HIGH-POWER ELECTRONICS

W. James Sarjeant

and

R.E. Dollinger

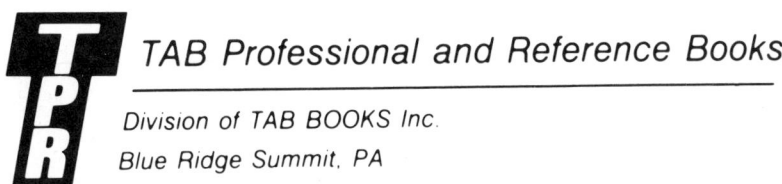

TAB Professional and Reference Books

Division of TAB BOOKS Inc.
Blue Ridge Summit, PA

Trademarks

Gap-Kap is a trademark of Centralab, Inc., Milwaukee, Wisconsin.
Teflon is a trademark of E.I. Dupont de Nemours, Wilmington, Delaware.
Freon is a trademark of E.I. Dupont de Nemours, Wilmington, Delaware.
Mylar is a trademark of E.I. Dupont de Nemours, Wilmington, Delaware.
Lexan is a trademark of General Electric Co., Fairfield, Connecticut.
Elkonite is a trademark of CMW Inc., Indianapolis, Indiana.
RTEMP is a trademark of RTE Corp, Brookfield, Wisconsin.

FIRST EDITION
FIRST PRINTING

Copyright © 1989 by TAB BOOKS Inc.
Printed in the United States of America

Reproduction or publication of the content in any manner, without express permission of the publisher, is prohibited. No liability is assumed with respect to the use of the information herein.

Library of Congress Cataloging-in-Publication Data

Sargeant, W. James.
   High-power electronics / by W. James Sargeant and R.E. Dollinger.
     p.   cm.
   Includes bibliographies and index.
   ISBN 0-8306-9094-8
   1. Pulse generators. 2. Pulse circuits. I. Dollinger, R.E.
II. Title.
TK7872.P8S27 1988
621.3815'34—dc19   88-7798
                                                                                 CIP

TAB BOOKS Inc. offers software for sale. For information and a catalog, please contact TAB Software Department,
Blue Ridge Summit, PA 17294-0850.

Questions regarding the content of this book
should be addressed to:

   Reader Inquiry Branch
   TAB BOOKS Inc.
   Blue Ridge Summit, PA 17294-0214

Edited by B.J. Peterson

Contents

	Introduction	vi
1	**Introduction to High-Power Electronics**	1

    1.1 Energy Storage Techniques—1.2 Characteristics of Pulse Shapes—
1.3 Switches—1.4 High-Power Electronics: Hard-Tube and Line-Type—
1.5 Impedance Matching—
1.6 Charging Systems for Pulse-Forming Networks—
1.7 Comparison of Hard-Tube and Line-Type High-Power Electronics—
1.8 Summary—1.9 References

2	**dc Power Supplies and Hard-Tube, High-Power Electronic Systems**	28

    2.1 dc Power Supplies—2.2 Hard-Tube, High-Power Electronics Systems—
2.3 References

3	**Pulse-Voltage Circuits**	87

    3.1 Marx Bank—3.2 Inversion Generator—3.3 Blumlein—
3.4 Spiral Generators—3.5 Coaxial Generators—3.6 References

4	**Transmission Lines and Pulse-Forming Networks**	117

    4.1 Transmission Lines—4.2 Pulse-Forming Networks—
4.3 References

5 Discharge Circuits and Loads — 137

5.1 Introduction—
5.2 Thyratrons, Ignitrons, Thyristors, and the Like—
5.3 Pulse Transformers—5.4 Switch Recovery and Resistive Effects—
5.5 Switch-Rise Time Effects in Ultrafast High-Power Electronics—
5.6 Effects of Interconnection Pulse Cables to Laser Loads—
5.7 Effects of Change in the Load Impedance on Circuit Performance—
5.8 Effect of Parasitics on System Performance—
5.9 Protecting Thyratrons from Excessive Voltages—
5.10 Effects of Load-Short Circuits—
5.11 Open-Circuit Protection with Cable Interconnections—
5.12 Laser Loads—5.13 Summary—5.14 References

6 Spark Gaps — 171

6.1 Gas Switches—6.2 Liquid Switches—6.3 Vacuum Gaps—
6.4 Solid-Dielectric Switches—6.5 References

7 Thyratrons and Ignitrons — 198

7.1 Triode and Tetrode Thyratrons—7.2 Glass Thyratron Construction—
7.3 Ceramic Thyratron Construction—
7.4 Thyratron Switching in Pulse-Forming Network Types of HPE—
7.5 Overview of Commercial and Developmental Thyratron Characteristics—
7.6 Development of High dI/dt Thyratrons—
7.7 Other Thyratron Applications—7.8 Ignitrons—7.9 References

8 Charging Systems — 239

8.1 The dc Charging Systems—
8.2 The ac Charging Systems—8.3 References

9 High-Voltage Air Core Pulse Transformers — 276

9.1 Discussion—9.2 Transformer Circuit Analysis—
9.3 Transformer Insulation—9.4 References

10 Measurement Techniques — 297

10.1 Voltage Measurements—10.2 Current Measurement—
10.3 References

11 Particular Applications — 317

11.1 Laser Loads—11.2 Summary—11.3 References

12 Energy-Storage Capacitors   335

12.1 Introduction—12.2 General Properties of Capacitors—
12.3 Capacitor Characteristics and Their Changes—
12.4 Applications of Energy-Storage Capacitors—12.5 Closing Note—
12.6 References

13 Grounding and Shielding in High-Power Electronics   371

13.1 Electrical Ground: Meaning and Interpretation—
13.2 Lightning Protection—13.3 Voltage Distribution and Skin Effects—
13.4 Grounding Schemes—13.5 Radiated Electromagnetic Noise—
13.6 Electrostatic Coupling—13.7 Shield Grounding—
13.8 Transformer Shielding—13.9 Shielding from Magnetic
Field—13.10 References

Index   387

Introduction

The purpose of this book is to present an overview of high-power electronics (HPE), particularly the power-conditioning aspects of it as applied to energy-transfer systems. The objective is to develop an understanding and appreciation of the physical processes that govern the performance of these systems, including the role of such elements as switches, capacitors, inductors, and resistors. The second objective is to discuss sources of information that can help solve problems when problem areas have been identified. A major effort in this text is to transfer this information in key areas to the reader as references. All the information is current and state of the art, and directed toward these pulse components, with particular emphasis on the area of high repetition rate systems, as these are taking on ever increasing significance.

In this book, W. Willis, R. Butcher, and W. Nunnally cover several areas with which they have considerable experience. J. Sarjeant and R. Dollinger handle text integration and discussions of the component technology base (for example, the present status of power conditioning systems and thyratron switching in general). As this text was created, it was arranged for several experienced individuals to participate in specialized topic areas, such as grounding and shielding, safety, transformers, and particular applications. G. Rohwein presents detailed information on pulse transformers and high-voltage insulation. A complete list of the authors of this book is:

Chapters 1 and 2—W.J. Sarjeant (State University of New York at Buffalo) and R.E. Dollinger (State University of New York at Buffalo)
Chapters 3, 6, and 10—W.L. Willis (Northrop)
Chapters 4 and 11—R.R. Butcher (XMR, Inc.)
Chapters 5 and 12—W.J. Sarjeant (State University of New York at Buffalo)
Chapter 7—W.J. Sarjeant (State University of New York at Buffalo) and D. Turnquist (Impulse Engineering)
Chapter 8—W. Nunnally (University of Texas at Arlington)
Chapter 9—G.J. Rohwein (Sandia National Laboratories)
Chapter 13—T.R. Burkes (Texas Technological University)

The first text covering this subject, now out of print for many years, is the World War II radar modulator text, *Pulse Generators* by Glasoe and Lebacqz. That book was primarily concerned with charging capacitors or pulse-forming networks (PFN) and discharging the energy stored therein at high repetition rates, either directly, or through a transformer, into a load. The secondary load was generally a radio frequency conversion device called a magnetron. For *High-Power Electronics,* the book by Glasoe and Lebacqz serves as a topic reference base, and technology development over the subsequent 30 years is molded around it.

To make the technical level of this book current, extensive reference information is used. References are listed at the end of each chapter, and where appropriate, footnote numbers in text indicate specific entries in the reference list.

This book is dedicated to
Jon Z. Farber
visionary and leader
in the field of
High-Power Electronics.

Chapter 1

Introduction to High-Power Electronics

by
W.J. Sarjeant and R.E. Dollinger
(State University of New York at Buffalo)

THE FIELD OF HIGH-POWER ELECTRONICS IS FRAUGHT WITH DIFFERING DEFINITIONS AS to what the term means, as well as to what the branching points are within each area. For the purpose of this book, define the concept of high-power electronics as the shaping of electrical power from the conventional 60 to 400 Hz (Hertz) mains, into temporally well-defined pulses of electrical energy having reproducible voltage and current time histories, (see Fig. 1-1). In this context, there is no specific mention of the repetition rate at which these pulses of energy are deposited into whatever the load might be. Normally, such systems are divided into single-shot (several times per hour or day) and repetitively pulsed (1 Hz upwards), primarily depending upon load requirements. A *high-power electronics (HPE) system* can then be defined as an energy-transfer system that stores energy, either mechanically or electrically, and then discharges a specific determined fraction of this stored energy as electrical energy into the load.

1.1 ENERGY STORAGE TECHNIQUES

There are two main types of energy storage techniques:

1. **Mechanical**—The mechanical energy, W_r, is stored in rotary motion of machinery, such as a flywheel homopolar dc source device or a pulsed compensated alternator:

$$W_r = 1/2\ I_o \omega^2 \tag{1-1}$$

2. **Electrical**—The energy is stored in electrostatic or magnetic fields.

$$\text{Electrostatic: } W_e = \tfrac{1}{2} \int_v \overline{D} \cdot \overline{E}\ dv \tag{1-2}$$

$$\text{Capacitor, for example, } W_e = \tfrac{1}{2}\ CV^2 \tag{1-3}$$

Introduction to High-Power Electronics

DEFINITION:
AN ENERGY TRANSFER SYSTEM THAT STORES ENERGY, EITHER MECHANICALLY OR ELECTRICALLY, AND THEN DISCHARGES A DETERMINED FRACTION OF THIS AS ELECTRICAL ENERGY INTO THE LOAD.

MECHANICAL
ENERGY, W_R, IS STORED IN ROTARY MOTION OF MACHINERY.
$$W_R = \tfrac{1}{2} I_0 \omega^2$$

ELECTRICAL
ENERGY IS STORED IN ELECTROSTATIC OR MAGNETIC FIELDS.

ELECTROSTATIC:
$$W_E = \tfrac{1}{2} \int_{volume} \vec{D} \cdot \vec{E}\, dV \qquad \text{E.G. FOR CAPACITOR OF CAPACITANCE C:} \quad W_E = \tfrac{1}{2} CV^2$$

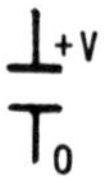

MAGNETIC:
$$W_M = \tfrac{1}{2} \int_{volume} \vec{B} \cdot \vec{H}\, dV \qquad \text{E.G. FOR INDUCTOR OF INDUCTANCE L:} \quad W_M = \tfrac{1}{2} LI^2$$

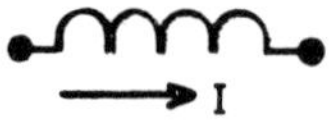

Fig. 1-1. Basic high-power electronics systems.

$$\text{Magnetic: } W_m = \tfrac{1}{2} \int_v \vec{B} \cdot \vec{H}\, dv \qquad (1\text{-}4)$$

$$\text{Inductor, for example, } W_m = \tfrac{1}{2} LI^2 \qquad (1\text{-}5)$$

1.1.1 Mechanical Energy Storage

These two classes of HPE, mechanical and electrical, are schematically illustrated in Fig. 1-2.

Mechanical energy HPE of both types A and B is being actively pursued at the Naval Research Laboratory (NRL) and the University of Texas at Austin, the latter under Department of Defense (DOD) and Department of Energy (DOE) support. It has been projected that these devices will be capable of transferring energies of up to 100 megajoules per unit in a time of several hundred microseconds for flashlamp pumping applications. It might be possible to extend the concept of a compensated pulsed alternator (type B in Fig. 1-2) to pulse-charging applications, up to 10 kHz repetition rates, for charging times from 10 to 1000 microseconds. If this extension is possible, these devices, following the normal time scale of high-technology development, might make excellent second-generation (1985–1990) charging units for fusion-reactor drivers. The units would charge high energy density transmission lines, which could then be discharged either in microseconds for laser drivers of the CO_2 type or in tens of nanoseconds for electron beam/light ion driven (EB/LI) or excimer (for example, KrF) laser-driven fusion systems. It is far from clear that the leakage flux of these rather low-voltage alternators can be reduced to the level that will allow efficient operation in the microsecond time scale. This point is currently under study at the University of Texas.

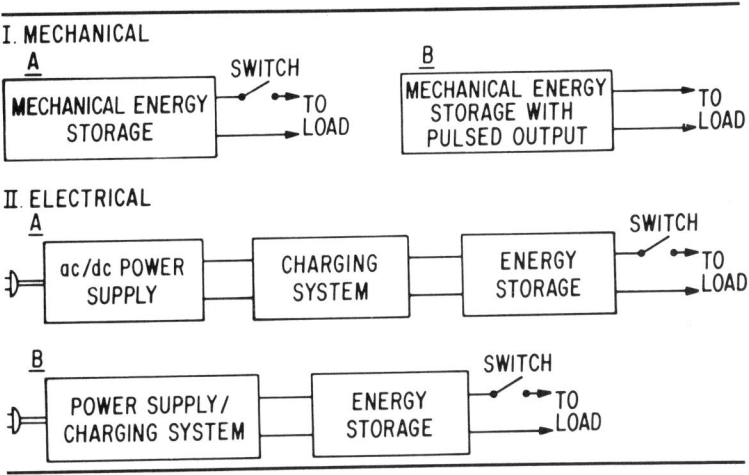

Fig. 1-2. High-power electronics systems.

1.1.2 Electrical Energy Storage

The electrical HPE fall into two categories, differing primarily in how the energy storage capacitors and/or transmission line are charged. For purposes of illustration, this discussion will include particular reference to some of the present repetitive HPE work underway at both Los Alamos National Laboratory (LANL) and Sandia Laboratory. In the type II-A of Fig. 1-1, a conventional dc power supply feeds energy into the energy-storage unit through the charging system. For single-shot systems, the latter might be only a string of charging resistors (giving a 50 percent energy-charging efficiency), but for repetitively pulsed applications, the unit is usually an inductive type of resonant charge network to ensure near-unity energy transfer efficiency. These inductive charging systems, with several high-regulation variations, are currently under development at LANL for laser isotope separation (LIS) applications, pulsed medical accelerators, the free-electron laser, and also are under study for use in the very large power conditioning systems (PCS) required for high-reliability fusion laser systems.

In addition to the above, pulse-charging transformers form a third charging technique used primarily where high-voltage gains and efficiencies are required. The development of large-scale, air-core, pulse-charging transformer techniques was pioneered by G. Rohwein at Sandia for high-voltage gain, fast charging (0.1 to 1 microsecond) and triggering applications. They have developed an extensive expertise in these air-core transformers, which are complementary to the iron-core units under research and development at LANL and in industry. These compact air-core transformers, designed with very high coupling coefficients and transfer efficiencies, have the potential of being vital components in future high-efficiency, cost-effective multikilohertz lasers, as well as in numerous related applications, such as research accelerators.

1.1.3 Repetitive High-Power Electronics System Technologies

The energy storage device required for high-reliability, long-life systems will likely continue to be capacitors for a considerable time until other technologies become sufficiently mature to displace them in a cost-effective way. At LANL and in industry, research and development is underway on capacitor structures directed toward satisfying the very long lifetime requirements in LIS, laser fusion, accelerators, and nuclear particle diagnostic systems. These efforts have clearly identified the need for a considerable technology-base development and effective laboratory-industry technology transfer in this area, which may soon take on considerable commercial significance. It has been recognized, through a number of LANL DOD and DOE programs, that the future systems evolving from the research described above will demand component performance levels vastly beyond those obtainable from current technology. For this reason, the research emphasis is being placed upon the two weakest areas in PCS: capacitors and switches for lifetimes in the region of 10^{10} to 10^{12} shots, as needed in the mature versions of all the above systems. In this field of high-repetition rate capacitor development, there has been exceedingly sparse data available in the area of long lifetime systems except for small radar modulator mica capacitors developed for the F111 fighter airplane ($>>10^{11}$ shot life) and polysulfone/polypropylene-silicone oil capacitors ($>10^{10}$ shot life) specifically created for driving industrial carbon dioxide (CO_2) transversely excited atmospheric-pressure (TEA) lasers.[4] The results of the work here have shown that the research and development approach to developing very long lifetime capacitor systems can be applied to HPE requirements for full-scale fusion laser driver applications.

In switching, several laboratories are working on complementary parallel efforts. At this time, it appears that the first feasibility demonstration fusion reactor systems, operating in the 1 to 10 Hz region, can be cost effectively switched utilizing spark gaps under development at Sandia and at LANL. At Sandia, components are currently being tested in the 30 kW range and will soon be into greater than 100 kW test-stand configurations. These are all Marx bank devices, and the Sandia facility will yield valuable scaling and lifetime data. Studies at LANL have shown that it is highly probable that alternative PCS configurations can be developed, utilizing extensions of adiabatic Blumlein techniques to continuous operation in the 10 to 40 Hz region as noted above, but yielding a mean time between failure (MTBF) of three to five years in contrast to months to a year (approximately 10^7 shots) for spark-gap systems.

For second-generation fusion drivers of the future, further development of switch and capacitor technology from the base currently being established through present programs is necessary to achieve industrial reliability levels for these systems.

The repetitive switching techniques undergoing research and development at LANL primarily utilize thyratron technology, because programmatic needs demand long lifetime operation in all of the high repetition rate systems (although some systems can function with spark-gap switches to meet near-term objectives). Through LANL contract support, the demonstration of current pulse-switching rates comparable to many single-channel spark gaps has been accomplished at over 10^{12} A/s. Scaling these new thyratron structures to exceedingly large power levels has been carefully studied.

It has been concluded that sufficiently large devices could be developed to allow a full-scale CO_2 fusion laser driver or EB/LIS pulse-charging unit to be switched with only a few thyratrons (two to six depending upon voltage needs), each passing multimegawatts of average power. Note that these thyratrons represent a direct scaling from megawatt devices produced through the ERADCOM program and the ultrafast tubes being developed for the LASL LIS program and show the significant offshoot benefits of both technology-base development programs.

As noted for thyratrons, burst mode tube developments are underway to meet adiabatic applications from 50 to 120 Hz repetition rates, for 60 to 90 seconds on-time, long off-times, and a limited number of total cycles (a few hundred). Thus, all components—thyratrons, capacitors, charging systems, and power supplies—need to be lightweight and of limited life. These conditions are diametrically opposed to the long-life needs of all long-life, industrial systems as well as numerous other program needs as described. The conditions dramatically illustrate the need for a research and technology base development program to create systems meeting these conditions that are so at variance with the burst-mode requirements.

In part II-B of Fig. 1-1, the power supply/charging system is shown as one block, connected directly to the mains. This type of configuration illustrates charging systems utilizing high-frequency inverters and polyphase cycloinverters, which directly charge the capacitor banks in times somewhat shorter than the pulse repetition period. In large-scale inverter technology, this is primarily at the research level (Gilmour, University of Buffalo). Cycloinverters are well established in the power industry (ac-dc and vice versa) and they may have application to single-shot systems. In most cases, their cost is so high that it is unclear just where they could be effectively deployed.

A high-power electronics (HPE) system stores electrical or mechanical energy and, on command, deposits a predetermined amount of this energy into a load (either a resistor, and inductor, or a more complex load system). As discussed, storage can be either electrical or mechanical, and either all or a portion of the energy can be discharged into the load. Storage elements include, for example, rotating machinery, pulsed alternators, and similar devices. In the mechanical sense, the energy is rotational energy, stored in rotary motion. In the electrical sense, there are two classes: (1) storing the energy as electrostatic energy in the dielectric of the medium and (2) in a magnetic case, where the magnetic energy is stored in the magnetic fields inside of the inductor. This discussion does not take into account time-varying capacitance or time-varying inductance. These relationships (Fig. 1-1) apply, then, to geometrically invariant systems, not varying in spatial dimensions or position with time. Otherwise, the C or L cannot be moved from under the integral sign. If dealing with nonlinear cases or time-varying C and L, return to the complete formulation of Maxwell's equations.

It is necessary to expand somewhat upon several facets of the pulse high-power electronic type systems. In mechanical systems, mechanical energy is often stored in a rotating machine and then discharged through a switch that couples the storage component to a load. A pulsed mechanical energy discharger, being developed at the University of Texas (Austin), is called a compulsator. It is essentially a magnetic-flux compressor. When the flux is compressed a maximum amount, a burst of energy is emitted into the load. The real advantage of this system is the enormous amount of energy that can be stored. At NRL, several megajoules have been stored routinely in a rotary machine. Mechanical stress analysis indicates that storage of up to several hundred megajoules in a superconducting rotary motion machine, or 100 megajoules in a straight mechanical wheel, is possible. The present difficulty with this system is the repetition rate for a fixed energy per pulse: the system is extremely slow to recharge, with bursts once every second to once every two minutes.

As discussed, electrical systems are of two major types as shown in Fig. 1-2.

In the first type (II-A of Fig. 1-2), the HPE can have an ac or dc power supply (ac is relatively rare), connected to a charging system that charges the energy storage device (capacitor, inductor, or transmission line) until the assigned stored energy is reached. At that time, the output switch is turned on, and the energy is dumped into the load.

The second electrical case (II-B of Fig. 1-2) is the high-frequency inverter power supply. It charges the energy system in a few milliseconds, permitting very high repetition rates. The system then charges the energy storage system in a very short period of time, the series tube turns off or is turned off, and the output switch closes, transferring the stored energy into the load. Other types of systems in this class are the constant-current dc supplies that use series vacuum tubes, and command-charging systems that use a series tube in the output or a thyratron turned full on to initiate the start of the (resonant) charge cycle.

If the energy storage device is charged rapidly, the probability of switch prefire is reduced, because the probability of prefire is proportional to the time the switch is at full voltage. When it is necessary to stress the switch to the maximum voltage, charging must take place very quickly (to decrease the probability of prefire). Corona (internal electrical unstable glow discharges—like St. Elmo's Fire) formation times are on the order of 1 to 10 microseconds. If the system is charged in times of that magnitude, the probability of the switch having a significant internal corona current, which gives a small electron density and thus a prefire, is quite small. The other advantage of pulse charging is that the dielectric strength of the energy storage medium is increased significantly (up to a factor of four higher for the same lifetime in a repetition-rate system). The device can then be made several times smaller for the same lifetime, appreciably reducing the system inductance. This has considerable advantage for airborne devices, portable devices, or in systems where extremely low internal impedance on the order of a few tenths of an ohm is required.

Each system has its own problem areas and performance limitations. When the switch closes, all the stored energy, if there is not perfect energy transfer, is not dissipated in the load. Then, when the switch tries to open again, at the zero of voltage, the small voltage perturbations from the energy stored in residual inductance and stray capacitance of the system generate very high frequency transients in the system discharge loop and in the energy storage system. These transients must be dampened, otherwise the switch will generally reclose and fault the system power supply. This is the main reason the early high repetition-rate TEA lasers were so very difficult to operate at repetitive rates above 1 kHz. There was enough energy stored in the loop inductances and stray capacitances in the system so that, whenever the switch started to open, there was sufficient inverse voltage (approximately 25 kV) applied to the thyratron switch to immediately fault it. It then acted as its own spark gap, tripping off the power supply. These transients can cause significant internal damage to triode thyratrons.

Tetrode tubes are far more immune to this fault. Spark gaps, on the other hand, if they turn on again, stay on, and fault the power supply, suffering no permanent damage. Ignitrons will also tolerate such treatment if they have proper anode and "keep-alive" auxiliary electrode materials and will then withstand significant current reversals. The ignitron auxiliary electrode is pulsed with a very large trigger pulse, and a relatively high energy is dumped into this auxiliary electrode (which keeps the pool ionized during the reversal period) so the mercury pool becomes the anode instead of the cathode.

The only disadvantage is that high rep-rates cause a tendency for the reverse arc to track along the surface of the anode insulator, destroying the tube hold-off capability and reducing lifetime.

1.2 CHARACTERISTICS OF PULSE SHAPES

In generating a voltage pulse, there are a number of parameters that are important in engineering, such as a peak value of the voltage and of the current through the load. These and other pulse-shape parameters are summarized in Fig. 1-3, and their determining factors, described in greater detail in the following sections, are shown in Fig. 1-4.

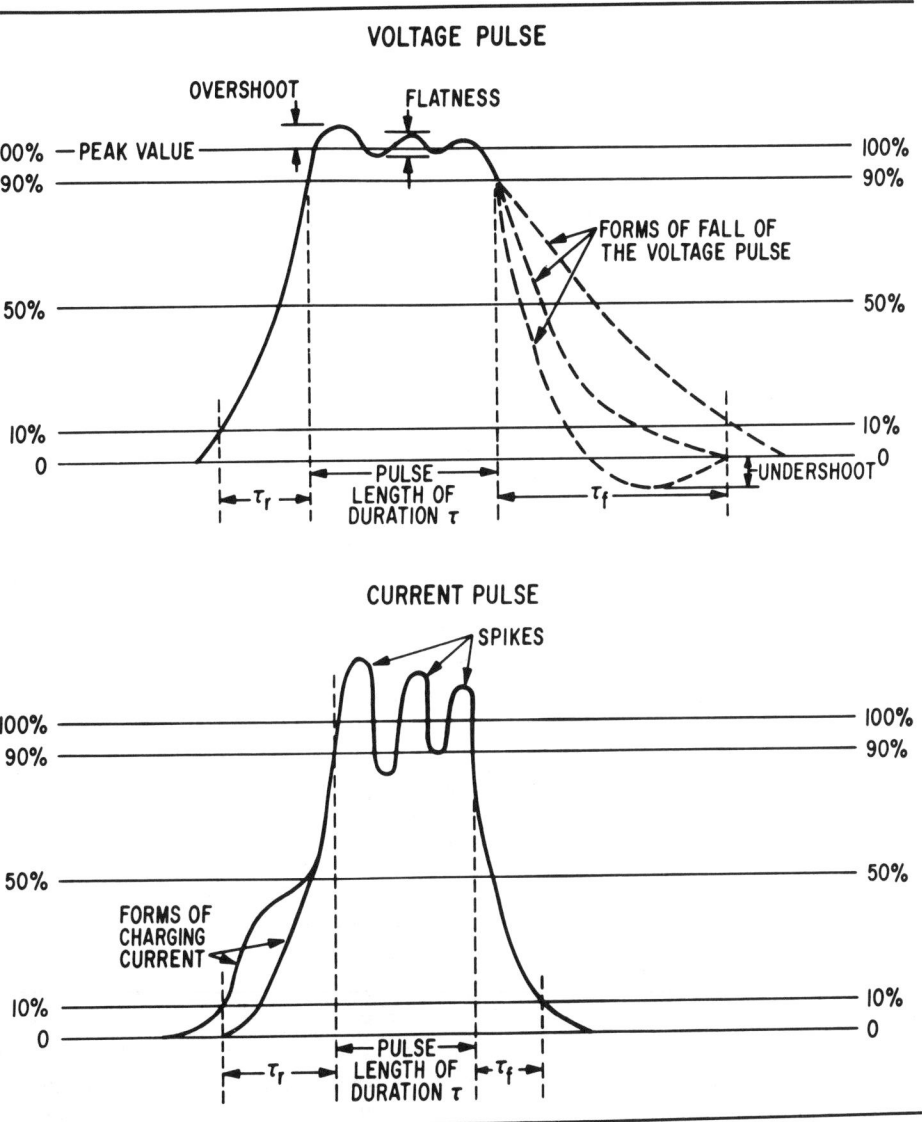

Fig. 1-3. Pulse shape parameters.

A. **FUNDAMENTAL**

1. **PULSE SHAPE:**
   A) τ_r: FROM 10-90%: RATE OF RISE
   B) SPIKES AND OVERSHOOT
   C) FLATNESS
   D) τ_f: FROM 90-10%: RATE OF FALL
   E) τ: DURATION AT 90% POINTS
   F) PEAK VALUE: V_{PK} AND I_{PK}

2. **PEAK PULSE POWER:**
   $$P_{PK} \equiv V_{PK} \times I_{PK}$$

3. **ENERGY:**
   ENERGY, W_L, DELIVERED INTO THE LOAD OVER THE PULSE DURATION VS THE TOTAL ENERGY DELIVERED INTO THE LOAD, W, DURING THE INTERPULSE INTERVAL.

4. **PULSE REPETITION FREQUENCY:**
   $PRF = \frac{1}{T}$: PROBLEM IS GENERALLY ONE OF INTERPULSE SWITCH RECOVERY TO PREDISCHARGE, QUIESCENT STATE.

B. **DERIVED**

5. **DUTY FACTOR** $\equiv \tau/T \equiv DF$
   (DUTY RATIO)

6. **AVERAGE POWER:**
   $$P_{AV} = \frac{\tau}{T} \times P_{PK}$$

7. **IMPEDANCE LEVELS:**
   ESPECIALLY IMPORTANT FOR TIME-VARYING LOADS IN ORDER TO MAINTAIN HIGH EFFICIENCIES.

Fig. 1-4. Pulse parameters describing high-power electronics systems.

1.2.1 Rise Times

One area of considerable interest is the rate of rise of the voltage pulse on the load τ_r; the second area is the rate of fall of the voltage, called τ_f. The required smoothness of the top of the voltage pulse is determined by the load characteristics. In the case of magnetrons, klystrons, and similar systems, this particular fluctuation of voltage and subsequent fluctuation in current causes a tendency toward frequency pulling in the device, and results in a time-varying radio frequency (RF) spectrum from the system.

Rate of rise can be defined in a number of ways. For these discussions, *rise time* will be defined as the time it takes the voltage to rise from 10 to 90 percent, rather than the *e-folding* time (that is, the time it takes the pulse, at an amplitude level of 1/e of its peak, to fold back fully upon itself to zero initial amplitude). The same applies to the *fall time*. When the output switch is turned on, the voltage rate of rise is determined not only by the properties of the HPE, but also by those of the load. In the case of a magnetron or laser, until the voltage rises to 50 or 60 percent of the peak value, the load is almost an open circuit. The slope of this voltage curve at the beginning can then determine how the current begins to rise through the system. This rise can have several forms, depending upon the stray capacitances in the system and the load impedance. For a magnetron or laser, when the thyratron is first turned on, the PCS discharges into an equivalent shunt capacity across the laser electrodes or magnetron, in series with some equivalent series inductance; this forms a series-resonant circuit with some damping in the equivalent inductance, allowing the output voltage to ring up rather quickly to values well in excess of the storage element charging voltage (*overshoot*). In fact, significant overvoltages can thus occur in these types of systems. Overvoltage is often of a considerable advantage in the actual operation of a laser system but generally causes additional insulation problems.

1.2.2 Fall Times

The falling characteristics of the pulse really depend upon the way in which the impedance of the load varies with time. The discussion up to this point generally assumes that there is a well-defined resistive impedance acting as the load for the HPE. As the voltage starts to decrease, for example, in the case of a magnetron, a point is reached at some voltage where suddenly the magnetron ceases to dray any current, and the load in the equivalent circuit for the HPE then becomes the shunt stray capacitance in series with the loop inductance, L. This resonant circuit has an energy of $1/2\ Li^2$ stored in this L at the moment of current decay that can then give rise to significant voltage ringing across the load. It is the damping of this ringing that is one of the most awkward problems in very high repetition-rate HPE design.

The pulse duration, τ, for this text is defined as the duration of the pulse at 90 percent of peak-pulse amplitude. Note that there is a τ for the current and for the voltage pulse, which are often significantly different.

The parameters that can be used to describe high power electronics systems can now be summarized. The fundamental parameters for the voltage and current pulses were previously discussed, for example, the rise time and also the rate of rise of voltage (of interest in the formation of glow discharges in lasers and in the initial turn-on conditions for klystrons and magnetrons) and are unique to each particular application. In the case of magnetrons, for example, the τ_r has to be slower than a critical value or a problem called *mode hopping* (a shift from one RF frequency to another during the time the current is building up) arises.

The second area of interest is spiking. Spikes are a major problem, even for relatively well-defined loads, such as the magnetrons and klystrons, as well as for laser systems, because these can give rise to arcing in the system when the voltage stresses are too high. If the current oscillations are excessively large, they can cause the discharge in some laser loads to quench effectively as the current goes down through zero. This produces a system in which the switch will try to reclose and no longer operate in the proper mode, generally causing the switch or laser to arc and fault the system.

1.2.3 Flatness

The pulse flatness is a primary consideration in applications requiring accurate voltage pulses for driving loads like Pockel's cells and other electro-optic light modulators. Flatness is not terribly important in many laser devices, which generally tend to behave as constant-voltage loads, smoothing out these peaks. As long as the energy is deposited into the resistive phase of the discharge in these systems (where the impedance is roughly constant), there is generally little advantage in efficiency for the applied voltage and current to have perfectly flat pulse shapes, provided the applied voltage is sufficiently above the threshold for efficient excitation laser kinetics. This has been a point of some considerable discussion and analytical work in CO_2 lasers in the last few years.

Work at LANL has shown that for an E-beam laser system with a very carefully controlled rectangular voltage and current pulse shape from the PCS into it, for which the gain is measured in the medium, this gain starts to grow to some peak value above which deactivation begins to dominate. In contrast, take the area under the power curve (which is some energy, W, deposited in the gas up to the gain peak) and then deposit the same energy in the same time, but without a sophisticated, complicated network. Rather, what is used is an inductor and a series capacitor. The energy is deposited in something like a $\sin^2 \omega\tau$ fashion, and measurement and calculation show that almost the same peak gain is obtained as for the carefully shaped flat pulse case. It is a very interesting observation for laser systems. As long as the energy is deposited into a constant-resistance system (glow-discharge) and the applied voltage is above some critical value, it appears that you do not need to be very sophisticated about the pulse shape. This applies to systems primarily where electron attachment and recombination are the dominant mode in which the electron density decreases during this portion of the pulse. In the case of some of the ultraviolet laser systems, that argument is not quite valid. This explains why you can get away with so much in driving most laser devices, in terms of pulse shape, and still have relatively high efficiency. In contrast, magnetrons and klystrons are very pulse-shape dependent for efficiency, particularly for very large pulsed klystrons.

1.2.4 Derived Parameters

There are three derived parameters in pulse-shape characteristics. One is the *duty factor*, the ratio of the duration of the pulse to the interperiod spacing. This is usually written as *DF* in most texts. If you multiply *DF* by 100, the result is the percentage of time a system is actively discharging energy. The duty factor can be used to calculate the average power, which is basically the duty factor times the peak power through the load per pulse, P_{pk}. The *total* energy discharged into the load per pulse is not necessarily useful energy. There is some type of time history of the power flowing into the load, and now all the energy that was stored in the energy storage device has been removed from the storage device. The characteristic of the load device (laser, magnetron, or whatever) might be such that the input energy, beyond some point, alters the load characteristics, and all this energy that is flowing into the system beyond this point in time is lost, so far as efficiency is concerned. This is a significant difference from the HPE requirements of yesterday.

In the 1950s, there were well-defined loads, such as magnetrons, and there were few efficiency problems. The model of a magnetron in this case was a resistor in series with a reverse-biased diode, which worked out quite accurately and efficiencies for these systems were often greater than 70 percent. Today, energy transfers in lasers are typically 2 to 50 percent. (That percent is the ratio of the energy stored in the HPE to the useful energy delivered into the load.) This fact can often be cause for considerable concern. The energy that is being delivered into the load that is useful energy and is producing photons, RF power, or the other output characteristics desired is often much less than the energy that is stored in the energy storage device of the PCS. Normally, the desired ratio is one (100 percent). For a number of new systems of considerable interest today, the ratio is much less than one. Therefore, the overall system efficiency that was formerly approximately 80 percent has decreased to a range of 20 to 35 per cent. When you consider devices where it is desired to put in hundreds of kilowatts of usable average power, this means that the total input power must be in megawatt range.

Another reason this point is becoming important is that a number of the load interface considerations in previous systems now change because there is a great deal of resistive power (Joule heating) to dissipate in the load. That thermal energy must be transferred somewhere else. This means either wasting energy in heating the gas laser or load medium or wasting it in the load device generating, say, RF radiation. The wastes causes a very severe thermal load problem. If, for example, 70 percent is lost energy, at a megawatt of average power flowing into the system, there is 700 kW (kilowatts) of undesired power in the load. The power must be disposed of; it is not useful power. Such thermal loading can cause significant problems.

The energy, W_L, is delivered into the load over the pulse duration, but it might not be the same as the total energy delivered into the load, W. That difference significantly affects the efficiency of the overall system.

The *peak pulse power*, P_{pk}, is the product of the voltage and the current at the point at which P_{pk} has a maximum value. More sophisticated definitions can be generated because of load characteristic shifts, but the convention suggested is that we look at P_{pk}, at the voltage peak, then we look at P_{pk} at the current peak, and define the peak power to be the larger number of the two.

The pulse repetition frequency (PRF), or sometimes pulse repetition rate (PRR), is a perfectly good parameter to use in systems that are repetitive at a given frequency (that is, in systems where there is a constant period of time from one pulse to the next, say 100 pulses per second). There are some difficulties when you consider, for example, modulators or pulsers with pulse repetition-rate agility. These are normally hard-tube (vacuum tube) switched systems. These systems generate a series of output pulses at an externally defined interpulse spacing that can be varying all the time. The pulse width also can be varying all the time according to some predisposed computer program. In that particular case, the concept of PRF does not apply.

The main problem in all of these high-power electronics systems is the one of interpulse recovery to the predischarge quiescent state. If there is some post-discharge energy left in the HPE, oscillations can arise in the discharge loop, and switches stay on that should not stay on, and components often become overstressed. What is desired is to recharge the energy storage elements after all switches are fully recovered.

1.2.5 Load Reproducibility Effects

If the load is reproducible on every pulse, the system output pulse shapes are perfectly reproducible on every pulse. For example, the impedance with time might have a specific shape, but in designing a system to work at even a single shot, you assume that this does not vary from pulse to pulse. The next pulse that comes along, then, varies the same way with time, providing that everything is charged to the same voltage.

In high repetition rate systems, particularly lasers and some RF tubes, this is not the case. A residual degree of ionization is left inside the system so the impedance profile per pulse might experience a different time history until equilibrium is reached. That is an important problem in very high repetition-rate systems. In small, high repetition-rate magnetrons, or in klystrons, it does happen, but to a lesser degree. In that particular case, it is primarily a result of the space charge saturation limitation in the cathode region. As cathode operating temperatures are reached from startup, there is a small shift of the available charge per pulse available from the cathode. The time history of impedance, then, changes slightly. It is meaningful to emphasize this point because, in

12 Introduction to High-Power Electronics

a number of the systems under consideration now, this will become a significant factor in HPE design for high efficiency of energy transfer.

In CO_2 TEA lasers operating at multikilohertz repetition rates, this particular impedance pulse shape is primarily determined by a combination of gas heating combined with the effects of residual interpulse ionization left in the system. Then it is necessary to choose which is the most efficient load profile in time to match into for maximum output energy for the specific experiment under consideration. If a very long operating time is not required, you could choose the startup load impedance, but if long runs are needed at maximum efficiency, it will be necessary to utilize a pulse-forming network that will maximize the energy transfer into the system at the equilibrium impedance level.

1.2.6 Load Characteristics and Their Effects on the PCS

These, then, are some of the different types of loads often found connected to HPE. A summary of their characteristics is presented in Fig. 1-5. Pure resistive, capacitive, and inductive loads are relatively straightforward to handle. The diode magnetron, on the other hand, has an equivalent circuit of a reversed-biased diode that turns on at some standard voltage. When it turns on, the voltage curve is almost flat with time. The dynamic impedance is quite low, basically like that of a glow-discharge, constant-voltage device.

In a magnetron, the dynamic (slope) impedance is about a tenth of the static value at the operating point. In lasers, there is generally a combination of this load behavior, with a time-varying component in the impedance. Figure 1-5 shows a sketch of what

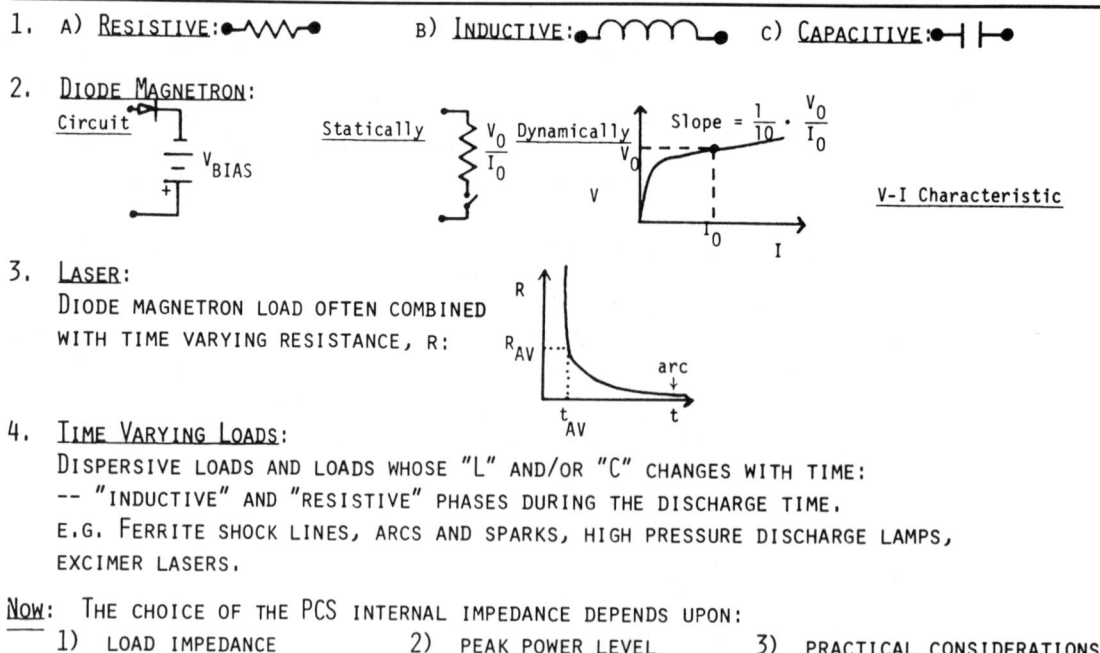

Fig. 1-5. Loads for high-power electronics systems.

the impedance with time looks like for an attachment-dominated excimer laser system (for example, KrF-, XeCl-, or XeF-laser). Evidently, the impedance decreases dramatically with time, but if most of the energy could be put into the system at times when the field intensity is sufficiently high for efficient pumping, then the laser system would work well and with high efficiency. Unfortunately, when this is attempted, a point is reached later in pumping time where the system arcs internally. This is normally caused by a thermally induced instability. In terms of load classes, one of the most difficult is the attachment-dominated laser discharge, which also happens to be rather interesting as an optical radiation source.

In the copper-vapor laser, for example, the impedance-with-time curve is somewhat shallower and alters dramatically as the repetition rate is increased. (The copper-vapor laser is a longitudinal, abnormal glow discharge excited with a ring electrode at each end of a ceramic tube.) Closing the switch in the HPE generates a voltage pulse that initiates a discharge, dumping the energy stored in a relatively small capacitor into the discharge tube at a repetition-rate of up to 25 kHz. By keeping the energy per pulse sufficiently low, it is possible to obtain a uniform, longitudinal discharge. Unfortunately, as the repetition rate is increased even further, thermal instabilities and kinetic effects constrict the discharge, raising the current and increasing the drop in impedance with time, which reduces system efficiency.

The fourth class of load characteristic is formed by time-varying loads. Dispersive loads change their permeability or dielectric constants with time as energy is deposited into them. An example is a shock-excited ferrite transmission line (for example, a coaxial transmission line with the center conductor covered by a ferrite material.) Driving the line could be a capacitor with a switch connected to the line input. Closing the switch starts energy flowing down the line, saturating the ferrite and steepening the rising edge of the voltage pulse as the wave front progresses down the line. As the ferrite saturates, the pulse sharpens. The sharpening is caused by the expenditure of energy in saturating the ferrite. Once the ferrite is saturated, the permeability is nearly one and, because the dielectric constant is rather low (12 to 15), the balance of the foreshortened pulse travels down a much higher impedance line. In a sense, such systems can be represented by lumped-element transmission lines, where the L and the C per unit length are time varying. This is a nonlinear problem and can be numerically solved with iterative solutions to Maxwell's equations. There is no analytical solution to the problem. With these systems, it has been fairly straightforward to generate pulses of 5 to 10 kV with rise times of 1 to 2 nanoseconds. With more optimal ferrite materials, it might be possible to reach 100 to 200 picosecond rise times. Unfortunately, there are serious reflection problems in these systems, making the generation of very accurate pulses a difficult and empirical design matter. This is very good technique for generating a very sharp pulse at the end of a coaxial cable.

Another type of load is the spark discharge used as a radiation source for short wavelength ultraviolet (UV) spectroscopy. The spark discharge is a point optical source, and it is excited by a switch discharging a charged, single-wire transmission line into the spark load. As the line capacitance is discharged, a wave is launched along this line (of about 150Ω impedance) providing the prompt energy for the spark discharge during the two-way transit time of the line. The spark is not a linear load with time and behaves similarly to the excimer laser loads discussed above. The difference in this case is that the arcing point is at the spark turn-on, and it is this arc that provides the optical radiation source. The last area is high-pressure discharge lamps, continuous-wave (CW) or pulse for driving glass lasers and UV process curing lamps, Hg discharge lamps and other such systems. These systems are operated in the glow (CW) or abnormal glow (pulsed) mode, all as optical radiation sources.

The internal HPE impedance selected depends primarily upon three parameters: the load impedance time history, the peak power level (and concomitant useful energy), and the practical considerations in selecting available circuit elements. Capacitors for low repetition rates are readily available. For high repetition rates (1 kHz and higher), low-inductance, high-energy, long-life capacitors are not yet available. For fast discharges (approximately 50 nanoseconds), capacitors generally behave as transmission lines. The liquid-impregnated, insulating-film capacitor is made of sections of paper, plastic film, and foil wrapped with connections on either end. At the self-resonant frequency, this is a parallel-plate transmission line with a surge impedance of a fraction of an ohm. For discharging this capacitor into a short circuit, the limiting transmission line discharge time for almost all such capacitors becomes approximately 150 nanoseconds. The only way to make faster discharge capacitors is to develop capacitor geometries or new types of energy-storage devices with shorter internal discharge times. Single-shot assemblies with discharge times of 40 to 50 nanoseconds have been developed, but they are presently unsuitable for high repetition rate operation.

There are two categories of high power electronics as shown in Fig. 1-6.

A) THOSE IN WHICH STORED ENERGY IS PARTIALLY TRANSFERRED EACH PULSE:

E.G., HARD-TUBE HPE, CROWBAR STORAGE BANKS, MECHANICAL HPE.

-- GENERALLY REQUIRE AN ANALOG SWITCH ELEMENT.

B) THOSE IN WHICH ALL STORED ENERGY IS TRANSFERRED EACH PULSE:

E.G., LINE-TYPE HPE, BLUMLEIN SYSTEMS, MARX BANKS.

Fig. 1-6. High-power electronics categories or classes.

1. Ones in which a small amount of energy is dumped from the storage device into the load. These are generally hard-tube HPE systems, series vacuum tube switches, crowbar storage devices, and mechanical HPE systems, like the compulsator. The advantage of these devices is that they generally utilize an analog switch element, so you can, if desired, modulate the energy flow into fixed and time-varying load impedances as a function of time. Note that the X-ray flux from the high vacuum systems at the 30 kV level and above can be quite considerable.
2. The second class is one in which all the stored energy is transferred into the load with every pulse. These are line-type HPE systems, Blumlein generators, Marx banks, etc.

1.3 SWITCHES

Normally, a switch is closed during discharge and open during the recharge time (Fig. 1-7). During recharge, a closed switch would short the charging system. The required characteristics of that switch depend on whether all the stored energy is dumped into the load in each pulse. If it is necessary to transfer all the energy out of the storage system at a constant ouptut voltage for a predetermined time, a transmission line or a pulse-forming network needs to be employed as the energy-storage device. Because the recharge time is almost always much longer than the discharge time, the discussion of the discharge part can be separated from that of the charging circuit. This approximation is not very good, however, for systems such as the copper-vapor laser with its 25 to 30 kHz operation—the discharge time is a fraction of a microsecond, and the recharge time is a microsecond or so. On that time frame, there are many other problems, particularly keeping the switch turned off during the recharge.

Ignitrons are much maligned switches, generally ideal for crowbar storage bank applications. There are modest repetition rate applications (less than 100 Hz) at high energies requiring an inexpensive switch that can handle substantial current reversal. This is a good application for ignitrons. One problem with them is the rather sophisticated trigger generators needed for reliable triggering. Normally, about a joule per pulse must be dumped into the igniter, basically through a constant-current source, causing some heating problems at high repetition rates.

- GENERALLY A SWITCH IS CLOSED DURING THE DISCHARGE AND OPEN DURING THE RECHARGE TIME.

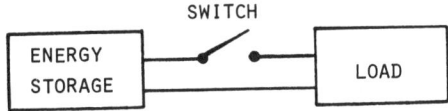

- SWITCH CHARACTERISTICS DEPEND UPON WHETHER OR NOT ALL THE STORED ENERGY IS TRANSFERRED PER PULSE.
  NOTE: PULSE SHAPING IS NECESSARY IN MOST DISCHARGE CIRCUITS WHERE ALL THE STORED ENERGY IS TO BE TRANSFERRED PER PULSE.

- NOTE ON CHARGING: SINCE THE INTERPULSE PERIOD IS MUCH LONGER THAN THE DISCHARGE TIME, IN GENERAL THE DISCHARGE CIRCUIT CAN BE LOGICALLY DISCUSSED SEPARATELY FROM THE CHARGING CIRCUIT.

Fig. 1-7. Switches.

1.4 HIGH-POWER ELECTRONICS: HARD-TUBE AND LINE-TYPE

1.4.1 Hard-Tube Systems

Hard-tube, high-power electronics systems are still alive and well. They are very useful in relatively small, low-voltage (less than 200 kV) systems. In a hard-tube HPE system (Fig. 1-8), a large energy-storage capacitor is charged from a dc supply through

16 Introduction to High-Power Electronics

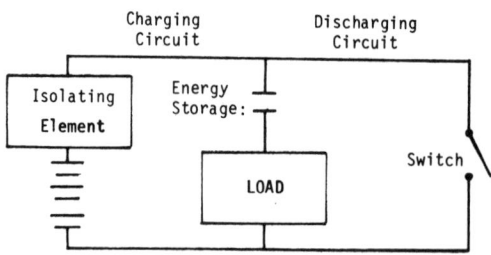

ENERGY STORAGE:

GENERALLY, ENERGY IS STORED IN A LARGE CAPACITOR AND ONLY A SMALL FRACTION OF THIS ENERGY IS TRANSFERRED TO THE LOAD PER PULSE.
-- THE SWITCH MUST HAVE VERY GOOD RECOVERY CHARACTERISTICS AS WELL AS HOLD-OFF CHARACTERISTICS: E.G., VACUUM TUBE.
NOTE: PULSE TAILORING CAN READILY BE PROVIDED WHEN USING A VACUUM TUBE SINCE THE GRID CONTROLS THE TUBE RESISTANCE.

Fig. 1-8. Hard-tube, high-power electronics systems.

element. The capacitor is connected to the load by a switch. If the switch is closed, energy is discharged into the load, and the isolating element serves to prevent the dc supply from discharging into the discharging circuit loop as well. The switch, however, opens to see almost the full potential of the dc power supply, because for good voltage regulation, only a small portion of the stored energy is discharged into the load on each pulse. The switch generally then needs to be a high-vacuum tube or a vacuum interrupter.

Hard-tube HPE systems are commonly used in the region of a few kilovolts at sub-nanosecond rise times. They have a great advantage in terms of pulse-width agility and repetition rate control. At high repetition rates, they are very useful, and the drive requirements are not severe. The real advantages of the hard-tube HPE are the variable pulse width and the high PRF capabilities at modest efficiencies.

Large hard-tube HPE systems (1000 A to 200 kV) exist. At these high voltages, the switching time is rather long, and the cost can be very high. They cannot generally be replaced with thyratrons, ignitrons, or spark gaps in these applications. On the other hand, they have limited di/dt capability because the vacuum tubes have a saturation resistance of a few hundred ohms, and that resistance limits the charging rate of whatever stray capacitances there are across the load. Their efficiency is nowhere near that of the line-type pulser (see Fig. 1-9).

1.4.2 Line-Type Systems

Line-type pulsers are quite different (Fig. 1-10). The energy is stored in a continuous or lumped-element transmission line, sometimes called an artificial transmission line, in the lumped-element case. Because this is the energy-storage element during the pulse discharge as well as the pulse shaper, the general designation of Pulse-Forming Network, (PFN), was coined by someone during World War II.

1.4 High-Power Electronics Systems: Hard-Tube and Line Type

ADVANTAGES:
- VERY HIGH PRF AS WELL AS APERIODIC OPERATION POSSIBLE.
- PULSE TAILORING.
- RAPID FAULT CLEARING

DISADVANTAGES:
- LIMITED DI/DT CAPABILITY BECAUSE OF RATHER HIGH TUBE-ON RESISTANCES ($\approx 100\Omega$).
- COST GENERALLY VERY HIGH COMPARED TO OTHER PCS.

Fig. 1-9. Characteristics of hard-tube high-power electronics systems.

There are two main classes of line-type HPE pulsers: voltage fed and current fed. In the voltage-fed HPE pulser, the energy, W_e, is stored in the total capacitance of the transmission line or PFN ($W_e = \frac{1}{2} CV^2$, where C is the total PFN capacitance and V is the charge voltage). In current-fed systems, a current is placed into the pulse-forming network and at the peak current value, $\frac{1}{2} Li^2_{pk}$ of energy is stored in the system. On the charging side is a closed series switch that can be opened at that point. As soon as the switch opens, it must withstand the full discharge voltage.

There are two general classes of voltage-fed, line-type HPE (Fig. 1-11). The first is a continuous transmission line that is charged from a voltage source. It has a switch at one end, and when the line is charged to voltage V, the switch is closed, discharging the energy stored in the transmission line into the load. If the load impedance equals the line impedance then $V_L = \frac{1}{2} V$. On the right side of Fig. 1-11 is sketched a time

1. THE ENERGY IS STORED IN A CONTINUOUS OR LUMPED ELEMENT (I.E., "ARTIFICIAL") TRANSMISSION LINE.
   - SINCE THIS SERVES AS THE ENERGY STORAGE ELEMENT DURING THE DISCHARGE PULSE AS WELL AS THE PULSE SHAPING ELEMENT, THE GENERAL DESIGNATION IS:

   "PULSE FORMING NETWORK" OR "PFN"

2. **CLASSES OF LINE TYPE SYSTEMS**

   THERE ARE TWO CLASSES:

   VOLTAGE FED
   - $W_e = \frac{1}{2} CV^2$
   - CLOSING SWITCH USED

   CURRENT FED
   - $W_e = \frac{1}{2} LI^2$
   - OPENING SWITCH USED

Fig. 1-10. Line-type, high-power electronics systems.

history of a typical charge and discharge cycle. The energy storage system is charged to the voltage, V, with a charging network. At the peak voltage, V, the switch is closed, and a pulse of energy flows out of the system and into the load. For a matched system, the pulse length is twice the one-way transit-time value of the continuous transmission line and all the stored energy is transferred to the load.

The second class of line-type HPE uses a pulse-forming network comprising a number of lumped elements of inductances and capacitors and a type E network as shown in Fig. 1-11. The designer chooses, basically, the L and C for the given load impedance and a number of sections in the line depending on the pulse fidelity required. With only one section, an LC discharge network is formed. As more sections are added, the second begins to discharge into the load concurrently with the first section, and a more rectangular pulse is formed in the load. With five or six sections, good pulse fidelities with ripples of about 4 to 5 percent are possible. Coupling between the inductances is possible, and some PFNs, such as the type E, are built that way. In this case, the different sections communicate in the flow of energy.

Some classes of time-varying loads can be accommodated by adding more coupling inductors between elements of the lines. These inductors are generally saturating inductors with ferrite cores. Years ago, there was some work done in far-infrared laser driver systems using that particular mode of operation. Such a direct, electron-pumped, far-infrared laser maintains high conversion efficiency as long as the voltage across the

---

TYPES

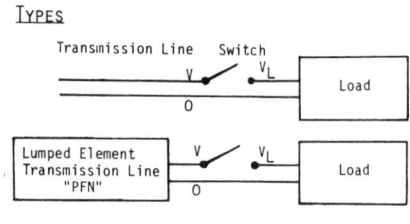

$W_e = \frac{1}{2} CV^2$

WHERE C IS TOTAL LINE CAPACITANCE AND V IS THE PEAK CHARGE VOLTAGE.

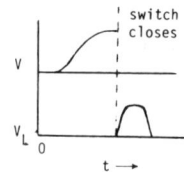

NOTES:

1. PFN CONSISTS OF A NUMBER OF CAPACITORS, C, AND INDUCTORS, L, AS LUMPED CIRCUIT ELEMENTS:

   TYPE E PFN:

2. VALUES AND NUMBER OF ELEMENTS IN PFN DEPEND UPON THE:
   A) DESIRED PULSE SHAPE
   B) LOAD IMPEDANCE
   C) COMPONENT AVAILABILITY

3. WHEN $Z_{LOAD} = Z_{PFN}$ (OR Z_{LINE}) THERE IS MAXIMUM ENERGY TRANSFER AND EFFICIENCY, ASSUMING A "PERFECT" SWITCH IS USED. OTHERWISE SOME ENERGY IS LEFT IN THE PFN AT THE END OF THE DISCHARGE TIME OF THE PFN CAUSING PROBLEMS IF THE MISSMATCH IS SEVERE; E.G., SWITCH RECOVERY, FAULTING, LOAD ARCS.

Fig. 1-11. Voltage-fed, high-power electronics systems.

laser load exceeds a critical value. To keep the electron temperature high, elements were chosen depending upon the laser impedance time history and component availability to preserve a rectangular voltage pulse shape.

Significant discussion of the current state of the art components follows, because they represent the major limiting factors in the technology base of HPE, particularly at high repetition rates and long lifetimes.

Current-fed, high-power electronics systems will be put to considerable use in the future as improved opening switches become available. In a current-fed HPE, in contrast to a voltage-fed pulser, the switch closes to allow current to flow through the battery (V_{dc}) to the energy storage (Fig. 1-12). When the peak current is reached, the switch opens and a negative pulse is generated across the load. The switch then experiences a very large voltage across it. Many current-fed HPE systems in small modulators are formed with relatively fast switching transistors, and they can be used to generate relatively high-voltage pulses with modest charging voltages. An inductor can be put in series with a transistor with a transmission line or capacitor across it. When the transistor is turned off, the energy stored in the inductor collapses and generates a voltage spike, which is useful for voltage multiplication. There are some very useful circuits in "Electronics Design" for voltage multiplying using this technique.

The pulse length is forced to be determined by the characteristics of the energy-storage device. Either a capacitor or a current-fed transmission line can be used. The latter has been studied extensively by a student of T.R. Burkes at Texas Tech University. Provided that the switch technology can be improved, this particular approach might be of some value in the 10 to 50 kV region. At peak current, the switch is opened, and

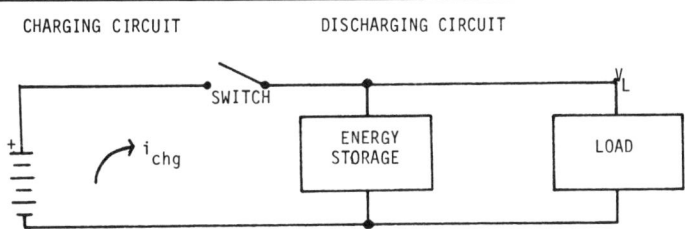

NOTES:

1) THE CURRENT BUILDS UP IN THE INDUCTANCE OF THE ENERGY STORAGE ELEMENT (TRANSMISSION LINE, PFN, OR TRANSFORMER).

2) AT THE PEAK CURRENT, THE SWITCH IS OPENED AND A HIGH VOLTAGE, V_L, APPEARS ACROSS THE LOAD.

$$V_L = \frac{Z_L}{2} \cdot i_{chg} [\text{PEAK VALUE}] : \text{FOR MATCHED IMPEDANCES}$$

3) THE SWITCH MUST RECOVER VERY QUICKLY TO HOLD OFF THE PEAK LOAD VOLTAGE.

Fig. 1-12. Current-fed, high-power electronics systems.

conservation of energy gives a voltage across the load impedance equal to the peak charge current times the load impedance over 2; but switch recovery is again a major problem. There is another problem in that when analysis of this sort of circuit is done, it is assumed that the switch is an ideal switch. If fact, what is normally used in this type of circuit is either a fast transistor, a series high-vacuum tube, or a vacuum interrupter. The equivalent resistance of the device increases as the current through it increases; in some cases, a nonlinear increase is found, which increases the charging losses. One other type of switch technique is to use a forced-commutation thyratron switch or a vacuum current interrupter. An auxiliary circuit can be placed across the switch, then it is discharged through the switch at the point of peak charging current so that the switch carries no net current and the switch then rapidly recovers. This is a method of turning off a switch that normally only recovers at zero current.

1.5 IMPEDANCE MATCHING

Matching the load impedance to the pulse forming network impedance is done primarily to obtain maximum power transfer (Fig. 1-13). The disadvantage, of course, is that only half the charging voltage is obtained. If the pulse-forming network is charged to 100 kV, only a 50 kV pulse is generated with typical efficiencies of 80 to 90 percent. The other point is that if the load and PFN impedances are not equal, reflections between the load and the PFN occur at the end of the discharge period of the PFN. If the load impedance is lower than the PFN impedance, a current reflection follows, and the current flow continues on for a very long time in an oscillatory manner. If the load impedance is larger than the PFN impedance, the load voltage, V_L is $V/2$ for the discharge time, τ, and depending upon the reflection coefficient (related to the ratio between the load impedance and the PFN impedance), a number of steps in the voltage pulse as a function of time can occur. Each step is τ seconds long.

---

1. NEARLY MATCHED IMPEDANCES ARE REQUIRED FOR
    A) MAXIMUM EFFICIENCY
    B) PULSE SHAPE FIDELITY
    C) MINIMUM POST-DISCHARGE VOLTAGE STRESSES ON SWITCH AND PFN.

2. IF $Z = Z(t)$ THEN THERE ARE PROBLEMS IN OBTAINING HIGH EFFICIENCIES.

3. IMPEDANCE INCREASES ARE OFTEN HANDLED BY PULSE TRANSFORMERS.
    - PRESENCE OF MANY 50 OHM COAXIAL CABLES TENDED TO MAKE THIS IMPEDANCE AN OFTEN-USED ONE FOR THE PFN.

4. FOR VOLTAGE-FED NETWORKS, THE VOLTAGE, UNDER MATCHED CONDITIONS, DECAYS TO ZERO AT THE END OF THE PULSE.
    - VERY GOOD SWITCH IS ONE THAT RECOVERS AT ZERO CURRENT AND NEAR ZERO VOLTAGE. E.G., SPARK GAP, IGNITRON, THYRATRON, SCR.
    - THE RECOVERY CHARACTERISTICS OF THIS SWITCH ALSO DETERMINE ALMOST EXCLUSIVELY THE NATURE OF THE INTERFACE TO THE CHARGING SYSTEM.

Fig. 1-13. Impedance matching of high-power electronics systems to their loads.

1.5.1 Effect of Load Mismatch on Switch Performance

This discussion points out one of the problems that occurs when the load impedance is much higher than the PFN impedance: there is current flowing through the switch for a relatively long period of time. It is a unidirectional current but takes a long time to die away. In such a case, switches such as spark gaps can revert from an arc to a glow and stay lit like a light bulb with milliamperes of current flowing through them. The actual power put into the switch can, operating under these conditions, be very significant. This is a strong argument for trying to match or slightly undermatch ($Z_L \lesssim Z_{PFN}$) the load to the PFN impedance.

If the load impedance is less than the pulse forming network impedance, the post discharge voltage pulse oscillates about zero. Then, the unidirectional switch turns off at the first voltage zero. A problem occurs if the load happens to short. This action generates a very large voltage reflection, which often exceeds the switch-recovery voltage, thus causing the switch to fault unless a voltage-limiting network immediately shunts the energy away from the switch.

A second class of problem arises when the load impedance is variable with time. A time-varying load impedance can move the operating point from a matched regime to an unmatched regime as the time increases. This is particularly true in lasers; it is also true in some high-current density discharge lamps. So, matching is not really practical and depositing the maximum amount of energy into the system in the relevant time is the most efficient approach. In designing for this regime of operation, it is far easier because of the voltage reversal across the switch, in the latter case. This is in contrast to the case ($Z_L > Z_{PFN}$) where a small unidirectional current is flowing through the switch for a very long period of time. The switch starts trying to recover in the zero current and zero voltage regime, and any small oscillations in the pulse-charging system will tend to inhibit this recovery. If the switch detects any appreciable positive voltage near the zero of current, it will not recover. It will fault on and dump the high-voltage power supply during recharge. This regime of operation was formerly referred to as *positive mismatch*. It was a favorite of many of the World War II radar modulator designers and is totally unnecessary today with the availability of modern inverse voltage control networks and considerably increased thyratron inverse-voltage ratings (up to -20 kV).

1.5.2 Transformers for Matching

There were a number of applications in the mid 1950s where relatively high voltages in the region of 100 kV or 200 kV were needed at tens to hundreds of amperes peak for driving large klystrons. (Klystrons needed physical length to accelerate the electrons versus the available electrical stress that the system permitted. The electrical stress was inside the system.) At that time, most switch tubes (thyratrons, ignitrons) were limited to the 20 to 30 kV range. The voltage, thus, had to be increased with a pulse transformer. The transformer technology for 1 to 5 microsecond pulse lengths is relatively well-established if the rise times are on the order of fractions of a microsecond. When rise times below 100 nanoseconds at significant average power levels are required, efficient design becomes very difficult. (With a sacrifice in efficiency, (20- to 30-nanosecond rise times are possible, but these are very sophisticated transformers with poor energy transfer—about 20 to 30 percent.) Few transformers are known that can handle significant energies at such rise times: one is a special version of a transformer developed by Gerry Rohwein at Sandia. It is a very low leakage, tape-wound, air-core transformer. The other was proposed by RCA to drive a very large load and to produce a quarter of a megavolt with an extremely fast rise time, at an average power of 30 MW. The primaries and secondaries were relatively thin copper foil, and water flowed through the copper foil, which was edge graded to prevent significant corona. The intention was to discharge 300 individual HPE pulses into the primary of that transformer.

Pulse transformers were used so much in modulator design because they were straightforward to design at the 50Ω primary impedance level. Fifty ohms was chosen, so the story goes, because there was 50Ω cable. This cable existed because it was determined in the 1940s that working at the 50Ω level gave the highest available stress in the cable system along with minimal dispersion. The cable was also easy to make.

1.5.3 Coaxial Cables

Cables normally have rubber, polyethylene, or air-gap dielectrics. For most HPE applications, air-gap dielectrics are not practical. In natural rubber dielectric cables, there are serious heating problems in short pulse length, high repetition rate applications. Natural rubber is not a good high-frequency insulator because it has considerable loss. In physical flexibility and mechanical lifetime, silicone rubber is still the best choice for general-purpose applications, with a design lifetime of many years at a modest PRF. Polyethylene cable is more fragile. It tends to have microbubbles filled with easily ionized water vapor. If these bubbles are undisturbed, they cause no damage. If the bubbles crack, they propagate; internal treeing and shorts develop.

In the matched loads, maximum energy transfer and efficiencies are possible, given a perfect switch (Fig. 1-13); however, very few switches are anywhere near perfect. An imperfect switch changes a linear circuit to a time-varying one. Spark gaps initially turn on as high-value resistors, but the resistance decreases dramatically with time, and as an arc column then forms, it expands or contracts with the current and acts then as an inductor. A switch can be a nonlinear element in the HPE. For short-discharge time, low-inductance systems, switch-limited pulse lengths are on the order of 20 to 100 nanoseconds single-channel devices (thyratrons, spark gaps, etc.). The switches become limiting components of the system. The alternate approach is to use a multichannel spark gap or the very large discharge area thyratrons currently under development.

If the switch is not perfect and even if the load impedance is matched to the generator impedance, some energy is left over at the end of the discharge pulse; this energy can cause oscillations in the system and give rise to switch-recovery problems. In other words, when you think you have zero current through the switch, you often do not have zero current.

1.6 CHARGING SYSTEMS FOR PULSE-FORMING NETWORKS

In the charging system, the switch is open during the charging period, and the charging time is much longer than the discharge time (Fig. 1-14). This is a general assumption applicable to most HPE. When the switch is open, a charge, q, is placed on the transmission line, charging up the total line capacitance, C, to voltage, V. There are now a number of choices for isolating elements.

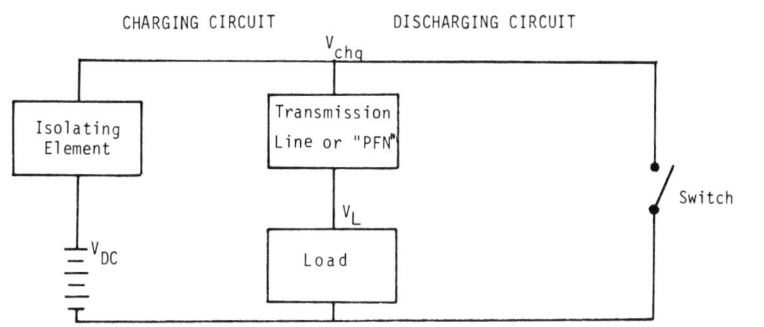

SINCE:

1. SWITCH IS AN OPEN CIRCUIT DURING CHARGING OF THE PFN OR TRANSMISSION LINE
2. THE CHARGING TIME IS MUCH LONGER THAN THE DISCHARGE TIME

THEN:

THE CHARGING SYSTEM CAN BE TREATED AS PLACING CHARGE ONTO THE LINE/PFN TOTAL TERMINAL-TO-TERMINAL CAPACITANCE, C.

ISOLATING ELEMENTS:

1. RESISTIVE: —WW— CHARGING INDUCTOR
2. INDUCTIVE: —mmm—▷|— CHARGING DIODE

Fig. 1-14. Charging systems.

1.6.1 Resistive and Resonant Charging

For a resistive-isolating element, generally used in single-shot HPE, the energy-transfer efficiency is always 50 percent or less. In inductive isolating element of inductance, L, the capacitor and inductor form a resonant circuit, and at the point of zero current through the charging loop, the charge voltage is equal to twice the dc voltage on the battery. The time for this to occur is $\pi\sqrt{LC}$, and the energy-transfer efficiency is very close to 100 percent. Most of the losses are resistive losses in the charging inductor; they can be made relatively small, with wall-plug efficiencies of 70-90 percent possible in the HPE. If the period is chosen to be equal to the above time, it is called the resonant-charging configuration, in which the discharge switch is turned on exactly at the zero charging current point. To work at other repetition rates, insert a blocking or charging diode in the charging loop. Once the voltage has come up to 2 Vdc, the current goes to zero, the voltage across the inductor collapses, the diode is reverse biased, and the 2 Vdc remains on the PFN to be discharged at will at any repetition rate up to $(\pi\sqrt{LC})^{-1}$. If this repetition rate is exceeded, there will be some dc current flowing through that choke at all times; that causes heating problems in choke design. There is also a net dc current through the discharge loop at all times, and the rate at which the voltage is reapplied to the switch and transmission line is quite a bit faster than in the resonant case. Switch recovery problems then emerge. Operating at these repetition rates is called linear charging. This operation had some stated advantages in early modulator design, but is not very useful today.

1.6.2 Repetition-Rate Factors

Constant-current or power-charging systems use a high-frequency inverter (usually a push-pull type) with a rectifier bridge (Fig. 1-15). The dc current, fed through an isolating inductor, charges the pulse-forming network; a closed-loop feedback system is used to control the frequency of the inverter. Frequency regulation thus provides very accurate charging currents and essentially very accurate voltages on the PFN. Charging units up to the order of a kilojoule per second can be bought as commercial units. They are small devices, but expensive.

Alternatively, in constant-current charging systems, a magnetic regulator can be used in series with the primary of the power supply transformer. For single-shot work, this was often used in some of the charging networks for fusion HPE. The current is controlled through the primary, charging the capacitor bank through an isolating network, and the bank final voltage can be held to a very accurate level. The other real advantage to this system is that, when it is first turned on, the peak current through the system is rather low. The equivalent circuit is a magnetic regulator in series with the high-voltage transformer, feeding a solid-state diode and a series resistor that connects the power to the capicitor bank. With the magnetic regulator in the primary, a high series impedance at turn-on is obtained; the impedance is much larger than the reflected load impedance, so the current flow is very small. The secondary charging current and voltage can be used to control the regulator. Magnetic regulators have a number of advantages in protecting the components, and they form an unusually rugged charging system.

Command charging uses a series switch in series with the battery or dc supply filter capacitor and the PFN. The switch is closed on command, and the PFN is charged up

---

A) CONSTANT CURRENT OR POWER

UTILIZES HIGH FREQUENCY INVERTER/RECTIFIER WITH CLOSED LOOP FEEDBACK:
- LEVELS UP TO 1 KJ/SEC CAN BE BOUGHT
- CAN ALSO USE A SERIES REGULATING SATURABLE REACTOR IN THE TRANSFORMER PRIMARY OF A NORMAL DC SUPPLY.

B) COMMAND CHARGING
- USES A SERIES SWITCH, EITHER VACUUM TUBE, IGNITRON, OR THYRATRON, TO CONNECT A RESONANT CHARGING NETWORK TO THE DC SUPPLY RESERVOIR CAPACITOR.
- ALLOWS MAIN PCS SWITCH TO FULLY RECOVER BEFORE CHARGING VOLTAGE IS APPLIED.

C) AC RESONANT CHARGING
- SATISFACTORY IF PRF IS LESS THAN TWICE THE AC FREQUENCY.
- TROUBLESOME FAULT MODES AND LITTLE PRF AGILITY LIMIT APPLICABILITY.
- LOWEST COST CHARGING SYSTEM AS ONLY A TRANSFORMER AND POSSIBLY SOME DIODES ARE REQUIRED.

Fig. 1-15. Other charging systems.

through some other isolating device—usually a series-resonant network. In that case, charging is completed to 2 Vdc, the series switch opens, and then the PFN is discharged through the load using the main shunt discharge switch. The real advantage of pulse charging is that the PFN discharge switch is allowed to fully recover before the voltage is reapplied to it and the pulse-forming network. This sequence ensures that the PFN discharge switch will experience an enormous decrease in its fault rate as the high-voltage on-time is reduced. The problem with pulse charging is that it normally requires a floating deck series switch with a large number of items placed at high voltage.

The last type of charging is ac resonant charging. Removing the diode and adding some extra inductance in a normal power supply (if desired), gives an LC resonant circuit with the PFN capacitance, C. The voltage on the PFN rings up and then is discharged into the load. The disadvantage is that the circuit must be used only at this resonant frequency. If lower frequencies are attempted, the fault modes are simply horrendous. This technique was, however, used in several fixed-PRF, World War II spark-gap pulsers. Some very large modulators were built that had no explicit resonant-charging network. They used a three-phase ac supply with adequate leakage inductance in the high-voltage transformer, and they charged three PFNs, using three switches to discharge them all into one load. The system for fixed-frequency operation can then be very compact.

1.7 COMPARISON OF HARD-TUBE AND LINE-TYPE HIGH-POWER ELECTRONICS

Figure 1-16 is a comparison between the hard-tube and the line-type HPE. The range in wall-plug efficiencies that the hard-tube systems work in is 10 to 15 percent, sometimes 20 percent. The first generation of line-type high-power electronics systems ran at 50 to 60 percent efficiency. A lot of power was lost in dissipation in the pulse-forming network components, in the spark gap switches, and even in the thyratron switches. Today, a typical figure for a complete line-type HPE system is an overall wall-plug efficiency in excess of 80 percent. The hard-tube HPE systems have increased their efficiency to about 20 percent because they then used three-phase or twelve-phase rectifiers with tubes requiring significant heater power. In addition, series high-vacuum switch tubes of somewhat higher efficiencies are available. The real advantage of the hard-tube HPE is its pulse shape and repetition-rate control. For varying load impedances, one can generate a well-controlled pulse shape from a hard-tube HPE. Currently, for small systems running around the 10 to 50Ω impedance range, the rise time is less that 1 nanosecond, increasing to around 20 nanosecond rise time in large systems. There are scalability limitations in the hard-tube HPE dictated by component availability for high average power systems.

1.8 SUMMARY

Pulse rise times and the flatness characteristics suffered significantly as a result of long resistive phase turn-ons of early switch tubes. What can be achieved to date in sub-ohm impedance circuits are current rise times of 20 nanoseconds at current rates of rise, circuit-limited, in excess of 1.2 MA/microsecond with new thyratrons under development. Five hundred kA/microsecond thyratron switch tubes are now available. One

Characteristics	Hard-tube HPE	Line-type HPE
Efficiency	Lower; more power lost in tube heating, and in dissipation in the switch tube.	High, >80%, particularly when the pulse-power output is high.
Pulse shape	Better rectangular pulses. Shape control by input pulse in analog fashion.	Poorer rectangular pulses, particularly through pulse transformer.
Impedance-matching	Wide range of mismatch permissible.	Small range of mismatch permissible ($\approx$20-30%). Pulse transformer will match any load, but power input to non-linear load cannot be varied over a wide range.
Interpulse interval	May be very short; as for coding beacons (i.e., $\approx$1 nsec).	Must be several times the deionization time of discharge tube (i.e., 2-300 μsec).
Voltage supply	High-voltage supply usually necessary $\approx V_L$.	Lower-voltage supply, particularly with resonant charging.
Change of pulse duration	Determined directly by input pulse shapes.	Requires high-voltage switching to a new network.
Time jitter	Negligible time jitter (<1 ns).	High-power line-type pulsers with a hydrogen thyratron give a time jitter of <2 ns.
Circuit complexity	Greater, leading to greater difficulty in servicing and high cost.	Less, permitting smaller size and weight.

Fig. 1-16. Comparison of the classes of repetitive high-power electronics.

now consider designing a number of line-type power high power electronics systems for long-life operation at kilohertz repetition rates that previously were thought of as impractical. The equivalent inductance of the switch thyratrons has concurrently been reduced by a factor of 20.

In impedance matching, the equivalent resistance of hard-tube modulators can be used in an analog fashion, so it is possible to accommodate very large impedance mismatches and time-varying loads. This was the reason for the first large-scale efforts to drive high repetition rate CO_2 systems with hard-tube HPE. (They can operate at megahertz repetition rates.) In this way, the decreasing discharge impedance with time could be accommodated. The mismatch range of 20 to 30 percent is one over which the efficiency of the system does not vary a lot. Going beyond that, for PFN-type HPE, all of the additional stored energy must be disposed of elsewhere if it is not deposited into the load in a time equal to one discharge time of the pulse-forming network. It is desired to deposit as much of that energy as possible in the discharge time of the PFN.

Typical recovery times today for small tetrode thyratron devices are 1 or 2 microseconds. The World War II design triode thyratrons that you can buy commonly now have a much longer recovery time. They also suffer recovery-time degradation with age because of their internal construction. It takes a long time for the plasma from the cathode to lose communication with the anode. Today, it is possible to go down to a microsecond recovery time in modest-size thyratrons discharging millicoulombs/pulse.

In terms of high-power spark gaps, recovery times are hundreds of microseconds and have not changed. If you want very fast recovery in spark gaps, it is possible to go to a device that adopts some internal quenching mechanism to get rid of the residual ionization. An example is the work done on hydrogen quench gaps, which are essentially a pair of electrodes separated by a number of fine meshes. Discharges through this device are at a modest pressure of hydrogen. The presence of the mesh cools the ionized gas in gap very quickly and enhances recombination around each mesh. This action causes little blobs of isolated plasma to form. These gaps can then work at a hundred kilohertz or higher repetition rate, discharging a few hundred picofarads capacitance per pulse; however, because all the current is passing through the screens, it is difficult to flow high average powers through the device.

1.9 REFERENCES

1. G.N. Glasoe and J.V. Lebacqz, "Pulse Generators," Dover Publications, Inc., New York, 1965.
2. A.E. Greenwood, "Electrical Transients in Power Systems," Wiley-Interscience, New York, 1971.
3. F.E. Terman, "Radio Engineers Handbook," McGraw-Hill, New York, 1943.
4. K.B. Riepe, "High-Voltage Microsecond Pulse-Forming Network," Rev. Scientific Instrum. Vol. 48, No. 8, August 1977, pp. 1028–1029.

Chapter 2

dc Power Supplies and Hard-Tube, High-Power Electronic Systems

by
W.J. Sarjeant and R.E. Dollinger
(State University of New York at Buffalo)

THIS CHAPTER COVERS SOME OF THE MAJOR CHARACTERISTICS OF DC POWER SUPPLIES and hard-tube pulsers, along with many of their HPE limiting properties and several of the more powerful vacuum tubes that are currently available for larger hard-tube pulser systems. The initial topic is elementary dc power supplies (Fig. 2-1), operating with standard 60 Hz ac input power. The full-wave bridge design is discussed in detail, illustrating the majority of the significant circuit parameters.

2.1 DC POWER SUPPLIES

If a simple dc power supply is snap started (suddenly energized to full voltage), and then the diodes, transformers, and sometimes the filter choke are damaged, what causes this damage is often a transient overvoltage effect rather than a short circuit on the output. These types of relatively low-frequency transients, which can cause damage and component breakdown in 60 to 400 Hz dc supplies, are the subject of considerable discussion in this section. The standard 60 Hz full-wave bridge geometry (FWB) will serve as the design center. The worst-case relationships for all the design parameters applying to the diode, transformers, and filter design and selection (that apply all the way from full-wave doublers to 6-phase and 12-phase Y-delta dc power supply systems) are compared and related to these FWB parameters. For all of these classes of supplies, the same type of transient analysis will apply to modeling the potential damage to components. A first-cut design of the power supply and the prerequisite damping networks can be obtained from this data and then, returning to the reference data that will be discussed, cost-benefit ratios for the supply design can be effected. In other words, what might come from the worst-case analysis here might be that the diode chosen has a higher current rating than that found necessary when a detailed circuit analysis has been executed. In this simplified, worst-case analysis, the design parameters determined will always err on the conservative side. One exception is the case in which the power-line

---

THIS SECTION CONSIDERS ELEMENTARY DC POWER SUPPLIES OPERATING FROM 60 Hz AC POWER. THE FULL-WAVE BRIDGE DESIGN WILL SERVE TO ILLUSTRATE IMPORTANT CIRCUIT PARAMETERS. THROUGHOUT THE DISCUSSION, RELATIONS APPLICABLE TO OTHER GROUPS OF DC SUPPLIES WILL BE CORRELATED WITH THE FULL-WAVE BRIDGE (FWB) PARAMETERS.

ELEMENTARY TYPES

A) HALF WAVE
B) FULL WAVE, CENTER-TAP
C) FULL-WAVE BRIDGE
D) FULL-WAVE VOLTAGE DOUBLER
E) VOLTAGE MULTIPLIER

Fig. 2-1. dc power supplies.

impedance is very low, potentially giving rise to enormous surge currents measuring hundreds of amps. For large, high-regulation, solid-state dc supplies of this type, it is necessary to obtain more accurate design data, which requires a return to the fundamental relations in the references. The best source for most of the diode and filter design data is the RCA thyristor manual. In the back of that manual is an application note that is a summary of the work by Shade in the 1960s, and it has almost all the design information needed to select a rectifier diode for the design. The other area discussed in the thyristor manual is how to execute relatively accurate ripple calculations when dealing with precision supply requirements.

2.1.1 Major Classes of dc Supplies

The five basic classes of elementary dc supplies are: half-wave, full-wave center tap, full-wave bridge, voltage doubler, and voltage multiplier. These types of supplies are described in detail in Terman. Although it deals with tubes, the Terman work is probably one of the better analyses of the general characteristics of these types of supplies for ripple and of the ratings the diodes must have. A summary of the different types of elementary supplies, A through E, is shown in Fig. 2-2.

- A. **Half-Wave**—The load is purely resistive. There is no filter capacitor included, giving bursts of voltage, one for each positive-going supply cycle.
- B. **Full-Wave with center-tap**—There is a diode added to the bottom half of the transformer that gives a ripple frequency of 120 Hz, or twice the input frequency.

C. **Full-Wave Bridge**—The main advantage of this supply is the high overall efficiency through full utilization of all the available transformer kVA or power rating.

D. **Voltage-Multiplying Circuits**—These are basically charge stacking systems that can be used to provide very low currents at relatively high voltage (they will not be discussed in detail). At 60 Hz, they tend to be very difficult to design for currents above 10 or 15 mA, so they are basically very low average current supplies. For low current requirements, on the order of milliamperes, they are an inexpensive voltage-multiplying technique. The slight disadvantage of using them is the relatively high RMS rating that the diodes and capacitors must have.

E. and F. **Cockroft-Walton**—This is the same type of voltage multiplier with a very high-frequency input. Then the reactances of the filter capacitors becomes very low, so that the amount of capacity needed to achieve very good regulation is small. Very large voltage multiplication factors with low ripples are obtainable.

2.1.2 Full-Wave Bridge dc Supply—Detailed Analysis with Choke Input Filter

Figure 2-3 shows a sketch of the full-wave bridge dc supply. On the left side is the main power transformer, fed power through a set of line filters to keep transients from passing through in either direction, and a circuit breaker that is rated to trip on severe overloads.

2.1.2.1 Circuit Operation.
If the power supply is reasonably large, the contactor shown is used to turn on the supply, and it is usually energized by a latching circuit. The transformer feeds the diode full-wave bridge, and the output voltage ripple is reduced by the LC filter. On a positive line cycle, current flows as shown by the arrows into the filter capacitors and back to the transformer. The transformer here is shown as having electrostatic shields, which are normally used in all sensitive circuitry or circuitry where there are relatively high voltages and transients. This arrangement shields the primary electrostatically from the secondary by a grounded metal structure. For single-phase designs, the shield cost is at most a low percent of the transformer cost, so there is no excuse not to ask the manufacturer to include them. Generally, for only a single shield, it is referred to the primary due to cost, because if a relatively high voltage is on the secondary, the added insulation for an electrostatic shield on the secondary is often expensive. The shield is usually a single layer of copper or aluminum foil over the primary, slit so the foil does not act as a shorted turn on the transformer, with a wire connecting the shield to the transformer case. That effectively puts an equipotential plane between the primary and the secondary as far as capacitive coupling is concerned. No matter how the transformer is wound, there is some capacity from the primary to the secondary, and what is desired is that any current that the secondary tries to induce from capacitive coupling to the primary is carried off to the ground via the shield.

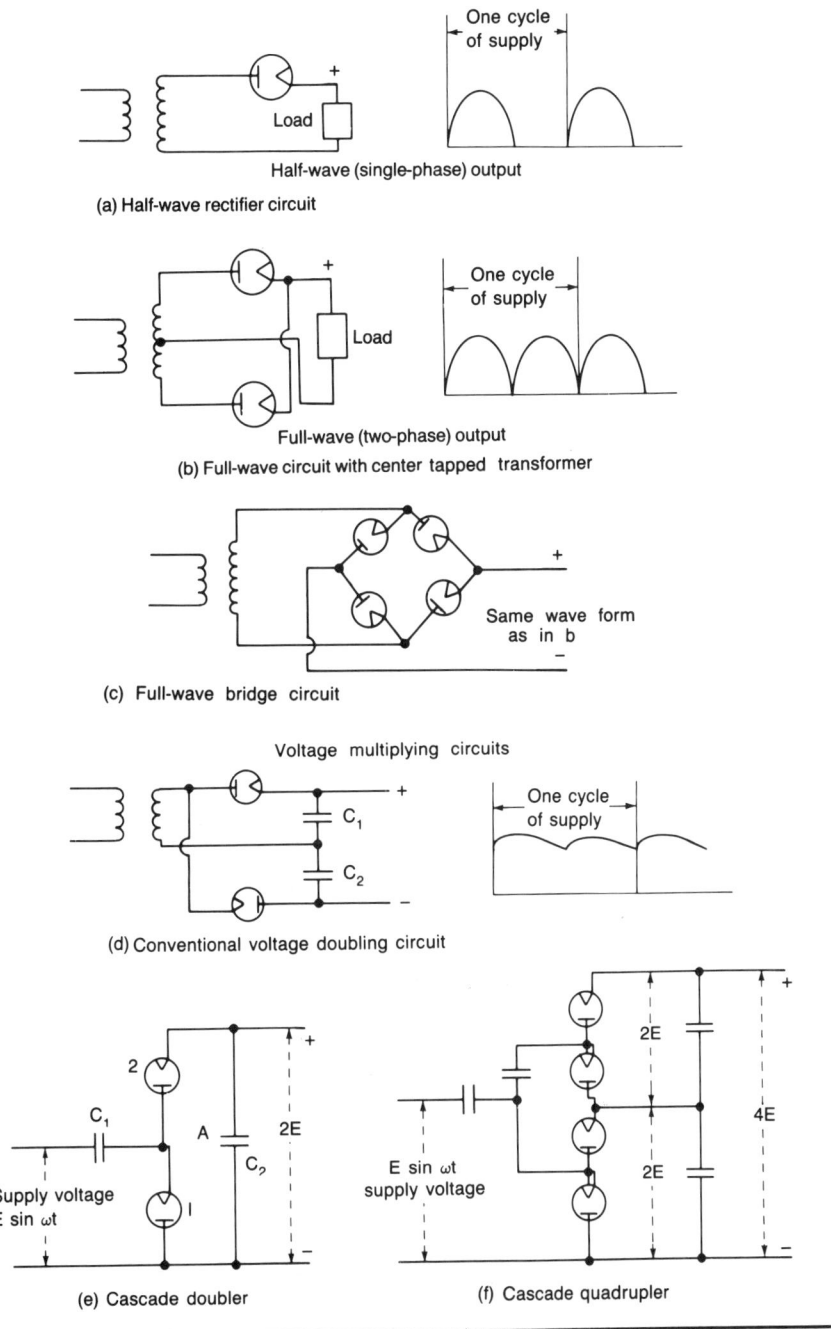

Fig. 2-2. Elementary dc power supplies. (Redrawn after Terman and used by permission of the Board of Trustees of the Leland Stanford Junior University).

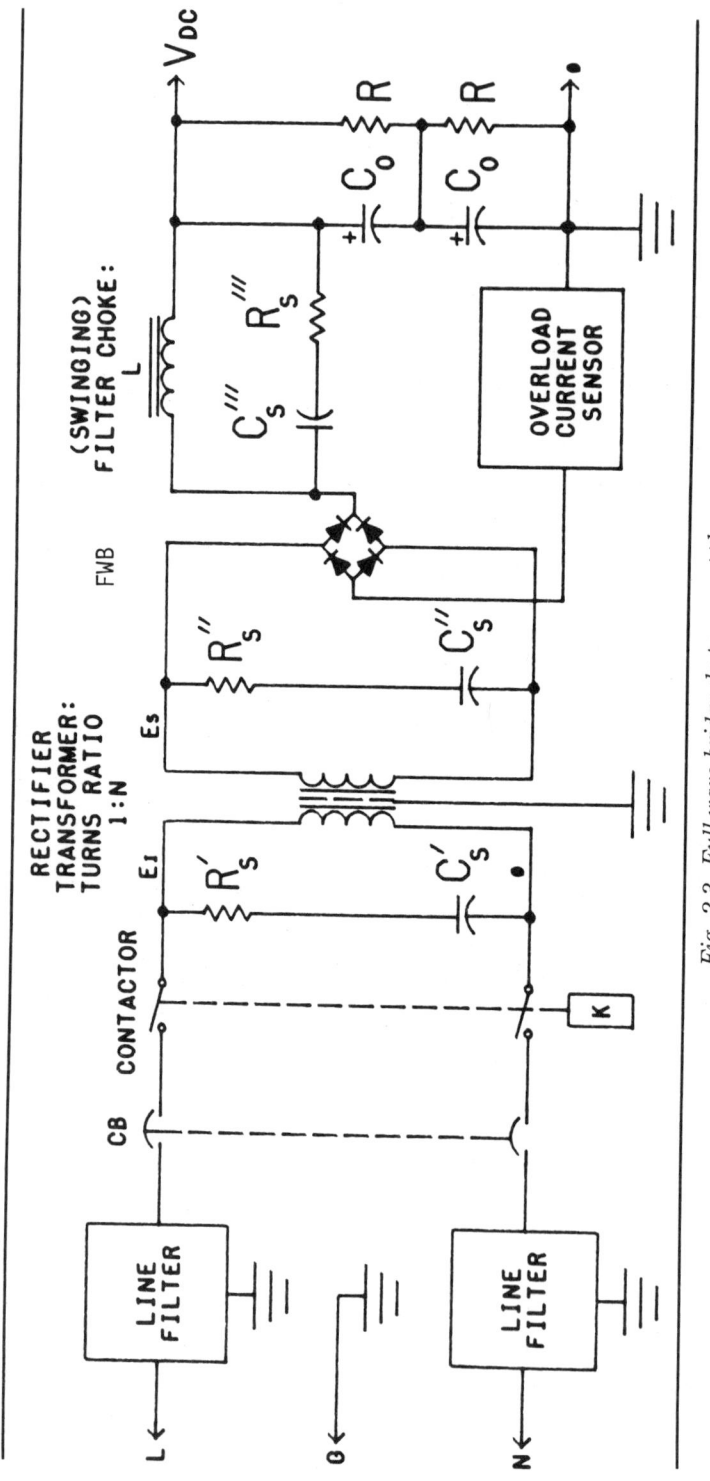

Fig. 2-3. Full-wave bridge dc power supply.

2.1.2.2 Transient Damping. On the primary side of the power transformer in the dc supply, when the contactor is energized, the equivalent circuit here has the minimum inductance in the transformer, giving rise to large inrush currents that must be taken into account in selecting circuit breaker and contactor surge ratings. In addition, a relatively large voltage transient can arise when the contactor is opened, caused by the collapse of the magnetic energy stored in the leakage and magnetizing inductances. The critical damping network, $R'_s\, C'_s$, is used here to dampen that transient when the contactor opens, particularly if it opens at maximum current into the transformer. The damping network, $R''_s\, C''_s$, on the secondary of the transformer, is used for protection from both rapid turn-on (if turn-on is at the peak of the primary voltage, this network dampens the oscillations in the secondary loop), and for turn-off of the system when the diode FWB is then decoupled from the load loop as the transformer secondary voltage drops to zero. The diodes then no longer conduct current and are no longer pushing charge into the filter capacitor, so the secondary is just an RLC oscillator of $R''_s\, C''_s$ and the transformer leakage and magnetizing inductances (referred to the secondary). Selection of R''_s and C''_s values for near critical damping then controls the oscillation amplitude and duration. The same argument applies to damping transients across the filter choke. All these situations use an analogous damping technique.

2.1.2.3 Filtering The output of the supply in this case is shown as two filter capacitors in series, which indicates that, when the voltage is so high that a single capacitor is unavailable, it is necessary to stack a number of them in series and to use the voltage-grading resistors shown. A grading resistor is usually incorporated to pass on the order of 8 to 10 times the maximum leakage current through the capacitor at the maximum operating temperature of the supply. This ensures no more than a 15 to 20 percent imbalance in the voltage division across the two capacitors shown.

2.1.2.4 Overload Protection The overload current sensor in Fig. 2-3 protects the diode bridge should the supply output be shorted to ground. It is timed to react within about 50 to 200 milliseconds (typically this opening time can be achieved with fast electromechanical relays or vacuum interrupters). So if there is a fault on the output, the pulse current will energize the sensor, which then trips open the primary contactor (in this case, the important cost parameter is the number of cycles the diodes can carry the fault current before the contactor opens).

2.1.2.5 Component Characteristics Next, the characteristics of each of the individual sub-components in the dc supply are discussed. There is an efficiency advantage of a full-wave bridge over the center-tapped geometry because the full-wave bridge constantly loads the secondary of the transformer, whereas the center-tapped geometry unloads the transformer on alternate half cycles when that particular portion of the winding is not conducting current. Also, during the recovery time of the diodes, it is possible for shock-excited oscillations to arise, damaging the transformer insulation or the reverse-biased diodes in the center-tapped geometry.

Figure 2-4 shows a summary from Mlynar of various design terminologies for diodes and transformers that is presented for general information. The key parameters of concern in dc supply design are the repetitive peak reverse voltage applied to the

VOLTAGE

1. **Design PRV** - the arithmetic sum of the individual peak reverse voltages of all seriesed rectifiers; the absolute maximum reverse voltage that the assembly can withstand without catastrophic failure.
2. **Rated PRV** - the maximum reverse voltage normally expected based on the actual circuit operating conditions; the design peak minus any safety factor.
3. **Transient PRV** - the maximum allowable transient reverse voltage that might be expected under normal operating conditions.
4.* **Blocking PRV** - the maximum value of repetitive peak reverse voltage rating permitted by the manufacturer under stated conditions.
5. **Output voltage** - normally applies to the DC voltage output of a complete rectifier assembly.
6. **AC voltage** - the RMS input value, either line to line or line to neutral, of voltage applied to the rectifier assembly or bridge.

CURRENT

1. **Rated current** - the average value of DC current, generally the output current of a rectifier assembly; however, can be the per-leg average output current.
2. **Short circuit current** - the value of current that the output load could experience under continued short circuit conditions, usually limited by circuit impedance.
3.* **Asymmetrical fault current** - the peak value of current delivered to the output load caused by overshoot of the short circuit current; typically 1.5 to 1.8 times the short circuit current allowed by circuit conditions.

KVA

1.* **Three-phase KVA using three single-phase transformers** - system KVA is equal to three times the individual KVA of the single-phase transformers; however the system impedance is 3 times the individual single-phase impedance value.
2. **Short circuit KVA** - maximum KVA available under short circuit conditions.

Fig. 2-4. Terminology—diodes and transformer (courtesy of Westinghouse).

diode, the worst-case asymmetrical fault current that the diode string can experience under start-up conditions or short-circuit faults in the load, and the kVA or power rating of the single-phase and three-phase power transformers. These are indicated by asterisks (*) in Fig. 2-4. In general, most of the needed design information can be obtained from the table from Mlynar shown in Fig. 2-5; however, the question of transients and some of the simplifications possible in designing for low ripple are not in the illustration. These questions are discussed in this book.

In multiphase power supplies, each phase has a unique role to play. For very high powers (megawatts at hundreds of kV), the most effective alternative is the six-phase series bridge (reference 12-1-1-B in Fig. 2-5) because the voltage stresses on the transformer can be kept quite low (essentially half), keeping the corona level low. Interphase

2.1 dc Power Supplies

Fig. 2-5. dc power supply data for resistive design loads (courtesy of Westinghouse).

CONFIGURATION	SINGLE PHASE HALF-WAVE	SINGLE PHASE FULL-WAVE CENTER TOP	SINGLE PHASE FULL-WAVE BRIDGE	THREE PHASE HALF-WAVE WYE	THREE PHASE FULL-WAVE BRIDGE
SCHEMATIC					
SYMBOLIC NOTATION	1-1-1-H	2-1-1-C	4-1-1-B	3-1-1-Y	6-1-1-B
OUTPUT WAVEFORM					
PEAK REVERSE VOLTS PER RECTIFIER LEG (WORKING PRV)	$1.414\,V_{AC}$	$2.828\,V_{AC}$	$1.414\,V_{AC}$	$2.45\,V_{AC}$	$1.414\,V_{AC}$
PEAK REVERSE VOLTS PER RECTIFIER LEG (WORKING PRV) RESTIVE OR INDUCTIVE LOAD ONLY	$3.14\,V_{DC}$	$3.14\,V_{DC}$	$1.57\,V_{DC}$	$2.09\,V_{DC}$	$1.05\,V_{DC}$
RMS VOLTS OUTPUT RESISTIVE OR INDUCTIVE LOAD ONLY	$1.57\,V_{DC}$	$1.11\,V_{DC}$	$1.11\,V_{DC}$	$1.02\,V_{DC}$	$1.00\,V_{DC}$
PEAK VOLTS OUTPUT RESISTIVE OR INDUCTIVE LOAD ONLY	$3.14\,V_{DC}$	$1.57\,V_{DC}$	$1.57\,V_{DC}$	$1.21\,V_{DC}$	$1.05\,V_{DC}$
NO LOAD RMS A.C. VOLTAGE [$V_{AC(NL)}$]	$2.22\,V_{DC(NL)}$	$2.22\,V_{DC(NL)}$	$1.11\,V_{DC(NL)}$	$1.48\,V_{DC(NL)}$	$.74\,V_{DC(NL)}$
AVERAGE D.C. OUTPUT CURRENT PER RECTIFIER LEG	$1.00\,I_{DC}$	$.5\,I_{DC}$	$.5\,I_{DC}$	$.333\,I_{DC}$	$.333\,I_{DC}$
RMS CURRENT PER RECTIFIER LEG	$1.57\,I_{DC}$	$.707\,I_{DC}$	$.707\,I_{DC}$	$.577\,I_{DC}$	$.577\,I_{DC}$
$I_{AC\,RMS}$ RMS CURRENT IN THE LINE BETWEEN THE TRANSFORMER AND THE RECTIFIER	$1.57\,I_{DC}$	$.707\,I_{DC}$	$1.00\,I_{DC}$	$.577\,I_{DC}$	$.816\,I_{DC}$
RATED POWER OF THE TRANSFORMER PRIMARY	$2.47\,I_{DC}V_{DC(NL)}$	$1.11\,I_{DC}V_{DC(NL)}$	$1.11\,I_{DC}V_{DC(NL)}$	$1.21\,I_{DC}V_{DC(NL)}$	$1.05\,I_{DC}V_{DC(NL)}$
RATED POWER OF THE TRANSFORMER SECONDARY	$3.5\,I_{DC}V_{DC(NL)}$	$1.57\,I_{DC}V_{DC(NL)}$	$1.11\,I_{DC}V_{DC(NL)}$	$1.48\,I_{DC}V_{DC(NL)}$	$1.05\,I_{DC}V_{DC(NL)}$
INTERPHASE TRANSFORMER RMS VOLTAGE (LINE TO NEUTRAL)	—	—	—	—	—
EQUIVALENT POWER RATING OF INTERPHASE TRANSFORMER	—	—	—	—	—

V_{AC} = RMS A.C. SECONDARY VOLTAGE.
V_{DC} = AVERAGE D.C. OUTPUT VOLTAGE OF RECTIFIER. PRV = $\sqrt{2}\,V_{AC(NL)}$.
I_{DC} = AVERAGE D.C. OUTPUT CURRENT OF RECTIFIER.
$V_{DC(NL)}$ = AVERAGE NO LOAD D.C. OUTPUT VOLTAGE.
$I_{DC}V_{DC(NL)}$ = IDEAL OUTPUT POWER OF THE RECTIFIER.

THREE PHASE DOUBLE WYE	FOUR PHASE FULL-WAVE CENTER TAP	FOUR PHASE FULL-WAVE PARALLEL BRIDGE	FOUR PHASE FULL-WAVE SERIES BRIDGE	SIX PHASE STAR (THREE PHASE DIAMETRIC)	SIX PHASE PARALLEL BRIDGE WITHOUT INTERPHASE TRANSFORMER	SIX PHASE PARALLEL BRIDGE WITH INTERPHASE TRANSFORMER	SIX PHASE SERIES BRIDGE
6-1-1-Y	4-1-1-C	2(4-1-1-B)	8-1-1-B	6-1-1-S	2(6-1-1-B)	12-1-1-B	12-1-1-B
2.45 V_{AC}	2.828 V_{AC}	1.414 V_{AC}	1.414 V_{AC}	2.828 V_{AC}	1.36 V_{AC}	1.414 V_{AC}	1.414 V_{AC}
2.09 V_{DC}	2.22 V_{DC}	1.11 V_{DC}	.785 V_{DC}	2.09 V_{DC}	1.01 V_{DC}	1.05 V_{DC}	.524 V_{DC}
1.00 V_{DC}	1.00 V_{DC}	1.00 V_{DC}	1.00 V_{DC}	1.00 V_{DC}	1.00 V_{DC}	1.00 V_{DC}	1.00 V_{DC}
1.05 V_{DC}	1.1 V_{DC}	1.11 V_{DC}	1.11 V_{DC}	1.05 V_{DC}	1.05 V_{DC}	1.05 V_{DC}	1.05 V_{DC}
1.71 $V_{DC(NL)}$	1.57 $V_{DC(NL)}$	.785 $V_{DC(NL)}$	.555 $V_{DC(NL)}$	1.48 $V_{DC(NL)}$	.715 $V_{DC(NL)}$	.74 $V_{DC(NL)}$	.37 $V_{DC(NL)}$
.167 I_{DC}	.25 I_{DC}	.25 I_{DC}	.5 I_{DC}	.167 I_{DC}	.167 I_{DC}	.167 I_{DC}	.333 I_{DC}
.289 I_{DC}	.50 I_{DC}	.50 I_{DC}	.707 I_{DC}	.408 I_{DC}	.408 I_{DC}	.289 I_{DC}	.577 I_{DC}
.289 I_{DC}	.50 I_{DC}	.707 I_{DC}	1.00 I_{DC}	.408 I_{DC}	.577 I_{DC}	.408 I_{DC}	.816 I_{DC}
1.05 $I_{DC} V_{DC(NL)}$	1.11 $I_{DC} V_{DC(NL)}$	1.11 $I_{DC} V_{DC(NL)}$	1.11 $I_{DC} V_{DC(NL)}$	1.28 $I_{DC} V_{DC(NL)}$	1.01 $I_{DC} V_{DC(NL)}$	1.01 $I_{DC} V_{DC(NL)}$	1.01 $I_{DC} V_{DC(NL)}$
1.48 $I_{DC} V_{DC(NL)}$	1.57 $I_{DC} V_{DC(NL)}$	1.57 $I_{DC} V_{DC(NL)}$	1.57 $I_{DC} V_{DC(NL)}$	1.81 $I_{DC} V_{DC(NL)}$	1.43 $I_{DC} V_{DC(NL)}$	1.05 $I_{DC} V_{DC(NL)}$	1.05 $I_{DC} V_{DC(NL)}$
.252 $V_{DC(NL)}$	—	—	—	—	—	.085 $V_{DC(NL)}$	—
.162 $I_{DC} V_{DC(NL)}$	—	—	—	—	—	.059 $I_{DC} V_{DC(NL)}$	—

Fig. 2-5. dc power supply design data for resistive loads (continued).

transformer power supplies in three-phase configurations can suffer from transient saturation effects when a pulse load is used (like a resonant-charging circuit or discharge networks).

Six-phase parallel operation (without interphase transformers) is acceptable, except that fairly fast recovery diodes are needed to avoid one diode starting to conduct while the other one is trying to turn off, producing a high reactive kilovar loading of the secondary of the transformer. The reactive power that flows through the winding and the ohmage loss in the winding create significant transformer heating. The three-phase, full-wave bridge is the best general-purpose, three-phase supply, with either a Y or a Delta secondary configuration.

Figure 2-6 illustrates some rectifier circuits. In this case, there is a large inductor in series with the rectifier assembly, and this inductance is chosen to be sufficiently large so that at all times there is a current flow through that inductor. As discussed later in this book, there is a critical value of the inductance, depending upon the minimum load current. If the current in the load becomes sufficiently small, a point is reached at which the inductor no longer conducts current continuously into the capacitance. There is thus no longer a unidirectional current flow through the circuit filter network so that the stored energy in the inductor can start impressing current into the filter capacitor if the net current flow is small enough, and the output voltage then rises to the peak voltage of the transformer. This is the turnover point in the regulation performance of the system, going from the choke input to capacitor input performance. This means that when a choke input supply is designed, there is a minimum load current that must flow through the output loop at all times. This is normally not a particularly difficult problem if the choke used has inductance that is a function of the current passing through it (swinging choke). The range in inductance from the minimum current to full-load current can normally exceed five to one, falling at the higher current as the choke core approaches saturation.

This summary has indicated some of the salient aspects of single and multiphase dc power supply design. The tables allow selection of the RMS value of the transformer secondary voltage and current ratings. The maximum diode inverse voltage, for example, for the double three-phase series connection was 1.05 and for the single-phase, full-wave center-tapped connection it ws 3.14. Analysis by Greenwood discloses that the peak overvoltage in any of these systems under stop-and-start conditions is about two. On rare occasions, peak overvoltages of three have been noticed. The usual engineering practice is to multiply diode inverse voltage requirements by two. This factor of two (a safety factor) almost guarantees that the device will not break down. However, for very high voltages (for example with 200 to 500 kV diodes), the ratio of cost to kV becomes larger and larger, making high kV ratings ever more expensive. In that particular case, the design is chosen to be as close to the safety margin as possible but, in general, not below 1.5 times the anticipated inverse voltage per leg is used. For example, for the above two cases, the inverse voltage ratings for the diode strings would be 1.6 and 4.7 times, respectively.

Returning to the full-wave bridge, its principal advantage is the superior transient suppression (Figs. 2-7 and 2-8) and more efficient use of transformer kVA. The center-tap design has a pair of solid-state diodes connected to a filter capacitor. During the transition when this diode becomes nonconducting, a small recombination charge current flows through it, leading to a step in the recovery response of the diode. When that

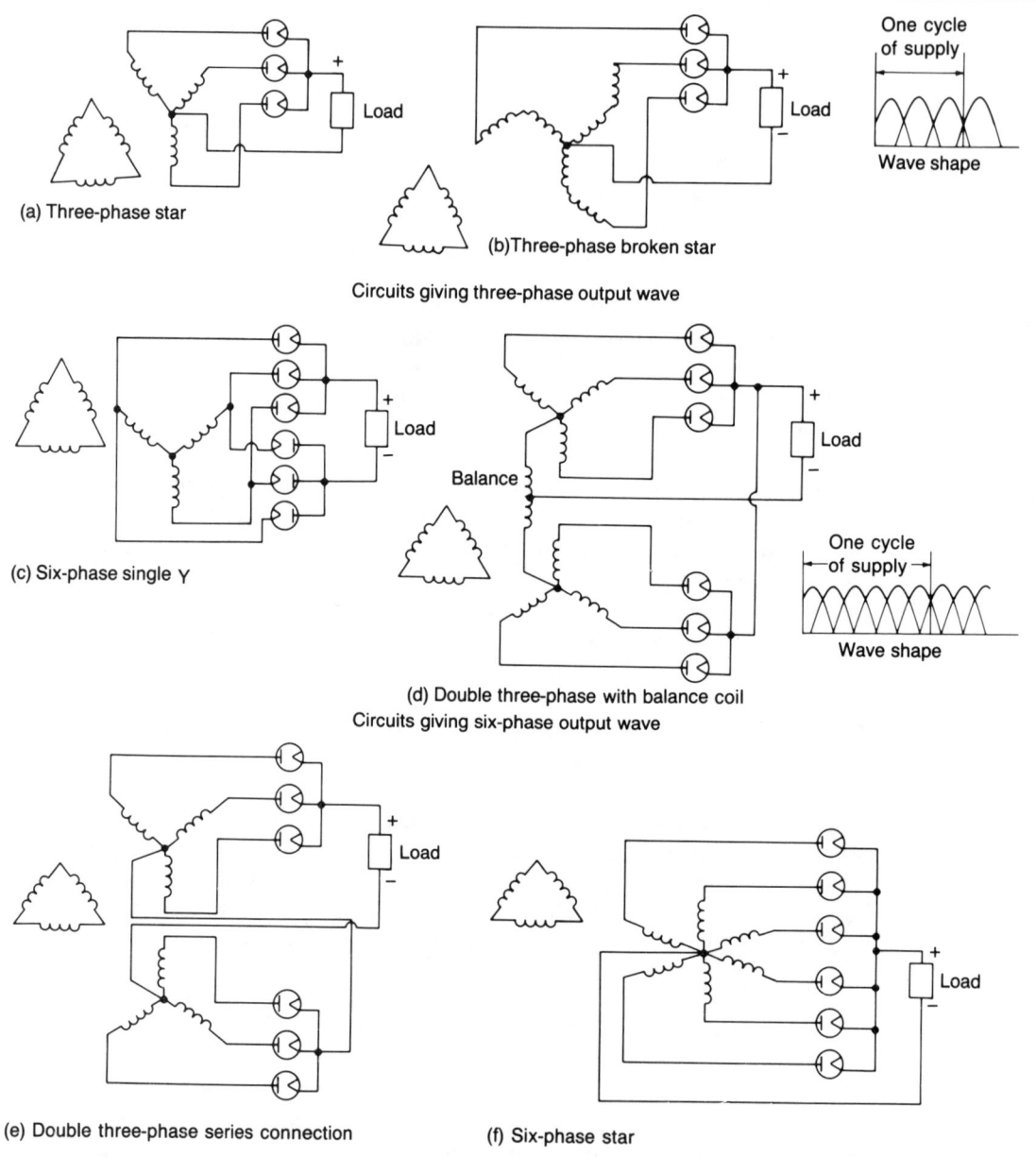

Fig. 2-6. *Resistive-load or choke-input rectifier circuits. (Redrawn after Terman and used by permission of the Board of Trustees of the Leland Stanford Junior University).*

happens, this current shock excites this part of the transformer into oscillation. An RC shunt network across the secondary will tend to dampen these oscillations. In the case of the full-wave bridge, the transformer is fully loaded for both cycles, minimizing the formation of such transients.

THE MAIN ADVANTAGE OVER THE CENTER TAP DESIGN IS THE FAR SUPERIOR TRANSIENT SUPPRESSION OF FWB DESIGN DURING RECTIFIER COMMUTATION. ALSO, THE MAXIMUM INVERSE VOLTAGE IN THE FWB CASE IS ONE-HALF THE CENTER TAP DESIGN.

<u>CHOKE INPUT</u>

LET V_{DC} BE THE (FULL LOAD) OUTPUT DC VOLTAGE AT CURRENT I_{DC}. DEFINE % REGULATION $= \dfrac{V_{NL} - V_{DC}}{V_{NL}}$

FOR A CHOKE INPUT FILTER THE MINIMUM INDUCTANCE AT THE MINIMUM LOAD CURRENT IS: $\dfrac{\omega L}{R_L^{max}} \geq \dfrac{E_{ripple}}{V_{DC}}$ $\quad \omega = 2\pi$ x ripple frequency $\quad E_{ripple}$ = peak ripple voltage

IF A LOWER CURRENT IS DRAWN, THEN THE CHOKE BECOMES INEFFECTIVE AND THE OUTPUT VOLTAGE RISES TO THE PEAK VOLTAGE OF THE TRANSFORMER SECONDARY.

<u>FOR FWB DESIGN</u>

$\dfrac{\omega L}{R_L^{max}} > 0.7 \quad \therefore \quad \dfrac{L}{R_L^{max}} > 10^{-3} \quad$ AND $\quad R_L = \dfrac{V_{DC}}{I_{DC}}$

<u>RIPPLE</u>

PEAK VALUE $\approx \dfrac{0.7\,V_{DC}}{\omega^2 LC} \qquad$ CF FOR RC FILTER $\approx \dfrac{0.7\,V_{DC}}{\omega RC}$

WHERE $\omega = 2\pi\,(2 \times 60)$

Fig. 2-7. Full-wave bridge (FWB) power supply.

2.1.2.6 Ripple Calculations and Filters In the choke input case for the FWB, if V_{dc} is the output dc voltage at I_{dc}, then if regulation is defined this way: the no-load voltage minus the dc voltage over the no-load voltage, then the choke input filter minimum inductance for a given R_L is:

$$\dfrac{\omega L}{R_L^{max}} \geq \dfrac{E_{ripple}}{V_{dc}}$$

The angular frequency, ω, (2π times the ripple frequency) times L (at the current V_{dc}/R_L^{max}) over the maximum value of the load resistor has then to be larger than the peak ripple voltage over the dc voltage. If a lower current is drawn, there is a point at which the choke is no longer effective (Fig. 2-9). If instead of an LC filter, an RC filter is used, the ripple at ω (the angular ripple frequency) would be considerably larger. From the calculation of the specified ripple required, work back and calculate the value of the LC product, under the constraint of the relations in Fig. 2-8. That capacity, C,

SINCE: $\dfrac{R_L^{max}}{L} \le 10^{-3}$

$E_{output}^{ripple} \approx 10^{-6} \cdot \dfrac{V_{DC}}{(LC)}$ FOR EXAMPLE, $C = \dfrac{C_o}{2}$ IN FIG. 3

COMPARE TO THREE-PHASE FWB WHERE:

$\dfrac{R_L^{max}}{L} < 40{,}000$ AND $E_{ripple}^{output} \approx 10^{-8} \cdot \dfrac{V_{DC}}{(LC)}$

THUS THE DESIRED RIPPLE AND LOAD RANGE DEFINES THE VALUE OF L. SWINGING CHOKES OFFER IMPROVED REGULATION FOR A LARGE RANGE IN R_L VALUES: A $\dfrac{\Delta L}{L}$ OF FIVE TIMES IS OBTAINABLE.

THE VOLTAGE RATING OF CHOKE = $2 \times V_{DC}$.

TRANSFORMER PARAMETERS IN THE FWB SUPPLY
___

1. $V_{RMS} = 1.11\, V_{DC}$
2. $KVA = 1.11$ DC OUTPUT POWER

} NEGLECTING CORE AND WINDING LOSSES

DIODE PARAMETERS IN THE FWB SUPPLY

$PRV \ge 2\, V_{DC}$ (x2 SAFETY FACTOR)

$I_{AV} = \dfrac{1}{2} I_{DC}$ AND $I_{PK} = I_{DC}$

R:
___

SELECT TO BE LESS THAN 10% OF THE LEAKAGE RESISTANCE OF CAPACITOR C_o AT THE MAXIMUM TEMPERATURE. THIS ENSURES AN EQUAL DIVISION OF VOLTAGE TO WITHIN ±20%.

Fig. 2-8. Full-wave bridge (FWB) power supply (continued).

is the total capacity on the output of the LC filter, but in the case of the full-wave bridge there are two capacitors, each of value C_o in series so that:

$$C = \dfrac{C_o}{2} \qquad (2\text{-}1)$$

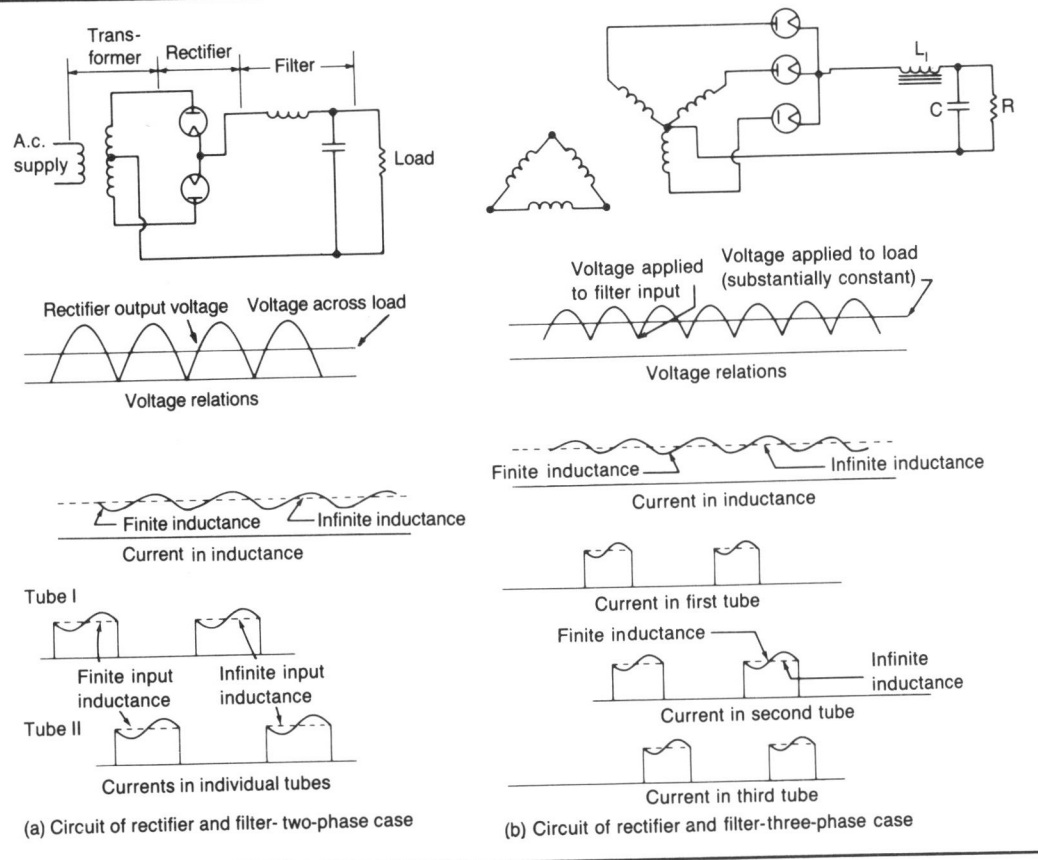

Fig. 2-9. Effect of a finite input inductance on current waveshapes. (Redrawn after Terman and used with permission of the Board of Trustees of the Leland Stanford Junior University).

If a broad range in output current is required. L_{max} can be derived from the amount of tolerable ripple, the minimum load current, and the swinging choke inductance ranges available. Given L and C, all the remaining design parameters in the system can be calculated.

In the full-wave bridge, the peak output ripple is:

$$10^{-6} \cdot \frac{V_{dc}}{LC} \qquad (2\text{-}2)$$

In this case, V_{dc} is the actual dc output voltage. This is to be compared to the three-phase supply, where the increased ripple frequency results in a factor of 100 times reduction in the ripple. That is one of the real advantages of three-phase over single-phase operation, other than regulation. It is either easier to filter to a given ripple level or, if a much smaller ripple is desired, the same size of inductors and capacitors can be used to gain several orders of magnitude reduction in the ripple. The major point about a swinging choke is that if there is a large range of load resistance to deal with, a choke whose inductance increases as the current decreases can keep the above relationships relatively constant. Now, a typical range in inductance is 5 to 1. Normally, in the design of filter choke assume that at turn-on there is zero voltage on the output, while peak volts are on the input. The insulation stress on the choke at 60 Hz is normally taken as twice that peak voltage—the safety margin being a factor of two. Recall that the safety margin in the diodes is also a factor of two. The rationale for that is in the economics of manufacture. It is relatively cost effective to design, up to voltages of 100 kV, with such an insulation and diode safety factor. The damping network controls the transient voltages to a factor of twice the peak voltage. This all fits in rather nicely with the design factor of two for the insulation level and defines the peak overvoltages to be experienced under worst-case conditions.

Where substantial high-voltage insulation is necessary in high-voltage chokes, the choke can cost as much as or more than a modest-sized power transformer. A typical choke, that might cost several hundred dollars to buy in an open mounting configuration, could cost a factor of two to three times more if it is hermetically sealed with a pair of high-voltage, feed-through bushings. So the optimal solution is generally to put all power supply components in one container, fill it with transformer oil, and put a top on it.

2.1.2.7 Capacitor Design and Voltage Equalization. A shunt resistance, R, is needed across each of the filter capacitors, C_o. Normally if R is chosen to be 10 percent of the leakage resistance at the maximum operating temperature, then the percent variation of the voltage across C_o under all conditions caused by temperature changes is ± 20 percent. The capacitor heats up and cools down (from the environment and ripple current heating), and the leakage resistance can vary widely. If the resistors were not there, it is conceivable that one of the capacitors could have a leakage resistance five to six times smaller than the other, causing the voltage rating of the latter to be dangerously exceeded.

2.1.2.8 Regulation Control with Bleeder Resistors. The third option in regulation control for the choke input dc supply when no-load to full-load operation is needed is to use a bleeder resistor. If the power supply is going to operate effectively from no load to some average current, this resistance can be chosen to draw sufficient current to ensure that the swinging choke is operating within the constraint of the relations in Fig. 2-8. The bleeder resistor can be combined with the shunt resistance, R, to simplify construction.

2.1.2.9 Diode Stack Voltage Grading. Consider the question of voltage division across the diode stacks as shown in Fig. 2-10. Consider a number (N) of diodes in series and a very sharp transient applied across them (the typical capacitances of each diode being 100 to 300 pF). The transient wavefront appears across the input diodes in proportion to the diode to distributed (C_p) capacitance ratio. It is quite conceivable that a particular diode experiences a transient voltage well in excess of its reverse voltage holdoff capability. What normally happens in a high-voltage stack is that this diode burns out, and with the next pulse the next diode burns out, then the next, and so forth all the way down the stack. The voltage across each of those diodes can be kept close to the

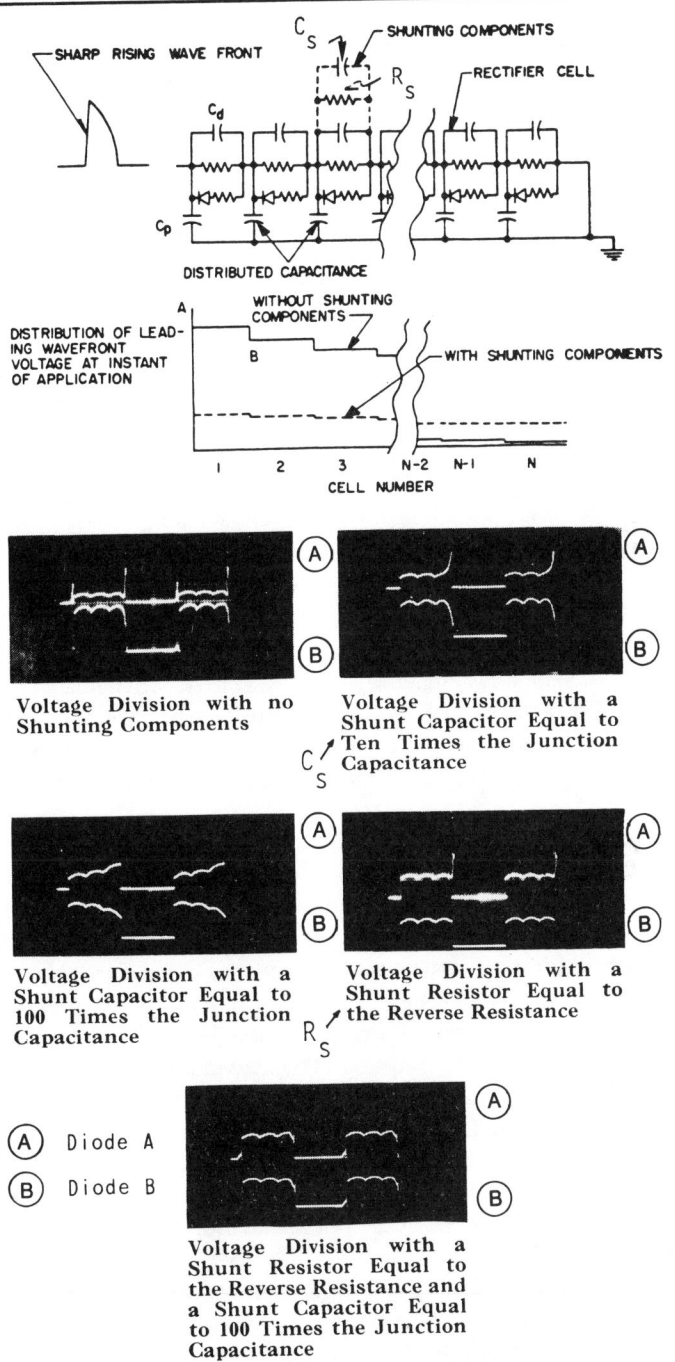

Fig. 2-10. Transient voltage division across diode stack elements (courtesy of Westinghouse).

peak voltage divided by N, by using relatively large capacitors across the diode (capacitors, in fact, that are much larger than any of the distributed capacitances or diode capacitances in the circuit). Under these conditions, the voltage grading becomes quite uniform and the transient voltage jumps across each of the divides are quite small. In Fig. 2-10 is an illustration of what happens when no shunting components are included in the system. There are very sharp spikes at the beginning and the ending of the commutation intervals. At the end of the interval where the diode becomes nonconducting, the energy stored in the diode and distributed capacitances has to discharge. Where does it go? Since the peak voltage can then exceed the diode PRV, the diode is forced back into conduction in the reverse direction, avalanches, and the current pulse travels up through the stack. That little current bump can cause significant damage by destroying the diodes during the avalanche breakdown. If an appropriate shunt resistor-capacitor network is inserted across each diode, the input and output become virtually identical. The shunt capacitances are normally in the range of 1 to 3 nF. Most of the modest-sized diode structures (1 to 5 A) can be fabricated up to the level of several hundred kV.

Experimental diode stacks can be built in the laboratory quite inexpensively. Avalanche-protected diodes are unnecessary. An avalanche-protected diode is one that behaves like a zener diode if a voltage in excess of its avalanche-rated inverse voltage is applied. That is supposed to protect it. There are a number of manufacturers who do not make RC-compensated diode stacks. Virtually everything they make is avalanche protected. The diode leakage is matched to within a few percent, and many of these diodes are stacked in series. A number of people who are using them end up incorporating RC compensation resistors and capacitors across each of those avalanche stacks. The point here is that the RC-compensated stack guarantees equal voltage division even for very fast-transients, particularly as experienced in laser systems.

2.1.2.10 High-Voltage Diode Stacks. Figure 2-11 shows several high-voltage assemblies made by Westinghouse, illustrating individually RC diode compensation. The heat sink is a large, thick aluminum plate between each of the diode assemblies. At the top is an illustration of one of the capacitor-resistor shunt assemblies for each of the 35 A diode elements. Westinghouse does make one now that uses this geometry but has a fully screened heat-sink arrangement so that they can be used in circuits where there are very steep transients on the leading edge. The reason that steep transients are a problem is that if a very fast transient is applied that rises more quickly than the propagation time of the stack, large transient overvoltages can occur and LC resonances can be shock excited, causing diode elements to avalanche and burn out. If the diode assembly is encased in a metal box with only a little slot in it, one side being the input and the other side being the output side of the diode, two equipotential planes are formed. What happens inside is screened from the outside world. These devices are made for use to 150 A at 500 kV.

Most of the small stacks that can be bought (average currents up to 1 A) generally use avalanche diodes. The solid-state world being what it is, it is straightforward at this current level; automated machines are available that select and grade according to reverse current leakage factors. The machines take the diodes in each group and stack them all together. To use one of these stacks in a fast circuit, the best thing to do is put a capacitive shunt across the total circuit. The shunt size can be estimated from the

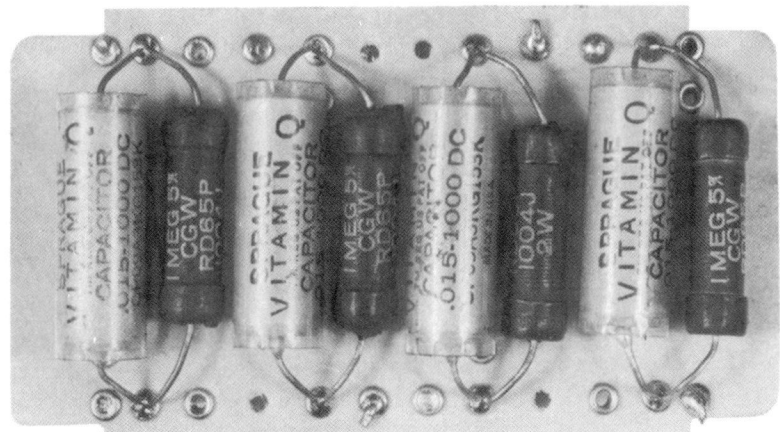

Back side of a module showing shunting components for four 35 ampere rectifiers

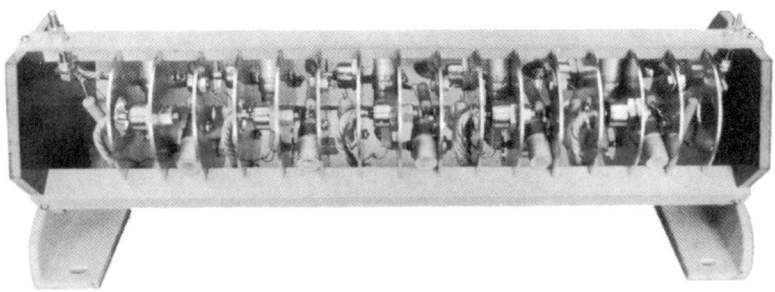

70 ampere high voltage assembly utilizing circular aluminum cooling fins

Fig. 2-11. High-voltage, RC-compensated diode stacks (courtesy of Westinghouse).

number of diode elements in the stack and the assumption that a shunt network of 3 nF would be correct for each individual diode.

2.1.2.11 Diode Lifetime and Reliability. Individual RC-compensated diodes can be bought from IRC and EDI, among others. There is a curve in Fig. 2-12 that gives an indication of diode reliability as a function of junction temperature, showing a very strong argument in favor of buying as much average current capability as can be afforded in a diode stack for maximum reliability. There has been a tendency in the high-voltage supply business over the years to use 1N4000 series diodes for all 1 A power supply stacks. Unfortunately, if the junction temperatures run relatively high, the percent failure might increase. For a given known probability of failure of the device, the weakest link in the system tends to be a solid-state diode running at a high junction temperature. The second weakest link tends to be transient suppression. If compensated properly and carefully, the devices that are statistically still most likely to fail at high temperatures in the system are the solid-state diodes. They are very good, but the hotter they run, the more problems they are likely to have. The power dissipated in the diode in watts (Fig. 2-12) is given in the manufacturer's data as a function of average current. The case temperature also goes up so many degrees for every watt of average power dissipated. That number is available from the manufacturer. To specify a certain failure probability, Fig. 2-12 gives the junction temperature, and the corresponding case temperature can be calculated, defining what the heat sink thermal resistance (degrees C/watt) rise above ambient has to be for that given case and ambient temperature. RCA has a set of notes in their audio amplifier applications brochure that does this calculation rather nicely. This is a representative diode curve in Fig. 2-12 for a 10 A silicon diode that Westinghouse made in the 1960s. It might be a lot better today. There is a reliability problem if the junction temperatures become too high. If it does not cost a great deal, it is advantageous from a reliability point of view to keep the junction temperature down. When it begins to cost a lot, it is necessary to make the heat sink capacity larger, reducing the junction temperature.

The manufacturer will generally provide his lifetime expectancy at least at the 99.5 percent confidence level. (There are some other applicable military standards for diodes.) Operating at the rated junction temperature (and therefore at a specific average current for the given thermal resistance) for a failure rate of 0.5 percent, only five of a thousand of these devices will fail in a thousand hours. That assumes that the distribution is normal. There is another way to make considerable improvement to this reliability, and that is to fold into that the number of hours the device is actually on versus on-off cycling. That is another component of reliability previously unmentioned. There is a report by the Boeing Corporation that provides additional information in this area. Formerly, it was assumed that the power supply will work for so many hours, and the diode will be on and under constant load. Cycling it on and off can buy lifetime. If it is on for a short period compared to thermal equilibrium time (adiabatic operation) and off for a couple of days, this curve is weighted, enabling the use of smaller devices. Experimentally, this appears to work, although the data base is sparse for high-power devices.

2.1.3 Capacitor Input FWB dc Power Supply

The other interesting power supply is the small FWB built in the laboratory. It serves as a full-wave bridge power supply but uses a capacitor input instead of an inductor input filter at the output of the FWB. The capacitor charges to the peak secondary voltage. Figure 2-13 illustrates the characteristics of the full-wave, capacitor-input power supply. In a FWB, the ripple frequency is doubled. There are some changes in the trans-

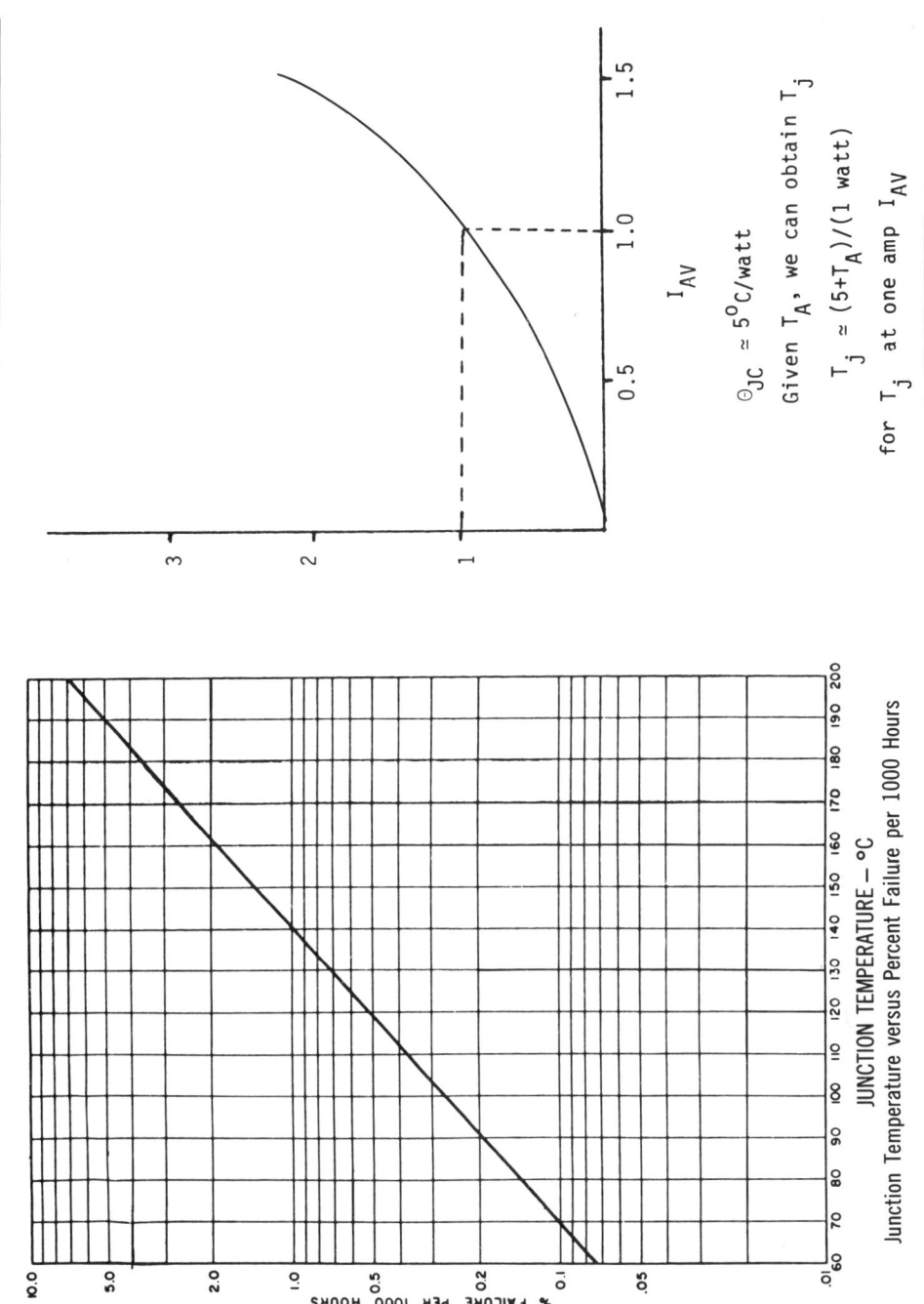

Fig. 2-12. Solid-state rectifier reliability (courtesy of Westinghouse).

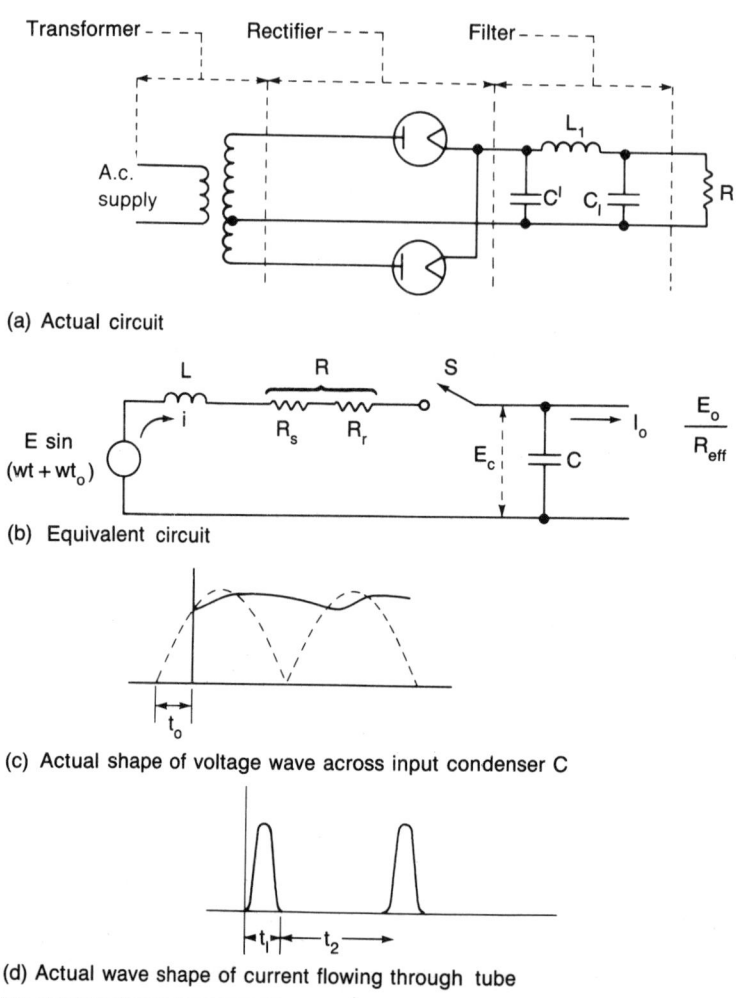

Fig. 2-13. Actual and equivalent circuits of a full-wave, capacitor-input power supply. (Redrawn after Terman and used by permission of the Board of Trustees of the Leland Stanford Junior University).

former requirements and the diode requirements. What is discussed here are worst cases based upon very good transformer performance—the lowest core losses and the lowest winding losses practical—and then some relations can be derived that will give an upper boundary to device requirements. This boundary will allow determination, for a diode with specific characteristics and a transformer with a given set of characteristics, of whether or not any of the components are going to burn out. In a transformer, the kVA or power requirement is as reported for the choke input plus the three loss terms (which the manufacturers provide) shown in Fig. 2-14. When dealing with a relatively low-current supply (an ampere or two at a few hundred volts), these losses matter little. These factors really become important when dealing with high currents or fairly high-voltage sup-

CAPACITOR INPUT

- Short out input inductor L in Fig. 2-3.
- $C = \dfrac{C_0}{2}$ now is charged to the peak secondary voltage.

(1) Transformer Parameters

A. $V_{RMS} = 0.7\ V_{DC}$

B. $KVA = 1.11 \times P_{DC}$ + winding losses + core losses + diode losses.

<u>Calculate</u>: $\left.\begin{array}{l} I_{RMS} \approx 5\ I_{DC} \\ I_{PK} \approx 40\ I_{DC} \end{array}\right\}$ from the diode curves the power loss is obtained. (This can be estimated as ≈ 1 watt per 2-A peak.)

(2) Diode Parameters

A. $PRV = 2\sqrt{2}\ V_{RMS}$ with a safety factor of 2
   - At very high voltages this is usually reduced to:
   $PRV = 1.5\sqrt{2}\ V_{RMS}$.

B. $I_{RMS} = 5\ I_{DC}$ $I_{PK} = 40\ I_{DC}$

C. The ripple specified determines the filtering needed which then defines C. Regulation defines the allowable series current limiting resistance:

For FWB: Good regulation and low ($\approx 20\%$) ripple demands

$\nu\, \omega\, C\, R_L = 50$ $\nu = 2$ $F = 60$ Hz

Note: for half-wave $\nu = 1$, voltage doubler $\nu = 1/2$: the following applies in general:

The ripple range is 2% (FWB) to 5% (FW doubler).

- Capacitor RMS ripple current $\approx 5\ I_{DC}$
- At 120 Hz ripple frequency for FWB
- At 60 Hz ripple frequency for FW doubler.

Fig. 2-14. FWB power supply.

plies where the transformers must be specified carefully for the best cost/performance tradeoff. During iteration with the manufacturer, there will be a cost tradeoff. The manufacturer can provide low copper losses that tend to improve regulation, but it will cost considerably more. As a rule of thumb, 2 to 10 percent is a reasonable regulation range to consider. Asking for 1 percent is very expensive for large supplies.

These are manufacturing restrictions, not fundamental engineering or physics restrictions. The diode losses were estimated from the curve on the reliability figure. The diode conducts such an average current and dissipates so many watts, and these are lost watts. They must be compensated for by increasing the power capability of the power transformer. From the diode curves in the RCA handbook, the worst-case RMS current rating for the transformer is five times I_{dc}, and I_{pk} is 40 I_{dc}. This indicates that another close look is needed because RMS current capability in a transformer has a large copper cost for a given percentage loss and this power loss must be folded into the total kVA lost in the transformer.

2.1.3.1 Transformer Design. In designing supplies where the transformer is to be bought, spend a considerable amount of time interfacing with the transformer manufacturer. The manufacturer can build into transformers a certain amount of reactive component that does not represent a real power loss but that does allow the limiting of the fault current in the secondary when it is shorted. Effectively, this puts an inductor in series with the transformer primary as a current-limiting element, physically, in the way the transformer is built. In addition, transients can be significantly reduced by the different ways in which the transformer windings are configured. A great deal depends on whether the primary is wound directly on the secondary, beside it, or on another leg of the core. By interfacing with the manufacturer, considerable money can often be saved for a given performance level requirement. A typical transformer cost saving on a large supply can be a factor of two or three, and for a transformer costing approximately 50,000 dollars, this can be worthwhile.

The most expensive approach is to prepare a set of design specifications, send them out for quotes, and wait for everybody to write back with the costs. It costs time and money. A more effective approach is to call the various people in the transformer-manufacturing field and talk with them. It might take a morning. When all that information is available, a more cost-effective design is achieved. In most cases, the cost of the telephone calls is more than recuperated. This is particularly true in cases of having to build very high kVA supplies at modest voltages (5 to 10 kV).

2.1.3.2 Diode Parameters. Diode costs can be staggering for the larger kVA high-current supplies. The parameter estimates given in Fig. 2-14 are a worst case and are inappropriate for very large (hundreds of kilowatts to megawatts) supplies. The analysis done in the RCA note advocates the selection of a PRV twice the peak voltage. This factor of 2 times is a reasonable approach to use for up to 100 kV supplies. Above 100 kV, it is probably too expensive and 1.3 to 1.5 times can be used, provided very careful fault-mode analysis modeling is executed. Westinghouse assures you with 90 percent confidence that 1.5 is sufficient PRV margin if a snubber is placed in the transformer primary and the secondary, as well as the filter choke. The worst-case RMS current rating of the diode should be five times the average current, and the peak current is

NOW THE DIODE SURGE RATINGS CAN BE ASSESSED:

LET $\underline{R}$ BE THE TOTAL SERIES RESISTANCE =

WINDING + EQUIVALENT DIODE RESISTANCE + RESISTIVE COMPONENT OF FILTER CAPACITOR REACTANCE.

$\underline{R}$ C = SURGE TIME CONSTANT

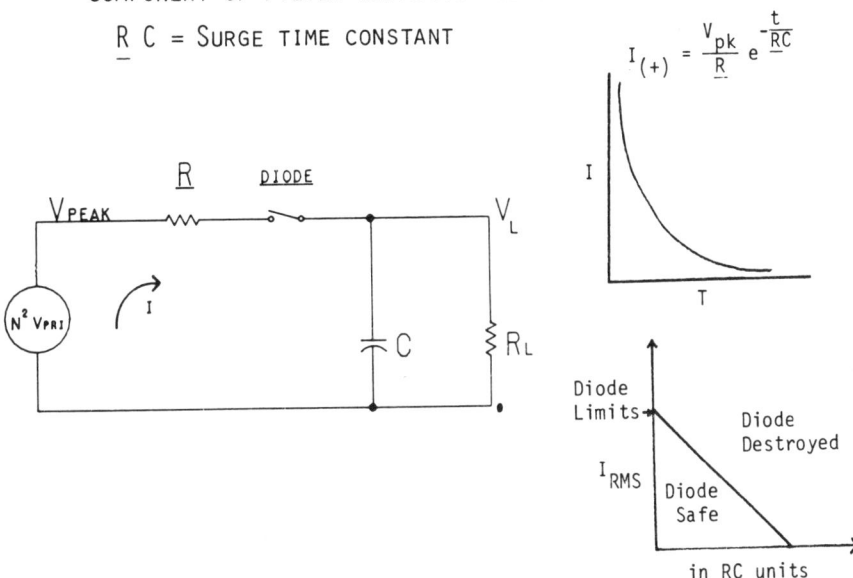

IF $\underline{R}$ IS NOT KNOWN THEN THE ITERATIVE TECHNIQUE MUST BE USED: ASSUME $\underline{R} \approx 0.01\ R_L$ (A WORST CASE). THEN, KNOWING FROM RCA APPLICATION NOTES:

$$I_{RMS} \cdot (\underline{R} \cdot C) = 0.7 \cdot V_{PEAK} \cdot C$$

CALCULATE I_{RMS} AND CHECK ON DIODE GRAPH. IF ABOVE THE LIMIT LINE, INCREASE $\underline{R}$. FOR A WORST-CASE ESTIMATE FOR $v \omega C R_L = 50$, $\underline{R} = 0.01\ R_L$ $I_{RMS} = 5\ I_{AV}$ $I_{PK} = 40\ I_{AV}$ (GENERALLY APPLICABLE)

Fig. 2-15. Diode surge ratings.

40 times the average current (Fig. 2-15). This can then be compared to detailed model predictions (using, for example, NET-2) in order to reduce the current requirements of the components to a safe minimum value.

2.1.3.3 Ripple Calculations. To obtain a degree of ripple reduction to 2 percent, find C from Fig. 2-14. Then the regulation-allowable voltage drop in the diode stack depends upon how much equivalent series resistance is present in each diode leg. Selecting, for 1 A dc, a 40 A peak diode, the output is faulted, and the peak current is obtained from the model (Figs. 2-15 and 2-16). For example, with a 100Ω load and a 1 A supply, faulting the output for a perfect transformer, diode, and zero input line impedance, there would be infinite fault current. There is, however, some internal resistance in the sup-

FOR A TRANSFORMER:

$$\% \text{ REGULATION} = \frac{V_{NL} - V_{FL}}{V_{FL}}$$

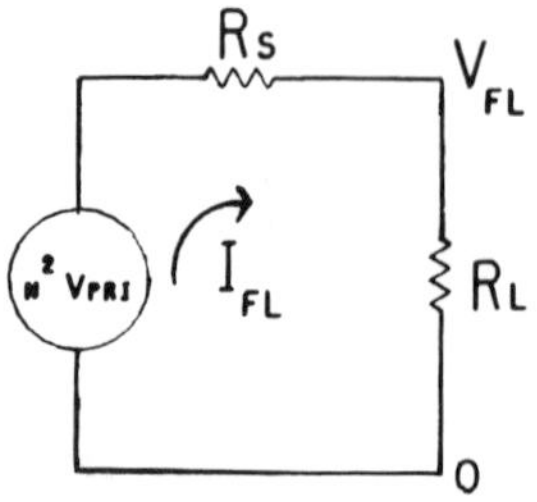

SINCE $V_{NL} - V_{FL} = I_{FL} R_S = \frac{V_{FL}}{R_L} R_S$

THEN $R_S = (\% \text{ REG}) R_L$

WHERE THE TRANSFORMER IS LOADED WITH A PURE RESISTANCE SO THAT SECONDARY POWER IS EQUAL TO DC POWER DESIRED PLUS LOSSES.

NOTE: GENERALLY $R_S \approx \underline{R}$ EXCEPT FOR LOW-POWER, HIGH-VOLTAGE UNITS.

FAULT CURRENT:

FOR THE WHOLE SUPPLY, MLYNAR HAS SHOWN:

$$I_{PEAK} \approx \frac{I_{DC} \times 1.8}{(\% \text{ REG})} \quad \text{ASYMMETRIC FAULT CURRENT}$$

— AVAILABLE ON TURN-ON INTO A SHORT CIRCUIT AT PEAK LINE VOLTAGE

Fig. 2-16. *Fault current at turn-on or output short circuit.*

ply. If the diode has a 40 A rating, this means that for a 100Ω load the internal supply resistance should be 100 over 40, which is 2.5Ω. With a 2.5Ω equivalent resistance inside the power supply, shorting the output of the power supply will keep the fault current to 40 A. Normally the transformer manufacturer can indicate the equivalent resistance of the transformer. This resistance is normally the dc copper resistance referred to the secondary of the transformer. There is also a small contribution from the diode stack. In this particular case for a 2.5Ω internal resistance, faulting the output using a 40 A diode, the circuit breaker would open or the fuse would blow, and the diode would survive.

These calculations can become more sophisticated for three-phase or more multiphase supplies or for complicated, very low ripple systems with a very high degree of regulation. In the full-wave bridge, if $\nu CR_L = 50$ is chosen, and $\nu = 2$ in this case, for good regulation the ripple is also low ($\cong 2\%$). In the half-wave supply $\nu = 1$, and for the case of a voltage doubler it is ½. This relationship allows the calculation of C and determination of the ripple range. The ripple range is 2 to 5 percent (peak to peak), in going from the FWB to voltage-doubler geometries. In attemps to reduce the amount of ripple further, the RCA curves reveal that peak current begins to rise enormously posing a clear trade-off situation. Should another LC filter be put in the supply? If desired to remain near the fault current levels of Fig. 2-15 for these three classes of supplies, it is almost always more cost effective to put another LC filter in the circuit. In fact, on very high-voltage supplies, the additional filters can be inserted into the negative line. This is a very inexpensive way to make a low-ripple, high-voltage supply with relatively inexpensive filter chokes. Most such chokes are rated for 1000 V RMS.

The inexpensive way to protect the chokes is to use a small spark gap (for example, Centralab Gap-Kap) that costs less than one dollar. It can be joined to a current-limiting resistor if desired. These gaps can handle about one joule of energy. For laboratory applications, they are good for a couple of hundred faults. They are not quite as effective as a nonlinear varistor that has a softer damping curve (so that as the voltage across it is increased, it gradually increases the current) but they are inexpensive. To stack 4 to 5 chokes like this in series (which, although not recommended, will work) the choke cores are floated, the core is tied to the next lower choke output lead (if two or more of them are in series), and the Gap-Kap placed across each set of choke leads.

2.1.3.4 Fault Currents and Overload Protection. During a fault in the output the chokes tend to develop the peak secondary voltage across them as they try to act as current regulators. The Gap-Kap shorts, essentially bypassing the fault current through the limiting internal resistance of the supply. There is another type of Gap-Kap that has a 0.01 μF capacitor inside it. These capacitors were extremely common when radios had tubes; the capacitors were used as protectors across the output transformers, particularly in radios that had remote extension speakers. Thyrites and metal-oxide varistors are much better but also far more expensive. For general laboratory applications, Gap-Kaps are most effective.

The relations in Fig. 2-14 allow selection of a value for C in the capacitor input case. Keeping $\nu\omega CR_L = 50$, in this ripple range of 2 to 5 percent, will cause the RMS and peak currents to obey the relations in Fig. 2-15. That is true for all types of supplies except in multiphase units. In general, in multiphase supplies, these would be very conservative design ratings. Additional design information is available in Mlynar and Terman.

2.1.3.5. Transformer Leakage Inductance. Consider the effects of leakage inductance with reference to Fig. 2-17. Leakage inductance just means that; between the primary and the secondary of the transformer, not all of the primary magnetic flux is coupled to the secondary. The loss arises primarily from geometrical factors. A small equivalent inductance is put in series with the ac driving voltage in the transformer equivalent circuit (Fig. 2-17). There is the resistance of the diode, R_D, and that of the transformer winding R_S. Consider each as an ideal diode switch closing once every 1/180 of a second. During each diode cycle, a large current pulse flows and puts a voltage on the load. Each diode is then sequentially reverse biased, and it is at these points during turn-on and turn-off that transient oscillations in the system can arise. The diode has to recover, it has stored charge in it, and the equivalent of an LRC generator in the supply circuit can then arise. The details of this are complicated and will not be discussed. Practically, all that is required is to add damping networks to the primary and secondary of the transformer. If the circuit does oscillate, nowhere in this voltage ringing will the voltage exceed twice the ac peak value from the transformer.

2.1.3.6 Filter Capacitor Heating. Large capacitors can have a significant RMS ripple currents passing through them. In the case of the full-wave bridge, the RMS current

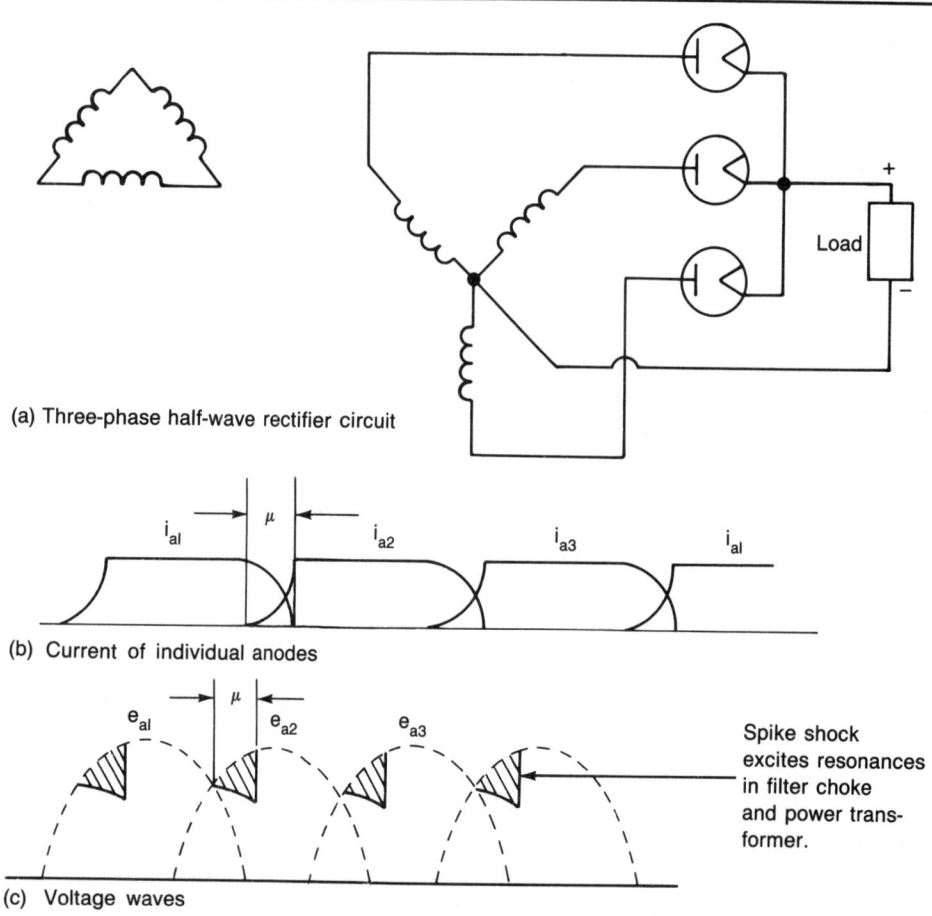

Fig. 2-17. Effects of leakage inductance on output voltage. (Redrawn after Terman and used by permission of the Board of Trustees of the Leland Stanford Junior University).

is about $5 \cdot I_{dc}$ at 120 Hz. It is just a point to be checked in the capacitor specifications to ensure that the capacitor will tolerate that amount of current heating. In the case of the surge ratings, this aspect is covered in more detail in the RCA notes. For a given filter capacitance, C and total internal resistance, R, of the power supply, the peak voltage instantaneously applied to this series RC network, through the diodes is N times the primary voltage of the transformer. Assuming an ideal diode, at the peak voltage the current flow is through an RC type of network. "R" in this case is this R. If the R is known, the peak current can be calculated and the proper diode selected. If the current is not known, estimate for practical transformers, assuming that R_{eq} is 1 percent of R_L (quite a good assumption) and, calculating from the application notes, this means the transformer regulation is 1 percent, which is a firm minimum number, (Fig. 2-15) so that:

$$I_{rms} \cdot R \cdot C = 0.7 \cdot V_{peak} \cdot C \qquad (2\text{-}3)$$

I_{rms} is calculated and checked against the diode curves available from the manufacturer. (Time is shown in RC units in the insert in Fig. 2-15.) If the number on the left side is above the limit line, the diode will not survive. What this means is that if the calculated RMS current is above the limit line, the diode will melt at turn-on. This is really an energy limit above which the chip melts.

Percent regulation has already been defined (Fig. 2-16). The voltage difference between the no-load and the full-load voltage is the full-load current times R_s, and since:

$$I_{FL} = \frac{V_{FL}}{R_L} \quad (2\text{-}4)$$

then the internal supply resistance, R_s is:

$$R_s = (\text{Percent Regulation}) \cdot R_L \quad (2\text{-}5)$$

For a given percentage regulation the equivalent resistance can thus be estimated.

2.1.3.7 Asymmetric Fault Currents. In this case, the equivalent loss resistive R_s is the dc (copper) resistive loss in the transformer plus the core loss in the transformer. The exception to this consideration arises in very large polyphase transformers with many amperes through the diodes where asymmetric effects in the multiphase transformer can predominate. In this case, the peak current is the dc current divided by the percent regulation. There is an asymmetric fault current factor ratio of 1.8 to be applied for the peak fault current as discussed in Mlynar. This is the peak current available when the whole supply is turned on to a short circuit peak line voltage. In conjunction with the transformer regulation and the average dc current, that will then indicate the absolute peak current to be expected, so far as the transformer or diode are concerned.

In the last part of this chapter, damping networks (what they do), transformer-equivalent circuits, and oil insulation are discussed. An introduction to hard-tube pulsers is also provided.

"High Voltage Engineering," by E. Kuffel and W. Zaengl (Pergamon Press, 1970) covers many interesting areas such as these. It is probably one of the best practical books published on spark breakdown, ionization, decay processes, electric breakdown of gases, and a most important area: measurement of high voltages. It introduces the literature of applied measurement techniques and forms a very valuable resource book. It actually covers practical areas, such as switch triggering techniques, and many types of high-voltage generators. The measurement section is comprehensive, starting at electrostatic volt meters and finishing with compensated probes.

2.1.4 Transient Damping

Throughout all pulse-power systems, there are potential transient problems. It is usually the transient phenomena that cause device breakdowns. Rarely in a well-engineered system, where these transients are under control, will the components fail. Systems usually fail when somebody turns on the power supply without the output load (or with the output load shorted); a transient is generated; a large peak voltage is generated in some series-resonant circuit; and a capacitor, choke, diode, or switch is lost. What will be discussed in this section is the problem of damping networks (Fig. 2-18) to control these transients. Essentially the problem is one of voltage damping in the series resonant circuit. When the power is initially applied, there is some inductance, L, there is a surge resistance, R, and the filter capacitance, C. If the power sine wave is applied at t = 0, the voltages will build up in the circuit until the equilibrium value is reached, generally requiring 10 to 30 periods at 60 Hz (that is ≈ 170 to ≈ 500 milliseconds).

For the purpose of this discussion, the fundamental concern is with the absolute maximum peak-to-peak secondary voltages across each high-voltage component, not with their growth or decay. Given that, the next question is how to ensure that this voltage can be kept within controlled limits so that the diodes, inductors, resistors, or capacitors do not break down.

2.1.4.1 Transients. For a full-wave bridge, when the power supply is turned on, the magnetic core of the transformer must be magnetized: the amount of current needed to do so is called the magnetizing current. The area inside the B-H curve is related to the core loss per cycle (it takes one full period to retrace the curve). The magnetizing current is essentially the amount of current applied that generates the magnetic field for zero secondary load power. In general terms, we can call it I_E. It is required to start the motion on this curve. The smaller this curve area, the smaller the core losses will

---

FOR THE SINGLE-PHASE FWB DC SUPPLY:

WHEN THE SUPPLY IS TURNED OFF AT THE PEAK OF THE LINE VOLTAGE AND THE LOAD IS NOT CONNECTED, THE PRIMARY CURRENT THROUGH THE MAGNETIZING INDUCTANCE "L_E" RAPIDLY COLLAPSES, GENERATING A HIGH VOLTAGE:

$$V_{pri}^{peak} \approx \left(\frac{L_E}{C_{shunt}^p} \right)^{\frac{1}{2}} I_{magnetizing}^{peak} \quad \text{:FROM CONSERVATION OF ENERGY}$$

UNLESS C_{shunt}^p IS SUFFICIENTLY LARGE, V_{pri}^{peak} CAN RISE TO LARGE VALUES, CAUSING PRIMARY INSULATION BREAKDOWN. TO CONTROL THIS, ADDITIONAL SHUNT CAPACITANCE IS ADDED AND, IF NECESSARY, SOME RESISTANCE IS PLACED IN SERIES WITH THE ADDED CAPACITANCE TO PROVIDE RAPID DAMPING OF THE TRANSIENT.

Fig. 2-18. Damping networks.

be. In traversing the curve counterclockwise, work is being done in moving the orientation of the ferromagnetic domains of the material as the coercive force, $H(I)$, varies. The thermal losses arising from this motion (core losses) change as the frequency (usually drastically increasing) increases and with the core material.

2.4.1.2. Transformers. When the power supply is turned on, current flows in the transformer primary even if the secondary is not connected. To measure I_E, an ammeter is put in series with the primary of the power transformer, the meter is shunted, and with the secondary open, power is applied. After a few seconds, the shunt is opened. The average value is measured with the milliammeter and that can be converted to the RMS equivalent. Given that, the energy stored in the equivalent magnetizing inductance L_E can be measured in the following way: for an ideal transformer with a magnetizing inductance, L_E, then the shunt damping capacitance C_s can be determined from conservation of energy:

$$\tfrac{1}{2} C_s \cdot V_{pk}^2 = \tfrac{1}{2} L_E \cdot I_{pk}^2 \qquad (2\text{-}6)$$

As before, V_{pk} is set equal to twice the peak line voltage and L_E obtained from $V_{pri} = \omega L_E I_{pri}$. Because this is an oscillatory RLC circuit, this energy $\tfrac{1}{2} L_E I_{pk}^2$ is dumped into the capacitor and oscillates repeatedly. This calculation can then give an estimate of the peak voltage that can be expected. Depending upon C_s and L_E, V_{pk} on the secondary can exceed the power supply peak voltage by a factor of 3 to 4. To dampen this voltage to twice the ac peak voltage applied, it will be shown that the:

$$Q = \frac{\omega L_E}{R_s} = 2 \qquad (2\text{-}7)$$

Now C_s and R_s in series with C_s are determined.

This magnetizing energy is not lost. It is in the iron, causing magnetic energy to be stored in the lumped equivalent inductance of the transformer primary. A current meter put into the circuit will measure a current flowing through the system with no load on the secondary. The transformer there is thus not ideal. If the current is measured, the value of L_E with no applied secondary load is obtained. It might not be a real inductor, but L_E is represented as a lumped equivalent inductance.

The problem, depending upon the transformer parameters, is that there can be significant transient voltages developed well above the line voltage. What is desired is to have a way of damping this large peak, irrespective of initial conditions. The same sort of argument applies to the damping of the transients across the choke in the filter (Fig. 2-19). The main factor in this arrangement is the distributed capacitances which is present from the coil to ground. This key parameter causes an oscillatory effect of this regenerative voltage and will produce a voltage of reverse polarity across the rectifier assemblies after the rectifiers have commutated the forward bias pulse. If this is an ideal transformer, it has a turns ratio of one to N so that if you apply 1 V ac to the input, N volts ac are at the output. This is what everybody would like to be the case, but alas the equivalent circuit is more complex and is sketched in Fig. 2-20. There is a magnetizing inductance, $L^p_{magnetizing} = L_E$, and a shunt resistance, $R^p_{core\ loss}$, representing the core loss (the work

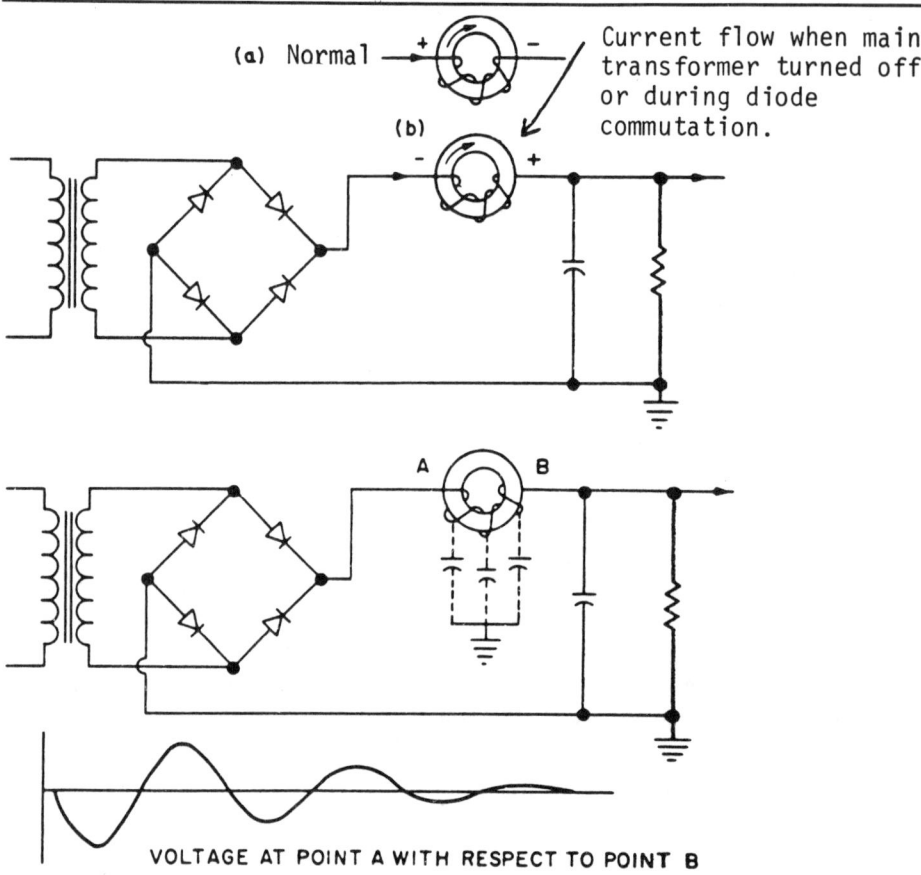

Fig. 2-19. Transient voltages generated across the input inductor during commutation of the diodes (courtesy of Westinghouse).

done in moving the ferro-magnetic domains around on the B-H curve). This series resistance, $R^p_{winding}$, is equal essentially to the resistance of the copper primary wire. The leakage inductance, $L^p_{leakage}$, which is shown lumped in the primary, represents essentially the percentage of the magnetic flux generated in a primary that is not fully coupled to the secondary. A rough number for that leakage inductance is 1 to 5 percent of the primary inductance. The transformer manufacturer can provide a more accurate number for a given design.

The equivalent elements on the secondary side are complementary. One small element that can become important is the primary to secondary capacity, C_{ps}. It is the real physical capacity from primary to secondary that one would measure with a capacitance bridge. In most circuits it is a relatively small number. The only time it becomes sufficiently large to be a problem is in multilayer, interleaved (multifilar) transformer designs popular in wideband pulse and audio transformers. For conventional pulse transformers in the microsecond region, it is preferable not to wind them with these interleaved structures, to minimize C_{ps} and reduce the coupling of transients through the transformer.

2.1.4.3 Equivalent Circuits. In this equivalent circuit in Fig. 2-20, all these lumped equivalent elements can be measured. For example, R_{gen} is the equivalent ac resistance in series with the generator; C^s_{shunt} is the shunt capacity of the secondary. It is a measure of the lumped shunt capacity across the secondary, and it is ascertained in the same way as for the primary (that is, by injecting a small current into the secondary and measuring the natural resonant frequency in resonance with the secondary leakage inductance with the primary shorted). Manufacturers of transformers for pulse applications can almost always give all these values to within 10 to 15 percent, which is generally satisfactory. If they cannot estimate them, consider whether to buy from that source. One very important point to note is that there is no electrostatic shield shown in Fig. 2-20. Because this represents an ideal transformer, there is no worry about that. It is suggested to put electrostatic shields on the primaries of all transformers. There is no reason the manufacturer cannot do it, but for separate primary and secondary bobbins, the individual can do it if desired. If the transformer primary is being referred to someplace other than ground, that electrostatic shield could quite legitimately be referred to that same place, recognizing that any current passing into it is going to traverse whatever other components are in that return loop to ground.

This complex circuit can be simplified. For high voltage, shielding is really intended to isolate the primary from fast transients that are being coupled back from the load, across the stray capacity of the filter inductor, back through the diodes (because they are shunted with capacitors), back through the transformer, and into its primary. The advantage of the electrostatic shield on the transformer is that these transient currents travel into the shield and are harmlessly directed off to the ground of the transformer assembly. This shield protects the primary from being exposed to such fast transients. Similarly, a fast transient coming into the primary side from the line cannot couple to the transformer secondary where it may become large enough to cause damage.

In the ideal transformer, because there is conservation of power, the output current is the input current over N (Fig. 2-21). In the real world, referred to the primary (because dissipated power is conserved when it is reflected from the secondary loss elements, all losses can be expressed as equivalent circuit elements in the primary, as illustrated in Figs. 2-21 and 2-22.

If the primary current is i_i and the secondary current is i_o, the same equivalent primary loss for power $i^2 R'_{equivalent}$ is sustained for both of these. So the overall equivalent resistance in the primary, taking into account the resistive loss in the secondary as well as the primary loss resistance gives:

$$R = R^p_{winding} + \frac{1}{N^2} R^s_{winding} \tag{2-8}$$

Whatever loss resistance is in the secondary can be referred back to the primary side, so lumped loss elements can be used for energy-dissipation calculations. This lumped equivalent model is a way of moving dissipative terms, capacitances, and inductances from one side to the other through ideal transformers to make calculations a little more straightforward. The power loss terms can be referred to the secondary or to the primary. This is an elementary way of referring the lumped elements to the primary side

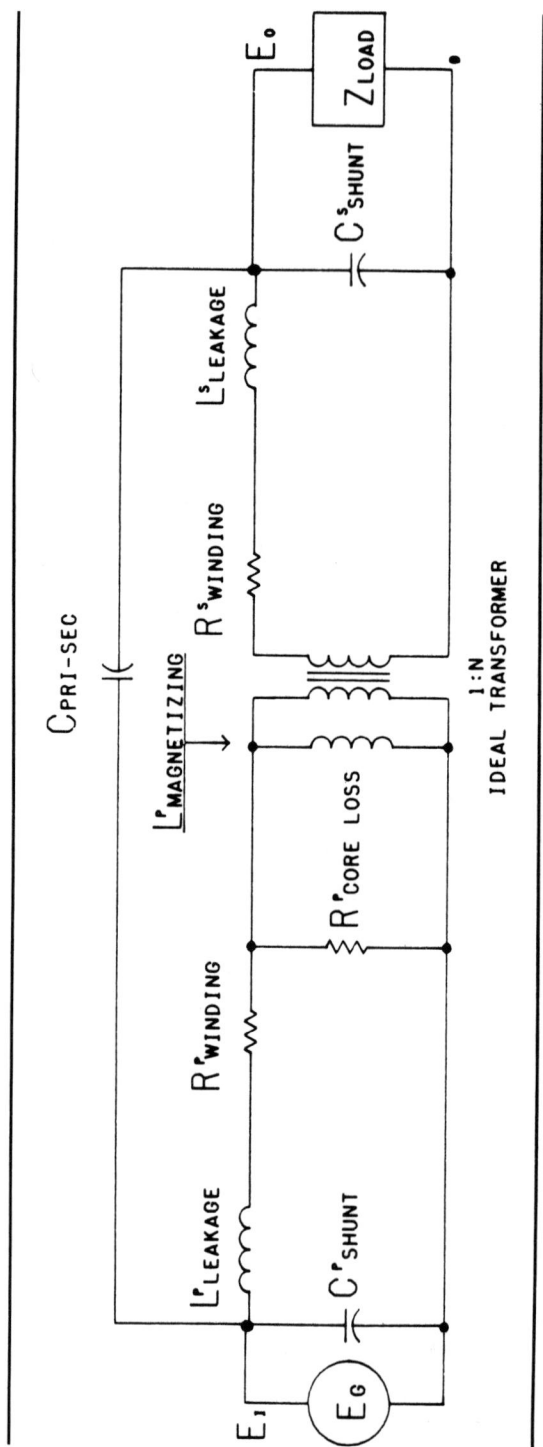

Fig. 2-20. General transformer equivalent circuit.

2.1 dc Power Supplies 61

IDEAL

$$E_i I_i = E_o I_o \qquad \therefore \text{Power conserved}$$

$$E_o = N E_i$$

$$I_o = \frac{I_i}{N} \qquad 1:N = \text{Turns ratio}$$

REAL: REFERRED TO PRIMARY

CONSERVING POWER DISSIPATED:

A. $I_i^2 R'_{equivalent} = I_o^2 R^s_{winding}$

$\therefore \underline{R'_{equivalent} = \frac{1}{N^2} R^s_{winding}}$

B. $\frac{1}{2} E_i^2 C'_{equivalent} = \frac{1}{2} E_o^2 C^s_{shunt}$

$\therefore \underline{C'_{equivalent} = N^2 C^2_{shunt}}$

C. $\frac{1}{2} I_i^2 L'_{equivalent} = \frac{1}{2} I_o^2 L^s_{magnetizing}$

$\therefore \underline{L'_{equivalent} = \frac{1}{N^2} L^s_{magnetizing}}$

D. $\dfrac{E_i^2}{Z'_{equiv}} = \dfrac{E_o^2}{Z_{load}}$

$\therefore \underline{Z'_{equiv} = \frac{1}{N^2} Z_{load}}$

Fig. 2-21. Simplified transformer equivalent circuits referred to primary.

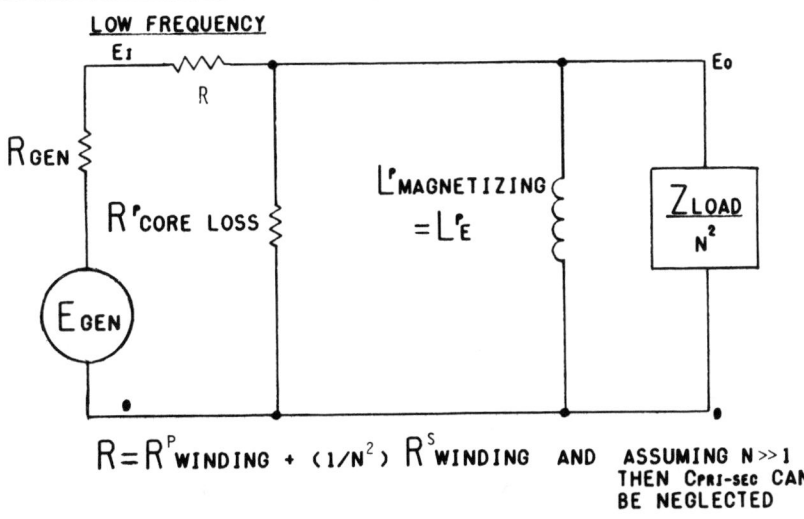

$R = R^p_{WINDING} + (1/N^2) R^s_{WINDING}$ AND ASSUMING $N \gg 1$ THEN $C_{PRI-SEC}$ CAN BE NEGLECTED

Fig. 2-22. Simplified low-frequency transformer equivalent circuit referred to primary.

from a loss point of view, an equivalent shunt capacity point of view, and from an equivalent inductance point of view. To reflect the other elements back into the primary, refer to Fig. 2-21 for the equivalents. This circuit comes from Terman, and there is a very good discussion in Fink's "Radio Electronic Engineering".

When all these lumped elements are reflected to the primary at low frequencies, the circuit of Fig. 2-22 results. At low frequencies, the capacitances are not required. R is the winding loss term for both primary and secondary losses and $R^p_{core\ loss}$ is the core loss term. The total magnetizing inductance is $L^p_{magnetizing}$, the load is Z_{load}/N^2, and this circuit equivalent is applicable from about 60 Hz to 5 kHz. It is a good circuit for the transformer, and it can be used for most power supply design.

At high frequencies, the second equivalent circuit (Fig. 2-23) is obtained. Shunt resistive and inductive components are no longer of concern, so further simplifications arise. The magnetizing inductance term was eliminated (the core loss still exists, but is relatively insignificant in terms of the energy stored in various shunt capacitances and leakage inductances). The L is the sum of all the leakage inductances. The winding loss and generator resistance remain although they might now be frequency dependent. The secondary shunt capacity is referred back to the right side of the equivalent leakage inductance as $N^2C^s_{shunt}$. That is important as now we have the potential for transient excitation of this dampened series RLC oscillator. The idea now is to dampen these oscillations and try to keep them under control so that the voltage nowhere exceeds twice the peak applied voltage. This amplitude limit is selected primarily because it gives a good cost/performance tradeoff.

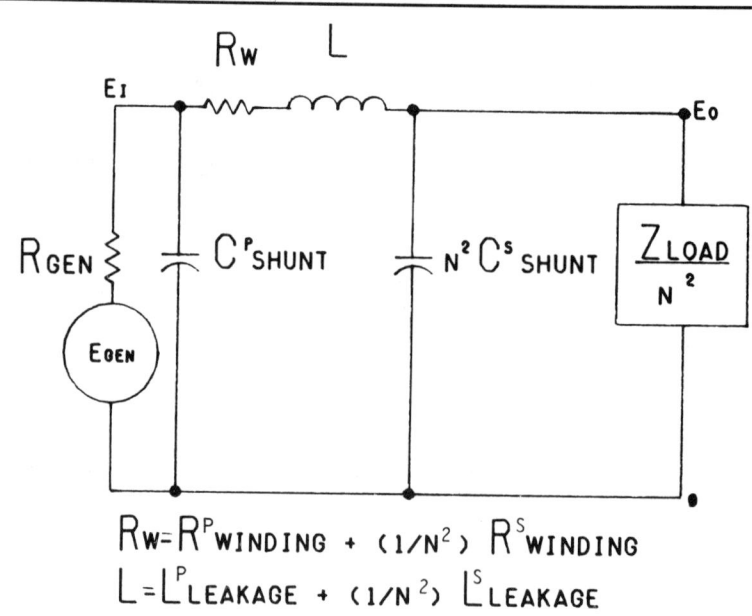

Fig. 2-23. High-frequency transformer equivalent circuit.

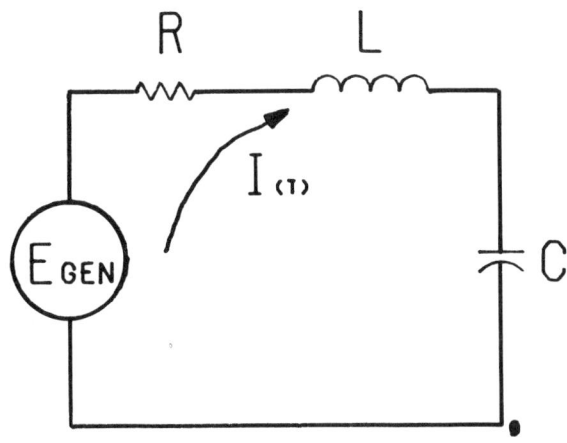

Fig. 2-24. The series-resonant circuit.

There is a resonant circuit as illustrated in Fig. 2-24 and 2-25, where the generatc voltage is E_{GEN}. Solving the circuit equations at the resonant frequency $\omega^2 LC = 1$ and $Q = \omega L/R$. At resonance, the circuit is pure resistive, the current is a maximum, so the voltage across the resistor is E_{GEN}.

What about the voltage across the inductor? If this voltage is desired to be twice E_{GEN}, then:

$$V_L = j\, Q\, E_{GEN} = j\, 2\, E_{GEN} \qquad (2\text{-}9)$$

Now for the capacitor, because $V_c = -j \cdot Q \cdot E_{GEN}$ or $= -j \cdot 2 \cdot E_{GEN}$ (for a $Q = 2$ circuit), the peak voltage across any element is never greater than twice the peak generator (or transient) voltage. That is all there is to snubbers. No matter what the initial conditions are, no matter what is done, there is no way for more voltage to appear across the inductor than twice the peak value of the ac voltage that is in the circuit. There is, however, one rare exception: there is a very special category of circuits called *two-frequency oscillatory discharge networks* and the one time they will be a problem is when there are ferroresonances in power transformers. With most of the problem areas covered in this book, there will never be a worry about this. It is a problem that comes about when circuit breakers do not always open simultaneously or they do not close simultaneously, or when power supplies are built that are three phase without phase drop-out detectors in them. When ferroresonance like that occurs, enormous voltages can be generated in the primary of the transformer. That is the only time not to believe the above analysis. Greenwood has an excellent discussion of this topic. To avoid that, just ask every power-supply supplier who is supplying three-phase supplies to put in a phase-drop or voltage detector. With such detection, the problem goes away.

2.1.4.4 Damping Networks. The damping network to be used for primary damping is shown in Fig. 2-26. The leakage inductance, L_E, is measured this way; $(E_i)_{rms}$ is applied and $(I_i)_{rms}$ measured for the secondary open circuit. From the previous analysis it can be shown that L_E can be calculated (Fig. 2-18). Just at turn-on there is no load

$$E_{GEN} = E_0 e^{j\omega t} \qquad \text{SET } I = I_0 e^{j[\omega t+\phi]}$$

Then, at time T for $I = 0$ and no "Q" on C:

$$E_{GEN} = IR + L\frac{dI}{dt} + \frac{1}{C}\int I\, DT$$

$$= I_0 e^{j[\omega t+\phi]} R + L I_0 e^{j[\omega t+\phi]} j\omega +$$

$$\frac{1}{C}\left[I_0 \int e^{j[\omega t+\phi]} DT \right]$$

$$\therefore \int e^{j[\omega t+\phi]} DT = \frac{1}{j\omega} e^{j[\omega t+\phi]} \Big| = \frac{1}{j\omega}\left[e^{j[\omega t+\phi]} - e^{j\phi} \right]$$

$$\therefore E_{GEN} = IR + j\omega LI + \frac{1}{j\omega C}\left\{ I - e^{j\phi} \right\}$$

Since at $T = 0$, $Q = 0$ $\therefore \phi = 0$:

$$\therefore E_{GEN} = Z I \text{ where } Z = \left\{ R + j(\omega L - \frac{1}{\omega C}) \right\}$$

Set $\omega^2 LC = 1 \longrightarrow Z = R \longrightarrow E_{GEN} = RI$

$\therefore$ CIRCUIT PURE RESISTIVE

Now $\quad V_L = j\omega L I = j\omega L \dfrac{E_{GEN}}{R} = j\dfrac{\omega L}{R} E_{GEN}$

Define $Q = \dfrac{\omega L}{R}$ $\therefore V_L = Q E_{GEN}$ i.e. V_L across the inductor leads E_{GEN} by 90 degrees and is "Q" times larger.

Now let us restrict the <u>peak value</u> of the voltage across the inductance to $2E_0$.

i.e. $(V_L)_{PEAK} = j Q E_0 = j2E_0 \rightarrow Q = 2$

Fig. 2-25. The series-resonant circuit

IN THIS CASE, THEN, $\omega L = 2R$

AND ACROSS CAPACITOR:

$$V_c = \frac{1}{j\omega C} I = \frac{1}{j\omega C} \frac{E_{GEN}}{R} = -j\frac{\omega L}{R} \cdot E_{GEN}$$

$$V_c = -jQ \cdot E_{GEN}$$

THUS THE PEAK VALUE HERE IS ALSO $2E_0$.

Fig. 2-25. The series-resonant circuit (continued).

on the power supply and the main inductance of the circuit is the magnetizing inductance. A series resonant circuit is then being excited, there is a shunt capacity C^p_{shunt} in series with the leakage and magnetizing inductances. The resonance can be dampened by inserting a shunt-loss term across the C^p_{shunt}. The resonant frequency of the circuit is 60 Hz. Now from $Q = 2$, the value of the shunt capacitance C'_s can be derived (Fig. 2-26) from $\omega^2 L_E C'_s = 1$ (generally $L_E >> L^p_{leakage}$). The core loss, $R^p_{core\,loss}$ is known, so from Fig. 2-26 the series damping resistance R'_s can be calculated. It turns out that C'_s is generally about 100 times bigger than C^p_{shunt}. So there is how a resonant network here, $C_s'' - L_E - (R^p_{core\,loss} + R'_s)$ resonating at 60 Hz, dampened with a Q of two so it cannot oscillate with a higher Q at the natural resonant frequency of $C^p_{shunt} - L_E$, which could be several tens of kiloHertz. Capacitors in the primary side, for C'_s are quite inexpensive. SCR (silicon controlled rectifier) snubber capacitors are generally used here because, although they cost about 20 percent more, they are of extended foil or multitab construction, and their design lifetime is an order of magnitude longer than the less expensive ones. An SCR snubber capacitor is a pulse-discharge capacitor made by the thousands. The people who make cycloinverters, for example, would be in serious difficulty if one of those failed. If one fails, the whole SCR stack can be destroyed very quickly. Capacitor manufacturers will sell them at a modest premium over the cheaper filter-capacitor design.

The secondary, depending upon the conduction angle of the diodes, for a very short time has a chance of being unloaded as the supply is turned on. That is one way of generating secondary transients, so a secondary snubber across the transformer is highly recommended. The same $Q = 2$ constraint will be applied to the secondary damping case. The secondary damping is described in Fig. 2-27. Figure 2-28 shows the equilvalent circuit and an analysis of damping characteristics by frequency shifting. If the line voltage is turned on faster than the diode turns on (and there is a good possibility of that happening—the coupling from primary to secondary via $L^s_{leakage}$ and C_{ps} tend to do that) oscillations in the secondary circuit can be shock excited at the resonant frequency of the transformer secondary with its shunt capacity. Experience has been that damping the secondary is far more important than damping the primary. However, it is more expensive. The difficulty is with the physical size of the capacitors needed in relatively

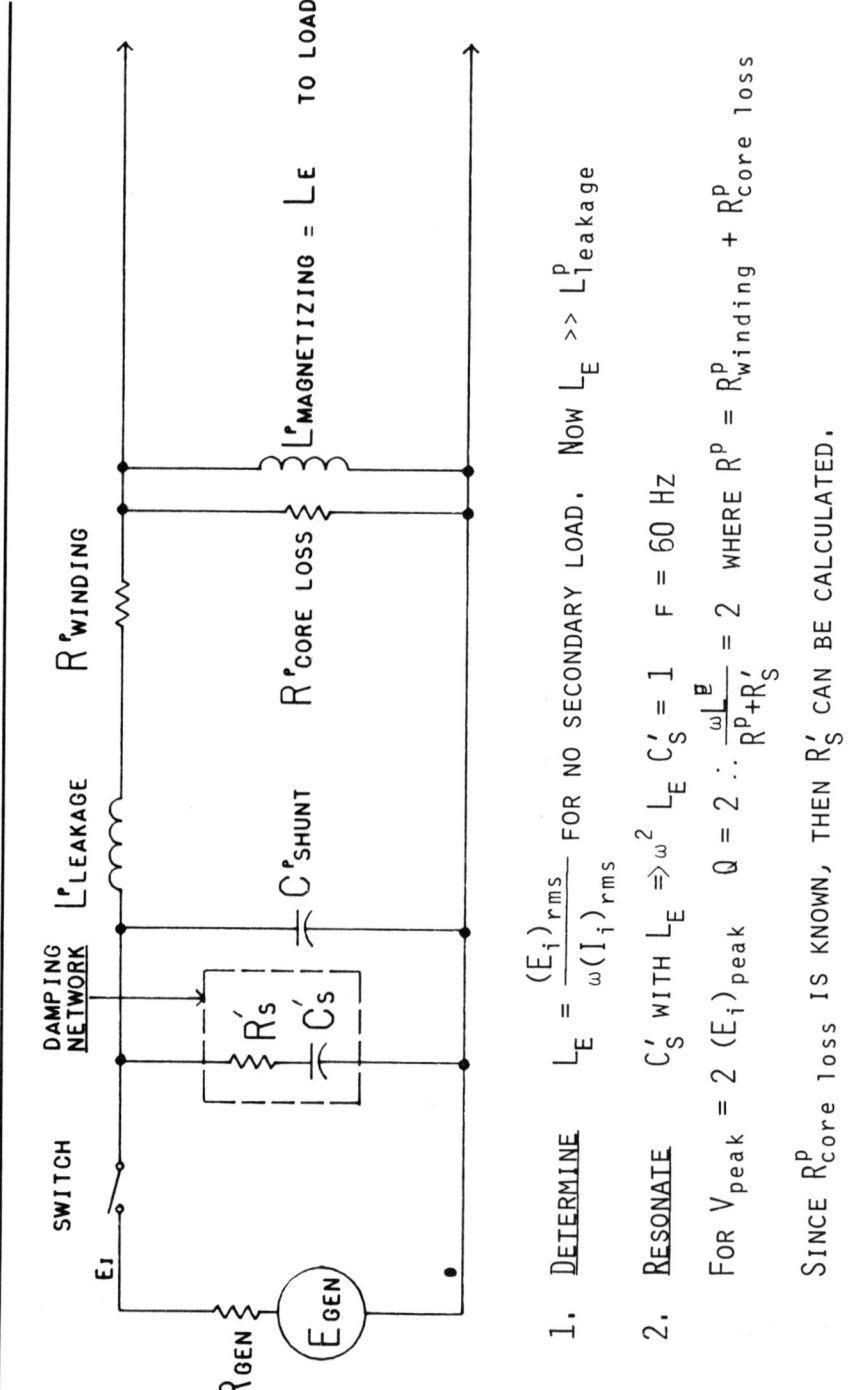

Fig. 2-26. Primary damping for a real transformer.

THE APPLICATION OF PEAK LINE VOLTAGE AT A RATE FASTER THAN DIODE TURN-ON (I.E. SUPPLY IS OPEN CIRCUITED) CAN GENERATE AN OSCILLATORY DISCHARGE IN THE SECONDARY CIRCUIT, ESSENTIALLY AT THE NATURAL RESONANT FREQUENCY OF THE TRANSFORMER (FROM 20,000 TO 100,000 Hz). THE FREQUENCY DEPENDS UPON THE SECONDARY "LEAKAGE" INDUCTANCE IN SERIES WITH INTERNAL SHUNT CAPACITANCE. THE CIRCUIT IS OF A HIGH Q AND CAN GENERATE VERY HIGH PEAK VOLTAGES ($\approx Q \cdot V_{SEC}^{PEAK}$). THESE TRANSIENT PEAKS CAN BE REDUCED TO REASONABLE AND CONTROLLABLE DESIGN LIMITS, WITHIN INSULATION RATINGS OF TRANSFORMERS (AND DIODES) TO, SAY, $2\, V_{SEC}^{PEAK}$, BY ADDING AN RC SNUBBER ACROSS THE SECONDARY OF THE TRANSFORMER. THE VALUES ARE OBTAINED FROM THE FOLLOWING MODEL, DERIVED FROM THE SERIES RESONANCE CIRCUIT THAT WAS BRIEFLY DISCUSSED BEFORE.

Fig. 2-27. Secondary damping.

high-voltage power supplies. For example, consider a 250 kW, 200 kV transformer system, that would probably be a six-phase Y-delta stacked configuration. (There is a Y secondary and FWB, and a delta secondary and FWB. The two are added in series.) In that type of circuit C_s'' of values up to 50 nF for the snubber are common. This represents a significant reactive power flow through the snubbers and, at these voltages, the damping networks become quite expensive.

The secondary resonant frequencies tend to be in the 20 to 100 kHz range. The resonance is between the leakage inductance and C^s_{shunt}, with resistive damping from $R^{s'}_{winding} + R^s_{core\ loss}$. There can be a large peak voltage—Q times the peak voltage in steady state at turn-on. Snubbers should be added to control these voltages and then values are derived from the model of the series resonant circuit, which is briefly discussed (Fig. 2-24).

This is interesting because this particular general circuit, with its damping snubbers, appears everywhere throughout HPE design. Snubbers are used on some laser systems, magnetrons (often for mode control), or klystrons, and it is always the same argument. This book attempts to find a general way of deriving the values for C_s'' and R_s'' (C_s' and R_s' having been determined in the preceeding—Fig. 2-26 analysis). There are a lot of cookbook prescriptions for these snubbers that are found in the various diode manuals or the transformer literature, but they tend to be overcompensation techniques, using large capacitors and modest values of resistors. This technique here allows considerable economy, because capacity in high-voltage capacitors costs money. If there is any significant RMS current (more than an ampere or two) it does cost a lot.

It can be shown that, starting from the $Q(\omega)$ analysis in Terman, if a snubber network is designed whose resonant frequency is ⅒ the natural resonant frequency, there

dc Power Supplies and Hard-Tube, High-Power Electronic Systems

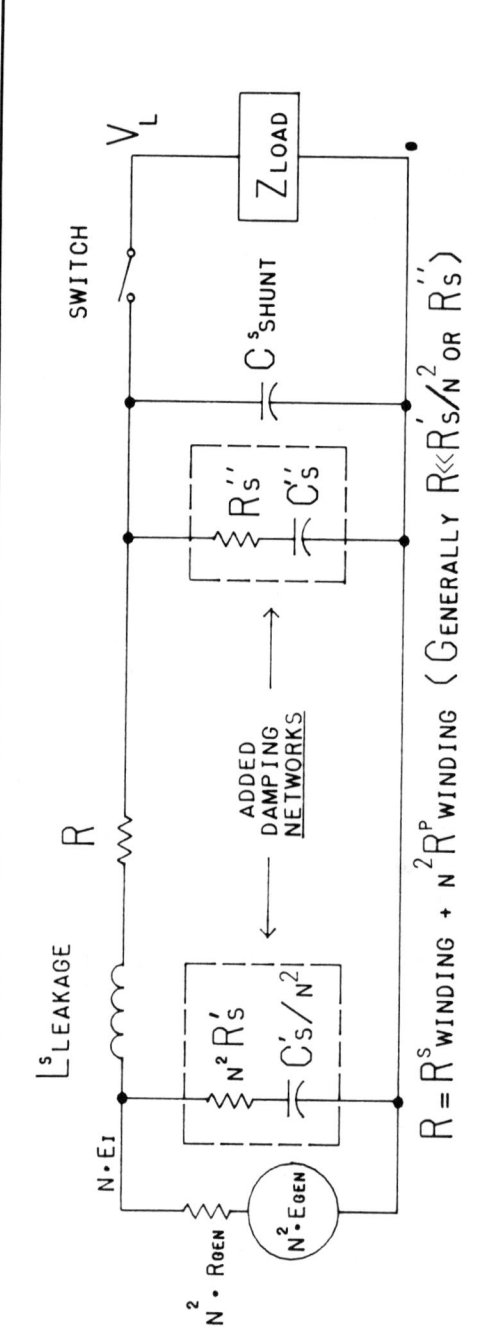

$R = R^s_{WINDING} + N^2 R^p_{WINDING}$ (Generally $R \ll R'_S/N^2$ or R''_S)

1. **Natural Self Resonant Frequency:** $\omega_R L^s C_T = 1$ where $C''_S = 0$; $C_T^{-1} = (C^S)^{-1} + \dfrac{C'_S}{N^2}^{-1}$

2. **Now:** For a peak overvoltage of 2x when switch opened then shift the resonance by N times in frequency (5x absolute minimum). In this case it can be shown that the peak voltage at ω_R remains below (N E_i) =

$$\left\{ \dfrac{1}{\dfrac{1}{C^S_{SHUNT}} + \dfrac{C''_S}{N^2} + \dfrac{C'_S}{N^2}} \div \dfrac{C^S_{SHUNT}}{1} = 100 \right.$$ Also, $2 R_T = \sqrt{\dfrac{L^S}{C_{T_O}}}$; $\therefore Q = \dfrac{\omega_0 L^S}{R_T} = 2$; $\omega_0^2 L^S C_{T_O} = 1$

$\dfrac{C_{T_O}}{C^S_{SHUNT}} = $ SHUNT

$R_T = R''_S + N^2 R'_S + R$, which defines R''_S.

Fig. 2-28. High-frequency secondary damping.

is no way that any Fourier components of any pulse of any shape that is applied to the circuit to generate a voltage across any component of more than twice the peak voltage of that Fourier component. The circuit lumped equivalent elements are all referred to the secondary to simplify the circuit in Fig. 2-28. Reference has been made to the primary damping capacitor, the primary damping resistor into the secondary. $L^s_{leakage}$ is the leakage inductance, R is the total winding loss, C^s_{shunt} is the shunt capacitance of the transformer. There is also an ouput switch: it could in fact be a diode. The power supply is turned on, the diode closes, and it conducts current into the load. The transformer load is in this case the filter inductor, the filter capacitor, and the load resistor that is attached across the supply.

The natural resonant frequency occurs when this $L^s_{leakage}$ is in resonance with the C_{To}, which is $(C^s_{shunt} + C_s'')$ in series with C'_s/N^2.

Initially, the resonant frequency ω_R is calculated for zero snubber (C_s'') capacitance. The analysis is shown in Fig. 2-28. For the peak overvoltage of twice the maximum secondary voltage when the switch is opened, proceed as before and shift the resonant frequency downwards by a factor of ten. Then the peak voltage at ω_o, the new resonant frequency, always remains below the peak secondary voltage. Shifting ω_R by only a factor of five means that the peak voltage at this new resonant frequency ω_o could be twice the peak secondary voltage. For a shift in resonant frequency of ten times, C_s'' and R_s'' are determined from the relations in Fig. 2-28. If the resonant frequency in the secondary is shifted by ten times, then it can be shown that nowhere can there be more peak volts than the peak volts from this new resonant frequency, even though as higher frequencies are reached higher Qs appear. Terman provides a lucid proof of this. This is important because it means no matter what happens there is no way the total peak voltage (the sum of these transformer and transient peak voltages) can then be more than twice the peak secondary voltage across any component. There are different self-resonant frequencies in the primary and the secondary of the transformer; normally they differ by a significant amount. They are determined primarily by the shunt capacities and the leakage inductances.

Everything discussed so far can also be used for designing pulse transformers. The same arguments apply for snubbing and control of oscillations in pulse transformers. Rise-time and fall-time aspects are considered in this book. If the resonant frequency is shifted by ten times, and if Q is equal to two, there is a new ω_o. C_{To} is known, then if that is divided by C^s_{shunt} and equated to 100 (this ensures that the resonant frequency is shifted by 10 times), then the resonances are going to be well-behaved and there will not be more than twice the peak applied volts. For $Q = 2$, then:

$$2R_T = \sqrt{\frac{L_s}{C_{To}}} \qquad (2\text{-}10)$$

With R_T known, then R_s'' can be calculated. C_{To} provides C_S'' and the snubber values are now known. Typical values are kiloohms and fractions of a nanofarad. Analysis reveals that the added KVAR loading from the snubber is about 5 percent for $Q = 2$.

If it is desired to eliminate the snubber capacitors entirely, they can be replaced with nonlinear elements called thyrites or metal-oxide varistors (MOV). What these are,

are nonlinear resistors that have a characteristic where above a certain critical voltage the current increases rapidly. They are essentially similar to symmetrical zener diodes, with a finite lifetime as a percentage of pulse overloads. The lifetime shortens drastically at high peak currents, making it necessary to control the peak fault currents through them very carefully in order to obtain reliable, long-life operation. The cheapest way to do this and the one that works very well is to put them across the snubber capacitors so you can reduce their capacitance value. Even if more voltage appears across that capacitor (as the Q is higher then), the varistors protect it by clamping the voltage level. They are difficult to use in large power supplies. They can only dissipate a few watts (up to 50 W), have significant capacitance, and are quite expensive. At higher voltages they are as expensive as the snubber capacitors. The normal MOV acts like a 0.01 to 0.1 μF capacitor, so there is a free damping capacitor. (One type that is rather useful is a zero-capacity MOV. Zero-capacity MOVs are valuable for pulse sharpening applications and have nanosecond rise times. They are made by a different process and have a 6th power law increase in current with voltage above the knee value, at a very low capacity. This characteristic makes them extremely useful for shunting transients and RFI in circuits when there are large peak-to-peak voltage problems at high frequencies. They can carry a fair amount of RMS current at high frequencies.)

This is how the transformer secondary can be dampened. The capacitor C_s'' resonates with the series RLC circuit at $\frac{1}{10}$ the natural resonant frequency, and the resistor R_s'' is selected so the Q is 2 (arbitrary choice to control the voltage to within a factor of two everywhere). The same argument applies to the filter choke damping. Typical values for the R_s'' of the right are kiloohms, and typical values for the C_s'' are fractions of a nanofarad. At the 100 kV level, a 1 nF capacitor is a size above which costs escalate rapidly. If 10 nF is desired, the price goes up very quickly. There is clearly a tradeoff required. Figure 2-17 illustrates the effect of leakage inductance in generating voltage transients in the secondary circuit.

For 60 to 360 Hz, kilovolt power supplies, thyrites and MOVs are quite useful devices. They can be used to build diode stacks in disaster situations to replace the capacitor/resistor combination in the compensative diode stacks just discussed. For 1 kV diodes, the 750 V standard MOV units can be used although they are rather expensive.

2.1.5 Oil Insulation

Figure 2-29 is a simplification of much of the insulation data in Mlynar's book. A study of these points might save a great deal of agony and lost time. Note that if there are even a few parts per million of suspended water in oil, the dielectric strength in kV per 0.1 inch drops drastically. People who make single-shot generators (for example, large Marx banks) really do not need to worry about that. When these large devices are assembled, the water sinks to the bottom. With small-size assemblies, particularly at high repetition rates, the ac corona component is significantly enhanced by the presence of water in the oil. Anyone contemplating high-voltage work is advised to acquire a small oil purifier. If at all possible, purify the oil whenever it is taken out. It is worth the trouble and the small investment in time. The typical oil sitting around the laboratory or outside, even in sealed containers rapidly (approximately six months) can become contaminated to such an extent that the dielectric strength is seriously degraded. The

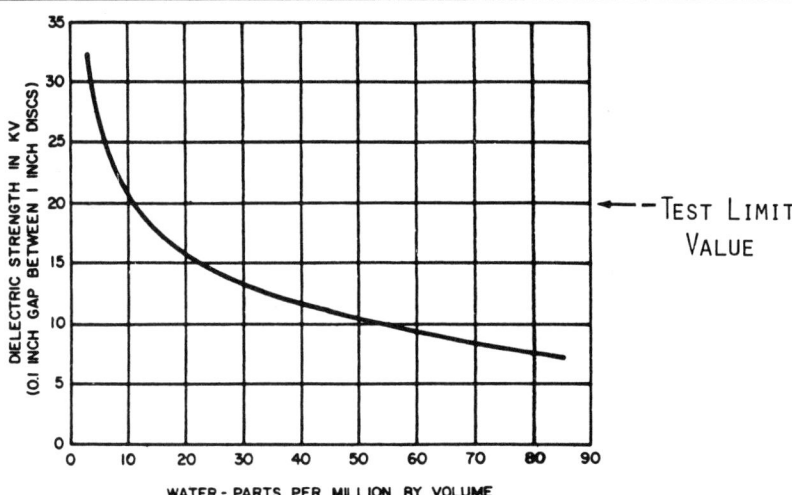

1. DESIGN FOR 30,000 VOLTS PEAK PER INCH.

2. CLOSER SPACINGS REQUIRE INSULATION BARRIERS BETWEEN HIGH-VOLTAGE POINT AND GROUND.

3. OIL EXPANSION WITH TEMPERATURE IS SIGNIFICANT.

4. FOREIGN MATTER CAN CONTAMINATE THE OIL (E.G. WAX-COVERED COMPONENTS, DIRT, PAINT, ETC.)
   - CAN REDUCE DIELECTRIC STRENGTH (AC, DC, OR PULSE)
   - CAN INCREASE LOSSES AT HIGH FREQUENCIES (E.G. RF CIRCUITS)

5. DISPLACE AIR IN TANKS WITH DRY NITROGEN OR ARGON. FREONS CAN BE USED IN EXPERIMENTAL SET-UPS (BUT FAR, FAR AWAY FROM CO_2 LASERS.)

6. OVERSTRESSES CAUSE WHIRLPOOLING AND LEAD TO ARCS.

Fig. 2-29. Oil-insulation data (courtesy of Westinghouse).

oil purifier is a small centrifuge device that separates impurities from the oil, possibly followed by an addition of a particulate and water filter.

Antioxidants put into high-voltage transformer oils react readily with a large number of components of which high-voltage assemblies are made, such as resistor varnishes and the wax coatings on cheap capacitors. These antioxidants and other materials dissolved in the oil can polymerize, allow tracking, and carbonize insulating surfaces. If components are to be used inside high-voltage assemblies, wax-covered components are not recommended unless specifically designed to work in oil (manufacturers provide this information). There are components designed for use in the standard high-voltage oils.

The recommended stress level near 60 Hz is 30,000 V peak per inch. At high frequencies, this decreases by a factor of 10 (at a few megahertz). With Shell Dialax, above a megahertz in RF systems becomes a problem. In dealing with high repetition rate circuits having very high frequency components in them, consider using the new silicone liquid dielectric insulating fluids from Dow Corning: DC 200, which has been available for years, and a new one, DC 561, that is about a factor of four more expensive than Shell Dialax and is quite impressive. It has a low corona inception point and a very low loss factor at high frequencies. Some people are sensitive to oils, so if you get them on your hands, wash often with pumice soap.

The antioxidants in high-voltage oils like Shell Dialax can degrade over a period of years under relatively high electric field stresses. Therefore, old oil should be rejuvenated with an additive before it is used. Over long periods of time, the hydrocarbons crosslink, combining to lose their ability to function as a high-quality insulator. They all depend upon molecular structure to absorb the gas molecules from corona areas, and this structure changes with time. If the oil is filtered and the antioxidants replenished, the oil can be rejuvenated. Oil should not be heated above 120° F. If it is, and then new transformers are impregnated with it, it is changed into something else that is not as good (its long-term antioxidation properties, its long-term breakdown, and its long-term corona resistance are all poor), and the system can fault. Silicone fluid can be heated (although there is no need to) to 240° F without damage; there is no damage until very high temperatures are reached. Mineral-based oils are also very stable and can be highly abused. The best oils for laboratory applications are Dow Corning 200 and 561. A resistor about 2 inches long that is normally rated for 10 kV dc, when it was put in Dow Corning 200, sustained 110 kV dc indefinitely. With Shell Dialax, the resistor tracked at about 30 kV. Dow Corning 200 is very useful material for lab applications, but may be too expensive for commercial applications. There are some insulating dielectrics better than the silicone fluids. They are very expensive and, for a given geometry, they can sustain a very high dc voltage before any significant corona occurs. That is what really is good about them. Their loss is generally high at high frequencies. The real advantage of silicone fluids is that they are very low loss up to 200 or 300 megahertz, and even then there is a transition to only a slightly higher loss. Silicone fluids (for example, diffusion pump oils) can be re-refined by the manufacturer and will work well afterwards.

Closer spacings and higher stresses than the above require insulation barriers between the high-voltage point and ground. If this is to be done, be sure that the insulating spacers do not suffer corona degradation or polymerize. One of the best materials known for this use, with high mechanical strength, is made by Permali. The possible exception to Permali might be in multimegahertz generators in which there are large RF voltages. Then it might tend to break down, and Teflon may be preferable.

Oil, unfortunately, expands with temperature. This is obvious, but once in a thyratron driver module made of ½-inch brass in a fully-sealed geometry, the leads were blown. Silicones are considerably better in this respect than standard transformer fluids.

Contaminants can dilute the dielectric strength of oils. Rosin in solders can polymerize and cause tracking. If possible, clean soldered items before putting them into oil (for example in chlorothene). Rosin solder is a very odd material. The nonactivated flux that Kester and others sell for soldering printed circuit boards, a very low viscosity material, is very good in terms of dc leakage and corona damage resistance. The standard rosin-core solder, particularly the activated types are to be avoided if fairly high-voltage stresses abound.

To minimize water vapor from the air passing into insulating fluids, the air in the high-voltage assembly is often displaced with dry nitrogen or argon (or Freon if there is no CO_2 laser around). Freon-12 dramatically suppresses surface tracking, keeps the water vapor out, displaces all the air, and will not allow air into the transformer.

Electrical overstressing causes whirlpooling in insulating fluids. Above certain stress points, the oil forms a whirlpool. For the laboratory application, it generally does not matter, but this level of stress will polymerize the oil and cause degradation in hold-off capability. It is not a recommended approach for long-life systems. This whirlpooling is caused by electron or ion-field emission from sharp edges where the ions push the molecules into a whirlpool. The high emission currents and fields also polymerize the oil. The whirlpool is simply an indication of other factors that are not obvious but are occurring simultaneously. A typical time limit in a test system at 100 kV was days. After weeks of operation, deposits occurred and caused numerous arcs. Whirlpooling also pushes the oil up the container surface and out onto the floor.

2.2 HARD-TUBE, HIGH-POWER ELECTRONICS SYSTEMS

2.2.1 Capacitive Storage System

A hard-tube pulser in high-power electronics can be represented as an analog switch in series with a battery. To a very good approximation (Fig. 2-30), the capacitive storage hard-tube pulser is a large reservoir capacitor with a series switch, some stray capacitance across the load, an isolating inductance for recharging, and some type of load. The inductive storage type is illustrated in Fig. 2-31. The balance of this discussion covers what happens when Switch 1 in Fig. 2-30 is opened and how to keep Switch 2 dissipation at a minimum. Typical switch resistance is several hundred ohms. That means a low-impedance Blumlein cannot be driven with a hard-tube pulser, so that hard-tube pulsers are very good to drive a high-impedance load. In the kilovolt range, small planar triode tubes achieve only 50Ω when saturated, and even the kiloamp tubes that cost thousands of dollars are 100Ω or so. That is the space-charge limited current the cathode can produce.

There are two switches in Fig. 2-30. If Switch 1 is closed and C is charged up, the RC charging time of that is limited by R_L. Opening Switch 1 and closing Switch 2, then the series resistance in the circuit is $R_L + R_p$. If Switch 2 is closed for a short time, τ, the change in the voltage across R_L is $V_{DC}\tau/RC$. The voltage is almost V_{DC}, but charge is being pulled out of this capacitor in an RC discharge. That is the prime switching cycle in a hard tube pulser.

When Switch 1 or 2 is closed, it is dissipating power and the power lost can be calculated. Given a voltage drop across the tube, V_p, at the peak current, I_p, then the energy lost in the switch per pulse is approximately $V_p I_p$. This times the PRF gives the average power dissipated, which can be a very large number. In a megawatt modulator, many kilowatts of average power can easily be lost into the switch.

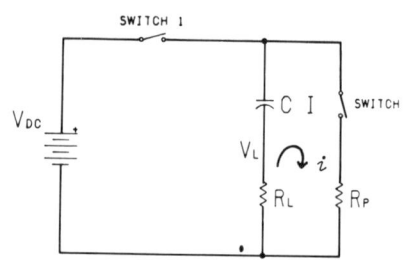

1. CLOSE SWITCH 1 TO CHARGE C; SWITCH 2 OPEN:

$$V_C(T) = V_{DC}(1 - e^{-t/R_L C}) - V_C(0)$$

2. OPEN SWITCH 1 AND CLOSE SWITCH 2:

$$V_L = V_{DC}\, e^{-t/RC} - V_P$$

SINCE $R = R_L + R_P$ AND $V_P = I_P R_P$, SWITCH VOLTAGE DROP = CONSTANT

NOW, LET SWITCH 2 BE CLOSED FOR TIME $\tau \ll RC$. $\quad \Delta V_L \approx \dfrac{V_{DC}}{RC} \cdot \tau$

DURING WHICH TIME SWITCH 2 DISSIPATES POWER.

FOR THE TUBE SWITCHED FULLY ON, THE TUBE DROP IS V_P AT PEAK CURRENT I_P. THE ENERGY LOSS PER PULSE IS $V_P \cdot I_P \cdot \tau$. AVERAGE POWER LOST PER PULSE IS THEN $P_{AV} = W_P \cdot \text{PRF}$.

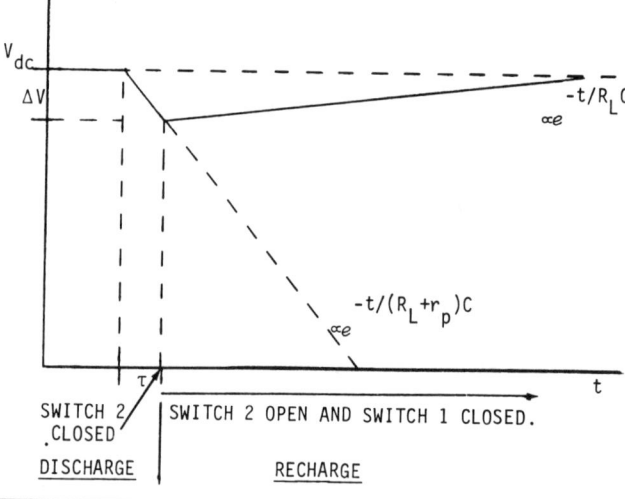

Fig. 2-30. Hard-tube pulsers.

Rise time is determined by the RC_{stray} time constant (Fig. 2-32) where R is a total equivalent resistance in series with C_{stray}, giving a rise time of that R times C_{stray}. Because generally for high-efficiency $R_p \ll R_L$, then $R \ll R_p$, which is the switch resistance. The high-power tubes are very large and have inductances of many microHenries. Rise time for hard-tube pulsers for such systems are several microseconds. This inductance does enter when the switch is opened in some circuits, sometimes giving rise to oscillations. If there is a switch with RC snubbers across it (added because it is misbehaving), there can be a RLC loop, in which the stored energy in that

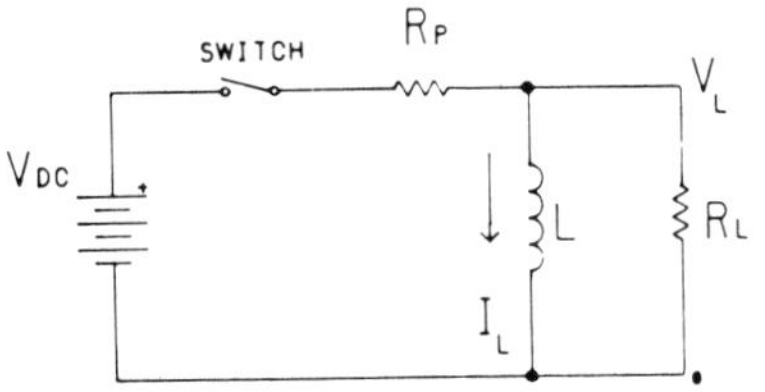

2.2 Hard-Tube, High-Power Electronics Systems

CLOSE THE SWITCH, THEN:

$$I_L(T) = \frac{V_{DC}}{R_P} \left(1 - e^{-\frac{R_P}{L}t}\right)$$

ASSUMING $R_P \ll R_L$ AND ZERO INDUCTOR RESISTANCE, AND OPENING THE SWITCH:

$$I_L(T) = \frac{V_{DC}}{R_P} \cdot e^{-\frac{R_L}{L}t}$$ (ASSUMING THE

SWITCH WAS OPENED AT PEAK CURRENT). THE PEAK LOAD VOLTAGE,

$V_L = \frac{V_{DC}}{R_P} \cdot R_L$, WHICH CAN BE VERY LARGE FOR $R_P \ll R_L$ AND THE

SWITCH MUST HOLD OFF THIS VOLTAGE AT THE INSTANT OF OPENING.

I.E. $V_L = -L\frac{di}{dt} = -\frac{V_{DC}}{R_P} \cdot R_L \cdot e^{-\frac{R_L}{L}t}$, AT T = 0,

$$V_L = -V_{DC}\left(\frac{R_L}{R_P}\right)$$

NOTE THAT THE SWITCH CURRENT REMAINS ON DURING THE CHARGE TIME SO THE POWER DISSIPATION IS HIGH.

Fig. 2-31. Power conditioning systems using inductive energy storage.

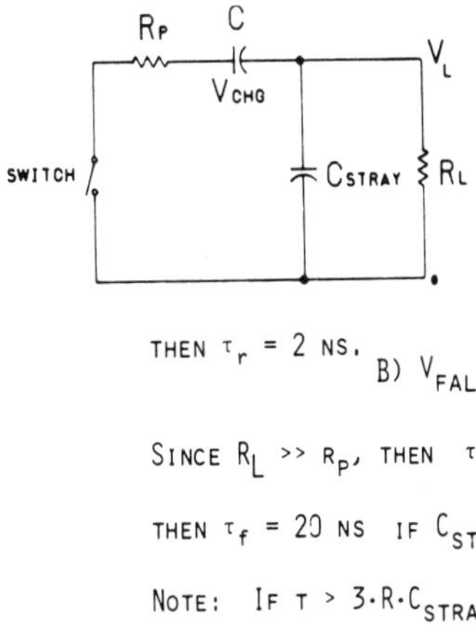

A) $V_{LEADING\ EDGE}(T) = \frac{V_{CHG}}{R_P} \cdot R \cdot (1 - e^{-\frac{t}{RC_s}})$

WHERE $\frac{1}{R} = \frac{1}{R_P} + \frac{1}{R_L}$ RISETIME = RC_{STRAY}
      $= \tau_r$

FOR $R_P \ll R_L$ THEN $\tau_r \approx R_P \cdot C_{STRAY}$

FOR EXAMPLE, IF $R_P = 100\ \Omega$ $C_{STRAY} = 20$ PF,

THEN $\tau_r = 2$ NS.   B) $V_{FALLING\ EDGE}(T) = V(\tau) e^{-\frac{(t-\tau)}{R_L \cdot C_s}}$

SINCE $R_L \gg R_P$, THEN $\tau_f = R_L C_s \gg \tau_r$ FOR EXAMPLE IF $R_L = 1000\ \Omega$

THEN $\tau_f = 20$ NS IF $C_{STRAY} = 20$ PF.

NOTE: IF $T > 3 \cdot R \cdot C_{STRAY}$, THEN $V(\tau) \approx \frac{V_{CHG}}{R_P} \cdot R$. ASSUMING THAT THERE IS SMALL VOLTAGE DROP ON "C".

NOTE: WHEN CONNECTED TO A CHARGING SUPPLY AT POINT A THROUGH A RESISTOR R_{CHG}, THEN ON CLOSURE, IF $R_{CHG} \gg R_P$, IT HAS NO EFFECT ON τ_r. NOW, WHEN THE SWITCH IS OPENED, R_{CHG} IS IN PARALLEL WITH R_L AND CAN BE USED TO REDUCE THE FALLTIME.

Fig. 2-32. Rise times and fall times in a hard-tube pulser.

gives rise to large oscillations, causing the tube to arc and fault the dc supply. On the falling edge, the switch is opened up and R_L is parallel with C_{stray}, revealing the falling-time constant.

When the power supply is added there is another resistive term to take into account. In inductive charging (Fig. 2-33) the inductor is placed across the load in series with Switch 2, forming the dc return. The inductor is added to reduce the recharging losses, and the output pulse appears on the load when the voltage exceeds the Switch 2 closure voltage. Switch 2 is then closed. (Switch 2 could represent a laser load that stays on while there are so many volts across it, and if the voltage on the load falls below the holding voltage suddenly Switch 2 opens, and the laser turns itself off. Magnetrons also behave that way.) When Switch 2 opens, there is another resonant circuit. If the Q of that is controlled correctly, there is no problem. If the Q is not low enough, a diode can be inserted across C_s as a shunt diode, clamping the inverse voltage as shown in Fig. 2-33. The conditions in here can be chosen for critical damping, as was previously described. A Q of 0.5 is chosen, and the critical damping is determined by resistor R_s + R_p in series with L_s.

The other possibility for fall-time control is to use a low-Q inductive resistor connected to the dc charging supply. The same argument then applies here. That R that we looked at before can be an inductive resistor. A lumped inductor and a resistor can be used to make an inductive resistor. The relation for the damping time is given in Fig. 2-34 as:

$$\text{DAMPING TIME} = \frac{6}{\dfrac{R_s}{L_s} + \dfrac{1}{R_p C_s}} \qquad (2\text{-}11)$$

Experience has shown that one of the best wire-wound resistors available is made by Ward-Leonard. They take an unusually high surface stress. Rod Carborundum resistors are capable of absorbing even higher energies and are available for operation up to 100 kV in air (and considerably more in oil).

Resistive charging is generally not used because the charging efficiency is only 50 percent, so inductive charging (Fig. 2-35) is required. If the damping components are chosen correctly, one can have critical damping. The capacitor, C, can be charged very quickly back to the dc voltage with a very small amount of overshoot. It can be shown that the overshoot on recharge is within 1 percent of the final voltage in the time $\pi/2 \sqrt{L_c C}$. If any significant overshoot on the pulse there is undersirable, a shunt diode can be put across R_L. It ensures that on the recharge, R_L does not carry any significant current. This is basically how high-efficiency, high-repetition rate, hard-tube pulsers are charged. This L_c is chosen to recharge C in a fraction of the PRF period.

2.2.2 Application of Pulse-Shaping Networks to Hard-Tube, High-Power Electronics Systems

This section is included because it has a significant potential for constant-current drive circuits (Figs. 2-36 and 2-37). The same arguments apply to using a thyratron or

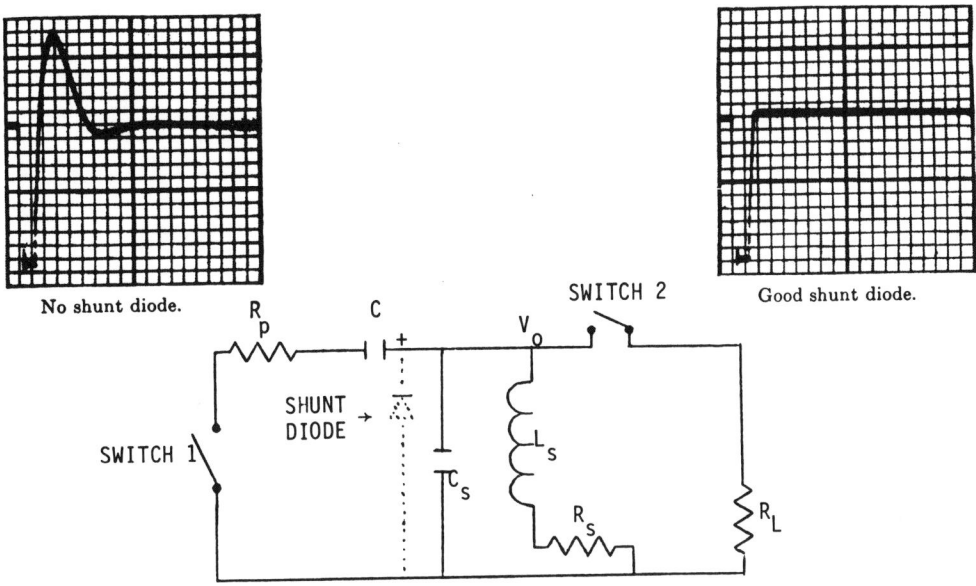

1. Close switches 1 and 2.
2. At the end of pulsetime, switch 1 is opened.
3. Voltage across R_L drops until switch 2 opens
   (e.g. laser load or magnetron)

Then: Since current is flowing in L_S a series resonant circuit of C_S, R_S, L_S is excited:

Oscillations arise if $\dfrac{R_S}{L_S} < \dfrac{2}{\sqrt{L_S C_S}}$ at the frequency $\omega^2 = \dfrac{1}{C_S L_S} - \dfrac{R_S^2}{4L_S^2}$

These oscillations are dampened and aperiodic if

$$\frac{R_S}{L_S} > \frac{2}{\sqrt{L_S C_S}}$$

For example, if $R_S = 50\ \Omega$ $L_S = 500\ \mu H$ $C_S = 50\ pF$

$\dfrac{R_S}{L_S} = 10^5$ $\dfrac{2}{L_S C_S} = 4 \times 10^8$ $\Rightarrow$ Oscillations

Fig. 2-33. Inductive and diode shunt elements.

An inductive resistor with $\dfrac{L_s}{R_s} \approx \dfrac{1}{2}\sqrt{L_s C_s}$ can be used to to obtain an aperiodic falltime $\propto e^{-At}$ then:

$$2A = \dfrac{R_s}{L_s} + \dfrac{1}{RC_s} \quad \text{AND} \quad \dfrac{1}{R} = \dfrac{1}{R_p} + \dfrac{1}{R_L}$$

Generally, $R_p \ll R_L$, so $R = R_p$ meaning

$$2A \simeq \dfrac{R_s}{L_s} + \dfrac{1}{R_p C_s}$$

Note: Damping is the same form for aperiodic or oscillatory case. Here the voltage ≈ 0 in a time $T_D = 3/A$ or "damping time" T_D where

$$T_D = \dfrac{6}{\dfrac{R_s}{L_s} + \dfrac{1}{R_p C_s}}$$

Fig. 2-34. *Inductive resistor used to charge energy storage capacitor and provide fall-time damping.*

any type of pulser switch imaginable. For the shunt LR network, if everything is correctly selected, and if $L = R^2 C$, it can be shown that the current under certain conditions remains relatively constant during the discharge interval. It is an effective driver, for example, for flash tubes and for certain classes of lasers (not KrF lasers but a number of other interesting glow-mode lasers). There is one important factor here, which affects the design of modulators. The current in this network can remain monotonic (that is it will not reverse) for $(\pi/2)\sqrt{LC}$ seconds. This is an important result for solid-state diode recovery. The current should remain monotonic through a diode while it is recovering. If this condition is chosen during snubber design, the result is the correct element to put in series with the diode so the diode cannot suffer a very rapid current reversal. This is a vital element in a number of high repetition rate modulator designs. Decoupling networks are needed to protect diodes in large modulators, as without them oscillations can ensue. This network will hold the current constant for that period of time, $(\pi/2)\sqrt{LC}$. As long as this time is longer than the diode recovery time, its precise value does not matter. If the diode can recover in three microseconds, the time is selected to be 5 to 10 microseconds. This is a very important point. One of these in series with diode bridge elements and recovery problems are negligible. One of these in series with the charging diode will keep the diodes from being damaged under any fault conditions except a prolonged dc fault. This network forces the current to remain monotonic during that short period of time.

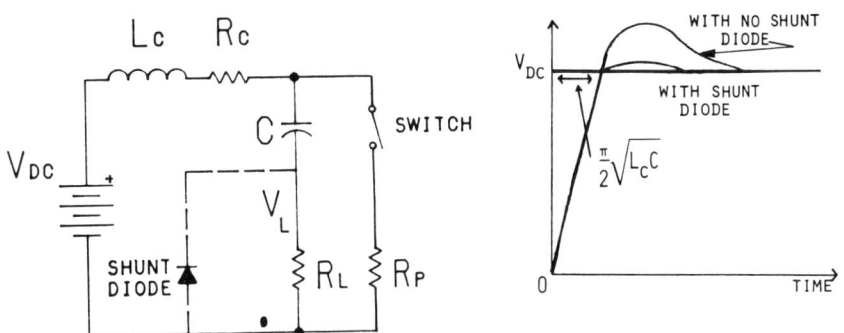

1. OPEN THE SWITCH.

   RECHARGE IS CRITICALLY DAMPENED IF $Q = 0.5$, I.E.

   $$\frac{\omega L_c}{R} = 0.5 \text{ AND } \omega^2 L_c \cdot C = 1 \text{ AND } R \equiv \{R_c + R_L\}$$

   THEREFORE: $\frac{L_c}{\sqrt{L_c C} \cdot R} = \frac{1}{2}$ OR $R = \frac{1}{2}\sqrt{\frac{L_c}{C}}$

   IN THIS CASE, OVERSHOOT ON RECHARGE IS NEGLIGIBLE AND RECHARGE TIME TO WITHIN 1% OF FINAL VOLTAGE IS

   $$\frac{\pi}{2}\sqrt{L_c C}$$

2. OSCILLATIONS MAY BE DAMPENED USING SHUNT DIODES ACROSS R_L TO CONTROL THE OVERSHOOT.

Fig. 2-35. Inductive charging elements.

2.2.3 Characteristics of Switch Tubes

A sketch of a vacuum-switch tube is shown in Fig. 2-38. All of the rest of this section is based on a large extent on T.R. Burkes' switch report. This sketch shows a profile of what tubes can be purchased (Fig. 2-39), to about 400 kV at modest peak currents. Three hundred megawatts peak power capability is available at just under 100 kV, although expensive to buy. Response time is generally in the area of microseconds because of the tube size. The peak current is space-charge limited, as it is cathode phenomena that limit all peak current.

Several points of interest are noted in Fig. 2-40.

1. **Peak Current**—Limited to 1 or 2 kA by available cathode-emission structures. Improving cathode emission would make more peak current become available.

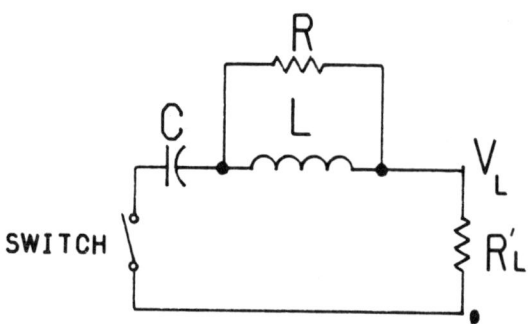

When $L = R^2 C$, the load current has zero slope at $T = 0$, implying a constant current source independent of circuit Q.

Note: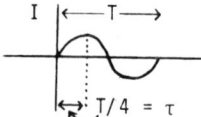

The current remains monotonic for a time $\tau \approx \dfrac{\pi}{2}\sqrt{LC}$

Important for solid state diode recovery, since current flow through them must be unidirectional until the diode recovers ($\approx$ 2-3 μs).

Fig. 2-36. Application of pulse shaping networks to hard-tube, high-power electronics.

2. **Lifetime Observations**—After 10,000 hours, the thoriated tungsten cathode emission drops. Oxide cathodes have shorter lifetimes than this by a factor of ten.
3. **Anode Voltage**—The basic limit is vacuum breakdown from induced field emission.
4. **X-Ray Radiation**—Especially troublesome for long pulse durations at high voltages and repetition rates.
5. **PRF**—Basically limited by anode and grid heating. The electrons are accelerated through the grids (each having a scattering crosssection) and some strike each grid; then they strike the anode, which heats them all up.
6. **Current Rate of Rise**—Fundamentally stray capacitance and tube saturation-resistance limited. To increase this, a saturating inductor can be used in the anode. Turning the tube on starts a little current flow through the large inductance. The ferrite then saturates, decreases the inductance, and a fast-rising output pulse is obtained. Unfortunately, when the tube is reopened, a long voltage decay ensues because the magnetic energy stored in the ferrite feeds energy back into the load.

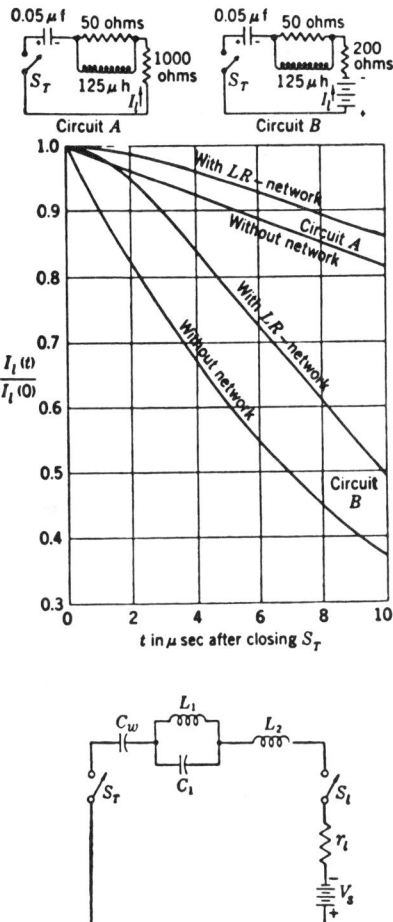

Output circuit of a hard-tube pulser with a network of passive elements to maintain approximately constant load current during the pulse.

Fig. 2-37. Constant-current, pulse-shaping networks (courtesy of G.N. Glasoe and J.V. Lebacqz, eds.: Pulse Generators, *Dover Publications, Inc. NY, 1965).*

7. **Pulse Width Limit**—Grid heating, a thermal effect limits the average power, except for oxide cathodes, where material sublimation limits temperatures because of enhanced field emission. Present limits are several seconds on time at the megawatt average power level.
8. **Delay**—Limited by RLC resonance effects in the external circuit and tube electron transit line.
9. **Jitter**—A couple of hundred picoseconds for planar triodes with dc heater power.

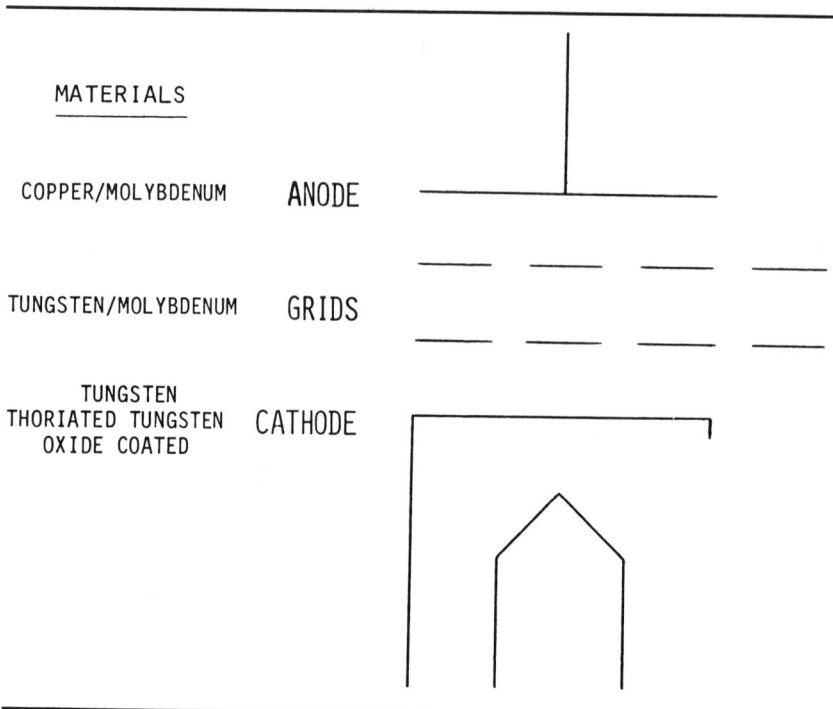

Fig. 2-38. Switch-tube materials.

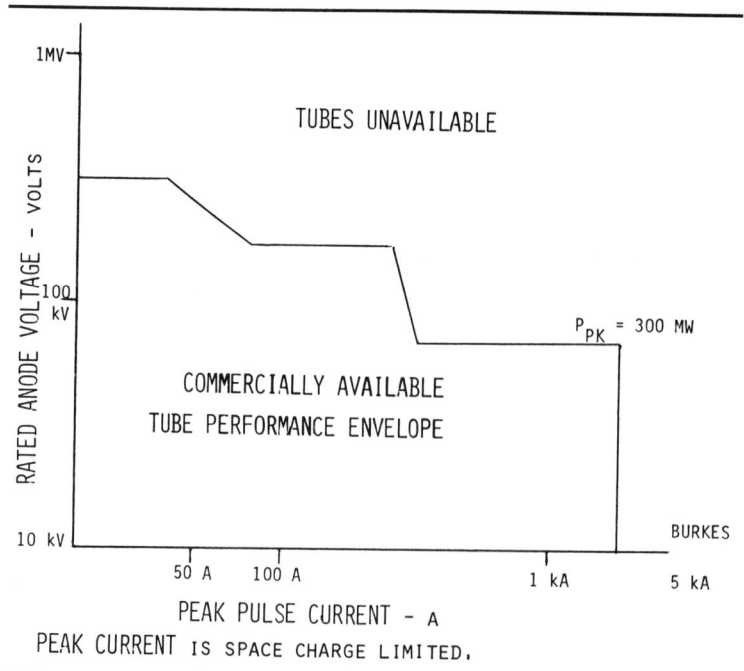

Fig. 2-39. Characteristics of high vacuum switch tubes. [5]

1. PEAK CURRENT: Limited by cathode emission and grid interception (and consequent heating) $\simeq 1000$ A.

2. LIFETIME: Oxide cathodes have short lives compared to thoriated tungsten. I.e. 10,000 vs 1000 hours as the emission drops.

3. ANODE VOLTAGE: Limited by vacuum breakdown from induced field emission and/or positive ion bombardment
   - Spacing/surface effects

4. X-RAY RADIATION: A significant problem at high repetition rates, voltages, and long pulse durations.
   - Anode phenomenon.

5. PRF: Limited primarily by anode and grid heating:
   $$I_{RMS}^2 = I_{PK}^2 \, (DF) \qquad DF = \text{Duty Factor}$$

6. CURRENT RATE OF RISE: If C_S is stray capacitance, then $R_S C_S$ is fastest risetime where R_S is tube (saturated) on resistance.

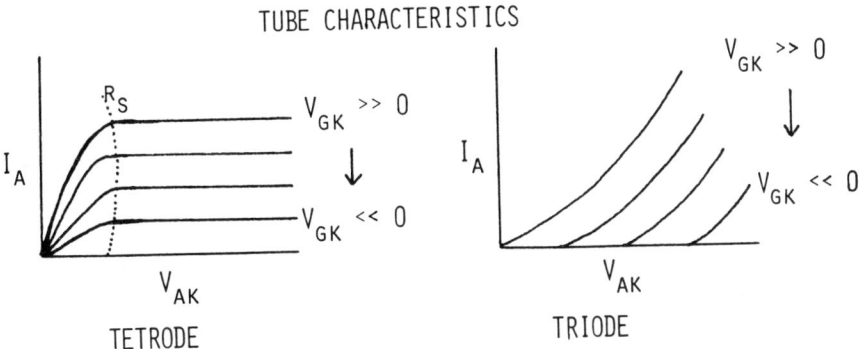

Note that resonant circuits have been used to improve risetimes, but generally at the expense of available peak pulse voltage

Fig. 2-40. Characteristics of switch tubes. [5]

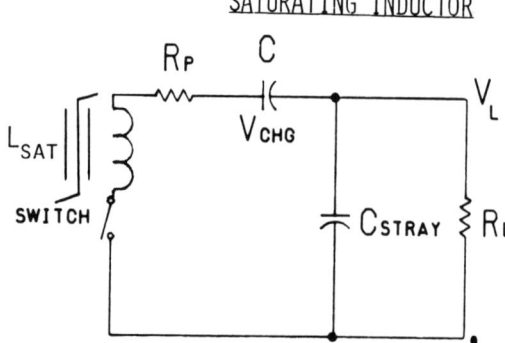
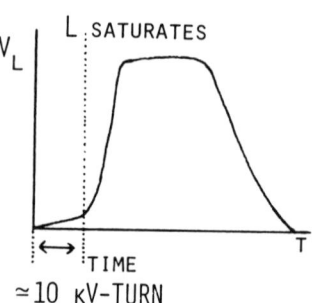

SATURATING INDUCTOR

≈ 10 kV-TURN

7. PULSE WIDTH: LIMITED BY GRID HEATING FOR ALL CATHODES; FOR OXIDE CATHODES SUBLIMATION OF OXIDE COATS GRIDS AND HENCE GRID EMISSION ENHANCED AT LOWER GRID OPERATING TEMPERATURES.
   - CURRENT LIMITS ARE SEVERAL SECONDS ON AT ≈1 MW LEVELS.

8. DELAY: LIMITED BY LRC RESONANCE AND TRANSMISSION LINE EFFECTS IN THE CASE OF CABLE STORAGE OR PFN SYSTEMS.

9. JITTER: USING DC ON HEATERS, JITTERS << 1 NS ACHIEVABLE.

Fig. 2-40. Characteristics of switch tubes (continued). [5]

2.2.4 Typical Hard-Tube, High-Power Electronics Systems

Figure 2-41 shows what a hard-tube pulser looks like. Believe it or not, they have not changed significantly since 1945. In this circuit there are a number of 6D21 tubes in parallel connected to a magnetron load through the two 0.01 µF storage capacitors. Turning the tubes on shorts the capacitors to ground, placing a negative pulse on the cathode of the magnetron. The pulse length is determined by the two-way transit time of the pulse cable connected to the triggering 4C35 thyratron. The pulse from this cable drives all the 6D21 grids positive, forming a fast driver that is contemporary today. The modulator produced 30 kV pulses at a 4 nanoseconds rise time.

2.3 REFERENCES

1. F.E. Terman, "Radio Engineers Handbook," (McGraw-Hill, New York, 1943).
2. P. Mlynar and G.C. Seaborn, "High Voltage Silicon Rectifier Designer's Handbook" (Westinghouse Electric Co., 1963).
3. RCA Thyristor Manual, Application Note AN-3659.

2.2 Hard-Tube, High-Power Electronics Systems

"The pulser was designed for a ... maximum voltage-pulse amplitude of about 30 kv, and a pulse current of about 35 amp. Since it was desired that the short pulses be nearly rectangular, the output circuit had to be designed for a very high rate of rise of voltage. The inductance and the shunt capacitance in the pulser circuit therefore had to be kept as small as possible. In order to minimize the inductance introduced by the circuit connections, the components were mounted as close together as their physical size would permit, allowing for the spacing necessary to prevent high-voltage flash-over. The importance of keeping the stray capacitance small is evident when

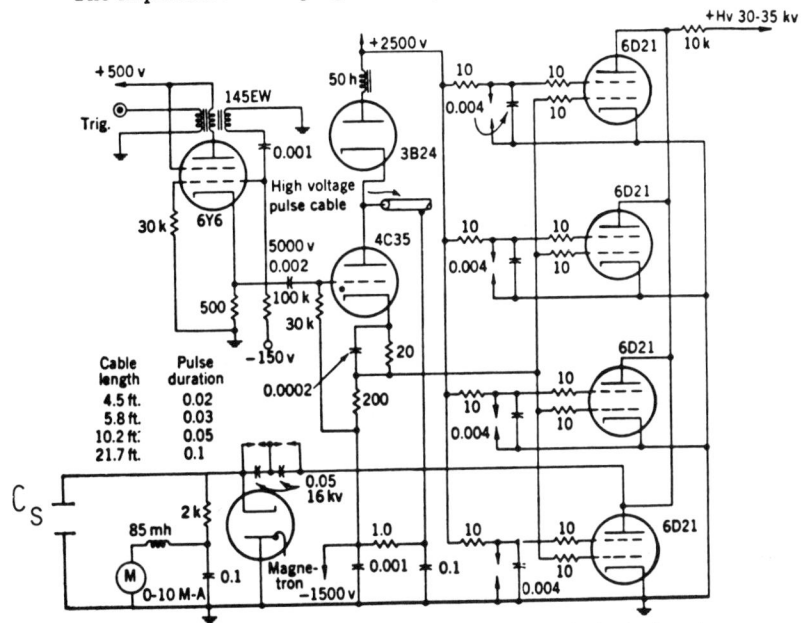

—Schematic diagram of the circuit for a high-power short-pulse hard-tube pulser.

the charging current, $C_s\, dv_l/dt$, for this capacitance is considered. For example, if the capacitance across the output terminals of the pulser is 20 $\mu\mu$f and the voltage pulse at the load is to rise to 30 kv in 0.01 μsec, the condenser-charging current is 60 amp. Since this current must flow through the switch tube, the maximum plate current needed to obtain a large value of dv/dt may be considerably greater than that needed to deliver the required pulse power to the load. The magnetron may have a capacitance of 10 to 15 $\mu\mu$f, between cathode and anode, so it is obvious that the additional capacitance introduced by the pulser circuit must be small."

6D21
    CATHODE: Thoriated Tungsten, 150 Watts Heating Power
    V_p(max):   38 kV
    I_p(max):   15 A

Fig. 2-41. High-power, short-pulse, hard-tube pulser (courtesy of G.N. Glasoe and J.V. Lebacqz, eds.: Pulse Generators, Dover Publications, Inc. NY, 1965).

4. W.G. Dunbar, "High Voltage Design for Airborne Equipment," Boeing Aerospace Co., Report AD-A029 268 (1976).
5. T.R. Burkes, "A Critical Analysis and Assessment of High Power Switches," Naval Surface Weapons Center, Report NP30/78, Sept. 1978.
6. A.E. Greenwood, "Electrical Transients in Power Systems," (Wiley-Interscience, New York, 1971).
7. G.N. Glasoe and J.V. Lebacqz, "Pulse Generators," Dover Publications, Inc., New York, 1965.
8. "What is a Fast Recovery Rectifier," Westinghouse Semiconductor Crusader, April 1979, p. 3.
9. "Characteristics of Common Rectifier Circuits," General Electric Technical Literature.

Chapter 3
Pulse-Voltage Circuits

by
W. L. Willis*

THE PROBLEMS ASSOCIATED WITH VERY HIGH VOLTAGES CAN BE AVOIDED IN PART BY using a variety of pulse-multiplication techniques. These methods are quite varied but share two features—the use of modest dc voltages and the use of switching techniques—to produce in some manner a vectorial transient addition of voltages.

Many such methods exist, but only the Marx bank, spiral generator, inversion generator, and Blumlein will be addressed in detail in this book. Two other methods will be given a brief discussion as matters of interest. An important system, the pulse transformer, is the subject of a separate discussion in Chapter 9 of this text. Also, some aspects of transmission lines will be taken for granted in this chapter and analyzed in detail elsewhere, particularly in Chapter 4.

3.1 MARX BANK

The history of the Marx bank generator is interesting, particularly to those who are actually struggling to build one. It is also interesting because the terminology describes circuits as an Erwin C_2 or Erwin R_3, recognizing Erwin Marx who described these generators in 1923. At about that time, quantum mechanics also began to appear in publication.

Rather good performance was obtained even in those days. In 1926 Burov, of Russia, described a 100 kV pulser with a 10 nanosecond duration. He did not describe the measurement technique. However, a 10 nanosecond duration pulse, even if it is triangular in waveform, yields a 5 nanosecond rise time. One hundred kilovolts in 5 nanoseconds is still respectable.

---

*Written while at the Los Alamos National Laboratory. Currently at the Northrop Research and Technology Center.

88 Pulse-Voltage Circuits

Burov used, incidentally, a Hertzian shortwave peaking gap, making this one of the earliest references for use of a peaking gap to sharpen the output of a Marx bank.

By 1938, Schering and Raske of Germany reported a 400 kV pulse with a 10 nanosecond rise time. This is 40 kV/nanosecond, which is also satisfactory. They described the five-stage Marx bank as using CO_2 pressurized spark gaps, and there was no mention of sharpening gaps to obtain that performance. The first reference in American literature seems to be R.C. Fletcher, in 1949, who used a gaseous peaking gap and reached 20 kV output, which is not particularly high voltage, but the 0.4 nanosecond rise time yields 50 kV/nanosecond rate of rise. Again, there was no mention of how the output was measured. Other information reveals that Fletcher built his own special spectroscope for doing so, but it was never described in print.

Figures 3-1 and 3-2, from J.C. Martin, "Nanosecond Techniques," show the state of the art through 1970.

There are two points that should be mentioned. First, these systems are not commercially available; somebody worked extremely hard for a special application. Second, numbers can be misleading; megavolt systems are required to get this kind of performance, not 100 kV or 50 kV systems, for reasons that will show up very quickly.

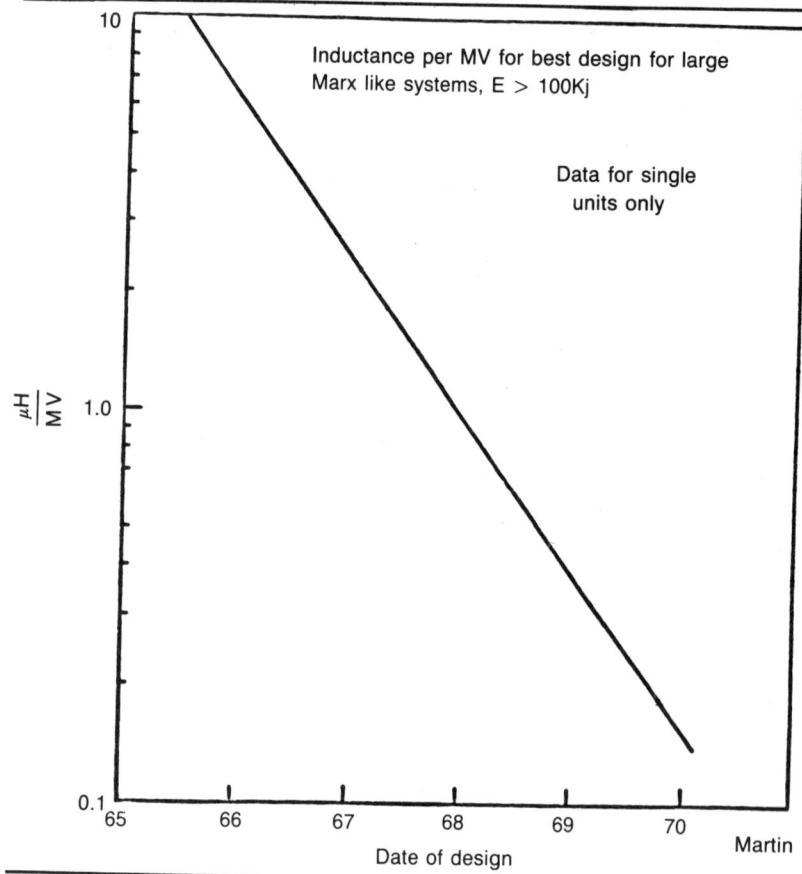

Fig. 3-1. Marx Bank state of the art. [1]

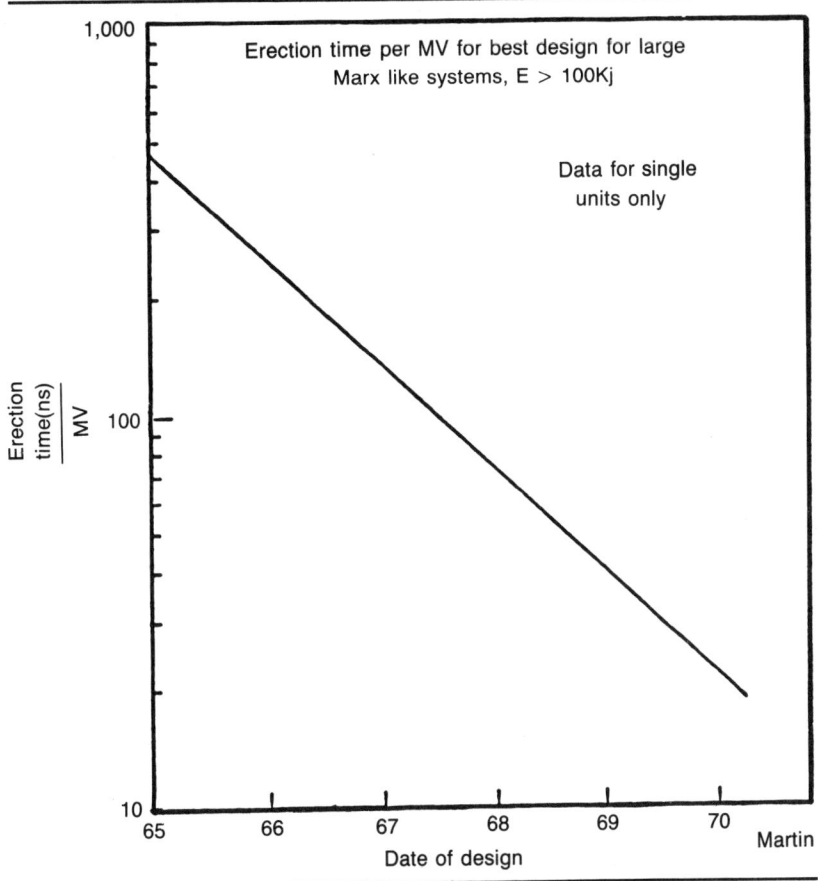

Fig. 3-2. Marx Bank erection time. [1]

Figure 3-2 shows the erection time per megavolt for the best design and the accent is on best. Erection time involves the manner in which Marx banks are fired. Firing is a critical parameter of Marx banks that has a lot to do with the mechanical layout and design. Erection time is not given by formulas very handily but is a good measure of how well a bank will work. Again, in Fig. 3-2 it is not stated whether or not the erection time includes a sharpening gap. As an example, consider a megavolt bank in terms of 100 kilojoules. Using the relationship for energy storage, and a simple calculation:

$$U = \tfrac{1}{2} C V^2 \quad (3\text{-}1)$$

so that

$$C = \frac{2U}{V^2} \quad (3\text{-}2)$$

which, at 1 MV, leads to

$$C = \frac{2 \cdot 10^5}{10^{12}} = 0.2 \ \mu F \qquad (3\text{-}2)$$

The Marx bank, means quite a few capacitors in a series, requires 100 nH in this system at 1970 performance levels shown in Fig. 3-1. That is impressive for a 10-stage Marx with 2 μF capacitors, only 10 nH per stage, including the inductance of the capacitor, the hookup, the spark gap, etc. If this is operated as a 100 kV system, the same relationship only allows 1 nH per Marx bank stage. That is why at this point the performance curve does not describe physically practical systems. Very high voltage Marx banks are required to yield this kind of performance.

3.1.1 Basic Configurations

The concept of the Marx bank is quite simple: charge capacitors in parallel and discharge them in series. Figure 3-3 shows three basic arrangements. Figure 3-3A is perhaps the most economical of parts because the output switch might be optional where a modest level of dc on the output is acceptable. If the load happens to be a gas discharge, and will operate with some dc on it, the designer can save a switch. It is not a usual case, but it is possible. The circuit gives the same voltage polarity out as it put in if polarity is of interest. Normally, this need is avoided by reversing the power supply, but sometimes it is difficult for one reason or another. It has a drawback with respect to triggering, which will be discussed later. Charging is of more interest at this point; however, the first gap in the string goes high upon firing, producing a high-voltage transient phase in the trigger system.

Figure 3-3B inverts the supply voltage sign and, in this case, the output is at ground potential except during the charging transient. One advantage is that one terminal of S1 is at ground potential. This might be an advantage in the triggering design because high-voltage isolation is not a problem. The Ion Physics MiniMarx line is of this design and employs a trigatron-type gap as the first switch. It generates a 100 kV pulse with a 2 nanosecond rise time. The trigatron is a type of spark gap that is usually triggered with respect to ground. Figures 3-3A and 3-3B are drawn in the ladder system. Figure 3-3C is somewhat different. The circuit is an even further step away from high-voltage dc supplies. It uses both a positive and a negative supply to charge, and there are only half as many switches as there are in the single-polarity circuit. Switching is not necessarily simpler. If the same voltage output is required as before, then the switch has to hold off twice as much voltage as in the other Marx banks, so it uses fewer switches but is a more difficult switch to build. Do not be misled by apparently concealed advantages. A drawback on the output point is that the full erected voltage appears across the output resistor. It is difficult to find a megohm resistor that will withstand a megavolt, even for a short time. One reason these circuits are typically built ladder style is that when they erect, the voltage across any one resistor is the stage voltage rather than the bank voltage. The resistor to ground on each stage does have advantages for charging speed. Visualize the charging components as inductors rather than resistors. There is an obvious charging time problem for banks of 10 or 12 stages. Marx banks typically

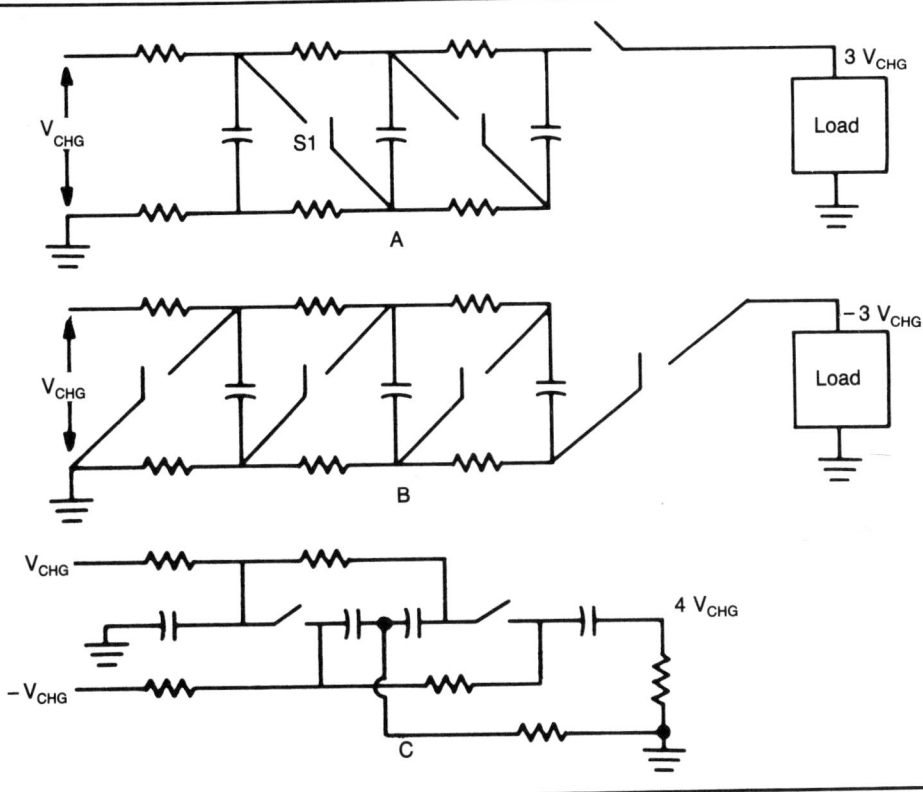

Fig. 3-3. Marx bank circuit configurations.

are not high repetition rate devices. The old timers were after very high voltages for things like E-beam systems, that tend to melt down if operated at a high repetition rate.

If a fast Marx bank is desired, do not use resistors. Serious inefficiency problems appear, for one thing. Another problem is getting the charge on the capacitors. Go to a set of inductors with another set of constraints; it is still possible to charge with a dc supply. In that case, the first inductor normally becomes very large with respect to the rest of the system components, so that when the bank is fired, it does not dump back into the power supply. It makes the calculation of how long it takes to charge much simpler than for the RC circuit. Also, there is no real reason why, given inductors and capacitors, the Marx bank could not be designed as a high-voltage, pulse-forming network (PFN).

3.1.2 Charging Methods

Figure 3-4 is an n-stage Marx bank with a step input U_t/n, where U_t is the desired output. For a charging analysis, it does not really matter whether the resistors are given as $R_o/2$ or are replaced by R_o with the lower elements eliminated. The output voltage available at any instant is, theoretically, the sum of the individual stage voltages. Thus, there is an RC line in which, except for the first stage, all forcing functions are time

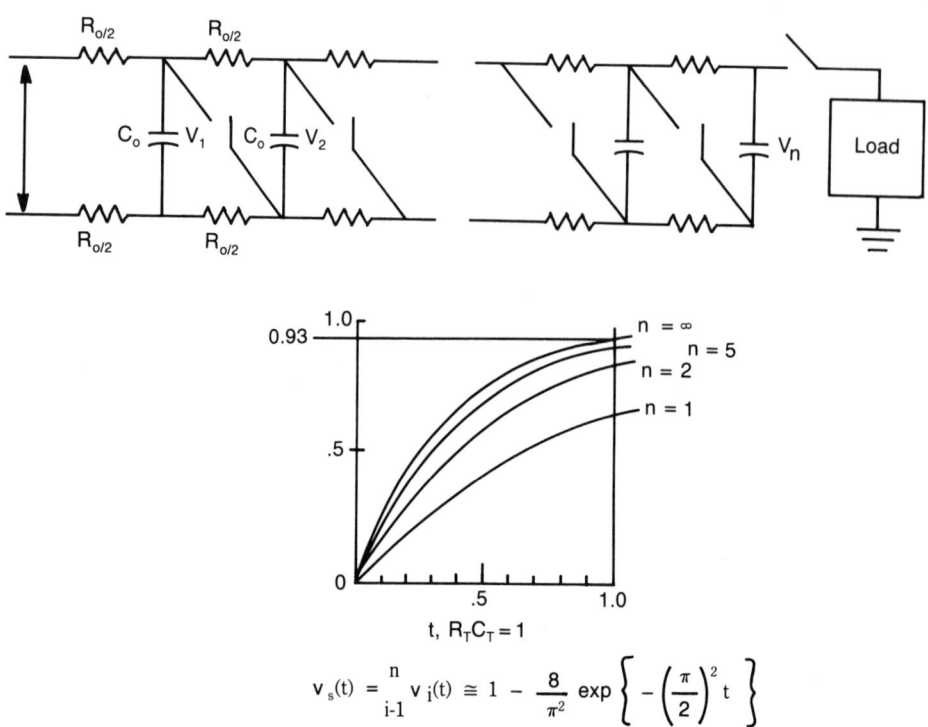

Fig. 3-4. Marx bank charging performance. (G.W. Swift, "Changing Time of a High-Voltage Impulse Generator," IEEE Electronics Letters, Vol. 5, No. 21, p. 534, October, 1969).

and position dependent. Two solutions are conveniently available. One relationship, according to Fitch, is:

$$T_{CH} = R_o C_o n^2 \qquad (3\text{-}3)$$

where T_{CH} is the lag of the nth stage with respect to the input stage.

G.W. Swift presents a power series analysis that he takes to the limit as n goes to infinity. This is the relationship shown in Fig. 3-4B with a few finite cases shown. The $R_t C_t$ time constant is the total R and total C of the circuit shown. The fact that five or more stages are effectively infinite makes this useful for calculation of banks of many stages. Note that the time t is given in terms of the $R_t C_t$ time constant. For a simple case, take 1 MΩ and 1 μF for R_o and C_o, which are typical numbers, and 10 stages. A 10-stage Marx of that type takes about 100 seconds after the first stage is charged before the last stage charges, almost 2 minutes. Is that consistent? In Swift's analysis in the same case, R_t is 10^7 because there are ten 1 MΩ resistors and 10 of C_o in parallel, for a total of 10^{-5} F. Thus, $t = R_t C_t$ and that still comes out to about 100 seconds on the normalized scale of Fig. 3-4B. Swift's analysis shows that the bank would be 93 percent charged at the end of that time constant. The agreement between the two methods is good—only a few percent difference. In either analysis, something better than 100 seconds is required to get the bank charged. The expression given in Fig. 3-4B is useful. Express time in terms of the RC time constants and solve for the time required to charge to 98 percent, 94 percent, or whatever is desired. The analysis does not describe the final stage. It says what the output would be if all these capacitors are in series with switches as shown and fired at a given time. In that curve, if the Marx is fired at 100 seconds, 93 percent is the total output because the input capacitors are charged but the ones behind it, although nearly charged, are charged slightly lower. It really describes the output voltage of the Marx bank, not the final stage voltage; however, a designer probably would not use an RC bank at a very high repetition rate but would use an inductive system.

In an LC circuit, recourse can be taken to the PFN characteristics that will be covered in a subsequent chapter. Ideally, such a network has a characteristic time given by:

$$\tau_{CHG} = \sqrt{L_{MARX} C_{MARX}} \tag{3-4}$$

and discharging into a matched impedance requires a time 2τ. In mismatched cases, oscillations occur that extend this time. More commonly, some command charge system is employed in which an external inductor, large with respect to the Marx inductors, is resonated against the total network capacitance. Charging is usually accomplished in a half cycle of the resonant frequency, leading to:

$$\tau_{CHG} = \pi \sqrt{L_{TOTAL} C_{MARX}} \tag{3-5}$$

Neglecting the switches, the point is that normally the pulse out is much, much faster, even at a kilohertz. There is perhaps 1 millisecond charge on the bank. The output falls typically in a microsecond or in that order, so that there are orders of magnitude differences between the discharge time and the charging time. That means the input inductance can be huge compared to the one in the Marx bank and is used for isolation. So, effectively, the Marx inductances almost are not there. Simply look at the bank as a total capacitor and a charging inductance. Put a voltage step on it, and it starts to ring in the classic resonating circuit fashion. It charges up in half a cycle at most—in a period that can be made fast compared to an RC time constant.

3.1.3 Discharging of the Marx

The inefficiencies of charging have a matching set of inefficiencies associated with the discharge. Figure 3-5 reveals that, in general, a stage capacitor is paralleled by two charging impedances, Z_o. In the resistive case, the self-time constant is just:

$$\tau_{DISCH} = \frac{R_o C_o}{Z} \tag{3-6}$$

and must be long compared to the output pulse for good efficiency. The two end cases are best, since each has only one resistor and therefore has twice the impedance. With many stages, that factor disappears in overall analysis. The point is the existence of a discharge time constant. The Marx discharges very quickly into the load with a much

94 Pulse-Voltage Circuits

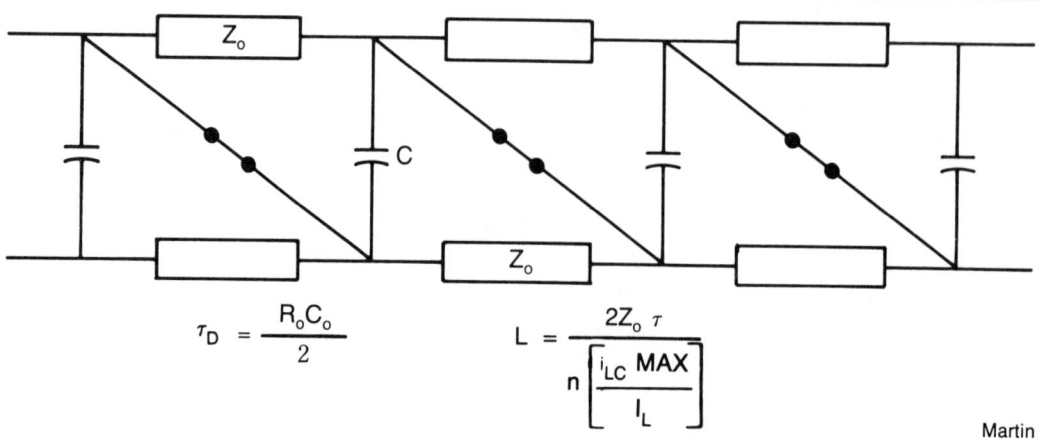

Fig. 3-5. Marx bank discharge relationships.

slower competing process because the impedances are large compared to the load impedance. Nonetheless the discharge is in there, drawing energy during the pulse and drawing current out of the capacitor back through the charging resistors (at the same time the bank is delivering current to the load). The competing process is an inefficiency. It has more than one repercussion, and it can involve heating problems in the system.

When inductances have been used as charging impedances, the behavior is similar except for the appearance of resonance in place of the simpler RC case. The analysis in Glasoe and Lebacqz is straightforward and summarized here.

A given mesh will attempt to resonate with a current given by:

$$i_{LC} = V_N \sqrt{\frac{C}{L}} \sin \frac{t}{\sqrt{LC}} \qquad (3\text{-}7)$$

where L and C are the mesh values of the Marx and V_N is the stage voltage. For good efficiency, the self-ring period, T, is chosen much greater than the discharge period, τ. Thus, the sine of the angle can be replaced with the angle and:

$$\begin{aligned} i_{LC} &\cong V_N \sqrt{\frac{C}{L}} \frac{t}{LC} \\ &= \frac{V_N t}{L} \end{aligned} \qquad (3\text{-}8)$$

and t has a maximum value of τ, because the process terminates when the bank is discharged.

If the Marx discharges into a matched impedance (PFN-like behavior), the Marx current is:

$$I_1 = \frac{n V_N}{2 Z_N} \qquad (3\text{-}9)$$

and the efficiency can be related by the ratio of i_{LC}/I_1.

This case is more apt to be discharge-inefficient than the resistive case. The resistive case normally has the loop currents swamped, so almost nothing goes into the resistors.

3.1.4 Marx Triggering and Erection

Up to this point, the Marx circuit has seemed quite simple and straightforward. The subtleties appear in the relationships between triggering, mechanical assembly, and erection. Very little has been mentioned about how the necessary switch closures are accomplished. Although other switches have been used, spark gaps are assumed for this discussion. (Two papers have been used as references for this discussion: "Marx and Marx-like High-Voltage Generators," by R.A. Fitch, and "Overvoltage and Breakdown Patterns of Fast Marx Generators" by R.W. Morrison and A.M. Smith.) This discussion gets into the part of the Marx bank that is both nonquantitative and frustrating. Discussion will start with the backbone circuit, Fig. 3-6. It is the relationship among the mechanical assembly, triggering system, and resultant erection (that is, sequential closing of all the switches) that yield a peak output voltage of N times the charging voltage per stage. Other switches have been used, but spark gaps are assumed all the way through these discussions. Very few Marx banks are not made with spark gaps, so there are not many exceptions to worry about. Assume now that somehow the first gap is triggered. Thus far there has been little discussion about the triggering. All the gaps here are two-element gaps, and the voltages can be increased until something collapses or the first one can have low pressure, or somehow be designed to be overvolted. It is possible to build a husky trigger system, pulse the first capacitor, overvolt that stage, and launch the bank that way. Argue that only the first gap is triggered on purpose.

When the first gap fires, all voltages around the loop must add up as before, to zero, by Kirchhoff's Law. (Gaps are opposite in sense to capacitors.)

$$N V_o = N V_{gap} \quad \text{initially} \tag{3-10}$$
$$N V_o = (N - 1) V_{gap} \quad \text{one gap fired} \tag{3-11}$$
$$N V_o = (N - 2) V_{gap} \quad \text{two gaps fired} \tag{3-12}$$

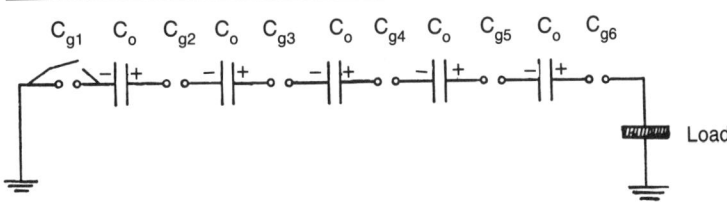

C_{g1} = gap electrode capacitance

C_o = stage energy storage capacitance

Fig. 3-6. Backbone circuit (courtesy of R.W. Morrison and A.M. Smith, "Overvoltage and Breakdown Patterns of Fast Marx Generators," IEEE Trans. Nucl. Sci., August 1972 © 1972 IEEE).

or, alternatively, the voltage across an unfired gap should be

$$V_{gap} = \frac{N V_o}{N - n} \qquad (3\text{-}13)$$

where n is the number of fired gaps and n is the number of stages in the Marx.

This model is, fortunately, too simple. It predicts an accelerating rate of overvoltage firing, which is not always observed, does not predict order of firing, and is pessimistic for many-stage Marx banks. The latter is due to the small overvoltage per stage produced by triggering one gap.

Figure 3-7 shows the physical layout and equivalent circuit of a simple stacked Marx. (In Fitch's notation, a C_1 Marx, that is, capacitively coupled, single column.) Some simplification is still present. C_g is the gap capacitance, C_c the capacitance between stage capacitors, and C_s the stray capacitance to ground.

Because of voltage holdoff and assembly considerations, C_s is usually very small, only a few picofarads. C_c can be quite different. In order to reduce height and inductance, the capacitors C_o may be quite close. Two Maxwell Type "S" pulse capacitors can exhibit up to 50 pF or so for C_c, and 1.25 μF, 4 kV CSI metal-cased units can have up to several hundred picofarads between the electrodes of adjacent units.

A trigger is shown in this particular case. That trigger is commonly brought in through a small capacitor for voltage isolation, and 250 pF is a typical number for the trigger. Assume a trigger is applied. In that case the lower electrode of C_{02} is driven with respect to ground to $+V$. Consequently, the upper electrode of C_{02} jumps to 2 V. It appears that this will overvolt gap C_{g2} because the charging string has the top of C_{g2} at

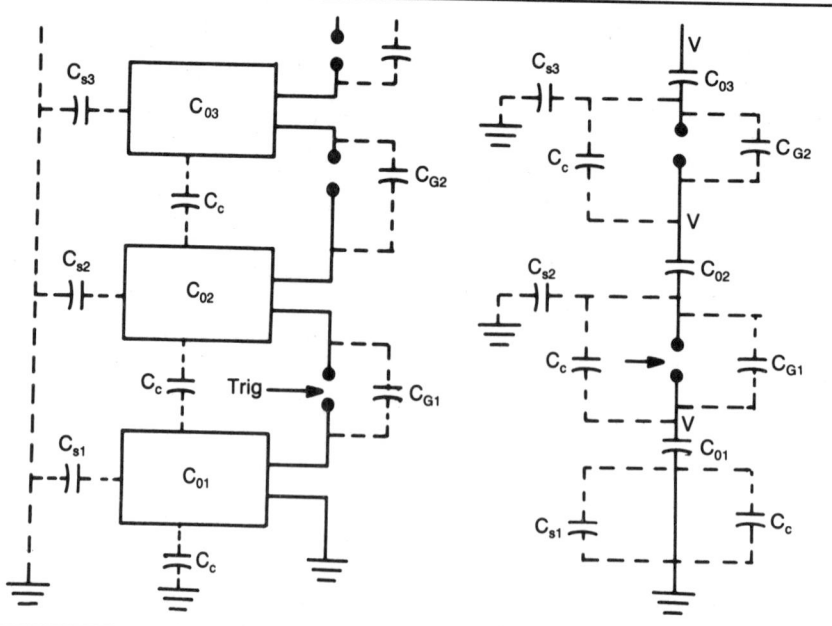

Fig. 3-7. Layout and circuit relationships.

dc ground before triggering. Unfortunately, the 2 V pulse sees divider C_c and C_{s3} to ground. The equation says that:

$$V_{cg2} = V\left(1 + \frac{C_{s3}}{C_{s3} + C_c}\right) \quad (3\text{-}14)$$

If the stray capacitance is small and the coupling capacitor is large, what happens is obvious. At first V_{cg2} is small, and the divider action takes it off. The step voltage will get there. The charging resistors are still present and eventually, because the charging resistor returns to ground, C_{s3} will discharge and the voltage will appear across C_c and C_{cg2}. Sooner or later the gap will go, but it goes with an RC delay, and the capacitive divider slows it down. Fitch comments that if the bank erects at 60 percent of the self-break on the gap it is a healthy bank. It should run dependably at 70 percent of self-break and up. But if the divider is bad enough, it may not even erect. The transients might all filter off on RC time constants down the line to where the bank ends up on the backbone circuit with the overvoltage distributed equally across all the gaps, and none of them fired. The solution for this is to go to what Fitch would call a C_2 Erwin Marx, as shown in Fig. 3-8. A zigzag pattern, mechanically, leaves the strays to ground from the capacitor the same as they always were. The stray coupling capacitor jumps stages, bridging two gaps in an alternating pattern. This can be done in a spiral of three and get a C_3 Marx, etc. In practice, three is about as high as people normally go. We will come later to the use of resistors instead of capacitors, and with physical components, start calling these R_1, R_2, and R_3 Marxes.

The behavior is a bit different in the C_2 case. When the first gap is triggered, the lower electrode of C_{02} jumps to 2 V, C_c is no longer a divider with respect to the next gap up. It does see C_s and has to charge it through the impedance of the stage. Basically, the voltage appears across C_{g2}, the capacitive divider has gone away, and the next gap fires quite nicely. In a Russian paper describing an Arcadev/Marx, quite large strays are inserted on purpose to hold the upper electrode of C_{g2} at ac ground and virtually the entire voltage appears across the second gap to fire it. This looks very elegant; it is going to go right up the line, firing 1, 2, 3, 4, and 5. Again life is not that simple. To preview coming attractions, look what happens when lower C_{g2} has been driven to 2 V, fired C_{g2}, and now lower C_{g2} tries to jump to 3 V. Unfortunately, C_c has coupled the 2 V point around to the top of C_{g3} and partially unloaded the gap intended to fire next. The interactions start to chase themselves up the ladder.

This problem has been analyzed in detail by Morrison and Smith. The complexity of the problem can be seen by examining their analysis of a 10-stage Marx in which the first gap is triggered and the interactions of the capacitive dividers are carefully studied (Fig. 3-8B). They have assumed ratios, by putting in a coupling capacitor large enough to give a ratio to the gap capacitor of 10 and a stray capacitance that is twice the gap capacitance. The stray is fairly easy to keep small, because normally the capacitors are well away from the ground plane. The coupling capacitor normally is not small because in fighting inductance the capacitors are close. These are given as $C_c/C_g = 10$ and $C_s/C_g = 2$. The analysis assumes firing the first gap and following the transients up a 10-stage Marx through all the capacitive dividers. After triggering gap 1, gap 2 is the most extremely overvolted of all gaps, so it will fire. When gap 2 breaks down, the voltages in the capacitive divider vary. At some points, for example, gap 4, they are inverted and at some they are not. But gap 3 is still normal and most overvolted. This action keeps on going to 3, then 4. When gap 4 breaks down, the problem is: "What Happens?" Gap 5 is high, and it goes. When 5 goes, 6 is high, 8 is high, 10 is high, and the question is: "What goes next?"

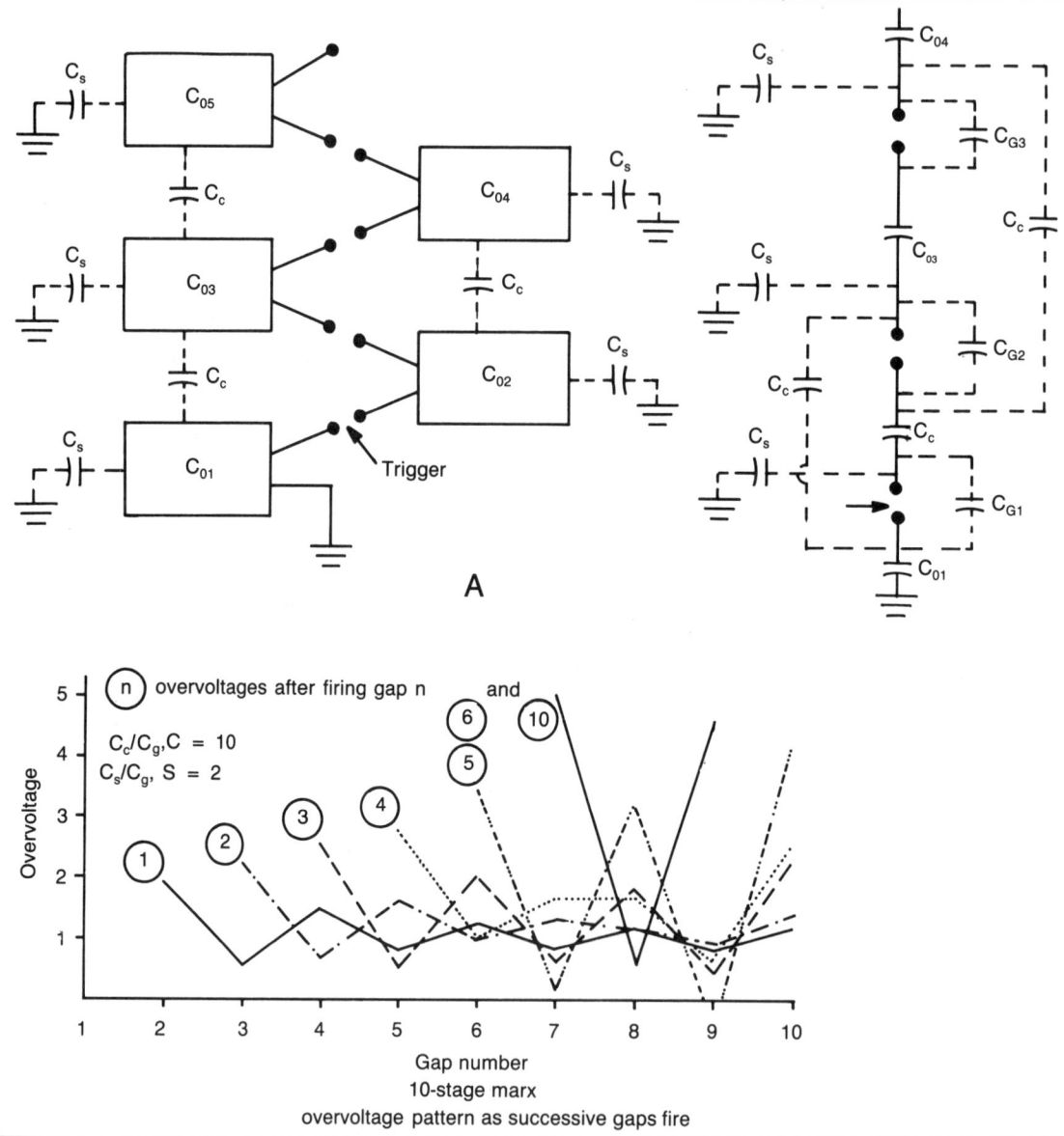

Fig. 3-8. Marx triggering sequence (courtesy of R.W. Morrison and A.M. Smith, "Overvoltage and Breakdown Patterns of Fast Marx Generators," IEEE Trans. Nucl. Sci., August 1972 © 1972 IEEE).

The outcome is that for different ratios in these dividers, the Marx bank might start erecting from the bottom, turn around and erect back down from the top and meet in the middle. Other kinds of unexpected patterns also can occur. Now that is not a trivial consideration. Take any two capacitors with a little space left between them. A charging current flows, and there is some inductance associated with that current. The bank has large capacitors with respect to the coupling values. This can result in a version of what will be encountered later, in resonant charging. The behavior depends on whether the capacitor has a metal can or dielectric can. But the circuit can swing up to approximately four times the stage voltage across that stray coupling capacitor. If it is closely spaced, arcs in the air or oil dielectric can occur. There will be arcs from stage to stage and firing patterns that are very hard to predict, in terms of voltages that are going to appear.

One consequence of this behavior is the use of midplane gaps in Marx banks. For triggering purposes, a nonsymmetric gap is desirable; however, this presupposes knowledge of the electrode voltages, a condition that is not usually met in banks of many stages. If the bank uses dielectric case capacitors, and they often do, this might punch right through the dielectric case. The Maxwells, for example, have a preferred side up from the ground. But if the preferred side up is *in* the overvolted stray side, and four times rated voltage occurs, punch-through might destroy it. So, how is a bank made to operate dependably? This question brings up the point of triggering.

The use of more than one triggered gap is common in many Marx banks in order to reduce jitter. This, combined with back coupling, yields a more predictable system. Figure 3-9 is an R_3 Martin Marx (named for J.C. Martin, who pioneered the use of three-electrode gaps in Marx banks) in which the first three gaps are externally triggered. This circuit depends upon triggering by maintaining the trigger electrode potential while the electrodes change voltage.

Up until now these have been called Erwin Marxes, because the Erwin Marx is a two-electrode system. The Martin Marx is characterized by spark gaps with trigger electrodes. In one of these, a trigger system fires the first three stages so that at the top of C_4 the voltage rises to 4 V for overvolting on subsequent stages. The gaps have trigger electrodes, and the circuit design reveals what is going on. The fourth gap trigger electrode is held down to the voltage of the bottom stage, which is going to be V_{CHG}, and the lower main electrode will jump to 4 V. What actually happens in this type of circuit is that the trigger voltage changes very little while the main electrode swings. In the first stages, swing the trigger electrode to fire the bank, and from there on hold the triggers down to the potential of preceding stages with discrete components, while the electrode itself swings away for triggering. These are quite dependable in jitter and triggering. There are a couple of penalties to pay. For example, shown is a very simple triggering with three input capacitors. When the plasma forms in the gap arc, the trigger electrodes are tied to stage voltages and are fed back down into the triggering system. It makes the triggering system rather complex. Also, voltages of 3 V, 2 V, and 1 V are summed, yielding miscellaneous sloshing in the trigger system. Typically, there is some resistance in series with each of the trigger capacitors to limit the surge currents that flow in the coupling capacitor. Only 50 Ω or so do not affect the system (because they do not draw current until the arc is actually formed, at which point the triggering is over). Do not, however, put the trigger resistor back at the other end of the cable because then the cable capacitance and resistor produce an RC circuit that will slow the

100 Pulse-Voltage Circuits

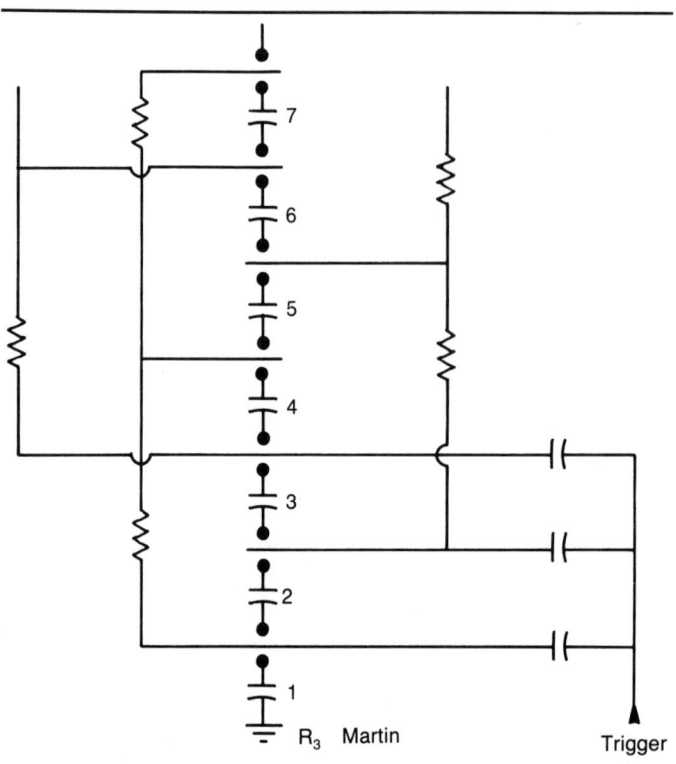

Fig. 3-9. R_3 Martin Marx (previously published by the Institution of Electrical Engineers, R.A. Fitch "Marx and Marx-like High-Voltage Generators" Proc. IEEE, 1972).

trigger pulse. Part of the precision of these systems is related to the rise time of the trigger generator.

There is yet another approach for which no figure is included. A high-energy Marx runs to large capacitors, and large strays are associated with them. It is not unusual to build another small Marx bank, which has the same basic characteristics, and cross-couple all the way up between stages. This small, very low energy Marx bank is very, very fast. It runs in parallel and the entire Marx bank is triggered with another Marx bank. That is an expensive and difficult way to go, but it can be the ultimate in precision, because the rise time for erection is set by a very small, fast Marx rather than a large sluggish Marx.

3.1.5 Output Considerations

G.A. Mesyats, in a translation of *Formation of Nanosecond Pulses of High Voltage*, presents an interesting analysis of the effect of the spark gaps on the Marx output waveform. This circuit model (Fig. 3-10) replaces the bank with C_o, an equivalent capacitor, which is the stage capacitance divided by the number of stages. An inductance associated with the Marx circuit is given. There are two resistors in the model. R_L is an equivalent fixed resistance. Sometimes a little resistance is added to a Marx bank

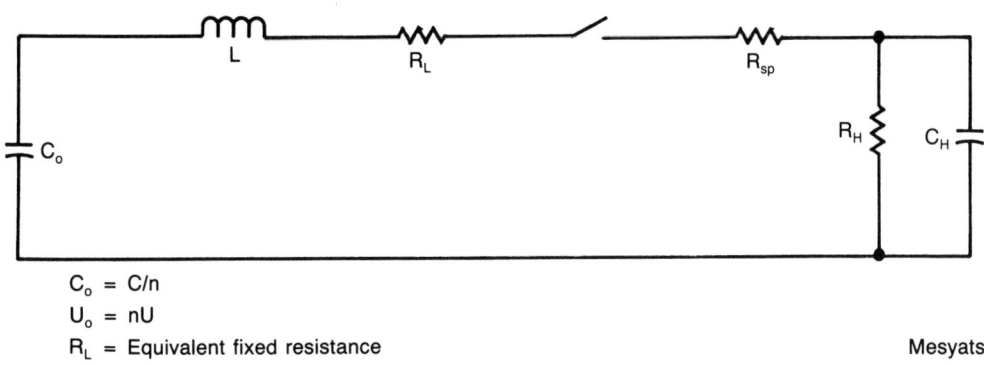

C_o = C/n
U_o = nU
R_L = Equivalent fixed resistance

Mesyats

Fig. 3-10. Mesyats circuit model of a Marx bank. [6]

purposes to prevent ringing. There is always some fixed resistance in the connections. R_{sp} is a spark gap resistance and there is an output load resistance paralleled with a load capacitance, probably a stray. When the gap resistance is pertinent, Mesyats approaches the problem with:

$$R^2_{sp}(t) = \frac{pd^2}{2a \int_0^t i^2 \, dt} \quad (3\text{-}15)$$

where $R_{sp}(t)$ is the gap resistance as a function of time, p is gap pressure in atmospheres, d is gap spacing in centimeters, i is current in amperes, and a is a gas constant:

$$a \cong 0.8 \text{ to } 1 \; \frac{\text{atm cm}^2}{V^2 \text{ sec}} \quad (3\text{-}16)$$

for air or nitrogen.

For a given breakdown voltage, as pressure goes up, R_{sp} goes down (Paschen's Law).

If the spark resistance is considered, the output pulse amplitude becomes:

$$U = U_o \cdot \left[1 - \left(\frac{1}{\sqrt{1+2B}} \right)^{3/2} \sqrt{\frac{1+2B}{2B}} \right] \quad (3\text{-}17)$$

where

$$B = \frac{R_{sp}C}{\Theta} \quad (3\text{-}18)$$

Pulse-Voltage Circuits

and

$$\Theta = \frac{2\,pd^2}{a\,U^2} \tag{3-19}$$

The pulse front width is defined as the ratio of pulse amplitude to maximum front curvature and is given by:

$$t_f = \frac{128}{27} \frac{\left(1-(2B)^{\frac{-1}{3}}\right)^{3/2}}{1 - \phi(A)} \quad \text{for } B \geq \tag{3-20}$$

where:

$$A = \frac{L}{\Theta R_H} \tag{3-21}$$

$$\phi(A) = 0.157A - 0.0108A + 0.00017A^3 \tag{3-22}$$

A small Θ (called the spark time constant) yields a larger, faster rising pulse. Examples are:

$$d \cong 1 \text{ centimeter at atmospheric pressure}$$
$$t_f \geq 10 \text{ nanosecond even with no circuit inductance}$$

When the gap pressure is high enough, the output is determined by the remaining parameters. If C_H, the load capacitance, is neglected:

$$t_f = 2.2\,\frac{L}{R_H} \quad 10 \text{ to } 90 \text{ percent} \tag{3-23}$$

Further, if R_H is $>> 2\sqrt{L/C_o}$, the half-width pulse height is $0.7\,R_H\,C_o$. This defines L in the sense that a 1 nanosecond rise time requires an L of $10^{-9}R_H$. For P, it is the spark resistance that has been calculated. The total bank capacitance for parameter Θ is C, and Θ, in turn, is derived out of PP_2.

Mesyats discusses the curvature of the front, where we would be more inclined to speak of the rise time. He calls Θ the spark time constant, and it is the key parameter of this discussion. The smaller Θ is, the faster the Marx bank is going to be. It does more or less agree with other relationships used to describe gap behavior. There is a relationship between geometry, pressure, and the voltage a gap will hold off. Going up in pressure increases Θ, but the down-quadratically-with voltage relationship agrees with other relationships that show a preference for small spacings and high pressures.

Mesyats worked out a couple of examples in his paper. He took a case where d was a centimeter and p was an atmosphere. This is about a 30 kV spark gap so it is not an unreasonable design. He goes through all these equations and comes out with 10 nanoseconds using that kind of a gap for the Marx rise time. This is the minimum time, even if the inductance for his model is zero. If the pressure is set high enough, the time can be reduced to 1 nanosecond. In this model, when the gap pressure is high enough, other parameters in the circuit determine rise time, which is the usual result in any of these spark gap switch systems. Increasing pressure and voltage with the spacing reduced generally yields a better system. If you are designing your own spark gaps, you are better off to increase pressures and voltages, and decrease spacing. The key relationship defines L and says that a nanosecond rise time requires inductance of 10^{-9} times the load impedance. If the load impedance is very low, then a fast rise time is difficult. Another point: can you neglect the stray capacitor C_H? Current flow is given by $C\ (dv/dt)$. This current drops some voltage back in the Marx impedance and in the case of very, very fast Marx banks, dv/dt can be a large number. Even small capacitors will start loading back in the system. There is an important comment related to that. It is easy, in these pulse systems, to come up with the situation where output voltage and output current waveforms do not remotely resemble one another. Sharpening gaps and so forth can produce a voltage pulse with virtually zero rise time. Then comes the discovery that the current pulse is performing differently because it is determined by the inductance of the system and the impedances scattered around. Power delivered to a load may be well out in time.

One way around this is to pulse charge a low impedance intermediate storage element with an output switch. If the bank is going to erect to 100 kV, try to set the output switch to break down at approximately 95 kV. The Marx pulse is rising in some fashion for 100 kV. At 95 kV, the sharpening switch closes and what the outside world sees is something that looks extremely sharp. This is very common practice in E-beam systems where there is usually some kind of transmission line that the Marx charges in 2 or 3 milliseconds. On the output, there is a spark gap that fires very quickly. When that breaks down, looking back into the system, you do not see the Marx bank, but the transition section, which has a very low impedance. This is capable of delivering a current pulse, similar to the voltage pulse, for many purposes. It is one way of getting around all of the inductance associated with the Marx bank and of stiffening up the rise time of the system.

3.2 INVERSION GENERATOR

Now consider the inversion generator shown in Fig. 3-11. It is related to the Marx bank and what Fitch calls a vector addition system. It has several advantages and disadvantages. It resembles a Marx bank but the switches are not in series. The charging circuit is not shown, but the capacitors are charged in alternating sequence, minus to plus, plus to minus, etc. One disadvantage is that there is no way to trigger one switch from the next, and there is no capacitive coupling between the stages. They do not see one another from a triggering standpoint, which means on this sort of circuit, triggering every switch must be accomplished somehow. There is no convenient way of triggering

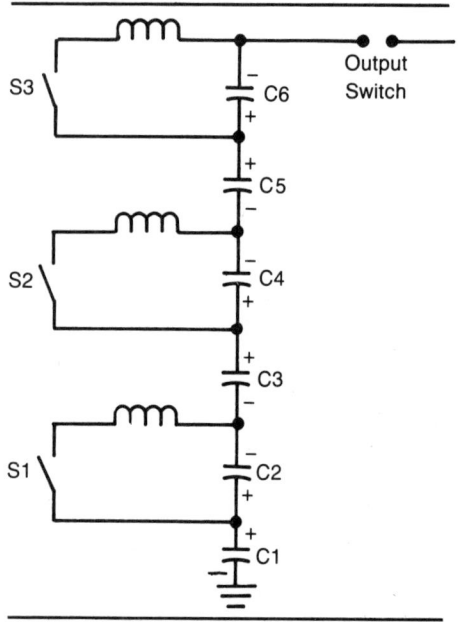

Fig. 3-11. Inversion generator circuit.

only one switch. Look at the advantages first. Erection of the inversion generator is quite straightforward. Closing a switch starts a capacitor-inductor circuit ringing according to:

$$I = V \sqrt{\frac{C}{L}} \sin \frac{t}{\sqrt{LC}} \qquad (3\text{-}24)$$

This L is fairly large relative to the value in the discharge circuit, and it is usually put in deliberately for timing purposes. Each circuit should ring with the same period. After a half period, $T = \pi \sqrt{LC}$, the capacitor has reversed polarity, adding to the unswitched units. Since some losses, R, are incurred in the switch and connections, the output is:

$$V_o(t) = \frac{nV}{2}\left(1 - e^{-\alpha t} \cos \frac{2t}{T}\right) \qquad (3\text{-}25)$$

where

$$\alpha = \frac{R}{2L} \qquad (3\text{-}26)$$

One thing is immediately apparent. The output voltage begins to appear at once with a rise time set by the ringing portion of the circuit. This is usually fairly long with respect to the desired output time in order to minimize losses in the switches. As a result, an output switch, commonly an overvolted gap, is normally a part of the circuit.

Unlike the Marx bank, load currents do not have anything to do with the switches. So all the switch loss, inductance, resistance, etc., are out of the way. The output is from a string of capacitors in series. A low-inductance, fast-current pulse can be produced by this generator. The generator has advantages and, buried away, a couple of big disadvantages. The triggering problem has been mentioned. The trigger must stand off the entire bank voltage on the upper stage. The really interesting fault occurs if one of the switches fails to go, for example, S_3, then C_2 and C_4 invert, and suddenly there is $+5$ V looking at a capacitance charged to -1 V. If the switch tries to hold, the capacitor is going to be highly stressed. High-voltage capacitors just do not like to be reversed. Even if an excessive overvoltage did not occur, damage could result from excessive ringing. Another difficulty occurs if the bank is charged up, and dumping becomes necessary. There is no really convenient way, in case of malfunctions, to save this circuit by dumping it in a hurry. Also, a prefire will ring one stage, probably without firing the other switches. The complexities of preventing these failures reduce the initial attractiveness of the LC inversion generator.

3.3 BLUMLEIN

The Blumlein rates about one page in Glasoe and Lebacqz, as a special case of a Darlington discharge circuit. In this case, the description is in terms of a PFN. In very fast discharge networks, the Blumlein is more nearly a single-stage LC inversion generator. Because the PFN mode is the normal one, the discussion will begin there.

Initially, all capacitors, C1 through C8 (Fig. 3-12), are charged to the same sign and potential. It should be noted that C5 through C8 must charge through the load. If the load is normally open, for example, a gas discharge, a parallel path must be provided. In many cases, the load is a pulse transformer primary for impedance conversion.

Something similar to this is used with some laser loads. The pulse transformer is hand for klystrons, magnetrons, and similar devices to provide an impedance match. This type of a PFN is used in rather broad pulse applications because it is difficult to get very low inductances and very low inductance capacitors. One side is charged to ground and the other side through the load. If the load happens to be a gas discharge, a bypass must be provided for charging C5 through C8. The bypass can be an inductor or a resistor.

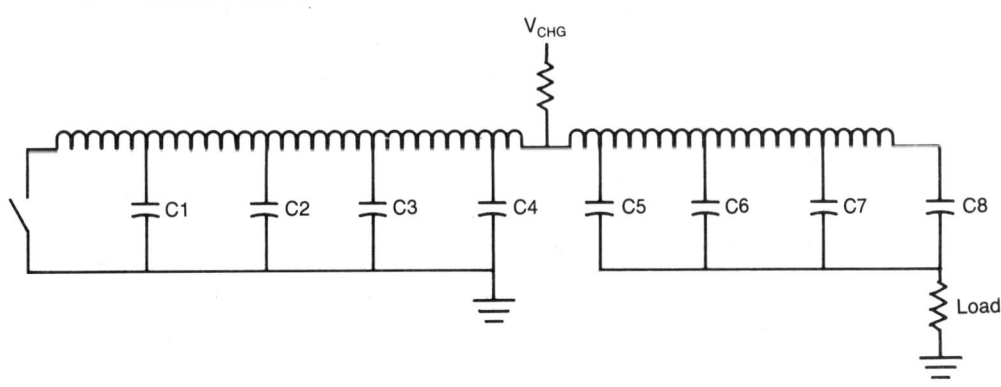

Fig. 3-12. PFN Blumlein.

When the switch is closed, the left stage, as a ringing stage, can ring until it has reversed the charge on C1. C1 is now plus on the bottom and minus on the top at which point C2 can ring and reverse. This ringing will progress down the line until eventually the charging capacitor C4 is charged inversely, and the whole line is effectively inverted. This reversal concept will be used several times in the rest of the discussion. The main point is that a short across the end of a transmission line causes the line to invert in polarity. At this point, the model of Fig. 3-13 shows a circuit in which effectively a battery that is inverse in sign to the original charge of transmission line impedance, the line impedances, and a battery of original polarity are across the load. It is now obvious that the plus to minus, plus to minus sequence yields 2 V across the load.

The transmission lines commonly built have all of the capacitors of the same size for procurement reasons, and the inductors are different because they are easier to tailor. There is a characteristic time given by $\sqrt{L_{tot} C_{tot}}$ that is the time it takes for a switching function to move one way down the line. It will turn out in most of these cases that the time of interest is down the line and back interval. With voltage doubling and 2 Z in the source, if the load impedance is 2 Z, the output voltage equals the charge voltage. The circuit doubles the voltage and then divides it again, a necessarily characteristic operation on these circuits. If it turns out that the load is not resistive but, in the case of common interest, some kind of a gaseous discharge, then this has a feature of ringing up to 2 V to break down the gaseous discharge, and then discharging current. If the overvoltage is needed for a sharpening gap—or something of that sort—the system doubles the charge voltage, and in that sense acts like an inversion generator.

The switch requirements are never greater than V charge, an advantage in many applications, such as with thyratrons. On the other hand, the switch shorts the left PFN so that the current is nominally V/Z or twice the matched load current.

In faster circuits, a PFN may not be practical. Then a distributed network, either cable or stripline, can be used. It is at this point that ambiguity begins. It is pointed out in the work by Mesyats that the switching time of a spark gap can affect the leading edge of a waveform in several nanoseconds. Also, the switch and connections will have an associated inductance. For a lumped PFN, this can be included in the PFN calculation. Here it leads instead to an:

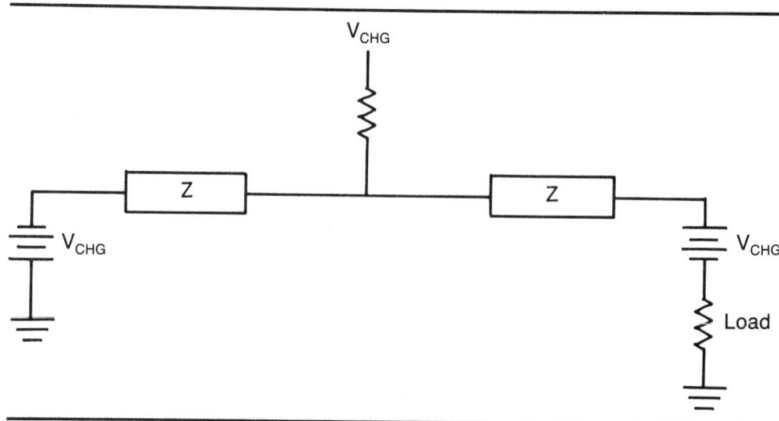

Fig. 3-13. *Blumlein transient equivalent.*

$$e^{-\frac{R}{L} \cdot t} \tag{3-27}$$

type of behavior with R as a function of time. The effect is to launch a wave down the Blumlein with a finite rise time. Since propagation velocities in lines tend to be 1.5 nanoseconds/ foot, a 10-foot cable will depart from a transmission line behavior and begin to behave more like a simple LC circuit with a ringing waveform. The analogy to an LC inversion generator then becomes apparent.

In Blumlein systems, operation ranges all the way from real transmission-line behavior to real LC inversion circuit technique, depending on the trade-off in time constants of the system. Several lasers are touted as a Blumlein system. In fact, the switch closure and gas breakdown occurs about half way through the inversion, and it does not look like a transmission line at all. It looks like lumped Ls and Cs. When someone talks about a Blumlein, look carefully at the closing time of the switch to see whether it is really a Blumlein or an LC inversion circuit.

A number of clever configurations that involve folding of the line have been developed. Figure 3-14 is a case in point. Initially the intermediate coaxial cylinder is charged as usual (upper half). When the switch is closed, the outer coaxial line undergoes the sign reversal from left to right previously described. At time τ, the reversal reaches the right end so that the situation is as shown in the lower half, with both lines adding across the load. The two lines then discharge in series according to the usual procedure. One necessary condition in this circuit is that the intermediate conductor is thick compared to the skin depth of the current that flows in the line. If that is not true, one side affects the other. This can also be accomplished with stripline geometries, and there are many, many geometries, Tom Burkes of Texas Tech has a number of these that he uses for transformers and stepups.

3.4 SPIRAL GENERATORS

Spiral generators are a little hard to understand. This material is directly from a paper by Fitch. It does not follow his notation because it is difficult to carry his notation

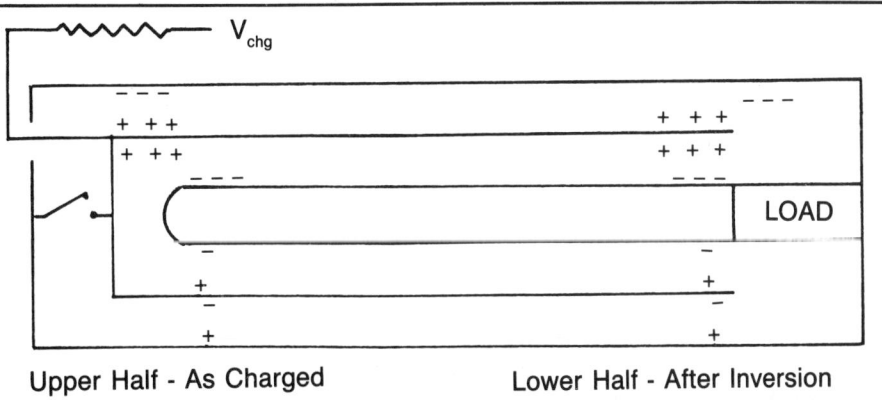

Fig. 3-14. Folded coaxial Blumlein (previously published by the Institution of Electrical Engineers, R.A. Fitch and V.T.S. Howell, "Novel Principles of Transient High-Voltage Generation," Proc. IEEE, Vol. IV, No. 4, April 1964).

108 Pulse-Voltage Circuits

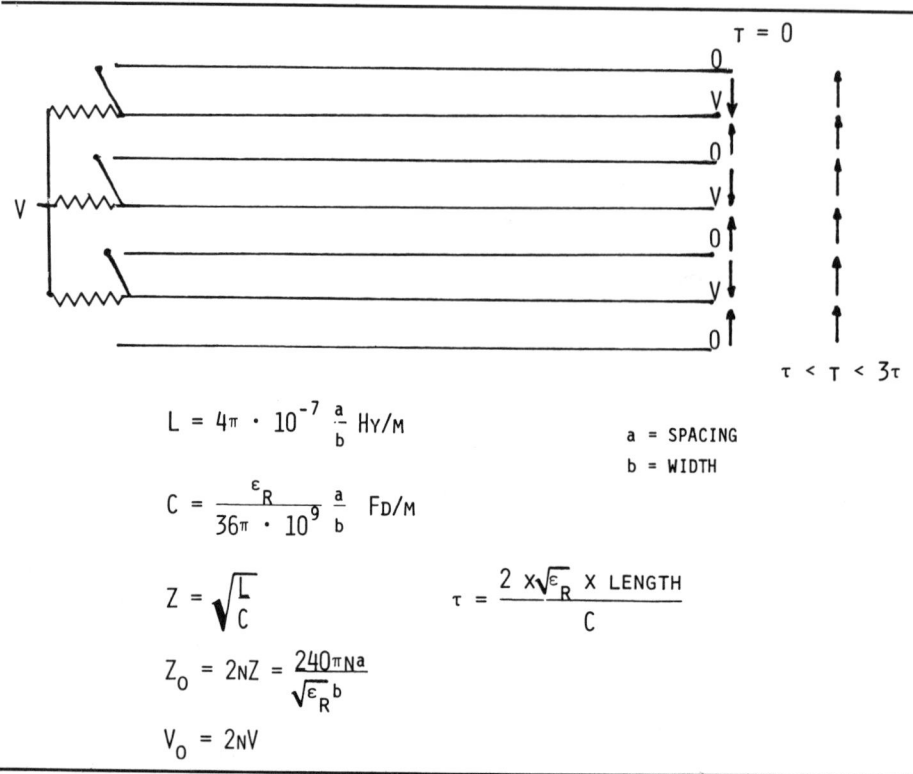

$$L = 4\pi \cdot 10^{-7} \frac{a}{b} \; \text{Hy/M}$$

$$C = \frac{\varepsilon_R}{36\pi \cdot 10^9} \frac{a}{b} \; \text{Fd/M}$$

$$Z = \sqrt{\frac{L}{C}}$$

$$Z_0 = 2nZ = \frac{240\pi n a}{\sqrt{\varepsilon_R} \, b}$$

$$V_0 = 2nV$$

a = SPACING
b = WIDTH

$$\tau = \frac{2 \times \sqrt{\varepsilon_R} \times \text{LENGTH}}{C}$$

Fig. 3-15. Stripline inversion circuit (previously published by the Institution of Electrical Engineers, R.A. Fitch and V.T.S. Howell, "Novel Principles of Transient High-Voltage Generation," Proc. IEEE, Vol. IV, No. 4, April 1964).

here. This is an approach to understanding two or three types of pulse generators other than Marx banks. In Fig. 3-15, if a set of striplines is charged in the fashion shown, alternating sign vectors zero to plus, as shown below $t = 0$ are established and the net output voltage is zero. The equations are for stripline. The inductance and the capacitance are in terms of spacing and width. From these arise the same impedance and transmission time relationships already mentioned. When the switches short the active lines in this system, at time τ, reflections occur on the right, the vectors are reversed on each line, and a 2 nV output voltage results of 2 nV. The impedance is 2 nZ because the lines appear to be in series electrically. The result is voltage multiplication from a relatively high-impedance source. Also, there is a delay after closing the switch of one transmission time. The circuit will stay inverted for two transmission times, so it is fairly easy to decide how long the lines have to be for a desired pulse width. What has been done in Fig. 3-16 is to go one step further and wrap the lines around a central core. Initially, these lines are charged from ground to positive. When the switches close, counterclockwise waves come around and invert in the same manner. The circuit approaches a spiral generator because the lines are wrapped around a common core. The principle is the same. Note that there are multiple switches. That is a complexity that will be eliminated.

3.4 Spiral Generators 109

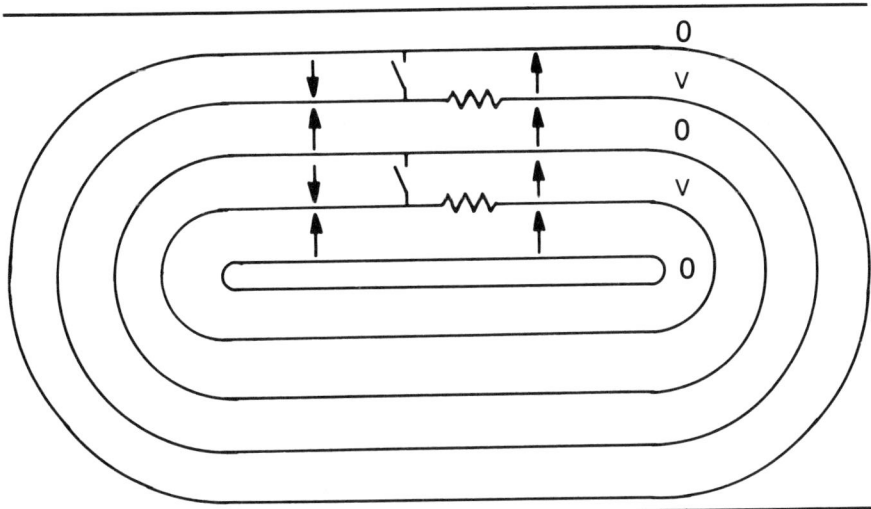

Fig. 3-16. Wrapped-line generator (previously published by the Institution of Electrical Engineers, R.A. Fitch and V.T.S. Howell, "Novel Principles of Transient High-Voltage Generation," Proc. IEEE, Vol. IV, No. 4, April 1964).

In Fig. 3-17, the active lines, the ones to be switched, are on the left. The passive lines are on the right. In a sense, this is Fig. 3-15 unfolded. The goal is to see what occurs and why these are not ideal. The switch shorts points A and C, launching an inverted vector down the line. This occurs on the other narrow lines so that there are six vectors stacked up. Unfortunately, there is another set of lines A-B, and these are shorted also. These vectors try to invert the passive vectors in the wrong sense. The impedance formula shows that the impedance depends upon the spacing. If the A-B spacing is large and the passive spacing is small, A-B is a high-impedance line, the passive line

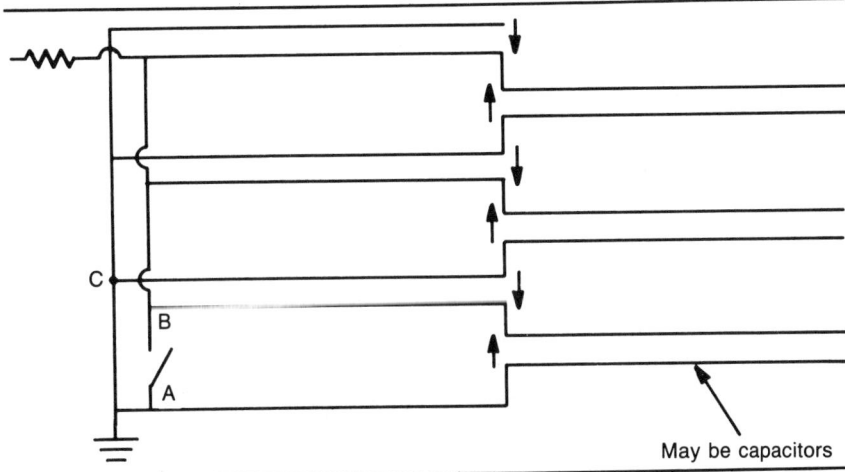

Fig. 3-17. Single-switch inversion circuit (previously published by the Institution of Electrical Engineers, R.A. Fitch and V.T.S. Howell, "Novel Principles of Transient High-Voltage Generation," Proc. IEEE, Vol. IV, No. 4, April 1964).

is a low-impedance line. When the vectors arrive, the circuit acts as a voltage divider that only cancels a portion of the vector. The formulas are worked out in terms of geometry.

A single-switch arrangement yields a loss mechanism that relates a number of parameters, and it comes out in a coerror function. Figure 3-18 shows this loss in the form of β. There is a different β for the spiral generator. This one is in terms of dielectric constant and length divided by spacing for copper. If the line is fairly long, the β will be quite decent. If the need is for a very fast line, then β will tend to be small. This expression will appear again in the spiral generator equations, but it is more complex. This is still the simple case of switching a set of parallel plate lines in parallel. In the spiral generator, the transmission line is wrapped on itself, and another layer of dielectric added so it does not short against itself. The result is a pair of lines back to back, or one line. One line is grounded and one charged as shown in Fig. 3-19. Use of a switch in the middle is a little easier to understand. Initially, there are alternating opposed vec-

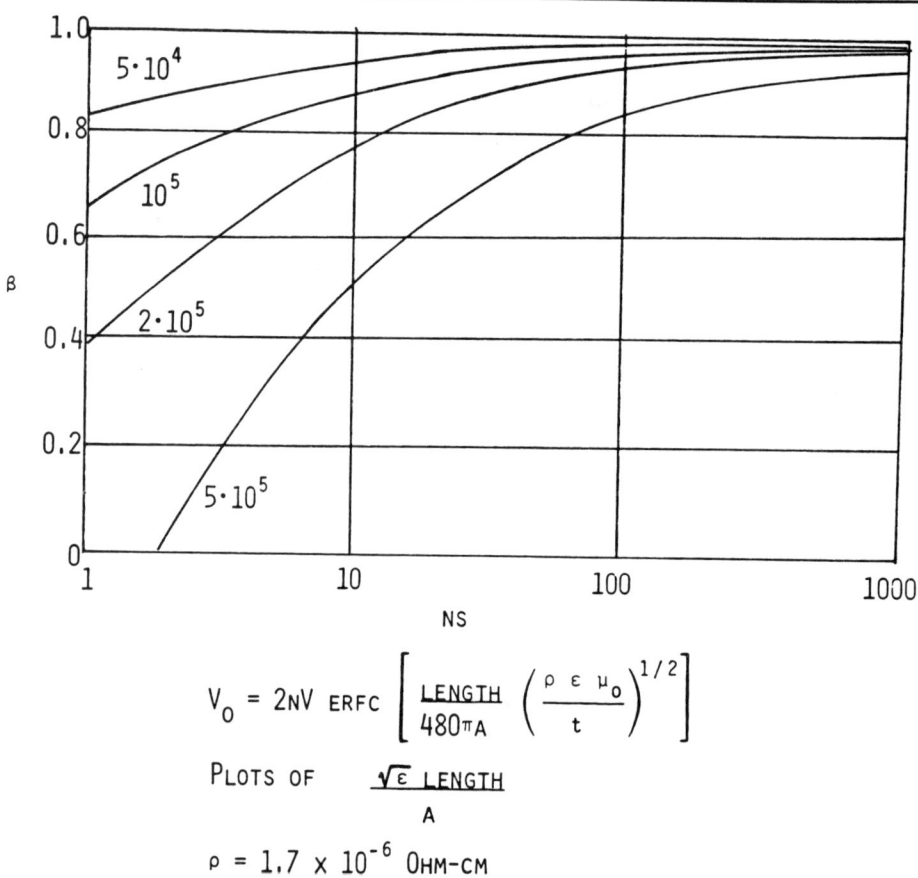

$$V_0 = 2NV \; \text{ERFC} \left[\frac{\text{LENGTH}}{480\pi A} \left(\frac{\rho \epsilon \mu_0}{t} \right)^{1/2} \right]$$

PLOTS OF $\dfrac{\sqrt{\epsilon} \; \text{LENGTH}}{A}$

$\rho = 1.7 \times 10^{-6}$ OHM-CM

Fig. 3-18. Inversion loss mechanisms (previously published by the Institution of Electrical Engineers, R.A. Fitch and V.T.S. Howell, "Novel Principles of Transient High-Voltage Generation," Proc. IEEE, Vol. IV, No. 4, April 1964).

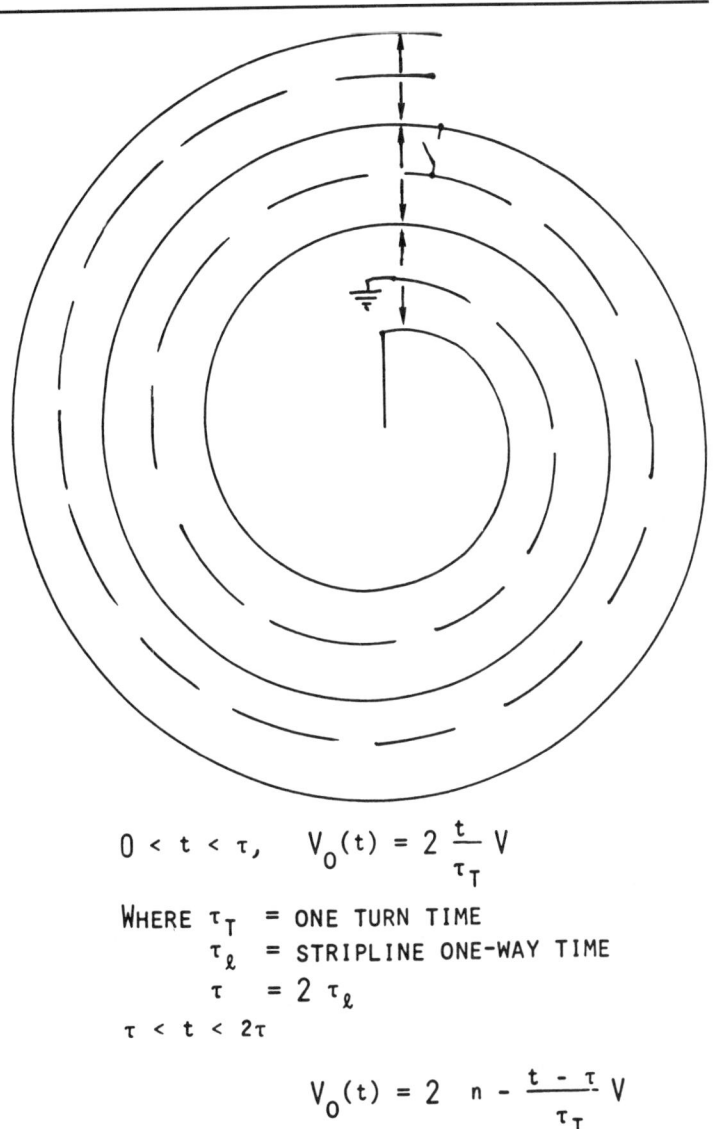

$0 < t < \tau$, $V_o(t) = 2 \dfrac{t}{\tau_T}$ V

WHERE τ_T = ONE TURN TIME
τ_ℓ = STRIPLINE ONE-WAY TIME
τ = $2\tau_\ell$

$\tau < t < 2\tau$

$$V_o(t) = 2\left(n - \dfrac{t - \tau}{\tau_T}\right) V$$

Fig. 3 19. Spiral generator (previously published by the Institution of Electrical Engineers, R.A. Fitch and V.T.S. Howell, "Novel Principles of Transient High-Voltage Generation," Proc. IEEE, Vol. IV, No. 4, April 1964).

tors. The switch is closed, and a vector is launched both ways from the switch. When these lines are shorted, the energy transfers from a voltage on a capacitor in an electrostatic field to a current in the inductor (a magnetic field). When the line reverses, the energy transfers back to the electrostatic field and progresses back down the line. In effect, the short launched by the switch runs around, unwinding the coils in both directions at first and cancels the upward-directed vector. As it goes around it generates a ramp. Eventually, the waves reach the two ends of the line, reflect, and start back inward. As they start back down, they begin to invert the cancelled vectors, and the voltage continues to climb. Finally, the transmission reflection arrives at the switch, reflects off the switches, starts in the other direction; the generator starts to wind down again in the same fashion, generating a saw-tooth output.

These systems are not normally high-energy systems; they are used in trigger systems. The rise time is relatively slow, particularly if large multiplication is desired. Many turns with a lot of voltage multiplication generates a long ramp in time. A sharpening gap is common in the output. Normally, the switch is not in the center because that is physically inconvenient, but at one end of the line and you launch only one vector. The commercial variety of these uses a krytron, the PT55. The krytron is a great device that runs for a few thousand shoes and wears out. The sharpening gap is electrically extremely noisy. The spiral generator does not produce a 2 nV output due to three loss mechanisms, switch impedance, resistive losses, and coupling effects.

Figure 3-20 treats the first case. If the switch closes exponentially with a response time τ, the voltage along the line at a given time t is described by $V_1(t)$ as shown. By manipulation, the voltage as a function of time at a position is then $V_t(1)$. Because the ideal switch charges the line to V, the area under the ideal voltage trace is just VL. Thus, the first loss is described by the area ratio defined as β_1 and plotted.

Resistive losses appear much the same as in the simple case treated before, except that the expression has been rearranged in Fig. 3-21. Again, area ratios with the ideal case lead to β_2, plotted for copper and aluminum. The axis is times a thousand. If the line is 6000 centimeters (200 feet $\cong$ 200-nanosecond rise time); this term alone is about 10 percent. This is worked out in terms of a spacing that is fairly realistic, 10 mils. Kapton has an ultimate failure of about 6 to 7 kV per mil and the failure point is 60 kV in that case.

There is a third β term difficult to describe mathematically. For example, the system behaves as a pulse transformer whose natural frequency depends upon the leakage inductance between the first turn and its closest neighbor. Second, the active and passive lines are always coupled by the inductance of one turn. The passive lines can discharge through the switch by the inductance of the midturn, and to an extent, the waves in the active lines enter the passive lines by inductance of inner and outer turn. This complex situation is not readily amenable to analysis. It should be possible to write an empirical formula. What Fitch concludes is that the overall data for the spiral generator are the product of the two—for which there is an analysis—plus a loss that is not readily analyzable. Numbers of around ½ for the product of the three terms are typical.

People have done better than that. A β of 0.9 has been built in specific cases. Maxwell has a paper on high-voltage spiral generators. They built a couple of them and have a couple of basic relationships on output capacitance, etc. They discuss the problems they have had. The problems typically are related to the fact that the generator is wound with a stripline. That means there is a nasty tendency to go around the edge or, more dramatically, right through because there is a high field stress right at the edge of the foil. Much work was done grading edges with resistive paper or conducting paints, water, and copper sulfate solutions. There are a number of techniques to keep the field concentrations down. The stripline, particularly narrowly spaced striplines, are subject

SWITCH EFFECTS

$$V_\ell(t) = V_\ell \left[1 - e^{-t/\tau_s} \right]$$

SO THAT

$$V_t(\ell) = V_t \left\{ 1 - e^{\frac{-(L-\ell)\sqrt{\epsilon}}{c\,\tau_s}} \right\}, \quad L = \text{TOTAL LENGTH}$$

$$\beta_1 = \frac{1}{VL} \int_0^L V_\tau(\ell)\, d\ell \qquad \text{OR}$$

$$= 1 - \frac{\tau_s}{\tau}\left[1 - e^{-\tau/\tau_s} \right]$$

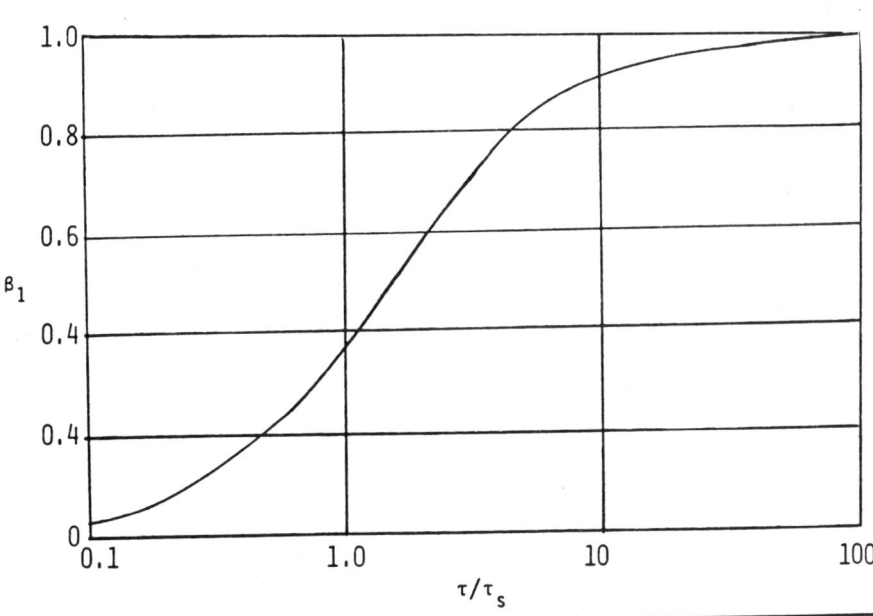

Fig. 3-20. Spiral generator switch loss (previously published by the Institution of Electrical Engineers, R.A. Fitch and V.T.S. Howell, "Novel Principles of Transient High-Voltage Generation," Proc. IEEE, Vol. IV, No. 4, April 1964).

to edge effects because of the high-field stress at the edge of the metal. Flashover is a serious problem. They quoted βs of about 0.6 or 0.7 on some units. They have generators 1½ feet in diameter and 3 feet overall length with two 10-mil conductors and 10-mil Mylar, vacuum impregnated and capable of repeated operation at about 700 kV. The 1 MV goal was obtained; however, edge failure prevented subsequent charge.

$$V'_\tau(\ell) = V_{eRFC} \left[\frac{1}{120\pi} (\rho \, \epsilon^{1/2} \, \mu_o \, c)^{1/2} \frac{\ell}{4a(L-\ell)^{1/2}} \right]$$

$$\beta_2 = \frac{1}{VL} \int_0^L V_\tau(\ell) \, d\ell$$

For $V_o = 2n \, \beta_2 \, V$
  $a = 0.25$ MM   (10 MILS)
  $\epsilon = 2.2$

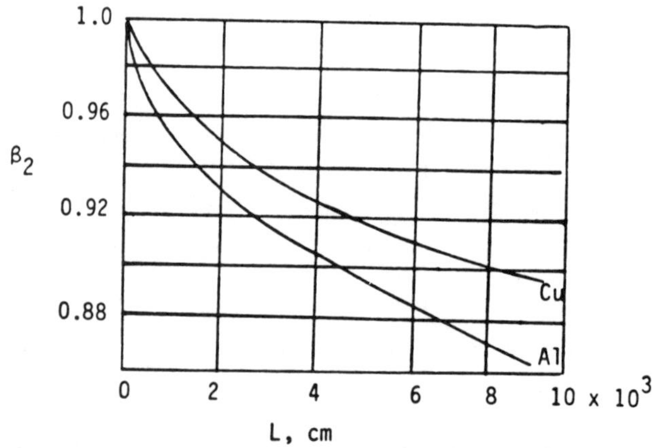

Fig. 3-21. *Spiral generator resistive losses-ideal switch (previously published by the Institution of Electrical Engineers, R.A. Fitch and V.T.S. Howell, "Novel Principles of Transient High-Voltage Generation," Proc. IEEE, Vol. IV, No. 4, April 1964).*

3.5 COAXIAL GENERATORS

Figure 3-22 has sketches of two coaxial generators. The first is connected so that the center conductors are bridged at the right end, and at the left end is a switch. This is not much different than the lines already discussed. Fitch reports an analysis that states that if 2 N is more than 9, performance degrades. That means about four cables are all that can be stacked. The limit arises because of the stray capacitance between the lines. The impedance of a coaxial cable compared to stripline is so high that the generator impedance increases rapidly with N. If this is 50 Ω cable, a 200 Ω source impedance results, with only four of them. It does not take much stray capacitance at 200 Ω to load the generator. These are typically relatively low-energy pulses. They might be used for trigger generators and applications of that sort.

The second circuit is electrically connected differently. It is called a time-isolation transformer. Instead of tying all the grounds together as in the first system, a center-conductor-to ground, center-conductor-to-ground pattern is used all the way up. All of

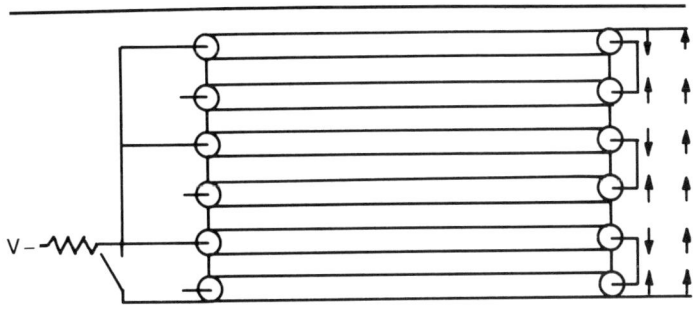

2n > 9   Due to Z, capacitance

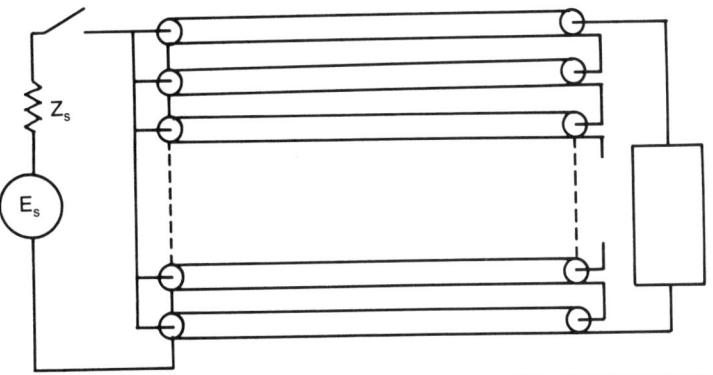

Fig. 3-22. Coaxial-pulse generators. (A.M. Chodorow, "The Time Isolation High-Voltage Impulse Generator," Proc. IEEE, July 1975 © 1975 IEEE).

the centers and all of the braids tie together at the left. Instead of charging the cable in this case, a fast pulse is launched down the cable. A 700 kV generator built this way had a very fast rise time. Again, it is a high-impedance source, but it goes along with Fitch's paper on the vector inversion concept. You are only limited by your ingenuity on how you can stack up elements and switch some of them to get the vectors all turned over and lined up for whatever period of time you need.

Chodorow quotes performance that is impressive. Chodorow's paper consists almost entirely of the mathematical analysis on how to fit all the various stray elements in order to optimize the system. It is mathematically beautiful, but there is not one word on how to physically implement the analysis.

3.6 REFERENCES

1. J.C. Martin, "Nanosecond Pulse Techniques," Circuit and Electromagnetic System Design Notes (AWRE), Note 4, April 1970.
2. G.W. Swift, "Charging Time of a High-Voltage Impulse Generator," IEEE Electronics Letters, Vol. 5, No. 21, p. 534, October, 1969
3. G.N. Glasoe and J.V. Lebacqz, "Pulse Generators," Dover Publications, Inc., New York, 1965.

4. R.W. Morrison and A.M. Smith, "Overvoltage and Breakdown Patterns of Fast Marx Generators," IEEE Trans. Nucl. Sci., August 1972, pp. 20-31.
5. R.A. Fitch, "Marx and Marx-Like High Voltage Generators," Proc. IEE, April 1972.
6. G.A. Mesyats, "Techniques of Shaping High-Voltage Nanosecond Pulses," March 1971, Foreign Technology Division, WPAFB, report FTD-HC-23-643-70.
7. R.A. Fitch and V.T.S. Howell, "Novel Principles of Transient High-Voltage Generation." Proc. IEE, Vol. IV, No. 4, April 1964
8. A.M. Chodorow, "The Time Isolation High-Voltage Impulse Generator," IEEE, July 1975, pp. 1082-1084.

Chapter 4

Transmission Lines and Pulse-Forming Networks

by
R.R. Butcher
(XMR, Inc.)

THE TOPIC OF THIS CHAPTER IS PULSE-FORMING NETWORKS. THE FIRST ITEM OF DISCUSsion is transmission lines because they are so prevalent, even if only in the form of coaxial cable. From there the subject proceeds to pulse-forming networks: the practical problems encountered with them, their advantages, and their disadvantages.

4.1 TRANSMISSION LINES

This chapter references two masters theses, and there is a bibliography available from Texas Tech University.[1,2] These theses present a good description of the transmission-line theory and more detail than can be presented in the limited space here.

Figure 4-1 shows the most familiar type of transmission line: a coaxial cable. By notation, there is an inner radius A and an outer radius B (which is actually the outer radius of the inner conductor and the inner radius of the outer conductor). This is an important point in that surfaces nearest each other are used because high-frequency currents flow near the surface.

Formulas for capacitance and inductance of coaxial transmission lines are:

$$C = \frac{2\pi \epsilon_o \epsilon_r}{\ln \frac{b}{a}} = \frac{5.56 \cdot 10^{11} \, \epsilon_r}{\ln \frac{b}{a}} \quad \frac{F}{M} \tag{4-1}$$

$$L = \frac{\mu_o \mu_r \ln \frac{b}{a}}{2\pi} = 22 \cdot 10^{-7} \mu_r \ln \frac{b}{a} \quad \frac{H}{m} \tag{4-2}$$

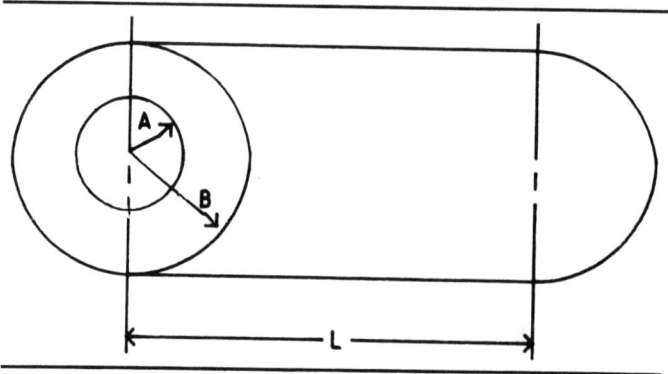

Fig. 4-1. Coaxial transmission line.

in units of farads and henries, per meter of length. Since most high-voltage transmission lines do not use ferrite or other magnetic material as a filler, the value of μ_r is usually taken as 1, and it is in the rest of this chapter. Given the inductance per unit length and the capacitance per unit length of a transmission line, the impedance of the transmission line can then be calculated:

$$Z_o = \sqrt{\frac{L}{C}} \qquad (4\text{-}3A)$$

When this is reduced, it becomes:

$$Z_o = \frac{377}{2\pi \epsilon_r} \ln \frac{b}{a} \; \Omega \qquad (4\text{-}3B)$$

where 377Ω is recognized as the impedance of free space. The relative dielectric constant is sometimes referred to as K, although occasionally the total dielectric constant is simply lumped as ϵ. Normally, break this down as ϵ_o times ϵ_r, where ϵ_o is the permittivity of free space.

The one-way transit time of a transmission line is calculated by:

$$\tau_{1\text{-way}} = \sqrt{LC} = \sqrt{\mu_o \epsilon_o \epsilon_r} = \frac{\sqrt{\epsilon_r}}{c} \quad s/m \qquad (4\text{-}4)$$

This is the time it takes the pulse to travel one meter along the transmission line. Notice that this reduces to the square root of the relative dielectric constant divided by the speed of light, c, and has units of seconds per meter.

If the total inductance and the total capacitance of the transmission line are known, the total one-way propagation time of that line can be found, or it can be done on a per-meter basis. Usually the numbers in tables are given on a per-foot or per-meter basis, and the impedance is a dimensionless quantity. The impedance does not depend on the length of the line; the length is assumed to be infinite.

The second common type of transmission line is the strip transmission line, which is characterized by two parallel plates (Fig. 4-2). This can be thought of as a coaxial cable that has been unfolded and laid out flat. In general, the thickness, t, is assumed to be much less than the width, w, of the line. The formulas for inductance and capacitance of the strip transmission lines are:

$$L = \mu_o \frac{t}{w} = 4\pi \cdot 10^{-7} \frac{t}{w} \qquad \frac{H}{m} \qquad (4\text{-}5)$$

$$C = \epsilon_o \epsilon_r \frac{w}{t} = 8.854 \cdot 10^{-12} \epsilon_r \frac{w}{t} \qquad \frac{F}{m} \qquad (4\text{-}6)$$

The impedance of the transmission line is:

$$Z_o = \sqrt{\frac{L}{C}} = \frac{377}{\sqrt{\epsilon_r}} \frac{t}{w} \qquad \Omega \qquad (4\text{-}7)$$

One-way transmission time is again as it was in the coaxial line:

$$\tau_{1\text{-way}} = \sqrt{LC} = \mu_o \epsilon_o \epsilon_r \frac{\sqrt{\epsilon_r}}{C} \qquad \frac{s}{m} \qquad (4\text{-}8)$$

A good approximation for impedance, if thickness is not much less than the width, is:

$$Z_o \cong \frac{377}{\sqrt{\epsilon_r}} \frac{t}{w} \;\|\; \frac{377}{\sqrt{\epsilon_r}} \qquad (4\text{-}9)$$

where $\|$ denotes parallel impedance. It is obvious that this type of transmission line precludes impedances greater than 377Ω. Even the 377Ω level is difficult to attain if the dielectric material used is other than vacuum or air. Twin-lead line at 300Ω is the highest value generally available.

Figure 4-3 shows a battery switched onto the end of an uncharged transmission line. This transmission line is characterized by impedance Z_1 and a one-way transit time, τ, and is terminated in an impedance Z_2. When the battery is switched into the line, a current of magnitude I_1 flows to the right. The arrows on the transmission line indicate that under certain conditions a reflected voltage and current wave, which are denoted

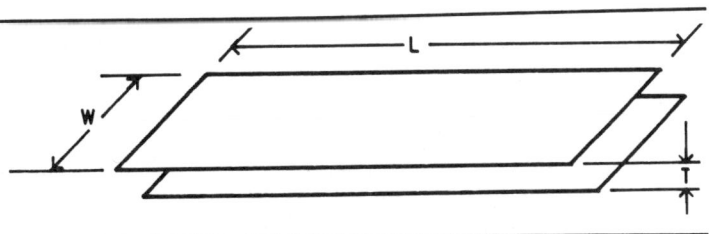

Fig. 4-2. Strip transmission line.

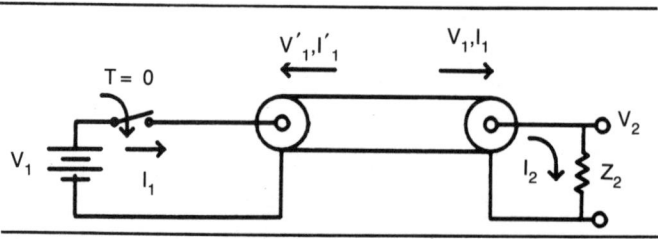

Fig. 4-3. Voltage and current waves in transmission lines.

V'_1 and I'_1, occur. These travel in the opposite direction down the transmission line (to the left in Fig. 4-3). At $t = 0$, when the switch is closed, a current of magnitude:

$$I_1 = \frac{V_1}{Z_1} \qquad (4\text{-}10)$$

begins to flow to the right in the line. When this wave reaches the end of the line (at time $t = \tau$), the voltage and current at the end of the transmission line must be continuous. The voltage and current are considered to be the algebraic sum of the wave traveling to the right and the wave traveling to the left. This can be written as:

$$V_2 = V_1 + V'_1 \qquad (4\text{-}11)$$

where V_2 is the voltage across Z_2. Currents can be calculated in the same way, resulting in:

$$I_2 = I_1 + I'_1 \qquad (4\text{-}12)$$

Ohm's Law declares:

$$I_2 = \frac{V_2}{Z_2} \qquad (4\text{-}13)$$

It also can be shown that the current traveling to the left is:

$$I'_1 = -\frac{V'_1}{Z_1} \qquad (4\text{-}14)$$

which is negative by convention, due to the direction of travel. If Equations (4-10), (4-12), and (4-14) are combined:

$$V_2 = \frac{V_1}{Z_1} - \frac{V'_1}{Z_1} \qquad (4\text{-}15)$$

Combing Equations (4-11) and (4-15) and defining:

$$K = \frac{Z_2}{Z_1} \tag{4-16}$$

allows derivation of the voltage reflection coefficient:

$$\frac{V'_1}{V_1} = \frac{Z_2 - Z_1}{Z_2 + Z_1} = \frac{1 - K}{1 + K} \tag{4-17}$$

The ratio of load current to line current is:

$$\frac{I_2}{I_1} = \frac{2Z_1}{Z_1 + Z_2} = \frac{2}{K + 1} \tag{4-18}$$

It is interesting to notice the effects of various configurations of transmission lines. In an open transmission line ($Z_2 = \infty$) that is initially uncharged (Fig. 4-4), when the switch is closed, current I_1 begins flowing into the line. Current travels down the line until it reaches time $t = \tau$ (the one-way transmission time of the line) the current must be continuous at the end of the line. Because the line is open, there can be no current flowing out of the line, and all the current must reflect and start back down the line in the opposite direction ($I'_1 = -I_1$). Given this situation and Equations (4-10), (4-11), and (4-14), the output voltage that results is:

$$V_2 = 2 V_1 \tag{4-19}$$

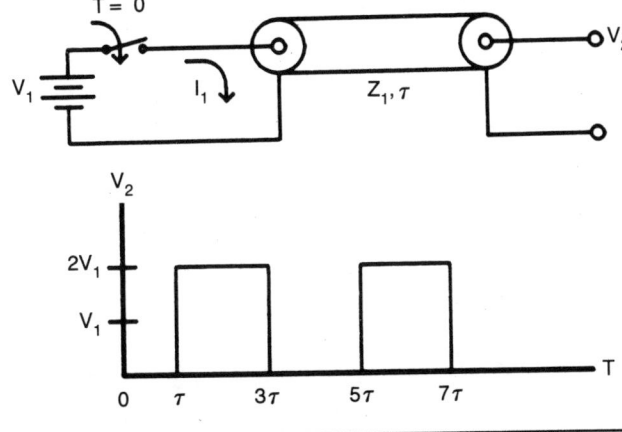

Fig. 4-4. *Open transmission line.*

or twice the initial voltage. This can sometimes be put to practical use producing a voltage-doubled pulse. These pulses are reflected back down the line, and current flows back into the battery. The net result is a voltage pulse of twice the initial amplitude that lasts for two transit times of the line. It is delayed by τ, and it lasts for 2τ. This can be either a help or a nuisance. It is disadvantageous when a fast wave is launched down a transmission line and hits the open end of the line. This action results in voltage doubling, which will track for tremendous distances on a coaxial cable in air. A resistor ($R \gg Z_o$) in series between the voltage source and transmission line prevents this doubling.

Marx banks are complicated by this effect. In Marx banks, there are several biasing resistors associated with each spark gap. If all the resistors and all the capacitors are mounted separately and then connected with a coaxial cable, pulses traveling back down the coaxial cables hitting the few-megohm resistors at the far end can cause serious tracking problems.

Figure 4-5 shows the case of a shorted transmission line. The switch is closed and the terminating impedance is zero, so the current builds up in a stairstep fashion. It goes to V_1/Z_1 and then—at time 2τ—it goes to three, then five, then seven times this value. This effect can be used to build a stairstep current generator.

Figure 4-6 shows a transmission line that is charged to some voltage V_1 and is shorted. The current initially flows in the negative direction. This is a perfect transmission line with no losses. The current in this transmission line will oscillate with a period of 4τ. If a terminating resistor, R, is inserted and the charged line is switched into this terminating resistor, the result is as shown in Fig. 4-7. If the resistance is matched to

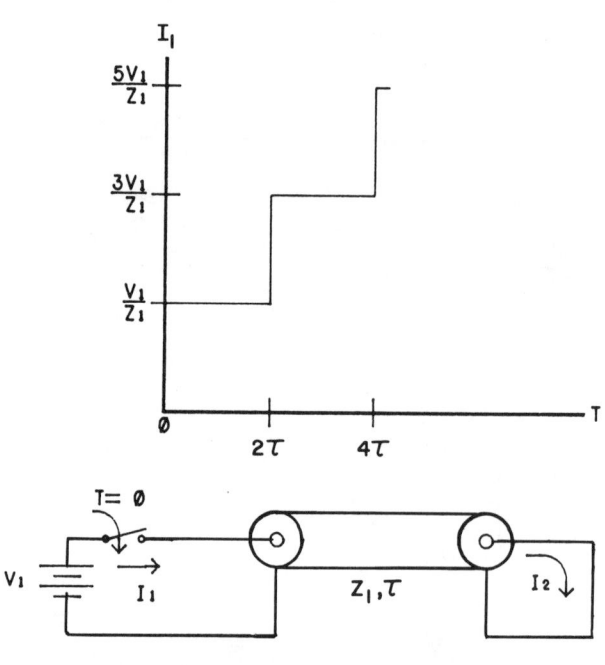

Fig. 4-5. Shorted transmission line.

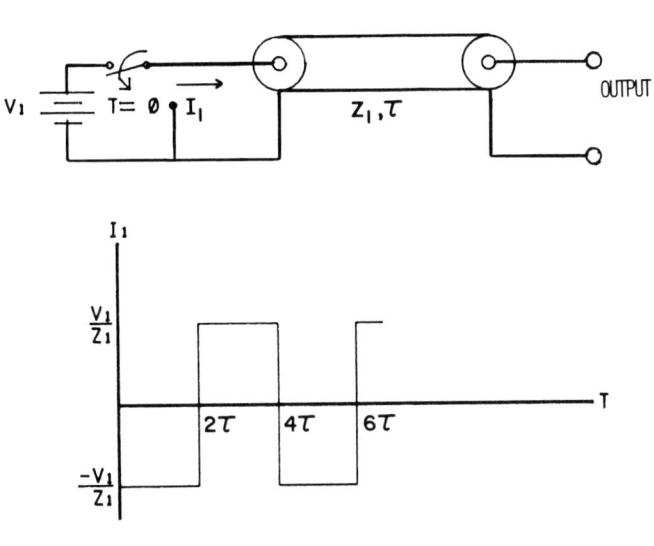

Fig. 4-6. Shorting of charged transmission line.

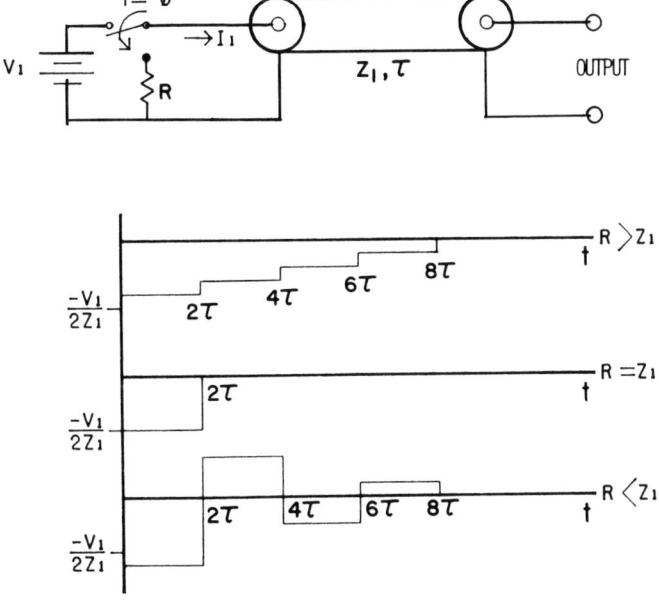

Fig. 4-7. Charged transmission line switched into a load.

the impedance in the transmission line, a square pulse of current will result with a magnitude of:

$$I_1 = \frac{-V_1}{R + Z_1} \qquad (4\text{-}20)$$

lasting for 2τ.

If the resistance is greater than the impedance of the transmission line, a stairstep generator results where the initial current is found from Equation (4-20). Then, because the energy was not dissipated in a single two-way transit time of the line, it will stairstep down.

If the terminating impedance is less than the characteristic impedance of the line, the initial current—from Equation (4-20)—will have somewhat greater magnitude than will matched-impedance, because R is smaller than Z_1, and it will tend to reverse. Therefore, there is both a current reversal and a voltage reversal in the load, and the circuit will tend to oscillate.

These cases are mostly of academic interest, because in reality there cannot be a perfectly resistive termination. There will always be some inductance in the termination that will round off the leading and trailing edges to make the waveform more sinusoidal. In general, most of the energy eventually ends up in the resistor.

Also, some of the energy is dissipated in the line, although a transmission line without loss is assumed. Real transmission lines include a resistive term in series with the line, R_S, (Fig. 4-8) caused by the resistance in the skin of the line, which can become appreciable at high frequencies. This resistance tends to cause dissipation of the energy traveling down the line. A recent measurement showed a loss of about 1 percent in a 10-foot length of cable at a frequency of 10 MHz; this loss is fairly typical of coaxial cables.

Another effect, which is usually small, is the parallel conductance or leakage resistance of the dielectric, R_L. Because dielectrics are not perfect, they do permit some losses. Usually this is a very large resistance, while the series resistance is fairly small. The resistance terms are ignored in the models; the analysis become very complicated if these terms are included.

A charged cable loses charge to leakage resistance or corona. If the cable were charged to a high enough voltage to produce corona effects, or if there were voids in the dielectric, corona losses would occur. These losses might be voltage dependent, and there might be a fairly large resistance up the point where corona begins. Then considerable losses occur. Such corona effects are detrimental to cable life.

Figure 4-9 shows a transmission line that becomes shorted when a battery is switched into an open transmission line. The current oscillates positive and negative with a period 2τ. This current will oscillate in the line, theoretically forever. But the resistive effects previously discussed eventually dissipate that energy, which will settle out at a zero cur-

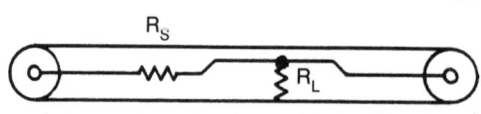

Fig. 4-8. Lumped loss terms in a nonideal transmission line.

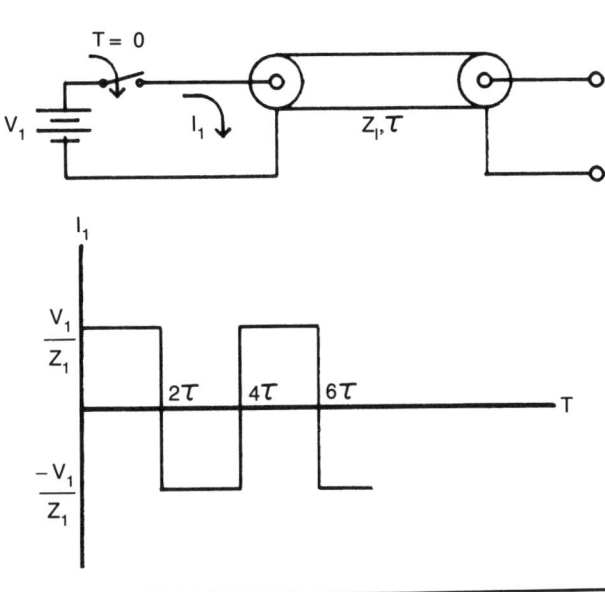

Fig. 4-9. *Current in open transmission line.*

rent value and a V_1 voltage value. This current waveform found when a battery is switched into an open transmission line is the same basic current waveform found when shorting a charged transmission line. It appears inverted, but this is a result of the sign convention.

4.2 PULSE-FORMING NETWORKS

If a shorted circuit produces a current waveform of a particular shape when the circuit is charged, it returns a current with the same shape when switched into a short circuit or a load resistor. This characteristic allows the design of a pulse-forming network in which the line is replaced with a lumped equivalent.

There are different types of pulse-forming networks. A current-fed line (Fig. 4-10) is of academic interest, but it is not of much practical use for high-voltage. A current-fed circuit can be made with transistors or vacuum tubes to a few hundred of volts (vacuum tubes permit kilovolt levels). A current source is switched into the shorted line, starting the current flowing into the line. At t = 0, the switch is opened. The line will then deliver a pulse into the load resistor. This is only of academic interest; although most

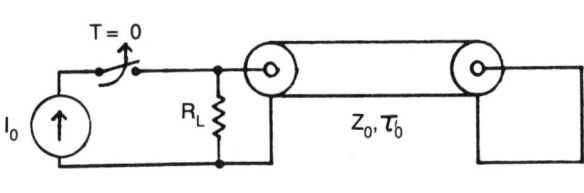

Fig. 4-10. *Current-fed line.*

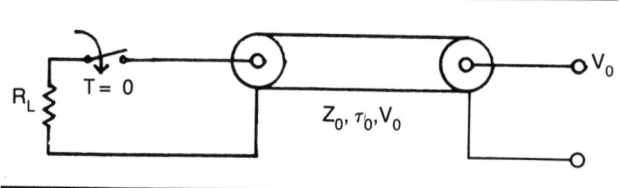

Fig. 4-11. Voltage-fed line.

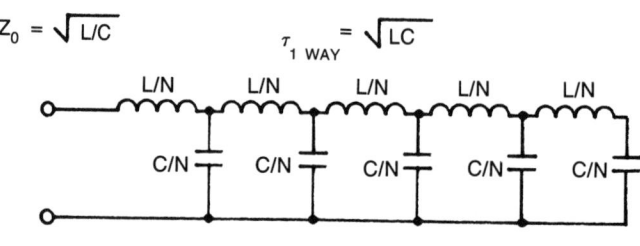

Fig. 4-12. N-section lumped equivalent line.

switches are able to close under high-voltage conditions, they are difficult to open. It is difficult and yet impractical to build a 50 kV switch that can open quickly. Vacuum interrupters open in a few microseconds or tens of microseconds, but these times are not considered fast. Like most switches, they prefer closing to opening.

The voltage-fed line is charged with some voltage V_o (Fig. 4-11). (Please notice the designation change.) The transmission line impedance is Z_o, the one-way length is τ, and the voltage is V_o. A closing switch switches into a resistive load at t = 0, producing a square pulse. The pulse will be similar to that of a matched load if the load resistance matches the impedance of the transmission line (that is, the voltage pulse will have a magnitude $V_o/2$ and width of 2τ).

Figure 4-12 shows how a lumped equivalent transmission line can be derived from series inductance and parallel capacitance. If the line is continued to the limit at which N approaches infinity and each of the inductances and the capacitances is very small, it looks like a true transmission line with distributed elements.

Again, the impedance of the line is the square root of L/C, and it can be either the total series inductance divided by the total parallel capacitance or L_N/C_N. Similarly, the one-way transit time of the line will be the square root of the product of LC, which is the total series L times the total parallel C.

A Fourier series approximation for a square-current waveform can be written as:

$$I(t) = I_m \left[\sin \omega t + \frac{1}{3} \sin 3\omega t + \frac{1}{5} \sin 5\omega t + \ldots \right] \quad (4\text{-}21)$$

This series has to extend to infinity to produce a true square pulse. The amplitudes of the higher harmonics diminish quite rapidly. The line is approximated by some finite series for pulse-forming networks, because it is awkward to have an infinite number of

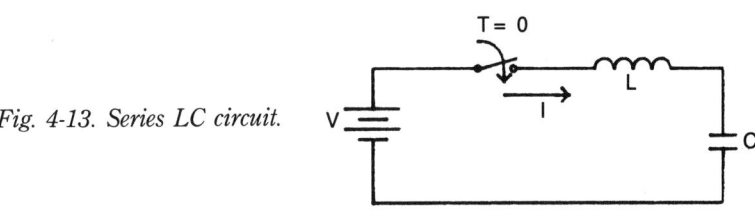

Fig. 4-13. Series LC circuit.

capacitors and inductors. If a battery is switched into a series inductance and capacitance (Fig. 4-13), the current will have the form:

$$I(t) = \frac{V_o}{Z_o} \sin\omega t \tag{4-22}$$

where Z_o is:

$$Z_o = \sqrt{\frac{L}{C}} \tag{4-23}$$

and ω is:

$$\omega = \frac{1}{\sqrt{LC}} = \frac{\pi}{\tau} \tag{4-24}$$

For a one-way transit time, τ, ω appears as a $1/\tau$ term, which leads to Guillemin networks.

The Type-C Guillemin network (Fig. 4-14) is the first item of interest. The battery is switched into a five-stage network, numbered C1, C3, C5, C7, and C9. These stages are numbered to produce odd harmonics (for example, the third harmonic is produced from the L3-C3 combination). For current oscillation as in a cable PFN, the proportionality is written:

$$I(t) \alpha \left[\frac{1}{Z_1} \sin\omega_1 t + \frac{1}{Z_3} \sin\omega_3 t + \frac{1}{Z_5} \sin\omega_5 t + \frac{1}{Z_7} \sin\omega_7 t + \frac{1}{Z_9} \sin\omega_9 t \right] \tag{4-25}$$

Notice that the I_{max} has been temporarily deleted. Comparing terms with Equations (4-21) and (4-25), it is evident that the conditions necessary for square-wave oscillations are:

$$Z_1 = 1, Z_3 = 3, Z_5 = 5, Z_7 = 7, Z_9 = 9$$
$$\omega_1 = \omega, \omega_3 = 3\omega, \omega_5 = 5\omega, \omega_7 = 7\omega, \text{ and } \omega_9 = 9\omega$$

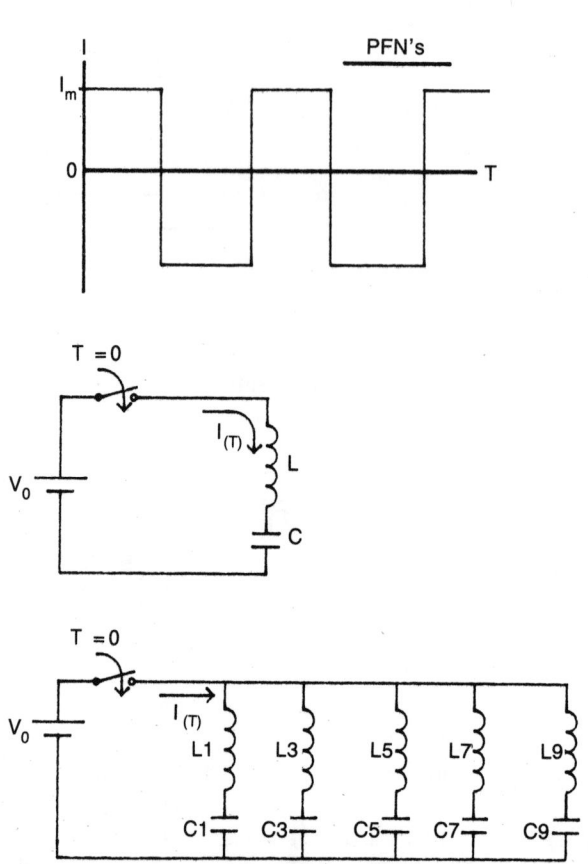

Fig. 4-14. *Five-stage Type-C Guillemin network.*

By comparison to Equations (4-23) and (4-24), generalize:

$$Z_n = n\sqrt{\frac{L_1}{C_1}} = \sqrt{\frac{L_n}{C_n}} \qquad (4\text{-}26)$$

and

$$\omega_n = \frac{1}{\sqrt{L_n C_n}} \qquad (4\text{-}27)$$

Now find the nth term by comparing to Equation (4-22) and by noting that all the impedances (Z_n) are in parallel:

$$Z_0 = Z_1 \parallel Z_3 \parallel Z_5 \parallel Z_7 \parallel Z_9 \qquad (4\text{-}28)$$

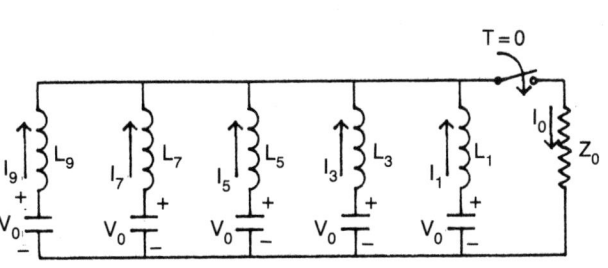

Fig. 4-15. Type-C Guillemin network with matched load.

One way of solving for the values in this circuit is to assume the capacitors are all charged to V_o and the circuit is switched into a load resistor of value $R = Z_o$ (Fig. 4-15). By analogy to the transmission line model, the current could be assumed to be a square pulse of amplitude of $V_o/2Z_o$ and of a duration τ. Now τ now is used as the full width of the pulse instead of the half width. This is done to be consistent with the text (see reference 3 at the end of this chapter). If there is matched impedance ($R = Z_o$), then all the energy stored in the PFN will be deposited in R during the pulse. This energy can be calculated from the definite integral of the power:

$$w = i^2 R = \left(\frac{V_o}{2Z_o}\right)^2 Z_o = \frac{V_o^2}{4Z} \tag{4-29}$$

which is written:

$$E = \int_0^\tau w \, dt = \frac{V_o^2 \tau}{4Z_o} \tag{4-30}$$

This energy can be equated to the energy stored on the total capacitance of the PFN:

$$E = \frac{1}{2} C_T V_o^2 \tag{4-31}$$

and solve for the total capacitance:

$$C_T = \frac{\tau}{2Z_o} \tag{4-32}$$

where Z_o is the load resistance in which the energy is deposited. In this type of network all the inductances are equal.

The difference in τ is visible in the capacitances of the coaxial lines. When τ is the one-way transit time, the capacitance of a coaxial or a regular transmission line is simply τ/Z_o.

Because there is not an infinite number of elements in this Fourier series, there will not be a perfectly square pulse, and the energy deposited will be slightly off because it is not a square pulse. In a five-stage, pulse-forming network, the energy will be about 4 percent off. If there are fewer stages, the discrepancy will be somewhat greater.

The pulse width, for this case can also be calculated from:

$$\tau = \frac{\pi}{\omega_1} \qquad (4\text{-}33)$$

this is from the fact that:

$$\omega_1 = 2\pi f_1 \qquad (4\text{-}34)$$

The period of the fundamental frequency is:

$$T = \frac{1}{f_1} = \frac{2\pi}{\omega_1} \qquad (4\text{-}35)$$

The pulse width τ, is the half period of the fundamental frequency, and the pulse width from the PFN will be:

$$\tau = \frac{T}{2} \qquad (4\text{-}36)$$

Rearranging Equation (4-33) results in:

$$\omega_1 = \frac{\pi}{\tau} \qquad (4\text{-}37)$$

The inductance can be calculated from:

$$L_n = L = \frac{1}{\omega_1^2 C_1} = \frac{1}{(\pi/\tau)^2 C_1} = \frac{\tau^2}{\pi^2 C_1} \qquad (4\text{-}38)$$

From Equation (4-26) find:

$$C_n = \frac{C_1}{n^2} \qquad (4\text{-}39)$$

which puts the capacitors in a ratio of 1:1/9:1/25:1/49:1/81. Therefore:

$$C_T = C_1 + \frac{C_1}{9} + \frac{C_1}{25} + \frac{C_1}{49} + \frac{C_1}{81} \qquad (4\text{-}40)$$

$$C_1 = \frac{C_T}{1.184} = \frac{\tau}{2.368Z_o} \tag{4-41}$$

From Equations (4-38) and (4-41):

$$L = \frac{2.368Z_o\tau}{\pi^2} \approx 0.24Z_o\tau \tag{4-42}$$

The remaining capacitors are found from Equations (4-39) and (4-41). Figure 4-16 shows a typical waveform for this type of circuit, done on the digital computer. It was calculated to produce a 1-microsecond pulse width with an amplitude of 10,000 A. The PFN is initially charged to 20,000 V, and it has a 1Ω impedance and is switched into a 1Ω resistor, so it should have about a 10 kA current for a microsecond. There is quite a bit of overshoot from the leading edge and the trailing edge because the eleventh harmonic is not there to reduce it. The ringing diminishes somewhat toward the middle of the pulse. As more stages are added to the pulse-forming network, there is less overshoot on corners and less ringing in the middle.

This effect led Guillemin to use the equation for a trapezoid and assume that he had a linear rise time, a flat pulse, and a linear fall time. He made the rise time some percentage, a, of the total pulse length (Fig. 4-17). With a being a percentage of the rise time, the requirements on a pulse-forming network can be relaxed. As the requirements are relaxed the a becomes larger, and the number of terms required in the pulse-forming network is reduced. A point is reached where the capacitor for the ninth harmonic becomes either zero or slightly negative, the inductance becomes slightly negative or infinite, and the inductance can be deleted. The contribution from those higher order terms are unnecessary to generate the trapezoidal waveforms.

Figure 4-18 shows another case examined by Guillemin: a flat-topped waveform with parabolic rise time and fall time. The total rise time is used as some percentage of the total pulse width, a, and the only differences are the parabolic rise times and fall times.

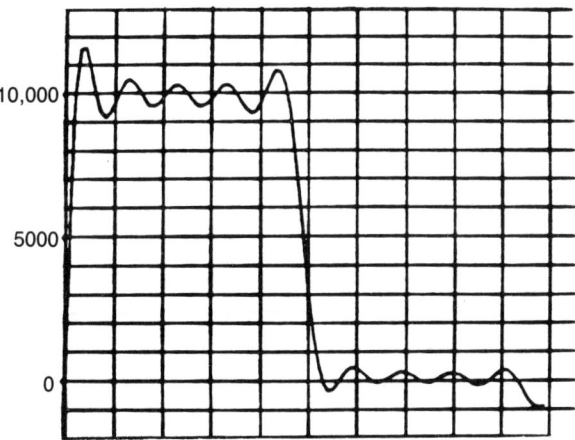

Fig. 4-16. *Type-C waveform, ideal case and resistive load.*

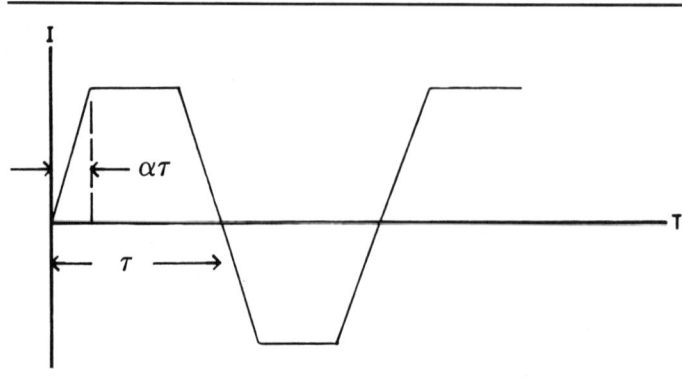

Fig. 4-17. Trapezoidal waveform—flat top with parabolic rise and fall.

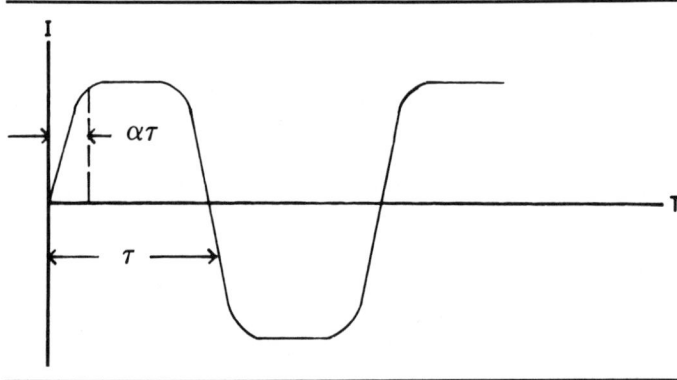

Fig. 4-18. Parabolic rise and fall.

Figure 4-19 shows some computer plots of the waveforms for the five-stage trapezoid and the five-stage parabolic generators terminated in the correct impedance. These waveforms are not ideal—in fact they look similar to Fig. 4-18. The main difference is that the overshoot is less. The few irregularities in the middle of the pulse are not exactly on the first, third, fifth, seventh, and ninth harmonics, but on some slight variation.

As the 10-percent trapezoidal waveform is approached, the pulse becomes smoother and as the 10-percent parabolic, it becomes lumpier. These are still better than the rectangular pulse in Fig. 4-18. Although they are meant to be the same pulse width, it is difficult to compare because they are plotted on graphs of different scales; however, the rise time is obviously good. A 100 nH inductance put in series at the load effectively increases the last inductor by 100 nH, producing an L/R rise time of 100 nanoseconds, which is a 10-percent value for the parabolic load. It rounds off the leading edge of the waveform, flattens most of the ripples, and rounds off the trailing edge. On all of these there is very little energy deposited in the resistance after the main pulse.

Guillemin found other circuits that produce the same general waveform (Fig. 4-20). Type A has parallel LC resonant circuits with one capacitor and one inductor in series.

All these circuits have contained capacitors of different values. Type B also has various inductances and capacitances, although it looks more like a classic transmission line.

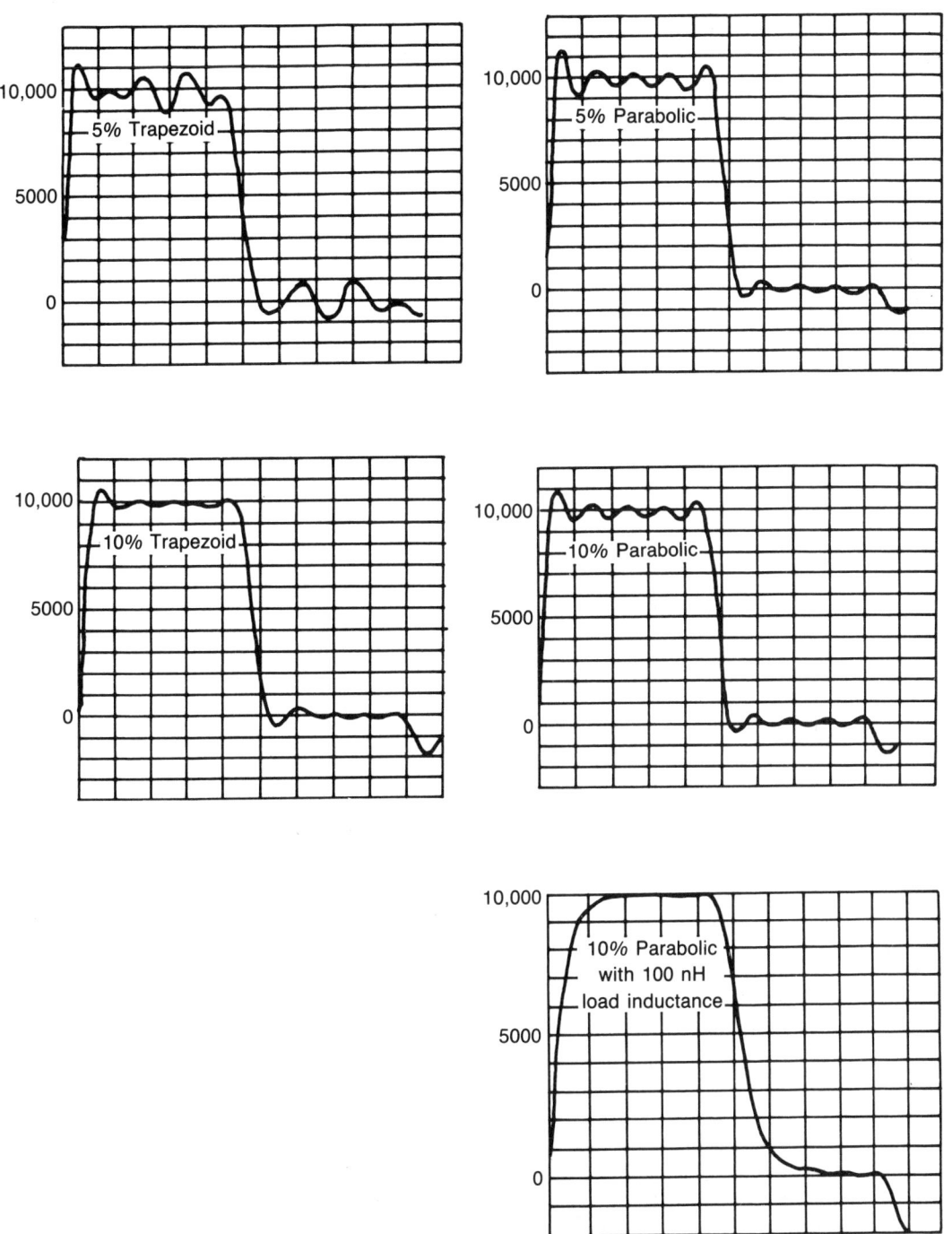

Fig. 4-19. Computer waveforms.

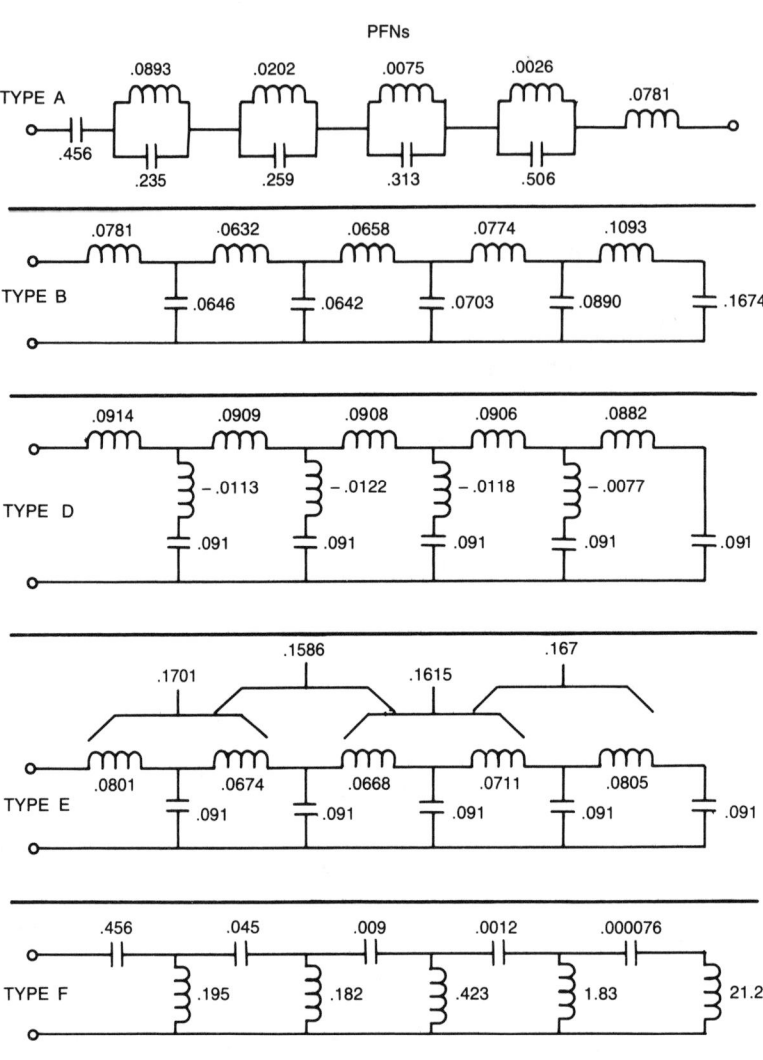

Fig. 4-20. Other Guillemin networks.

This configuration is requisite to generate the trapezoidal pulse with a nominal 8-percent rise time.

The Type F has all the capacitors in series in a transposition of the Type-B network, replacing all the capacitors with inductors, etc. Unfortunately, these capacitors have some very unusual values.

The Type-D network uses capacitors of equal value, but it has negative values of inductance in series with several of the capacitors. Coupling between inductors can realize this physically and produce the Type-E Guillemin network. In this network, there is a coefficient of coupling between the various inductors, and all the capacitors are of equal value.

Total inductances desired to produce these mutual inductances can be found from:

$$L_T = \frac{\tau Z_o}{2} \qquad (4\text{-}43)$$

and the capacitances are derived from:

$$C_T = \frac{\tau}{2Z_o} \qquad (4\text{-}44)$$

This capacitance is divided equally among the number of stages. Then a coil is wound on a long cylindrical form having the total desired inductance, with 20 to 30 percent additional inductance added to the coils on each end of this form. The ratio of length to diameter should be chosen to provide mutual inductance of about 15 percent of the self-inductance in each center section. Soldering the capacitors on results in the correct diameter-to-length ratio, winds the coil, and taps it at the appropriate number of turns.

There are certain minimum inductance limitations in the various types of networks. Some PFNs have inductors in series with capacitors; others, like the Type-C network with its coil wound on a form, require a somewhat larger minimum inductance value. The importance of this minimum inductance number can be seen from rearranging Equation (4-43) to calculate a minimum pulse width-impedance:

$$\frac{\tau Z_o}{2} \geq L_{min} \qquad (4\text{-}45)$$

On a Type-A network, if the minimum inductance is reduced to 100 nH, the Z_o product must be greater than or equal to 38½ times 10^6. Because the Type-E network has mutual inductance, the absolute minimum inductance is about a microhenry, which produces a 1×10^6 τZ_o product. Type-C networks can have minimum inductances to the level found in the capacitors plus the connections.

Capacitors can be bought of almost any inductance, but vary greatly. A practical lower limit does exist for capacitors. Maxwell Series S capacitors have about 20 nH in the capacitor and can usually be connected into a circuit in which the inductance of connections can be kept down to 20 nH. An inductance of 40 nH produces a τZ_o of 0.16 microseconds-ohms.

This relationship is plotted on a graph in Fig. 4-21. If the desired pulse width and impedance of load are known, this circuit will provide some idea of the practical minimums available in Guillemin-type networks. Notice that at impedances of 1Ω, pulse width remains around a microsecond with a Type-C Guillemin, but that generally low impedances dictate long pulse widths. High impedances allow short pulse widths. Ten nanoseconds is a reasonable minimum pulse width.

Also available on the graph is the useful range of distributed pulse-forming networks (coaxial cables). With enough coaxial cables in parallel, impedance can be reduced to 1Ω. Because coaxial cables with a polyethylene dielectric have a one-way propagation time of about 1 nanosecond per foot, given a fast enough switch, a 1-nanosecond pulse width can be attained on a cable or transmission line. Strip lines provide low impedances

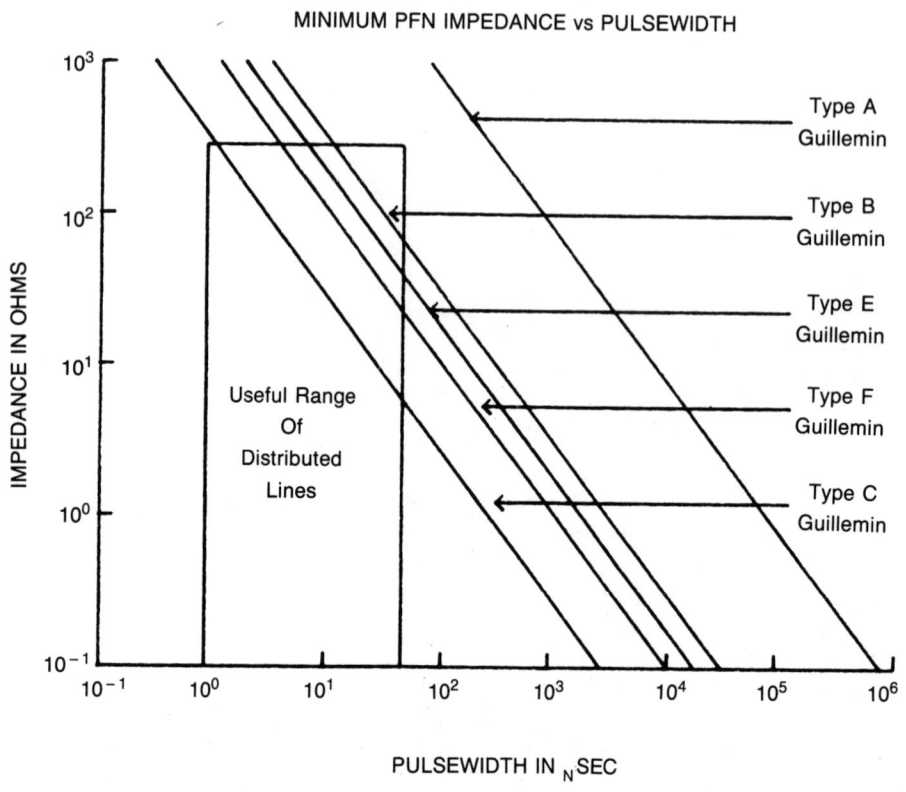

Fig. 4-21. PFN parameters determined by minimum attainable inductance.

but are difficult to switch. Neither cable nor lumped PFNs will generally deliver the desired pulse unless the switch closes much faster than the desired pulse width. Switching, therefore, presents a problem.

4.3 REFERENCES

1. E.G. Cook and T.R. Burkes, "Pulse Forming Network Investigation," Dept. of Electrical Engineering Report, Texas Tech University, Lubbock, Texas, August 1975.
2. D.G. Ball and T.R. Burkes, Report "Pulse Generation for Time-Varying Loads," Dept. of Electrical Engineering, Texas Tech University, Lubbock, Texas, August 1975.
3. G.N. Glasoe and J.V. Lebacqz, "Pulse Generators," Dover Publications, Inc., New York, Chapter 6, 1965.

Chapter 5

Discharge Circuits and Loads

by

W.J. Sarjeant
(State University of New York at Buffalo)

IN THIS CHAPTER, SOME OF THE GENERAL PROPERTIES OF LOADS ARE EXAMINED:

1. Their interface with the energy storage and switching devices.
2. General problems encountered with different types of loads.
3. How load behavior and fault modes can impact on the design of high-power electronics (HPE).

5.1 INTRODUCTION

5.1.1 General Properties of the Discharge Circuit and Load

Consider, in Figs. 5-1 and 5-2, a transmission line with some impedance, Z_o, and a one-way transit time, δ, connected in series with an ideal switch, and a load resistor, R_L. Voltage V_o is the voltage to which the line is charged. When the ideal switch is closed, a current[1] equal to:

$$\frac{V_o}{R_L + Z_o} \tag{5-1}$$

flows for a time $\tau = 2\delta$. For this discussion, let us accept this as an experimental fact. The pulse width τ, base to base, is then 2δ for a load impedance that equals the impedance of the transmission line.

For maximum energy transfer, it can be shown that there is a relationship between the energy stored in the line and the energy delivered to the load. There is a broad maximum in the energy transfer efficiency (Fig. 5-1). Even with a significant impedance mismatch, there can still be a quite high transfer of efficiency in the system. From 0.5 to 1.5 in the ratio of R_L/Z_o, there is a transfer efficiency of about 80 percent. The real

General Properties:

- Basic characteristics determined by circuit elements: switch, PFN, and load.
- Pulse cable can introduce some pulse shape degradation, amplitude attenuation, and average power loss.

Ideal Discharging Circuit

Transmission Line:
(impedance is Z_o)

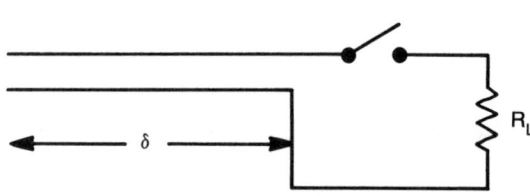

$$I(t) = \frac{V_o}{R_L + Z_o} \quad 0 \leq t \leq 2$$

$$= 0 \quad t > 2\delta \text{ if } R_L = Z_o$$

and $2d = \tau =$ PULSE WIDTH
where δ is the one-way transient time.

For Maximum Energy Transfer, $R_L = Z_o$
As R_L varies from Z_o:

$$P_{R_L}/P_{Z_o} = 4 \frac{R_L}{Z_o} \cdot \left[\frac{1}{(1 + R_L/Z_o)^2}\right]$$

$$P_{R_L} = \left[\frac{V_o^2}{(R_L + Z_o)^2}\right] \cdot R_L$$

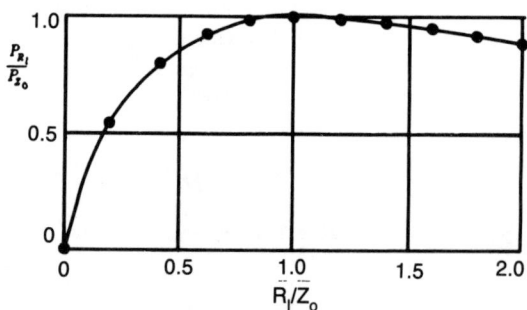

Fig. 5-1. General properties of the discharge circuit and loads (courtesy of G.N. Glasoe and J.V. Lebacqz, eds.: Pulse Generators, *Dover Publications, Inc. NY, 1965).*

reason that the line impedance is matched to load impedance is to control the voltage left on the line at the end of the discharge cycle. If an 80 to 85 percent efficiency is acceptable, there can be a large mismatch.

When the load impedance equals Z_o, the impedance of the transmission line, the energy stored in the line is one-half $C_o V_o^2$ and it is all discharged into the resistive load R_L in the same time, τ. It can be shown (Fig. 5-2) that the energy deposited into the load is the integral of the instantaneous voltage times the current over the discharge time. The discharge time is twice the line capacitance times the line impedance. This is a useful relationship to keep in mind.

The pulse rise-time problem in short-pulse duration discharge circuits is due to the parasitic capacitance across the load. The parasitic capacitance times the impedance of

For $Z_o = R_L$, all the energy in the transmission line is dissipated in R_L:

Let C_o = Line capacitance

$$\therefore \frac{1}{2} C_o V_o^2 = \frac{V_o^2}{4Z_o} \tau = \frac{V_o}{2} I_o \tau$$

$$\text{I.E.} = \frac{V_o}{2} I\tau = \frac{V_o}{2} \cdot \frac{V_o}{2Z_o} \tau = \frac{V_o^2}{4Z_o} \tau$$

or $\tau = 2C_o Z_o$ = Pulse duration

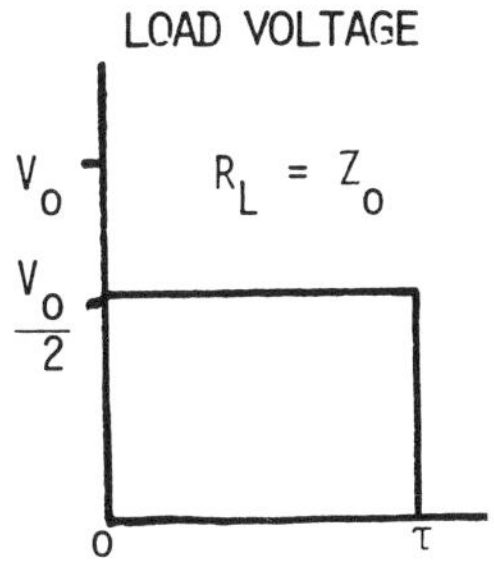

Note that stray capacitances may degrade very short pulse length pulse shapes.

Fig. 5-2. Impedance matching effects on the general properties of the discharge circuit.

the transmission line gives a fairly good measure of what the ultimate rise time will be if the capacitance is fairly large. That is normally the property of the real-world system that degrades the voltage rise time on the load. The current rise time is determined primarily by the inductance in the switch-load interface area.

5.1.2 Pulse-Forming Network Systems

For pulse-forming network (PFN) systems, Fig. 5-3 shows an example of a system having a unidirectional switch, a Type-E PFN (of four sections) discharging into an impedance equal to the PFN characteristic impedance. The PFN discharge current in each section is shown to illustrate the different current pulses flowing through the different capacitances in the line. The last capacitor in the line sees twice the peak current. The RMS current through each is about the same, but the peak current in the final capacitor is definitely higher.[1] This means that during the design of a PFN, it is usually more cost effective to provide the capacitor supplier with a sketch of the network along with the capacitor values. Inform the supplier of the load peak current. This will allow the supplier to design the capacitors taking into account the current ratios in the various PFN capacitors.

For very short-duration current pulses in the multikiloamp range, the PFN capacitor is preferably of the extended-foil type to eliminate internal arcing and tab interface power losses.[2] If the current rise time required in the load is less than $0.1\,\tau$, there is a way of achieving this, which was started years ago in gas discharge systems. Given an ideal switch in a four- to eight-section transmission line, the first four capacitances are selected in the ratio of 1:2:4:8 and with the same impedance per section, $Z_o = \sqrt{L_N C_N}$. This does not produce a trapezoidal shaped pulse, but it gives a very fast rise time pulse. In this configuration, water vapor lasers were driven with relatively steep rising current pulses. Because they were direct electron-pumped systems, this increased the electron temperature significantly and increased their laser efficiency. The rise time

PFN: Very fast risetimes and falltimes demand tapered PFNs. Low ripple demands many sections.

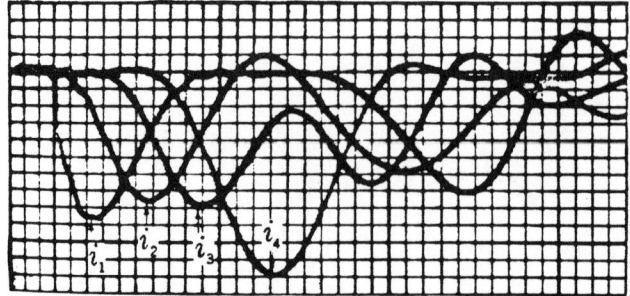

Condenser-discharge currents in the four-section type-E network

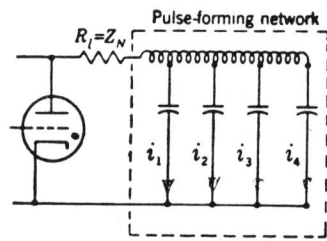

Fig. 5-3. Pulse-forming network characteristics (courtesy of G.N. Glasoe and J.V. Lebacqz, eds.: Pulse Generators, Dover Publications, Inc. NY, 1965).

is effectively the discharge time for the very first stage provided the PFN-load interface inductance L is small enough so that L/R_L is less than the first PFN section discharge time. This is a useful technique for generating a steep wave-front pulse. Time-varying load impedances can be roughly matched for these lasers by setting a ratio for the impedances of Z_o to 2:4:8 for the first three sections, creating what is called a tapered transmission line. This can cause reflection problems, but it is a rather unique and efficient way of driving such loads as hydrogen fluoride (HF) lasers and CO_2 transversely KrF excimer systems, whose impedance decreases with time. This is a crude but economical approximation to the general time-varying PFN problem.

For very low ripple on the top of the pulse, there must be many sections in the PFN. A 20-section PFN can be built to have a peak-to-peak ripple of one percent, which appears to be a practical limit. It is very difficult to achieve less ripple than this.

5.1.3 Parasitic Capacitance Effects

A major problem in all these systems can be parasitic capacitance and its effects on rise time (Fig. 5-4). In most cases, two of these parasitics predominate: there is one across the switch, and there is another across the load. Both are difficult to avoid. The one across the switch is discharged at an extremely high frequency when the switch closes. This could be used to advantage in spark gaps to shorten the resistive phase. The one across the load slows the rate of rise of voltage, creating problems in very fast circuits.

For most applications, stray capacitances in the power conditioning system fast discharge portion can be lumped into a "C_s'" at the output of the PFN.

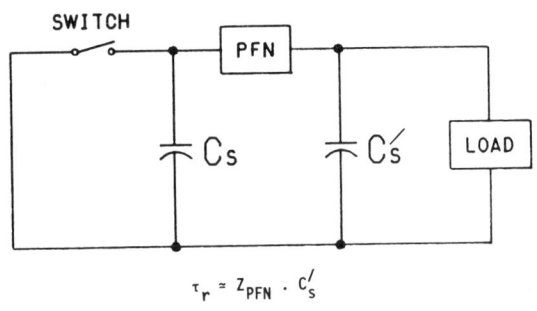

Fig. 5-4. *Parasitic capacitances and their effects on pulse rise time.*

5.2 THYRATRONS, IGNITRONS, THYRISTORS, AND THE LIKE

5.2.1 Switch Common Areas

This section also discusses thyratrons, ignitrons, thyristors (silicon-controlled rectifiers), or any bulk ionization device where everything is controlled by avalanche ionization and the appropriate analogue in the solid-state case. EG&G has been studying thyratrons for the Los Alamos National Laboratory AP Division to determine the limitations in their switching properties, the maximum speed at which they can be made to switch, and the physical model that correlates with the real world. It turns out that they can be made very fast switches indeed. Ristic shows that in spark gaps,[3] for the case of nitrogen, the gap voltage decreases with time and Ristic found a t^{-3} dependence. There is a great deal of discussion about this time dependence; referring to Chapter 6 on spark gaps for additional information.[4]

5.2.2 Thyratron Switch Model

For thyratrons, a trigger pulse is applied, and there is then an avalanche growth of electron density forming a glow discharge that may tend towards an abnormal glow for high-peak currents. All switches of this class work this way. At $t = 0$, $-V_o$ is added in series with the switch so that no current flows. As time passes, the voltage source decreases in amplitude, and current flows through the system. The current flow is limited in its rise time by the lumped inductance L, which couples the switch and the load. The voltage drop decreases exponentially to the constant switch drop level for the duration of the pulse. The best values experimentally achieved for τ_F are 1 nanosecond for thyratrons at kiloamp peak current. That is a factor of 10 better than the initial value some 20 years ago.[1]

For thyratrons, the model that gives very good results in comparison to the data, with the discharge times of 20 to 100 nanoseconds, replaces the thyratron with a voltage source (Fig. 5-5). When the switch is closed, it has a series inductance (from the

142 Discharge Circuits and Loads

Most current model:

Let V(t) be the voltage across the switch.

Then:

$$V(t) = V_0(+e^{-t/\tau}) + \text{switch drop of} \cong 150 \text{ volts}$$

$3\tau \cong \tau_F$ = Resistive phase falltime (i.e., 100 to 0% voltage falltime)

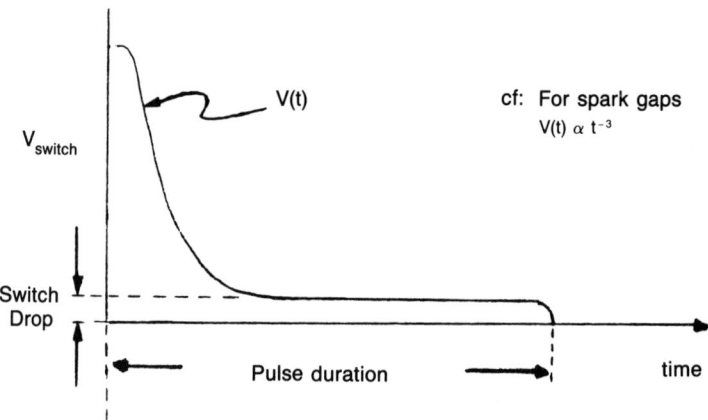

cf: For spark gaps
$V(t) \propto t^{-3}$

Note: That for most switches the drop is nearly independent of current.

Sarjeant and Nunnally

Fig. 5-5. Thyratron switch model.

thyratron and its connections), a voltage source in series with it, nd the switch drop (which is an invariant 100 to 200 V for almost all gas switches). The τ that gives the best agreement with the data is, if τ_F is the total fall time from top to bottom, $3\tau = \tau_F$. That gives agreement, within 5 to 10 percent, with the data for all types of tubes. The examples shown using this model predict somewhat slower rates of rise than are observed with present thyratron switch tubes.[5]

5.3 PULSE TRANSFORMERS

5.3.1 Pulse Transformer Models

Pulse transformers (Fig. 5-6), referring everything to the secondary, have a magnetizing inductance L_E', a shunt capacitance C_c, primary to ground converted to secondary, the secondary-to-ground capacitance C_D, the leakage inductance L_L, the core loss R_E', the load resistance R_L, and the voltage, NV_o. The switch, actually on the primary, is closed with everything referred to the secondary. The diagram then shows an LRC circuit with the rise-time $L_L - C_D$ limited.

High-pulse fidelity pulse-transformer design is still a compromise of efficiency against pulse fidelity (Fig. 5-7). Flatness is predominantly controlled by three factors: resonances in the transformer structure, flux saturation level and the time taken to reach saturation (which is material dependent), and the amount of energy that flows into the magnetizing inductance. For very long pulses, it is the magnetizing inductance that shunts away the desired energy from the PFN. The cost of large, high-quality pulse transformers goes up very rapidly with pulse width. Unless absolutely necessary, pulse transformers are an expensive route for pulses longer than 20 to 25 microseconds.

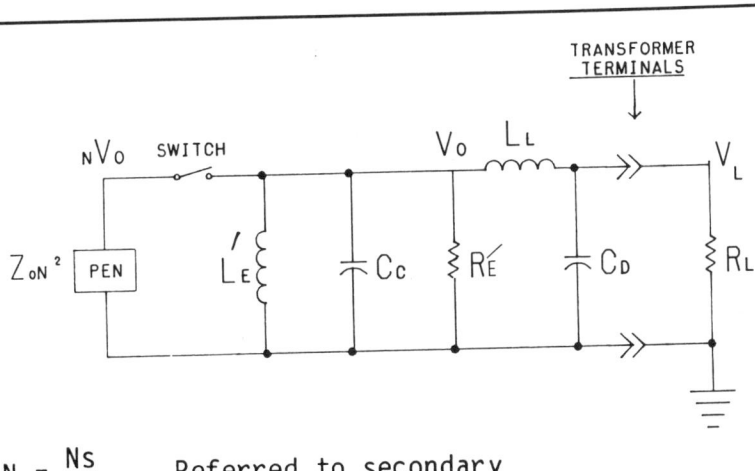

$N = \dfrac{N_s}{N_p}$ Referred to secondary

L'_E: Magnetizing inductance $\simeq L_E N^2$

R'_E: Core loss $\simeq N^2 R_E$

L_L: Leakage inductance $= L_L^{sec} + N^2 L_L^{pri}$

C_D: Shunt capacitance-secondary to ground

C_C: Shunt capacitance-primary to ground $= C_p/N^2$

Fig. 5-6. Pulse transformer model.

144 Discharge Circuits and Loads

Main problem areas are:

1. $\tau_R \simeq \pi \sqrt{L_L C_D}$: Risetime limit

2. Flatness - determined primarily by resonances in $L_L C_D$, and flux "saturation" level and time, and energy in L_E'.

3. $\tau_F \simeq \pi \sqrt{(L_E' + L_L) C_D}$: Falltime limit - especially if R_L is biased diode load.

   $\simeq \pi \sqrt{L_E' C_D}$

   since $L_E' \gg L_L$ in general

4. Reversal depends upon matching and Q of circuit.

Fig. 5-7. Pulse transformer limitations.

5.3.2 Pulse Transformer Limitations

Given a load like a magnetron or a direct discharge pumped gas laser, as the current and the voltage start to decrease, the sustaining voltage is passed, the current suddenly ceases to flow in the load, except for the current flow in the recharging inductance or resistance across the load. The fall time limit is then determined by the total loop inductance, which for pulse transformer drive can easily be 400 to 500 nH, and the PFN impedance, resulting in fairly long fall times. When a laser switches off, the energy stored in this shunt inductance and in the transformer leakage inductance discharges in a resonant fashion into the inductance/capacitance loop (Fig. 5-7). Another problem is that because this is a resonant discharge, there can be a voltage reversal at the load point. The amount can be determined through a nonlinear analysis for high-power pulse transformers, but it is difficult to do. The amount of reversal is not constant as the load impedance changes. The load and the pulse transformer are intimately interconnected.[1]

For very fast pulses, coaxial-cable transformers can be used to provide voltage gains of two to four times, but they are always rather high impedance. Cable transformers experimentally give much lower multiplications than the theoretical formula predicts. For two or three times, they are an economical way of stepping up voltages. All comments made for the same transformer using ferrite cores to increase permeability also apply to coaxial-cable transformers.

For overall pulser efficiencies, a rough number is about 60 to 80 percent, which has not changed significantly since the 1940s. A repetition-rate system, even at 1 to 2 kHz, has a given input power of about 70 percent of a given output power for the system. This is due to heat lost in the diodes and in the charging unit, and to switch losses. This applies to ignitrons, thyratrons, vacuum-arc devices, and some thyristor (silicon-controlled rectifier) modulators. Spark gaps in repetitive circuits can have quite large losses, and overall system efficiencies can fall well below 60 percent at kilohertz repetition rates.[4]

5.4 SWITCH RECOVERY AND RESISTIVE EFFECTS

When a load turns off, the current decays towards zero. The resonance effect in all the inductors with the stray capacitances and the recharge circuit tries to put charge back on the PFN. The recovery time for the switch device, whatever it is, is over the time so marked in Fig. 5-8. The area of the first negative pulse represents a positive ion energy deposited inside the thyratron gas switch tube (gas cleanup), which can, in effect, increase the heat load on the tube and severely limit the lifetime.[1] In large, high repetition rate systems, where the switch tube is being pushed to the limit of its performance, an enormous amount of positive ion heating of the anode of the tube can occur, easily equaling the anode heating of the tube during the switch turn-on time. A similar behavior can apply to spark gaps. Energy is deposited in several areas inside the gap until it fully recovers. Just how important that is depends on the individual circuit.

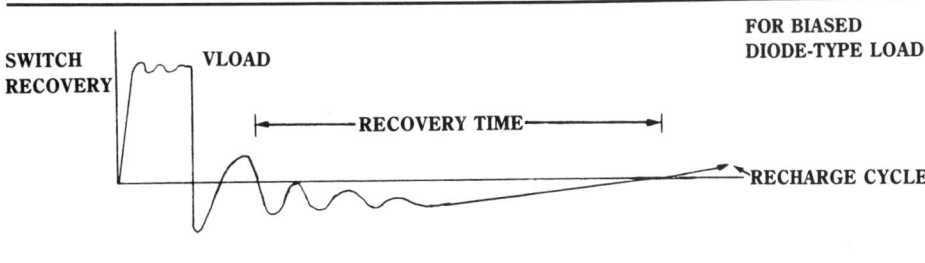

Fig. 5-8. Switch recovery and resistive effects upon circuit performance.

5.5 SWITCH RISE-TIME EFFECTS IN ULTRAFAST HIGH-POWER ELECTRONICS

The switch can degrade the rise time. This is illustrated by considering a circuit in Fig. 5-9 proposed for a midplane spark-gap trigger.[5] It is necessary to put a negative bias on the midplane through a coaxial cable. The inductor to the bias line decouples the midplane during the trigger-pulse duration. The line of length W_1 is resonantly charged to voltage V_o. The switch is closed and allows this line to discharge into the second one of length W_2. There is a negative-going pulse on the second line, and the peak value equals the bias voltage plus the absolute value of V_o. The particular circuit under consideration has a 23Ω line, 50 nanoseconds long, charging another 23Ω line. The second line, being a long cable, eliminates reflections during the spark gap turn-on phase. The ground current through the coaxial outer conductor is decoupled with ferrite toroids

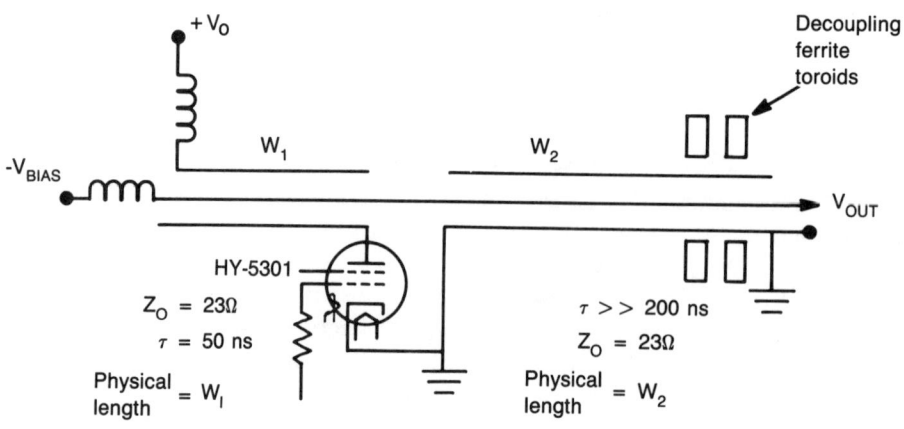

V_{out} as a function of time, using NET 2 program and thyratron model as shown in Fig. 5-5:

For thyratron:
    Tube drop ≅ 150 V
    τ_F = 10 ns ∴ τ = 3.3 ns = ⅓ · τ_F
    and L = 20 nH = tube plus mounting inductance

Sarjeant and Nunnally

Fig. 5-9. *Ultrahigh-speed spark gap trigger generator.*

that have been shown to work very well. There are several ferrite toroids that work particularly well for fast-pulse circuitry (for example, Ferroxcube Type 144T500). About 20 to 30, spaced along the line, will provide a great deal of increased shield inductance. They increase the effective inductance between the grounds shown to minimize current flow through the cable during discharge.

5.5.1 Switch Modeling

Bill Nunnally (University of Texas at Arlington) has modeled this with a switch-resistive phase time, τ_F, of 10 nanoseconds, a τ of 3.3 nanoseconds, and a lumped switch plus interconnection inductance of 20 nH.[5] This is comparable to state-of-the-art systems. (Such a thyratron switch tube is about 1 inch high, and with a current return shroud attached, a total inductance of 10 to 15 nH is measured.) For an ordinary thyratron of conventional design, which is much larger, the inductance can be 200 to 300 nH. In the thyratron trigger, for conventional tetrode tubes, τ_F is ≅ 40 nanoseconds, and τ_F is 80 nanoseconds for triodes. The tube model is illustrated in Fig. 5-10.

Initially, the thyratron is at +50 kV, and the bias on the spark gap at −50 kV. A 50 pF shunt load is assumed across the end of that line. The 23 and 50Ω cables were selected and were found to have little effect on rise time. The switch is a HY-5301, and

System Initial Conditions

V_{bias} = -50 kV
V_o = +50 kV
C_L = 1 pf and 50 pf: Load shunt capacitance

Use thyratron model: For HY-5301 tube

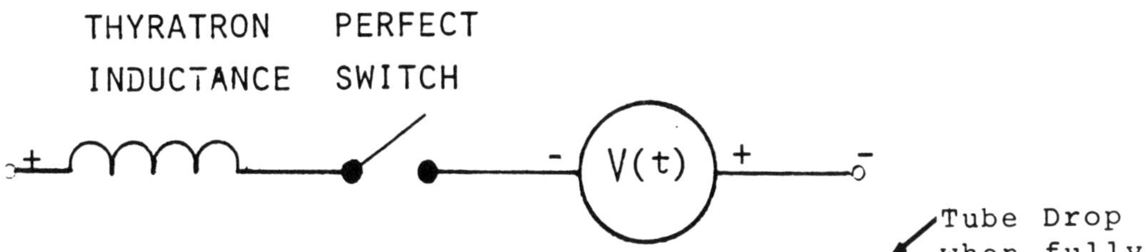

```
THYRATRON      PERFECT
INDUCTANCE     SWITCH
```

$$V(t) = -(50{,}000)\, e^{-(t-t_o)3/\tau_R} + 150 \quad \text{Tube Drop when fully conducting}$$

V_o = 50,000 = peak voltage across tube
20 nH = tube inductance
t_o = time at which perfect switch is instantly closed
τ_F = resistive phase time for thyratron
≈ 10 ns   Note: For conventional tubes
≈ 3τ τ_F ≈ 40 ns resulting in some risetime degradation time

where: τ is 1/e time

Sarjeant and Nunnally

Fig. 5-10. HY-5301 thyratron model and trigger-circuit initial conditions.

it gives an output pulse (Fig. 5-11) rate of rise of 7.5 kV/nanosecond, which is more than adequate for multichanneling most spark gaps that require 6 kV/nanosecond. An ordinary thyratron provides only 1.5 to 2 kV/nanosecond. This is a relatively compact switch device that can be used to provide fast trigger systems.

5.5.2 Trigger Generator for Multichannel Spark Gaps

Another tube examined was a developmental 100 kV device (HY-5323), with a 15-nanosecond resistive phase fall time and 30 nH total inductance. The advantage of

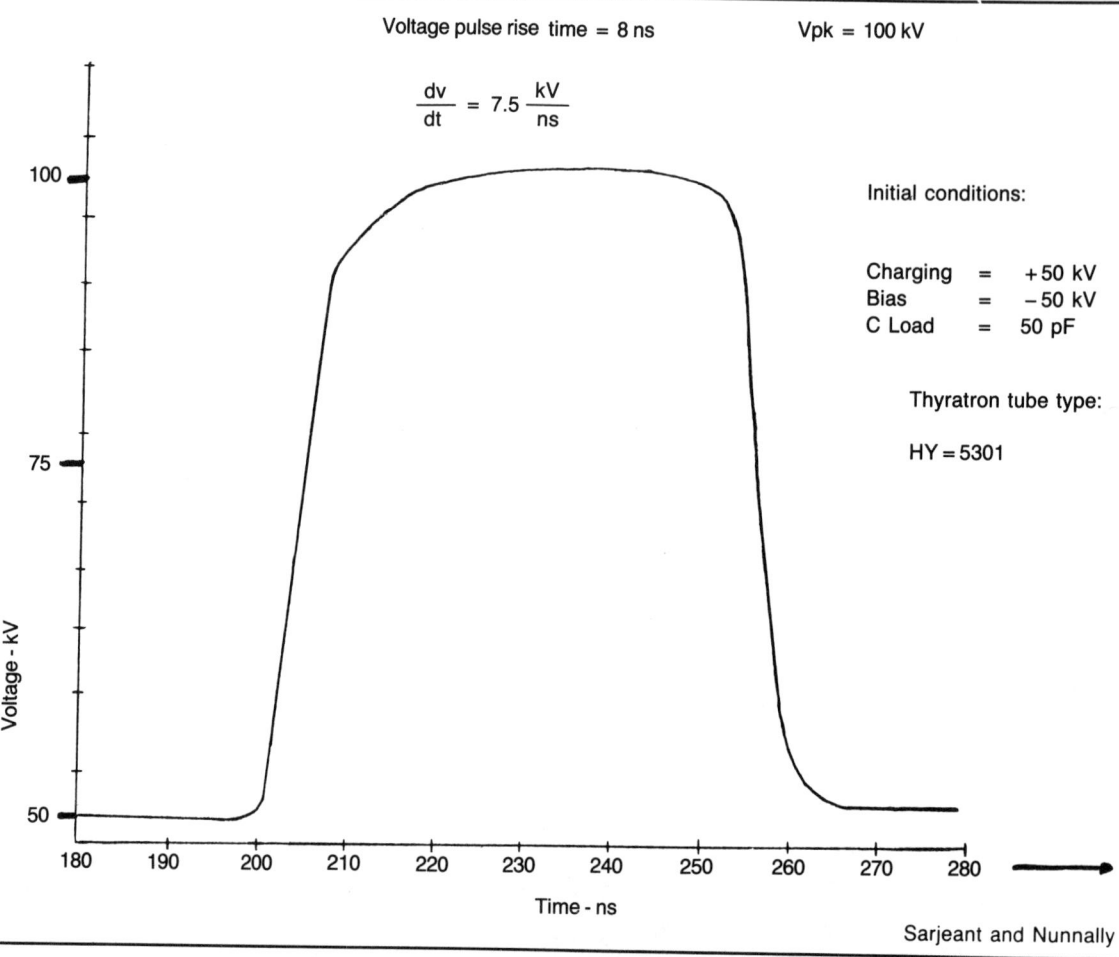

Fig. 5-11. Predicted trigger-pulse shape using an HY-5301 thyratron switch.

going to the higher voltage is the steeper slope (Fig. 5-12). For multichanneling, the calculated 11 kV/nanosecond is more than adequate. This pulse can be generated at repetition rates of several kilohertz with low delay times and extremely low jitter ($<<1RS$).

5.6 EFFECTS OF INTERCONNECTION PULSE CABLES TO LASER LOADS

There are four points to note on the effects of pulse cables connecting to laser loads (Fig. 5-13):

1. Given a piece of cable connecting the laser to the pulser, the laser impedance being usually time-varying, reflections can travel back and forth on the cable, causing oscillations on top of the voltage pulse.
2. Coaxial cables are not perfect cables; they have leakage inductances of about 5 percent and resistive losses. To derive the same peak power from the system, additional power must be put in the input. To minimize additional power,

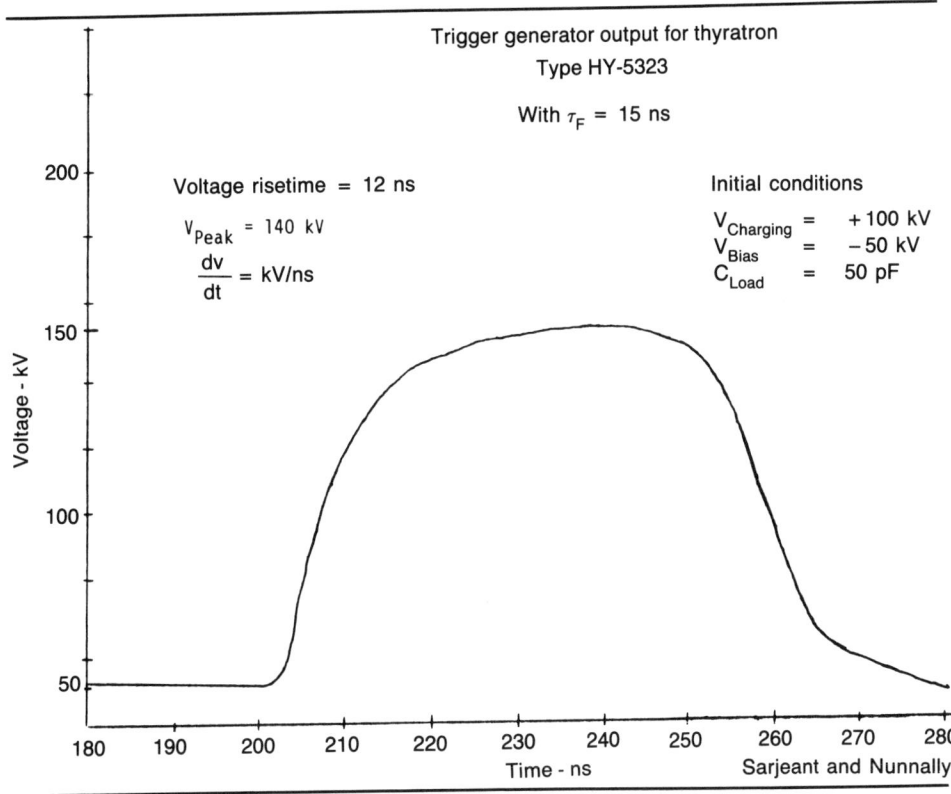

Fig. 5-12. Predicted trigger-pulse shape using an HY-5323 developmental thyratron switch.

special totally shielded cable can be purchased, or high-voltage coaxial cable for underground power distribution systems is readily available at less cost. This cable works well.

3. When the load faults, an opposite polarity voltage pulse travels back along the cable. Depending on the timing of the fault, enormous voltages can accumulate at the input of the cable. This problem is discussed in detail in Greenwood,[6] (Chapter 3). If the cable is designed to handle three times the pulse load voltage, faults will not destroy the system. Most faults will not exceed three times the load voltage.

4. For fast pulses of about 1 nanosecond rise time, there is very little current penetration into the conductors. The voltage pulse shape rapidly degrades substantially as it travels down the cable. The skin effect is well discussed in the book by Metzger and Vabre.[7] The problem was minimized in one case by keeping the cables shorter than 20 feet for 6 kV, 120 picosecond pulses. One-nanosecond pulses can be propagated for 100 feet over RG-17 and -19 uncompensated cables, if the cable is kept free of disturbances and physical damage. (The fork-lift degradation factor is 0.1 and 5 percent reflection added per runover.) Foam cables are quite good, but should be strung overhead to avoid physical damage. (RG-8 foam will operate to 6 kV pulse at low repetition rates.)

For laser loads, pulse cables create:

1. Reflection oscillations on pulse crest

2. Current increases because of cable resistive losses and radiation (shields not perfect).

3. For load faults, cable can propagate more than $2V_0$ reflections depending upon total circuit Q.

4. For fast pulses ($\simeq$ nanosecond risetime) skin effect degrades risetime over $\simeq$ 30 m for RG-17U.

Note that the cable looks like:

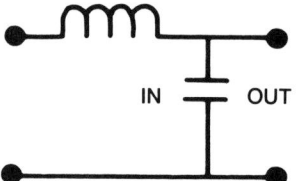

for pulse durations much longer than its two-way transit time. Then this cable acts as a resonant circuit in conjunction with PFN and/or transformer stray C and L. This can give rise to post-discharge oscillations that may be damaging to the switch or load. Damping shunt L-R networks in series with the coaxial cable center conductor at both ends of the cable generally control this.

Fig. 5-13. Effect of pulse cable between the fast-discharge circuit and a laser load.

In times much longer than the two-way transit time of the cable, a resonant circuit with a series inductance and a shunt capacitance develops. Pulse discharge oscillations can be dampened in many cases by adding LR networks in series with the center conductors at both ends of the cable; this circuit is discussed in detail in this book.

5.7 EFFECTS OF CHANGE IN THE LOAD IMPEDANCE ON CIRCUIT PERFORMANCE

Difficulties arise with a time-varying load or one that can have pulse-to-pulse fluctuations (Fig. 5-14). The TEA CO_2 and the HF lasers have built-in fluctuations

For rapid impedance changes during a pulse or pulse-to-pulse variation, usually caused by load faults, the designer must protect pulser against prolonged short or open circuits in the load loop.

FOR IDEAL PULSER DISCHARGING A PFN OF IMPEDANCE Z_N INTO RESISTIVE LOAD R_L:

$$I_L = \frac{V_N}{R_L + Z_N} \qquad V_N = \text{PFN charge voltage}$$

$$V_L = \frac{V_N}{R_L + Z_N} \times R_L$$

Voltage left on the network at the end of the pulse is[1]

$$V_{N-1} = \frac{R_L - Z_N}{R_L + Z_N} \times V_N$$

∴ % Voltage Reversal on PFN For BIDIRECTIONAL SWITCH:

$$100 \times \frac{V_N - V_{N-1}}{V_N} = \frac{2Z_N}{R_L + Z_N}$$

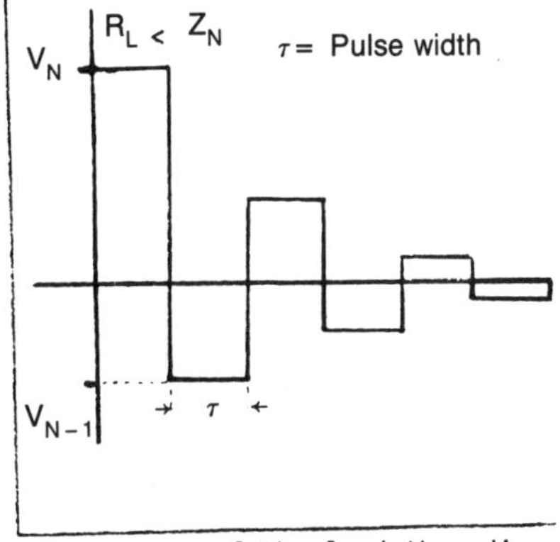

$R_L < Z_N$ $\tau =$ Pulse width

NOTE: For biased diode load, if $V_{N-1} < V_{threshold}$ of the load then discharge is through recharge elements in parallel with the load.

Glasoe and Lebacqz

Fig. 5-14. *Effects of change in load impedance on the performance of line-type power conditioning systems.*

that are not necessarily repeatable or constant, while vacuum devices like magnetrons and klystrons have almost perfect reproduceability. Many of the fluctuations are caused when the loads arc, posing the problem of protecting the HPE from serious damage.

5.7.1 Voltage Reversal After PFN Discharge

An ideal switch discharging a PFN has a charge voltage, V_N, a current through the load, a voltage across the load, an impedance, Z_N (of the PFN), and a load resistance, R_L, on the output. At the end of the discharge time, a voltage can be left on the line. The amount of reversal is:[1]

$$\frac{V_{N-1}}{V_N} = \frac{R_L - Z_N}{R_L + Z_N} \tag{5-2}$$

where V_{N-1} is the peak voltage of the first reversal.

When R_L is smaller than Z_N, which is unfortunately the case for most lasers, there can be a significant degree of voltage reversal. With a unidirectional switch, the charging circuit gradually puts current into the PFN and starts recharging. This negative voltage must be kept within the inverse ratings of the switch device. A typical number for ignitrons is 15 kV, for newer thyratrons 15 to 20 kV, and for older thyratrons 5 kV. If this number is exceeded, the device arcs. Arc damage depends solely upon the properties of the switch tube. Ignitrons, even at high repetition rates, can handle some intermittent arcing. Thyratrons usually include protective snubber devices to keep the voltage under control. If the load does arc and it is a tetrode thyratron tube, nothing happens to it because there is the forward bias plasma in the cathode region. When the switch breaks down from excessive inverse voltage, it tends to do so in a bulk ionization mode. Spark gaps of course are fully bidirectional and thus can handle inverse currents with no problem.

If the inverse voltage is less than whatever is needed to turn the load back on in the reverse bias direction, the PFN discharges through the shunt element that is in parallel with the load, which is either the inductor or resistor used to allow the PFN to recharge. The negative voltage just discharges through those elements.

When the load resistance is greater than Z_N, a step results (Fig. 5-15). When lasers or magnetrons turn off, the impedance becomes very large and a long time to zero can follow. Spark gaps and ignitrons have no problem handling this, but thyratrons do. Older modulators were designed using this positive mismatch to avoid inverse voltage on tubes. It was felt that there must always be a time at which the voltage and current must reach zero to let the tube recover. Today, the best way to achieve this goal is to pulse charge the PFN. Instead of only a resonant or resistive charging unit, a switch is put in series with them and turned on to recharge the PFN well after the output switch has recovered.

5.7.2 Voltage Reversal During Load Faults

If there is a short-circuit fault in the load during the discharge, it almost instantaneously puts $-V_N$ across the PFN. If the switch recovers, it starts charging from $-V_N$ and resonantly charges up to four times the peak dc (Fig. 5-15). A general property of resonant circuits[1] is that they will charge up to twice the sum of the absolute value of the voltage below the zero line plus the dc voltage. At high voltages, the switch tube faults as it exceeds its forward hold-off voltage, and the protection circuits in the system must come into play.

5.8 EFFECT OF PARASITICS ON SYSTEM PERFORMANCE

A sketch of a typical line-type HPE is given in Fig. 5-16, with several stray capacitances shown. The inverse diode D_2 is used to clamp any negative voltage left on the PFN at the end of the discharge cycle or during load faults. The charging diode is D_1. The World War II modulator designs used the vacuum diodes shown, which could be replaced with solid-state diodes. There are still, however, stray capacitances. When the switch tube is turned on, C_s and C_s' contain stored energy, and being small, force the tube to shunt this energy to ground at a very high rate of rise of current; oscillations occur. In system design, a current-viewing resistor (for example, T & M, Inc., Type

5.8 Effect of Parasitics on System Performance

NOTE: $R_L < Z_N$ case with unidirectional switch: switch opens when voltage across it passes through zero. Then PFN has V_{N-1} on it at beginning of next charge cycle. For example, in resonant charging:

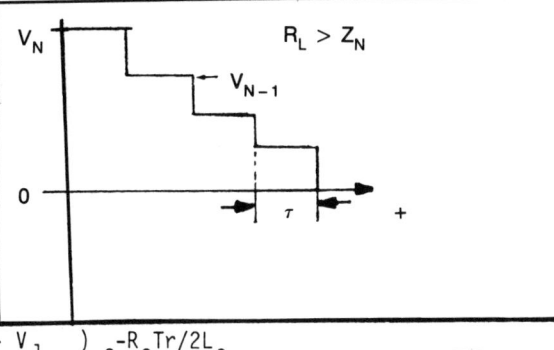

V_{Nth} charging period $= V_{DC} + (V_{DC} - V_{J_{N-1}}) e^{-R_c T_r / 2 L_c}$

$V_{J_{N-1}}$ = voltage left on PFN after N-1 discharge

$T_r = \pi \sqrt{L_L C_{PFN}}$
L_C = Charge L
R_C = Loss resistance in L

For example, let the load fault with unidirectional switch leaving $-V_N$ upon PFN

$\therefore V_1 = V_{DC} + (V_{DC} + V_N) e^{-\pi/2Q}$ and $V_N \approx 2V_{DC}$

If $Q > 10 \therefore e^{-\pi/2Q} \approx 1$

$\therefore V_1 = V_{DC} + (V_{DC} + 2V_{DC}) = 4V_{DC} \gg V_N$

GENERALLY SWITCH FAULTS AND PULSER OVERVOLTAGE/CURRENT SENSORS TURN HV OFF

Cases of interest for PFN charging voltages:

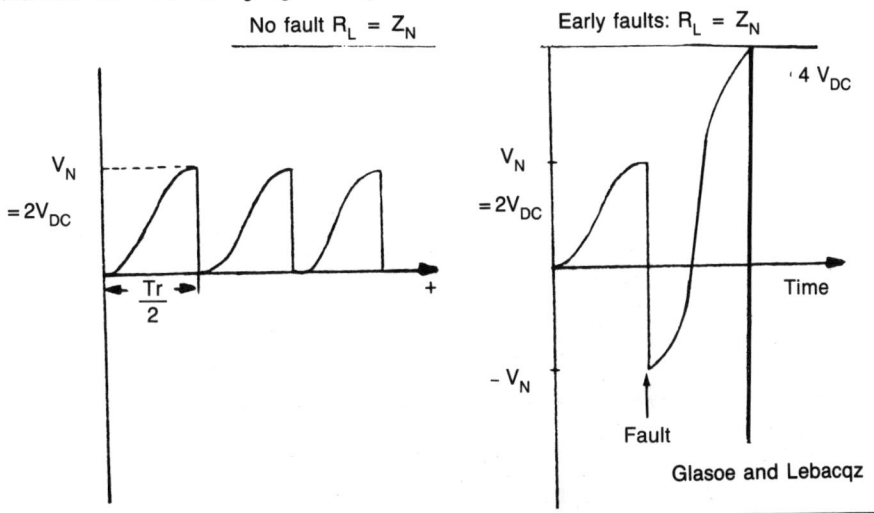

Glasoe and Lebacqz

Fig. 5-15. Further effects in change in load impedance on the performance of line-type power conditioning systems.

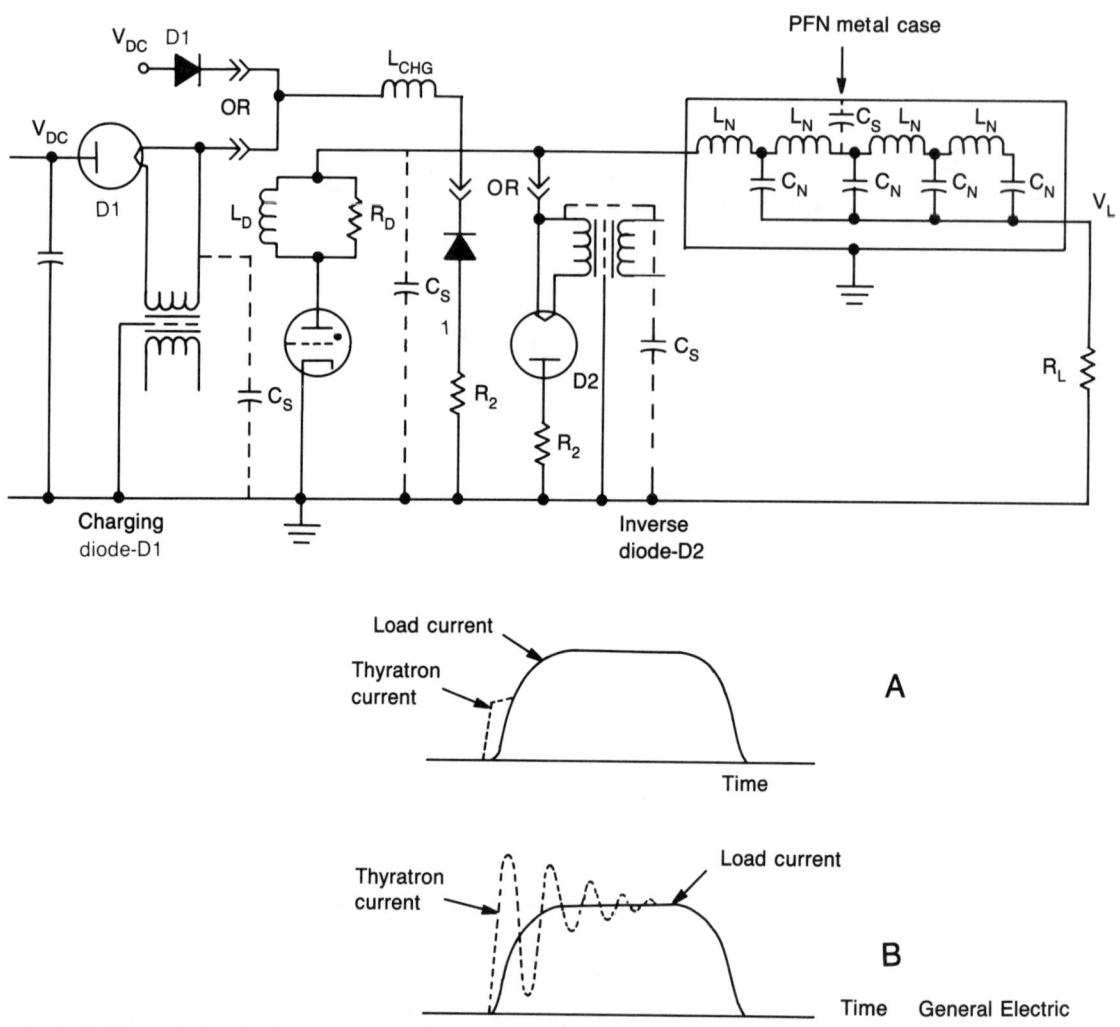

Fig. 5-16. (A) Typical line-type HPE and (B) Waveforms showing load and thyratron current in a line-type power conditioning system with distributed capacitances.

W) should always be inserted in the circuit in the cathode of the thyratron, ignitron, krytron, etc. (Current transformers are difficult to use in very fast circuits: they have a non-Gaussian response and give rise to shock-excited oscillations with a driving current waveform that has a front rise time faster than their effective rise time.) If oscillations are apparent, the lifetime of the device in the HPE usually will be shortened. During the rapid time-varying resistive turn-on phase, the instantaneous power dissipated in the device can be excessive.

Oscillations can be dampened with a shunt LR network in the thyratron anode lead. Choose a Q less than 2 if possible. Ohmite makes a parasitic suppressor used for suppressor grid and neutralization oscillation suppression. It is a 50Ω resistor in parallel with a silver-plated inductor of 0.3 μH inductance and is inexpensive. They can be used in series with the anode of the tube and are effective.

Spark gaps also dissipate power during the turn-on time. That is a waste of the capabilities of the device, and the losses should be minimized by reducing the turn-on time.

5.9 PROTECTING THYRATRONS FROM EXCESSIVE VOLTAGES

If there is a short on the load (Figs. 5-17 and 5-18) and if a resistor equal to the impedance of the load could be instantaneously put in parallel with the switch tube, the maximum apparent voltage would be $-V_{dc}$. With a noninductive resistor inserted in the circuit during a short, a tube charged to $+50$ kV will see -25 kV on the anode, and the tube could arc back. With a tetrode tube, the arc damage from occasional faults is not significant. To really protect the tube, a shunt network could be added from anode to ground. Adding a diode-resistor end-of-line clipper across the PFN ($R'_D = Z_o$). Then 12.5 kV will result, and the tube will be safe.[1] The biggest problem is the turn-on time of the diode. Many diodes have fast turn-on times, but are capacitively graded and have very little protection from stray electromagnetic fields that couple to them and destroy the junctions. For research applications, one means of obtaining fast recovery is to make a coaxial array of very fast recovery diodes and to put them all around the thyratron.

Even is everything is matched, a shunt inductor, L_S, not equal to zero is included in the system. When the load is a laser and it turns off with some voltage across it, some current has been flowing through L_S at that point. A voltage reversal may result, depending upon the ratio of the stored energies in the PFN and the shunt inductance (Fig. 5-18). Unfortunately, L_S, the inductance across the load, is usually fairly large to keep the efficiency high; thus, even though the system has perfect matching there might be some need for inverse diodes to protect the tube from load faults.

The resistor R_D is normally quite a large value (several hundreds of ohms), and it is chosen to be large enough to keep the voltage across the switch tube negative until recovery is assured. For small tubes that can be 2 microseconds, and for larger tubes it can be 50 to 250 microseconds.

5.10 EFFECTS OF LOAD-SHORT CIRCUITS

Faulting in the load during discharge (Fig. 5-19) can dramatically increase the voltage to which the PFN is charged in the following recharge. Only by providing an impedance-matched, inverse-diode network can this be avoided under all conditions. With present solid-state diodes and hydrogen diodes, such a low-impedance inverse network is almost always possible. These comments pertain specifically to very long-lifetime systems (5 to 10 years) at kilohertz repetition rates with no component failure desired.

The network, consisting of a diode and a resistor, also can be connected across the PFN. When it is, it is called an end-of-line clipper; it serves primarily to clamp the PFN voltage reversal to a low value and protect the PFN capacitors. When the load shorts, there is a negative-going pulse, and the energy represented by the hatched area under the curve is deposited in $R_D = Z_N$, and recharge then occurs. R_D should be slightly smaller than Z_N so that sufficient negative voltage recovery time is available.

Note that the switch still must hold off $V_N/2$ during the remainder of the discharge time (Fig. 5-20). Normally, restrikes of the switch will occur infrequently during these

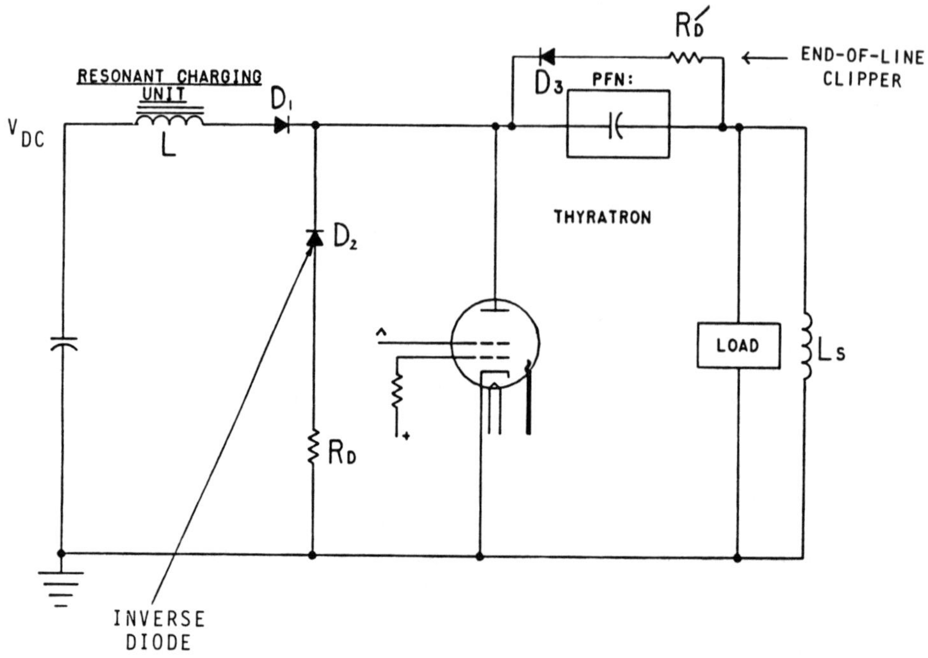

156 Discharge Circuits and Loads

[Circuit diagram showing Resonant Charging Unit with V_{DC}, inductor L, diode D_1, inverse diode D_2, resistor R_D, thyratron, PFN with diode D_3 and resistor R_D' labeled as End-of-Line Clipper, Load, and L_S.]

Now the "inverse diode" conducts during fault-induced voltage reversal on PFN, removing PFN voltage to zero before resonant recharging begins. If $R_D = Z_N = R_D'$, all voltage, V_{N-1}, is removed in the first reversal pulse.

NOTE: Even for $Z_L = Z_N$, for $L_S \neq 0$ there is some current flow in L_S, thus at end of pulse:

$$i_{L_S} = \frac{V_L}{L_S} \qquad \int_0^\tau V_L = -\int_0^\tau L_S \frac{di}{dt}$$

$$V_L \Delta T \simeq L_S \Delta I$$
$$= L_S i_{L_S}$$

Can show: $\dfrac{V_{N-1}}{V_N} \simeq -\sqrt{\dfrac{L_{PFN}}{L_S}}$ and $\Delta T = \tau$

∴ may need D_2 even if $Z_L = Z_N$

Fig. 5-17. Inverse voltage removal during load faults.

conditions. Adding an end-of-line clipper reduces the voltage by another factor of 2, clamping the PFN reversal to less than $V_{dc}/2$. With fast circuitry, they fail rather frequently and an inverse diode network only is used.

What is normally meant by tetrode thyratrons is an anode, a control grid, a cathode structure, and an envelope with some hydrogen inside. Normally there is a tiny pin stuck

At end of pulse of length τ, for laser or magnetron loads, the load open circuits when the voltage is below the holding voltage (Sustaining Voltage) V_S. Energy stored in L_S charges C_N capacitance in PFN:

$$\therefore \tfrac{1}{2} L_S (I_{L_S})^2 = \frac{V_L^2 \tau^2}{2 L_S} \qquad I_{L_S} = \frac{V_L \tau}{L_S}$$

And
$$= \tfrac{1}{2} C_N V_{N-1}^2$$

But Since $Z_N \simeq Z_L$ $\qquad V_L = \dfrac{V_N}{2}$

$$\therefore \tfrac{1}{2} C_N V_{N-1}^2 = \tfrac{1}{8} C_N V_N^2 = V_N^2 = \frac{V_N^2 \tau^2}{8 L_S}$$

$$\left(\frac{V_{N-1}}{V_N}\right)^2 = \frac{\tau^2}{4 L_S C_N}$$

But
$$\tfrac{1}{2} C_N V_N^2 = \frac{V_L^2}{Z_L} \tau \qquad V_L = \frac{V_N}{2}\ \&\ Z_L = Z_N$$

$$\therefore \tfrac{1}{2} C_N V_N = \frac{V_N^2 \tau}{4 Z_L} \qquad Z_L = \sqrt{L_N/C_N}$$

$$\tau = 2 C_N Z_L = 2 C_N \sqrt{L_N/C_N} = 2 \sqrt{L_N C_N}$$

$$\therefore \left(\frac{V_{N-1}}{V_N}\right)^2 \simeq \tfrac{1}{4} \times \frac{4 L_N C_N}{L_S C_N}$$

Or
$$\boxed{\left(\frac{V_{N-1}}{V_N}\right) = \pm \sqrt{\frac{L_N C_N}{L_S C_N}} = \pm \sqrt{\frac{L_N}{L_S}}}$$

WHERE THE SIGN DEPENDS UPON CURRENT FLOW DIRECTION: e.g., If I_{L_S} is the same sign I_L was, then negative sign is used since $V_{N-1} \therefore \dfrac{dI_L}{dt} < 0$.

Fig. 5-18. Inverse voltage calculation and removal during load faults.

in the side of the grid structure; when forward biased, it will reduce the jitter. That is not a tetrode tube; that is a tube that has a pre-ionizer electrode in it. This tube otherwise behaves like an ordinary triode thyratron. A true tetrode thyratron actually contains another grid. An electron cloud is generated through forward biasing this first grid, and the potential well from the reverse-biased upper grid prevents these electrons from seeing the accelerating anode potential. Note that true tetrode (with two grids) thyratrons are highly damage resistant to inverse internal arcing. They should be used whenever possible.

158 Discharge Circuits and Loads

As shown in Fig. 5-15, faulting in the load during discharge can dramatically increase the voltage the PFN is charged to in the following recharge. Only by providing an impedance-matched inverse diode network can this be avoided under all conditions. With present solid-state diodes and hydrogen diodes such a low impedance inverse network is almost always possible.

This network can also be connected across the PFN called an "end-of-line clipper" and serves primarily to clamp the PFN voltage reversal to low value and protect PFN capacitors.

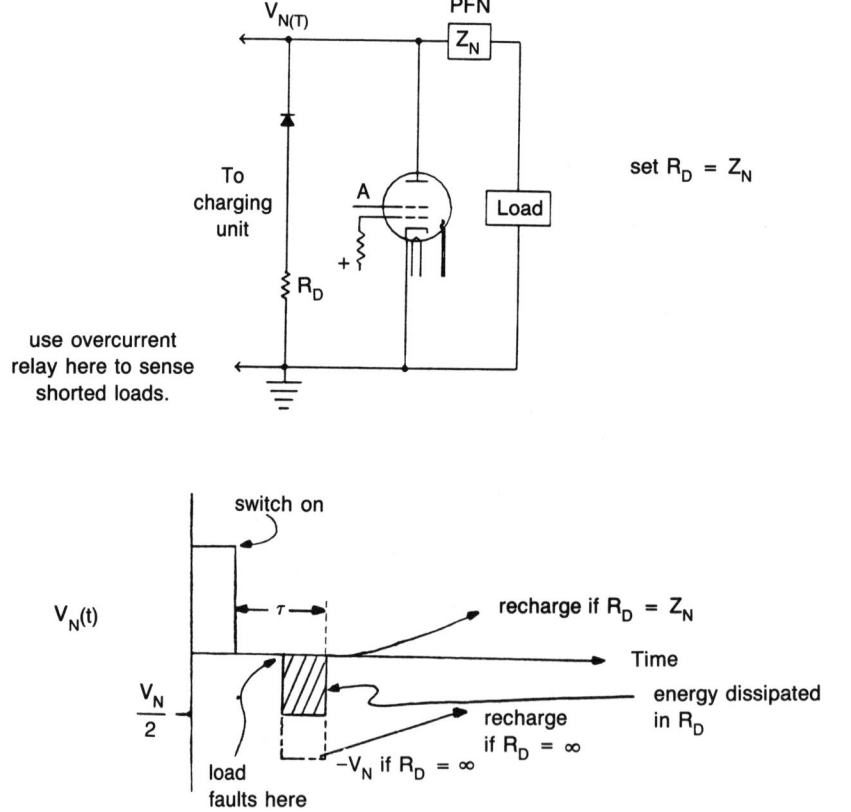

Fig. 5-19. Load short-circuit faults.

In a large reliable system, fault current should be controlled with an inverse network. Current could be put through a time-varying resistor to keep the inverse voltage initially low and gradually discharge the PFN, reducing premature failure. This can be done with thyrite varistors. They are highly capacitive and difficult to use in kilohertz repetition rate circuits.

NOTE: The switch still must hold off $\approx V_N/2$ during the remainder of the time τ. Normally restrikes of the switch occur infrequently during these conditions. Adding an "end-of-line clipper" reduces the voltage by another factor of 2, clamping the PFN reversal to $\approx V_{dc}/2$. With fast circuitry they are "often replaced" and an inverse diode network only is used.

NOTE: True tetrode thyratrons are highly damage resistant to inverse internal arcing. They should be used whenever possible.

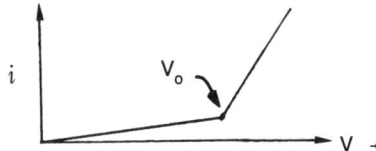

NONLINEAR CIRCUITS: The "mov" and "thyrite" materials have $I \propto (V - V_o)^6$ so that they can replace all or a portion of R_D to reduce diode $\frac{di}{dt}$ and allow larger peak currents:

$$\frac{di}{dt} \leq 3 \, \frac{KA}{\mu s} \quad \text{(Westinghouse 20 A stacks)}$$

$$\leq 1 \, \frac{KA}{\mu s} \quad \text{(Westinghouse 5 A stacks)}$$

→ CONSULT MFR FOR PEAK CURRENT LIMITATIONS: Generally $I^2_{pk} t$ energy limit

Fig. 5-20. Inverse voltage appearing during load faults and nonlinear circuits to control this voltage.

The metal-oxide varistor and thyrite materials have $I \propto (V - V_o)^6$, (V_o is the turn-on voltage and V the applied voltage) so that they can replace all or a portion of R_D to reduce di/dt and still allow larger peak currents. Keeping di/dt less than 3 kA/microseconds for 20 A stacks and about 1 kA/microsecond for 5 A diode stacks is recommended by Westinghouse.

5.11 OPEN-CIRCUIT PROTECTION WITH CABLE INTERCONNECTIONS

Referring to Fig. 5-21, if the load is shorted, a wave is sent back and is multiplied repeatedly until the cable breaks down. If L and R are correctly chosen, the voltage pulse can be reduced significantly. In large systems with very high voltages, there is no choice but to do this. In single-shot systems, a single point-plane spark gap connected to ground through a copper sulphate resistor is used to offer additional protection. If the laser faults, it sends back a $-V$, and the point-to-plane gap across the cable input breaks down controlling this voltage. Such spark gaps can be used for high repetition rate systems, but erosion and gas flow needs may pose problems. In another circuit, the switch can be self-triggered. If relatively long pulses are used, a triggered spark

160 Discharge Circuits and Loads

For cable connections to the load, use a shunt spark gap to ground or a shunt L/R network at each end of cable to protect from overvoltages.

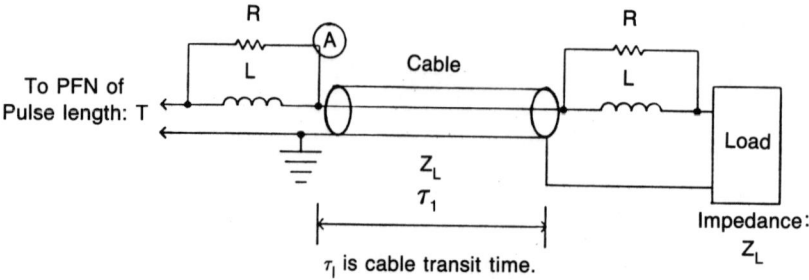

τ_1 is cable transit time.

L/R < τ_R AND L/R >> $2\tau_1$ for proper damping. If L/R = $2\tau_1$ then at time $t = 2\tau_1$ voltage at (A) is $\alpha(e^{-L/R \cdot t})V_N/2$
For load shorted: $V_{(A)} = V_N/2 \times |1/e^2|$ (i.e., 90% damping) pick $Q = \frac{\omega L}{R} = 0.5$ to quench oscillations; therefore voltage stress at (A) has been reduced from $\simeq 2 V_N$ to $.05 V_N$ and only small oscillations are allowed.

Fig. 5-21. *Open-circuit protection with cable interconnections.*

gap protector can be added. When the wave comes back, a capacitive divider reduces the voltage and (fed through a diode) triggers the spark gap.

For cable connections to the load, either a shunt spark gap to ground or a shunt LR network at each end of the cable is suggested to protect from overvoltages.

Figure 5-22 is a useful compilation of information on inductive loads prepared by Bill Nunnally.

5.12 LASER LOADS

5.12.1 Direct-Discharge Pumped Excimer Laser Loads

Rare-gas halogen lasers, of considerable interest as sources of intense ultraviolet energy, represent one of the most challenging time-varying loads to come into existence in the last decade (Fig. 5-23). One of the difficulties in these systems is their time-varying nature.

It is not unusual to see them go from many ohms to milliohms, making the maintenance of a constant V_L (which determines the excitation efficiency of the system) extremely difficult. Unlike the carbon dioxide laser, these discharge-pumped systems are quite unstable plasmas, so that the impedance of the krypton (or argon, xenon, etc.), helium, and fluorine mixtures decreases monotonically with time. These notes highlight the most extensive engineering study performed to date on the electrical characteristics of these lasers,[9] and space permits sketching only some of the interesting properties of these loads.

Figure 5-24 schematically illustrates the type of excimer laser of interest here. Energy is stored in the driver capacitor D and transferred into the cable capacitance C during

5.12 Laser Loads

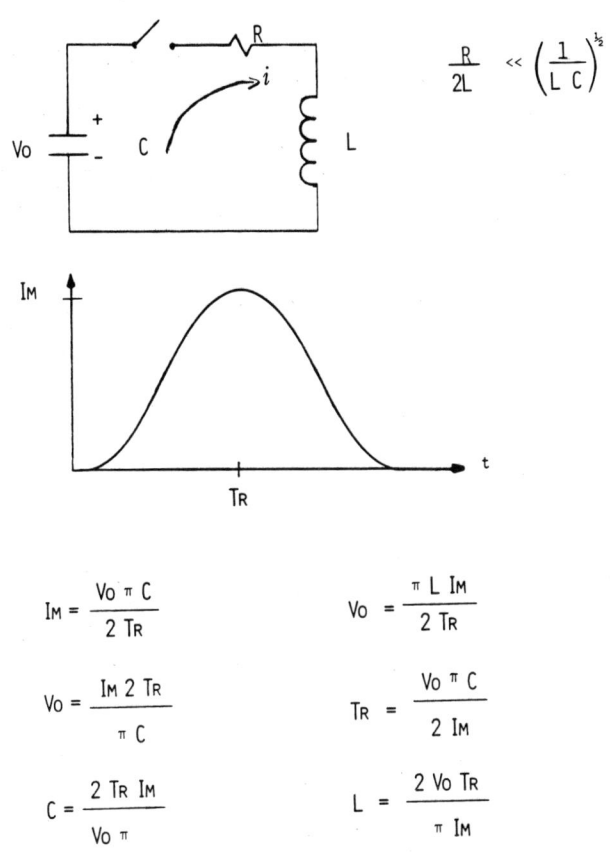

Fig. 5-22. Circuit information for inductive loads.

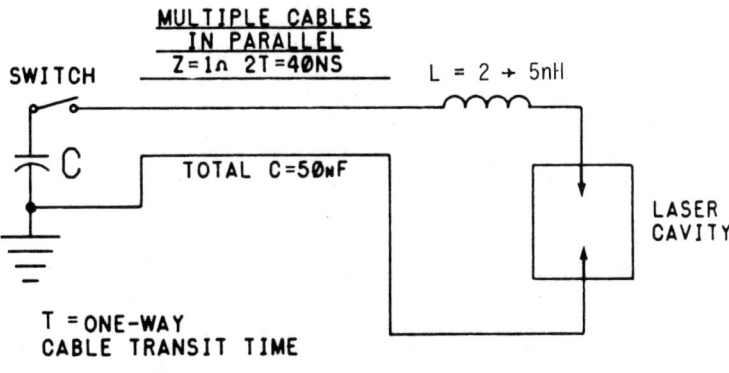

Fig. 5-23. Simplified excimer laser driver. [9]

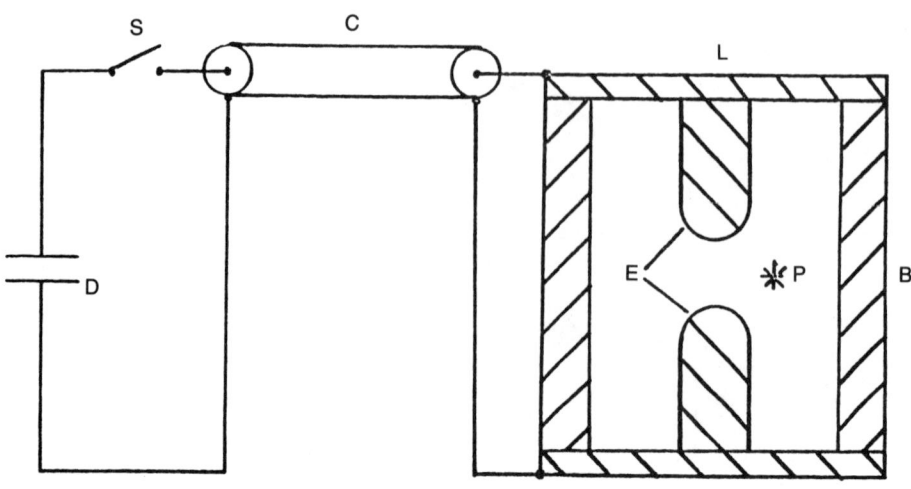

Typical excimer laser. D—driver capacitor; S—switch; C—cable PFN; L—aluminum plate; E—electrodes; P—pre-ionization source; B—rigid dielectric body

Butcher

Fig. 5-24. Typical excimer laser. [7]

and after the closure of the spark gap or thyratron switch S. The cable PFN generally has an impedance of 0.5 to 1Ω and a discharge time approximately equal to the sum of the cavity time to break down plus the time to termination of the laser pulse. (This is close to the time the current takes to decay through zero.) As the switch closes, there is a permissible build-up time. Initially, the gas mixture is pre-ionized by one of several means, here by an array of sparks P along the electrodes E. After a short delay, the switch S is closed, and the voltage on the electrodes builds up from zero to break down in about 80 nanoseconds. Then a large current flows through the low-inductance load (L is approximately 2 nH). The most significant problem in this system is the rapid decrease in the load impedance with time, clearly indicating that conventional PFN design techniques are inappropriate to achieve the theoretically high efficiencies.

Efficiencies in excess of 1 percent have been achieved at optral energies of $\cong 1$ J. The major difficulty with this system is that when the switch closes, it does so with a time constant, so the voltage on the electrodes has some build-up profile (either inductance limited or switch resistive phase limited). Another major difficulty here is the current unavailability of lumped elements with parasitic inductances sufficiently low to allow the synthesis of time-varying PFNs required in this application. Devising techniques to effect such networks or transmission lines, whatever may be the highest pumping efficiency drivers for these systems, remains one of the more difficult power conditioning engineering problems today.

In these systems, the closure times of switches currently available are around 20 nanoseconds, significantly affecting the build-up times on the cavity electrodes and

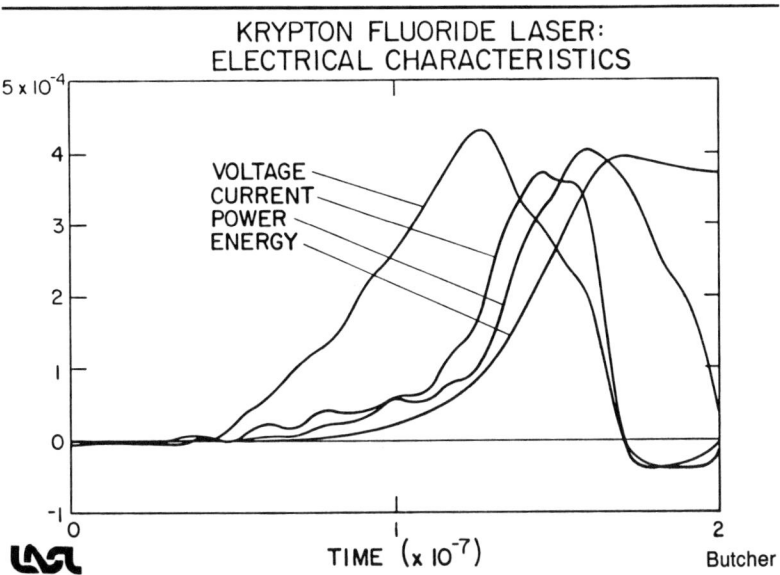

Fig. 5-25. Time history of circuit and system parameters in the discharge-pumped KrF laser.*

adding to the energy losses in the discharge loop. Advances in pulse-charged, very low inductance thyratrons and spark gap switches might well make long-life lasers of this class a reality. The remaining challenge will be to design and construct low-inductance, inverse-diode systems to protect the thyratron switches under conditions of laser-cavity arcing.

Figure 5-25 shows that for a KrF laser there is a voltage buildup until the discharge turns on. The current builds up to peak value while the voltage is decreasing. This is undersirable because the voltage must stay above a threshold level for optimal excitation kinetics. This threshold level is, unfortunately, rather high. Most of the energy deposited in the system is inefficient in pumping the laser. Ideally, a time-varying PFN is needed that has the reciprocal of the impedance of the discharge with time down to levels of $\cong 50$ mΩ. In contrast to this, typical extended-foil capacitors have an internal resistance of $\cong 25$ mΩ.

Figure 5-26 also shows the predicted power versus time and the predicted impedance verses time.[9] The agreement with the model is quite good.

5.12.2 Flashlamp Loads

A flashlamp is a glass tube with two electrodes, containing xenon, krypton, or some other gas mixture at rather low pressures (Fig. 5-27). Assume a length of 1 and a diameter of d. Normal flashlamps are driven by long (multimicrosecond) pulses; there is very little concern about switch losses, and they can even be driven with silicon-controlled rectifiers. Empirically, it has been shown that the lamp voltage is some constant times the square root of the current through it.[10]

Capacitor cost can be reduced by allowing a small 20 percent reversal and designing for maximum energy transfer in the first current half cycle. If the light output of the

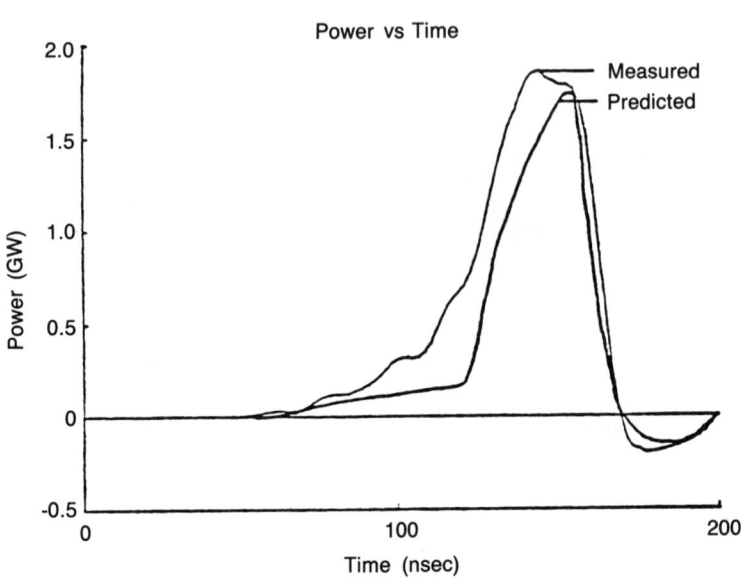

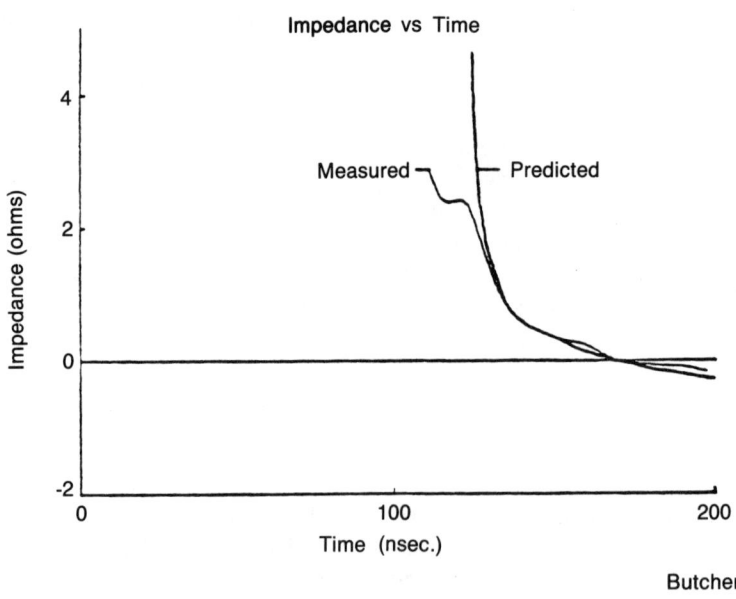

Fig. 5-26. Comparison of predicted and measured powers and impedances in the discharge-pumped KrF* laser.

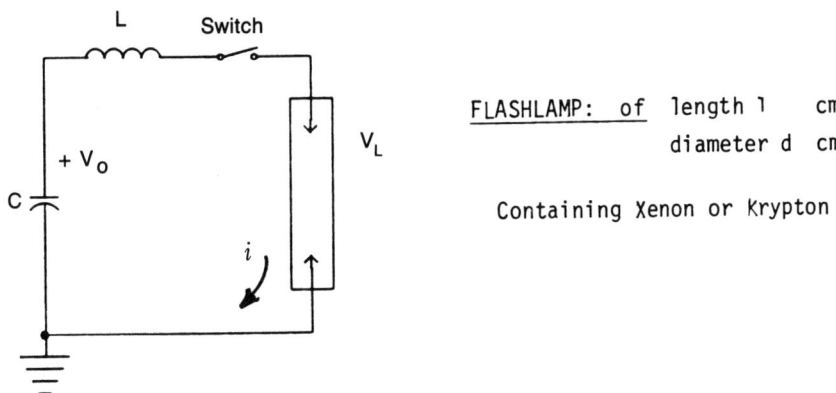

Pulse widths are typically 5 to 125 μs so that switch losses can be neglected.

$$V_L = \pm k |i|^{1/2}: \text{ Empirically determined}$$

For small voltage reversals on C and maximum energy transfer into the lamp in the first current ½ cycle: (say ≈ 20% reversal as determined by the cost for C for a given lifetime)

Then:

1. $K = 1.3 \frac{1}{d} \left(\frac{P}{P_1}\right)^2$ P_1 = 450 torr for xenon

    P = Actual xenon pressure in flashlamp

2. $U = \frac{1}{2} C V_0^2$: Energy stored in capacitor.

3. Let τ = Pulse width (not FWHM but zero to zero of current).
   $\tau = \pi \sqrt{LC}$

4. $C^3 \simeq \frac{0.5}{9} \frac{U \tau^2}{K^4} \simeq \frac{0.05 \, U \tau^2}{K^4}$

5. Check U explosion = $6.8 \times 10^4 \, ld \, (LC)^{1/4}$:   50% probability
   (τ>10μs)   = $3.8 \, ld \sqrt{\tau}$ of explosion   Willis

Fig. 5-27. *Flashlamp loads.*

lamp and the energy to be deposited in the lamp are known, the explosion limits for relatively long pulses can be determined. When the energy stored in the capacitor is equal to the determined energy, there is a 50 percent chance that the lamp will blow up. In most circuits, it has been found that the longer lamps can be operated near the explosion limit.[10]

The most convenient way to trigger a flashlamp is with a series injection trigger transformer, usually on the ground end. This generates a pulse that exceeds the self-break voltage of the lamp and the lamp turns on. The fluctuations in the breakdown time can be significantly reduced by overvolting the lamp by a factor of two.

5.12.3 Carbon Dioxide Laser Load

There are two kinds of CO_2 lasers: electron beam lasers and TEA lasers. The latter are similar to the excimer laser, wherein their impedance decreases with time, but at a slower rate. The CO_2 lasers with electron beam controlled discharge systems, for example, basically act as resistive loads with some turn-on voltage. The gas discharge voltage can be provided by a Marx bank. If the pumping voltage wave form is altered to rise more quickly, the gain rises faster; this rise permits more energy to flow out of the system for shorter optical pulses into it. Experimenters have looked into a Type-C PFN, with two sections (two L and C, one each per section in the PFN) and have derived the voltage pulse shown in Fig. 5-28. This worked very well.[11]

The TEA lasers are like HF and KrF lasers that are discharge pumped. They can be made to work very well at high repetition rates. Everything that has been said about the KrF laser can be said about TEA lasers, multiplying the time-scale of excitation by a factor of 10. The TEA lasers also have a smaller decrease in impedance with time; compared to KrF lasers the ratio is considerably less.

5.12.4 Discharge-Pumped Hydrogen Fluoride Laser Load

The HF lasers are attachment dominated, and the impedance decreases with time (Fig. 5-29). Impedance was measured as a function of current in kiloamps for different pressures of the gas (Fig. 5-30). To design this particular system, choose a pulse width less than the arcing pulse duration ($\cong 0.25$ microseconds). Even if the electric field is kept constant, the gas heats with time until it experiences thermal breakdown. This is a common characteristic power-loading limit of all lasers.

5.13 SUMMARY

With the exception of electron-beam systems, lasers all have decreasing impedances with time. This presents a challenging area for research to provide high-efficiency lasers with electrical PCS drivers optimized to meet the load excitation kinetics requirements.

5.14 REFERENCES

1. G.N. Glasoe and J.V. Lebacqz, "Pulse Generators" Dover Publications, Inc., New York, 1965.
2. W.J. Sarjeant, "Energy Storage Capacitors," LASL Report No. LA-UR-79-1044, August 1979.
3. T.P. Sorensen and V.M. Ristic, "Risetime and Time-Dependent Spark-Gap Resistance in Nitrogen and Helium," Journal of Applied Physics, Vol. 48, No. 1, January 1977, pp. 114-117.

Electron beam controlled discharge
- pure resistive load compared to diode magnetron type of load
- can use type C PFN in Marx stages
- large units: $Z \cong 3\ \Omega$ for 2.5 μS discharge times

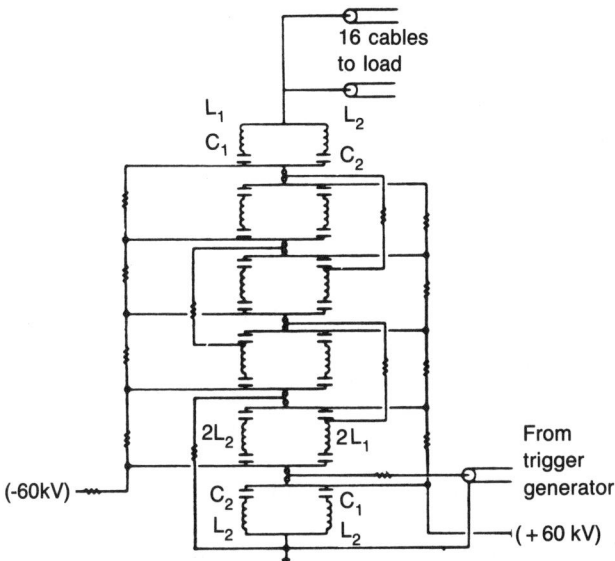

600-kV, 2.5-Ω type-C Guillemin-Marx network. This is the ideal circuit, ignoring stray inductance. Circuit element values are discussed in the text. A "stage" is considered to be a spark gap, a set of positively charged capacitors, and a set of negatively charged capacitors.

Per stage: type C PFN

$L_1 = 0.63\ C_1 Z_o^2$

$L_2 = 6.5\ C_2 Z_o^2$

$C_1 = 0.40\ \dfrac{\tau}{Z_o}$

$C_2 = 0.042\ \dfrac{\tau}{Z_o}$

Defining τ, V_{pk} and U stored defines all the particular components.

then $C_1 = 8.6\ \mu F$
$C_2 = 0.58\ \mu F$ For
$L_1 = 220\ \mu H$ $Z_o = 2.5\ \Omega$
$L_2 = 235\ \mu H$

Advantages for Resistive Load:
1. V_{pk} per stage times $n = V_L$
2. Stray L in connection loop affects risetime but not peak power.
3. Waveform readily changed.
   $\therefore$ Can use with time varying loads.

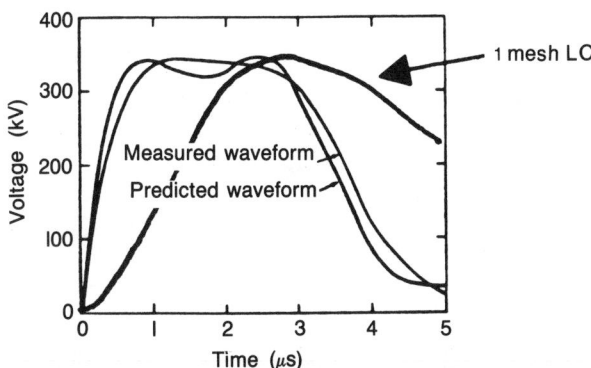

Predicted and actual waveforms into a 3-Ω dummy load in the PFN tank.

Fig. 5-28. *Carbon dioxide laser loads. (Courtesy of K.B. Riepe, "High-Voltage Microsecond Pulse-Forming Network, Rev. Sci. Intrum, Vol. 48, No. 8, aug. 1977, © 1972 AIP).*

1. Electrical drive similar to KrF* laser discussed before.
2. From experiment: Impedance decreases with increasing discharge current: A time-varying load.

   –Take Z at peak current to design PFN with peak voltage determined from a scaling experiment.
   –Design PFN width for optimum pumping and pulse termination before the onset of arcing (300 ns for a laser electrode spacing of 5 cm)

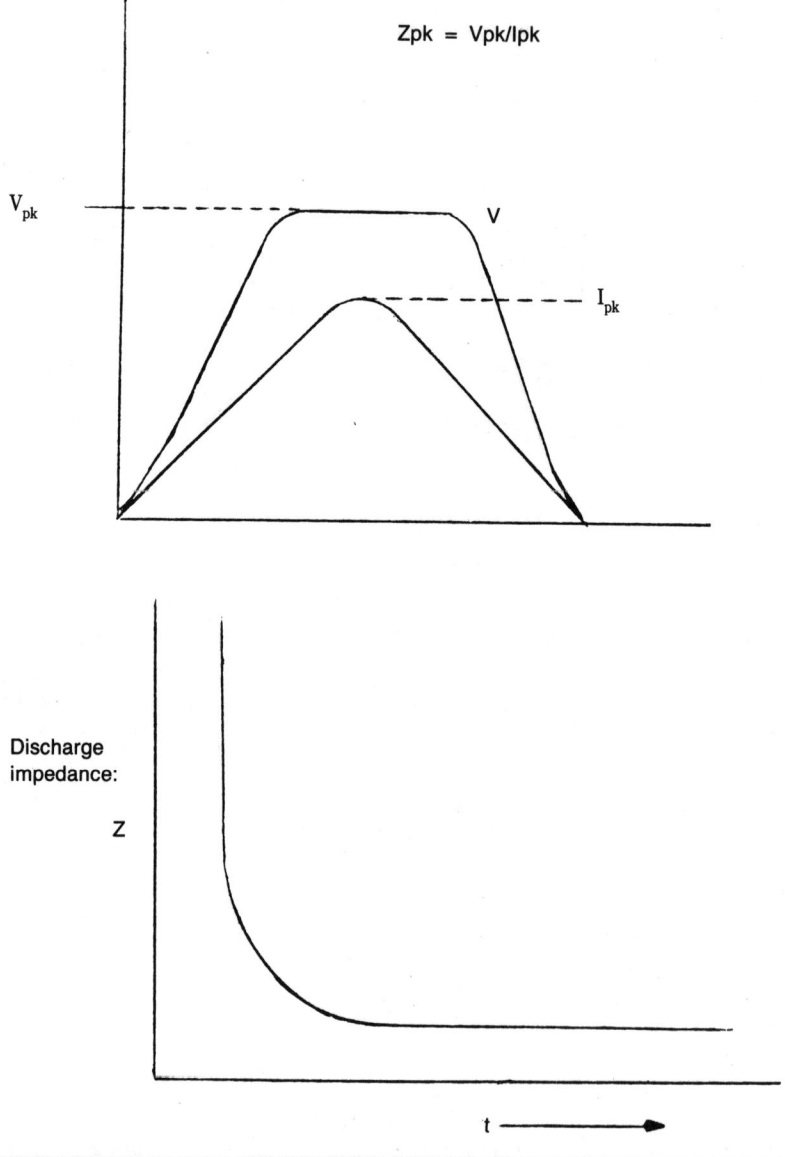

Fig. 5-29. *Discharge-pumped hydrogen fluoride laser load.*

3. Results of the calculations

   -Used NET 2 program and incorporated of L of each capacitor as part of the PFN. Type E-C hybrid PFN could be designed analytically but far more cost effective to use a computer.

4. Short circuit ringing tests:
   a. short cavity
   b. from ringing, following may be determined.

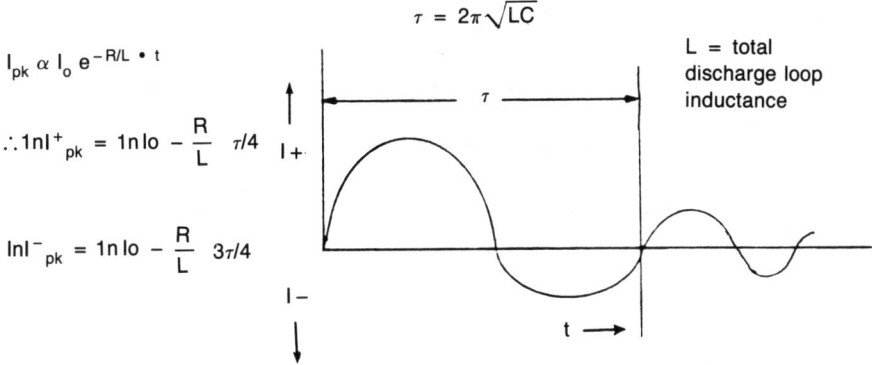

$I_{pk} \alpha I_o e^{-R/L \cdot t}$

$\therefore \ln I^+_{pk} = \ln I_o - \frac{R}{L} \tau/4$

$\ln I^-_{pk} = \ln I_o - \frac{R}{L} 3\tau/4$

$\tau = 2\pi\sqrt{LC}$

L = total discharge loop inductance

Subtract

$\ln I^-_{pk} - \ln I^+_{pk} = \ln(I^-_{pk}/I^+_{pk}) = \frac{R}{4L} \cdot (3-1)\tau = R\tau/2L$

$= \frac{R}{2L} \cdot 2\pi\sqrt{LC} = \pi R\sqrt{C/L} = \ln$ (Damping ratio)

$\therefore$ obtain "L" and "R"

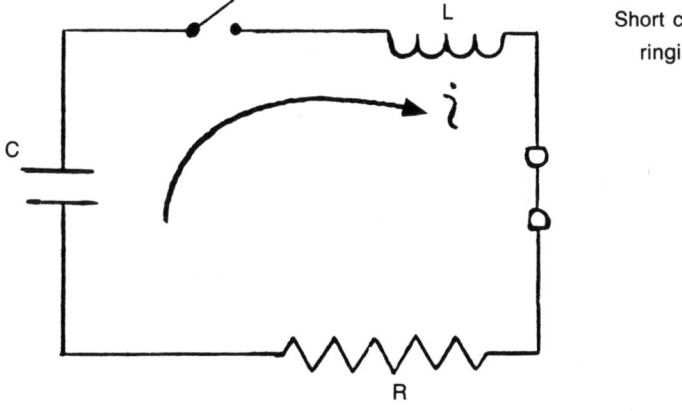

Short circuit laser cavity ringing experiment

Willis

Fig. 5-29. Discharge-pumped hydrogen fluoride laser load (continued). [10]

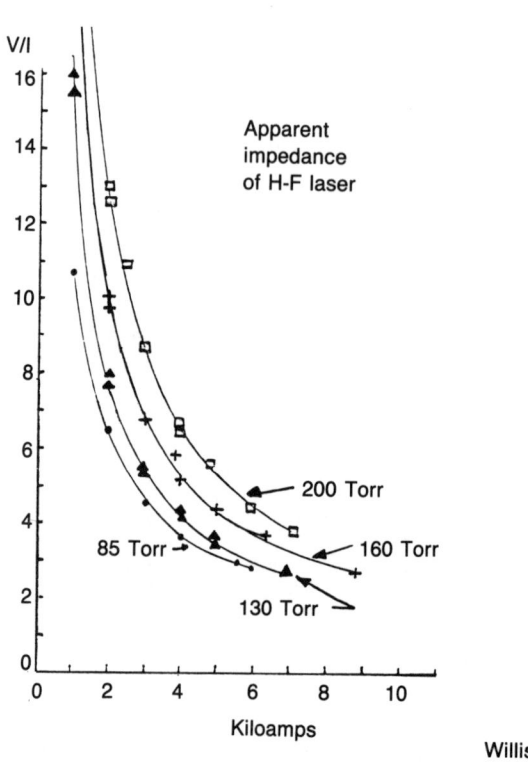

Fig. 5-30. Apparent impedance of the discharge-pumped hydrogen fluoride laser load.

4. W.L. Willis, "Spark Gaps," Los Alamos Scientific Laboratory Report No. LA-UR-80-634.
5. W.J. Sarjeant and W.C. Nunnally, "A Kilohertz Trigger System for Multichanneling a 100-kV Midplane Spark Gap," Los Alamos Scientific Laboratory Memorandum No. E-4-79-101, March 5, 1979.
6. A.E. Greenwood, "Electrical Transients in Power Systems," (Wiley-Interscience, New York, 1971.)
7. G. Metzger and J.P. Vabre, "Transmission Lines with Pulse Excitation" (Academic Press, New York, 1969).
8. "Hydrogen Thyratrons—Theory and Applications," General Electric Technical Literature.
9. R.R. Butcher, "A Comprehensive Study of Excimer Laser Systems," Los Alamos Scientific Laboratory Thesis Report LA-7329-T, June 1978.
10. W.L. Willis, unpublished data.
11. K.B. Riepe, "High-Voltage Microsecond Pulse-Forming Network," Rev. Scientific Instrumentation, Vol. 48, No. 8, August 1977, pp. 1028-1029.

Chapter 6

Spark Gaps

by
W.L. Willis*

CONCEPTUALLY, A SPARK GAP IS A SIMPLE DEVICE CONSISTING OF TWO CONDUCTORS AND an overstressed dielectric. The dielectric can be any insulator, gas, solid, liquid, or vacuum. A spark gap is typically used as a high-voltage closing switch with the requirement that the applied voltage must be removed for recovery to the open state.

6.1 GAS SWITCHES

Practically, spark gaps are quite complex, particularly if close control of closure time, inductance, losses, and lifetime is required. As a direct consequence, geometry, electrode materials, insulating structures, and the overstressed dielectric all interact. Gas dielectrics are most commonly used in the latter function and the greatest amount of information exists for this type of gap. Most of this detailed analysis covers the gas type of gap, but other kinds of gaps also are discussed.

Even with this restriction, several basic types are involved, including two electrode gaps, trigatrons, field distortion gaps, rail gaps, and more complex geometries.

6.1.1 Gas Breakdown

The details of gas-insulated gaps depend strongly upon the breakdown mechanisms of the gas involved. There are typically two stages, avalanche and streamer formation, although a thorough analysis includes more complex stages.

J.S. Townsend ("Electricity in Gases," 1914) did the basic work in this area. A sidelight to his work was the discovery of cosmic rays in order to account for the observed conduction in gases. When an electric field exists in a gaseous medium, a small current can be observed due to available free electrons resulting from ionizing radiation.

---

*Written while at the Los Alamos National Laboratory. Currently at the Northrop Research and Technology Center.

As the field is increased, electrons begin to acquire enough energy between collisions with gas molecules to produce secondary ionization upon impact. Townsend defined α as the number of ionizing collisions per centimeter in the field direction produced by a single initiating electron. This leads to:

$$N = N_o e^{\alpha d} \tag{6-1}$$

as the description of the current reaching the anode.

As the field is further increased, a second mechanism takes effect: generation of electrons at the cathode due to positive ion bombardment. This has a coefficient that relates ion current to electron generation. When this term is added, the current takes the form:

$$I = I_o \frac{e^{\alpha d}}{1 - \gamma(e^{\alpha d} - 1)} \tag{6-2}$$

when N is related to I by taking into account the electronic charge of each electron. At some point, the denominator approaches zero so that:

$$1 - \gamma(e^{\alpha d} - 1) = 0$$

or, approximately, when $e^{\alpha d} >> 1$:

$$\gamma e^{\alpha d} = 1$$

which justifies Paschen's Law, to be discussed later.

When the term $e^{\alpha d}$ is on the order of 20, transition from avalanche to streamer takes place. One consequence of this is that the dielectric strength for small (less than 1 centimeter) spacings is greater than for larger gaps.

Figure 6-1 is a typical voltage-current relationship for a gas in a uniform field. The behavior, after reaching breakdown, depends upon the gas. In general, a sharp drop in voltage occurs. Figure 6-2 represents the growth of a single electron in avalanche mode with transition to streamer mode. Note that a negative space charge builds up due to the relatively immobile ions. Eventually, a virtual cathode forms out in space, and it tends to produce secondary structures. A physical difference between avalanche and streamer mode is that avalanche is invisible, but streamers are marked by photo-ionization and photoemission and are brightly luminous. Also, the velocity of propagation is different. A velocity of 10^7 centimeters/second is accepted for avalanche, 10^8 centimeters/second or greater is typical of streamers.

There are two other mechanisms of importance to consider in gas breakdown: electronegativity and the Penning effect. Some monovalent gases, such as fluorine, and some more complex gaseous molecules, such as SF_6, have outer rings deficient in one or two electrons. These tend to capture or attach free electrons to form negative ions. The low mobility of such ions effectively removes the electron from the avalanche process and reduces the first Townsend coefficient, alpha. If this attachment coefficient is given

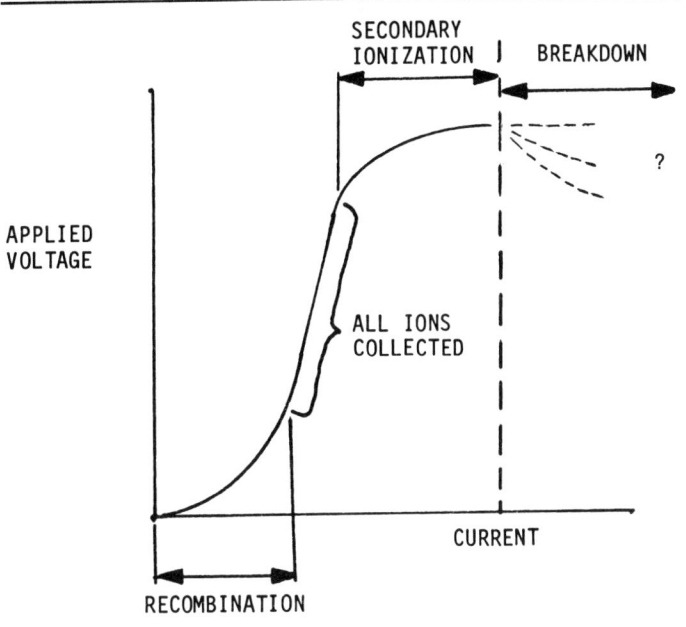

Fig. 6-1. V-I characteristic for a gas in a uniform electric field.

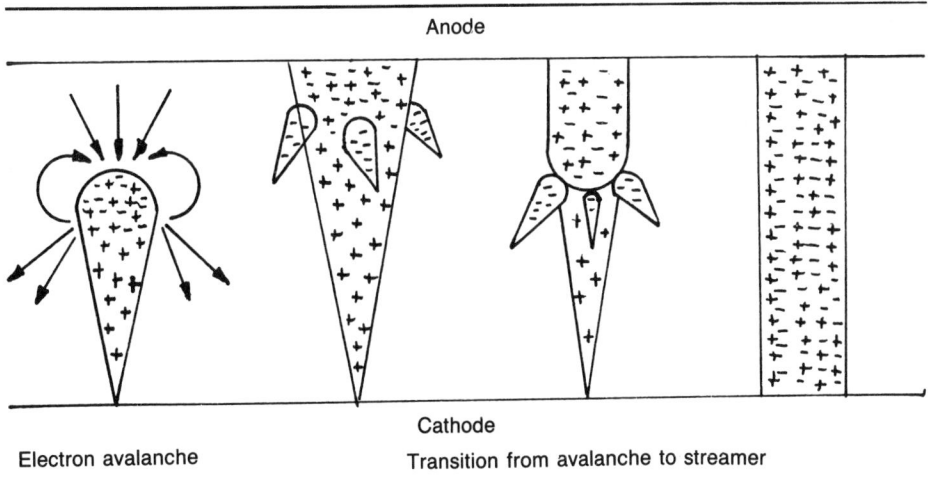

Fig. 6-2. Breakdown in gaseous dielectric.

by n, the breakdown criterion becomes:

$$\frac{\gamma^\alpha}{\alpha^{-n}} \left[e^{(\alpha-n)\gamma} - 1 \right] = 1 \tag{6-3}$$

174 Spark Gaps

Table 6-1. Electronegative Gases.

GAS	SYMBOL
Sulfur Hexafluoride	SF_6
Dichlorodifluoromethane	CCl_2F_2
Perfluoromethane	C_3F_8
Perfluorobutane	C_4F_{10}
Hexafluoroethene	C_2F_6
Chloropentafluoroethene	C_2ClF_5
Dichlorotetrafluoroethane	$C_2Cl_2F_4$
Tetrafluoromethane	CF_4

The Penning effect, on the other hand, reduces breakdown strength. If, for example, a trace (1 percent) of argon is added to neon, a large reduction in breakdown strength occurs. Several mixes exhibit this effect including helium-argon, neon-argon, helium-mercury, and argon-iodine.

Table 6-2. Breakdown Voltage Between Bare Electrodes 1 Centimeter Apart at One Atmosphere.

Gas	He	Ar	H_2	NH_4	N_2	Air	CO_2	SF_6
KV dc	1.3	3.4	12	18.5	22.8	23	24	67

At low frequencies, for example, 60 Hz power, the same voltages are observed, provided that the factor for peak rather than RMS is taken into consideration. When the voltage pulse is fast compared to propagation times in the gas, the situation changes. Figure 6-3 shows a particular case. When the formative time is on the order of 10 microseconds, the impulse ratio approaches one; that is, the dc breakdown value is observed. When the pulse is 10 nanoseconds, the breakdown strength is about 1.85 times the dc value.

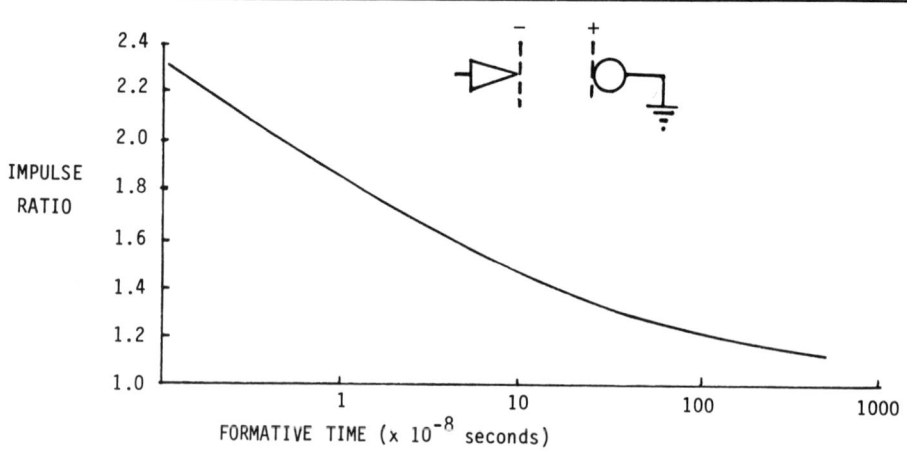

Fig. 6-3. Relation between formative time and impulse ratio for a negative point-sphere gap in air.

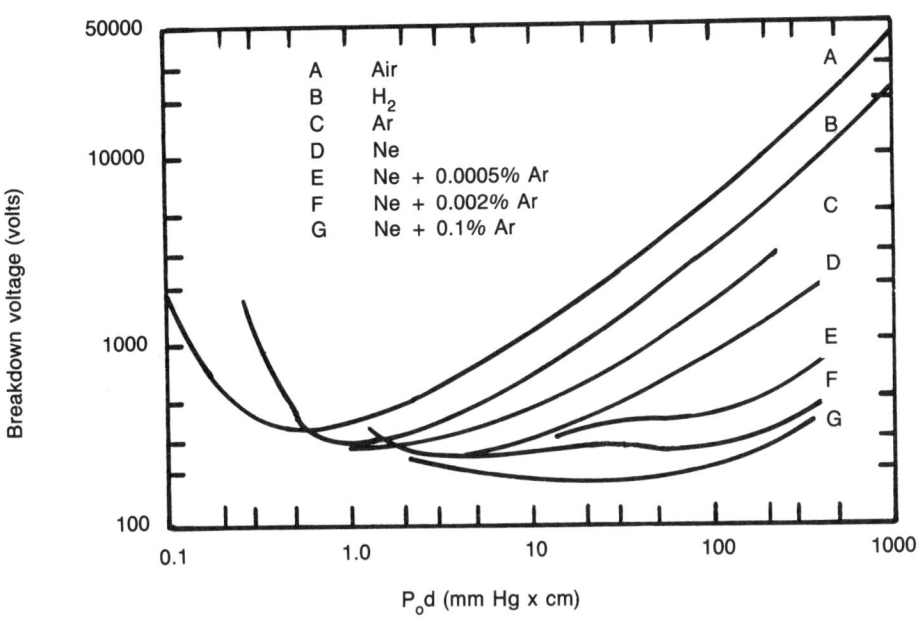

Fig. 6-4. Typical breakdown voltage curves for different gases between parallel-plate electrodes. [2]

The Paschen curve for several gases is shown in Fig. 6-4. Note the Penning effect on neon. Also, note the minimum point. For a given spacing, as pressure drops, so does the probability of an electron-gas molecule collision. A point is reached where mean free paths correspond to electrode separation, and the drops. At this point, the apparent dielectric strength increases again.

A final point before proceeding to the specifics of gap design pertains to field enhancement factors. Most of the data so far pertains to uniform field conditions or to specific geometries. In gap design, the infinite parallel plane case is seldom of interest. Other geometries, however, produce field distortion with locally more intense electric fields. These geometries can markedly affect the gap behavior. Refer to Fig. 6-5 and note that potential is given in terms of U, not V.

This gives the maximum field strength, not the effective field strength. This distinction is important, as is shown by some calculations in the next section.

Also, only breakdown (switch closure) is discussed in the preceding discussion. Switch recovery is more complex but will be discussed in this chapter.

6.1.2 Self Break

This discussion might begin by quoting J.C. Martin: "One of the minor irritations of my life has been the fact that while approximate calculations of the breakdown volts of practical sphere/sphere and cylinder/cylinder gaps give reasonable agreement to 10–20 percent, when more accurate attempts are made to compare experiment with crude theory, reality seems strangely perverse."

Spark Gaps

CONFIGURATION		FORMULA FOR E	EXAMPLE
Two parallel plane plates	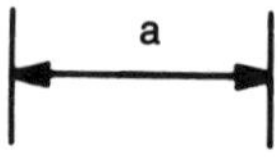	$\dfrac{U}{a}$	$U = 100$ kV, $a = 2$ cm, $E = 50$ kV/cm.
Two concentric spheres	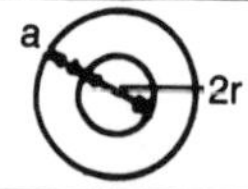	$\dfrac{U}{a} \cdot \dfrac{r+a}{r}$	$U = 150$ kV, $r = 3$ cm, $a = 2$ cm, $E = 125$ kV/cm.
Sphere and plane plate	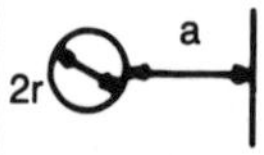	$0.9 \dfrac{U}{a} \cdot \dfrac{r+a}{r}$	$U = 200$ kV, $r = 5$ cm, $a = 8$ cm, $E = 58.5$ kV/cm.
Two spheres at a distance a from each other	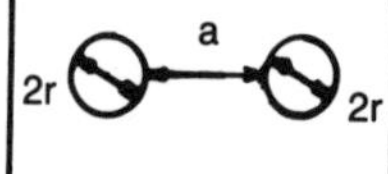	$0.9 \dfrac{U}{a} \cdot \dfrac{r+a/2}{r}$	$U = 200$ kV, $r = 5$ cm, $a = 12$ cm, $E = 33$ kV/cm.
Two coaxial cylinders	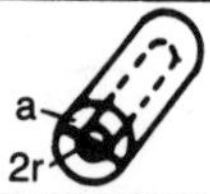	$\dfrac{U}{2.3\, r\, \lg \dfrac{r+a}{r}}$	$U = 100$ kV, $r = 5$ cm, $a = 7$ cm, $E = 22.9$ kV/cm.
Cylinder parallel to plane plate	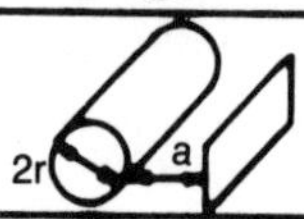	$0.9 \dfrac{U}{2.3\, r\, \lg \dfrac{r+a}{r}}$	$U = 200$ kV, $r = 5$ cm, $a = 10$ cm, $E = 32.8$ kV/cm.
Two parallel cylinders		$0.9 \dfrac{U/2}{2.3\, r\, \lg \dfrac{r+a/2}{r}}$	$U = 150$ kV, $r = 6$ cm, $a = 20$ cm, $E = 11.5$ kV/cm.
Two perpendicular cylinders	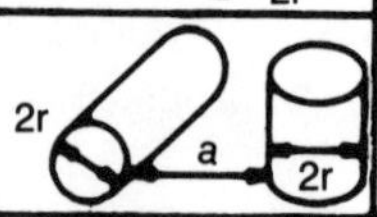	$0.9 \dfrac{U/2}{2.3\, r\, \lg \dfrac{r+a/2}{r}}$	$U = 200$ kV, $r = 10$ cm, $a = 10$ cm, $E = 22.2$ kV/cm.
Hemisphere on one of two parallel plane plates	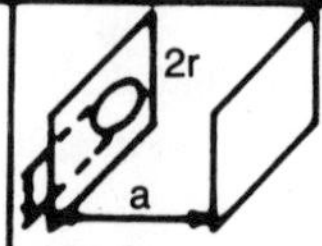	$\dfrac{3U}{a}$; $(a \gg r)$	$U = 100$ kV, $a = 10$ cm, $E = 30$ kV/cm.
Semicylinder on one of two parallel plane plates	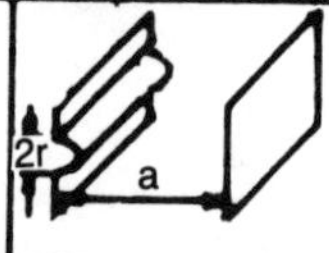	$\dfrac{2U}{a}$; $(a \gg r)$	$U = 200$ kV, $a = 12$ cm, $E = 33.3$ kV/cm.
Two dielectrics between plane plates ($a_1 > a_2$)	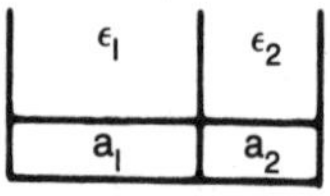	$\dfrac{U \epsilon_1}{a_1 \epsilon_2 + a_2 \epsilon_1}$	$U = 200$ kV, $\epsilon_1 = 2, \epsilon_2 = 4$, $a_1 = 6$ cm, $a_2 = 5$ cm, $E = 11.8$ kV/cm.

Fig. 6-5. Maximum field strength, E, with a potential difference, U, between the electrodes for different electrode configurations. (A. Bouwers and P.G. Cath, "The Maximum Electrical Field Strength for Several Simple Electrode Configurations," Philips Technical Review, Vol. 6, No. 9, with permission.)

6.1 Gas Switches

In this section, calculations are presented for the case of a Maxwell 100 kV gap as a practical case. Figure 6-6 presents the geometry and field enhancement factor equation for the two-sphere case. The midplane is ignored for the calculation. (Actually, you can do sphere-to-plane and double the result—the answers are nearly identical.)

First, determine the field enhancement factor (from Fig. 6-5, fourth case):

$$E = 0.9 \frac{U}{0.754} \frac{0.794 + 0.75/2}{0.794} = 1.76 \, U$$

Now, watch closely. The uniform field would be:

$$\frac{U}{0.754} = 1.326 \, U$$

Thus the FEF (*field enhancement factor*) is the ratio, or:

$$FEF = \frac{1.76 \, U}{1.326 \, U} = 1.327$$

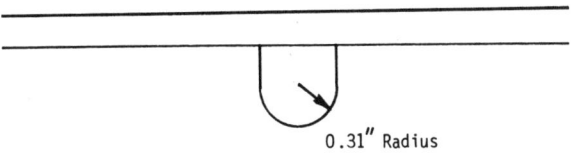

0.31" Radius

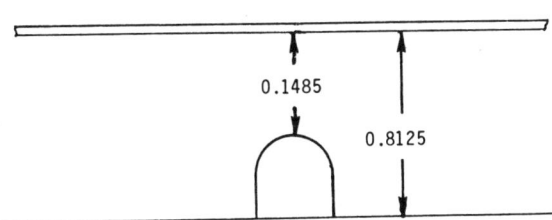

0.1485

0.8125

$$0.9 \, \frac{U}{S} \, \frac{R + \frac{S}{2}}{R} = E$$

U = kV   E = kV/cm

$$R = \frac{5''}{16} = 0.794 \text{ cm}$$

S = 2 × 0.1485 × 2.54 = 0.754 cm

Fig. 6-6. Maxwell 100 kV gap geometry.

Although it seems a shame to cloud the issue here, we now proceed down the path by which J.C. Martin was troubled first.

Figure 6-7 is a reasonably accurate Paschen curve for air. Figure 6-8 is the Maxwell published curve for the 100 kV gap. Examine the correlation you can obtain from this data.

Take the left intercept of Fig. 6-8. This appears to be, quite closely, 25 kV at 0 psig. The gap is 0.754 centimeters so that the pressure-spacing product at one atmosphere is 573 Torr-centimeter. Entering the Paschen curve at 573 Torr-centimeter, the breakdown is about 23 kV for air, uniform field.

If this is reduced by the FEF, the expected breakdown would be:

$$\frac{23 \text{ kV}}{1.327} = 17.3 \text{ kV}$$

Not too bad, but not right. J.C. Martin worked out corrections for this. First, take the uniform field equation for air. (This is the equation for Fig. 6-7):

$$E = 24.5 \, p + 6.7 \, \beta \, \frac{p^{1/2}}{S^{1/2}} \text{ kv/centimeter} \qquad (6\text{-}4)$$

where E is the breakdown field, p is pressure in atmospheres, and S is the spacing in centimeters.

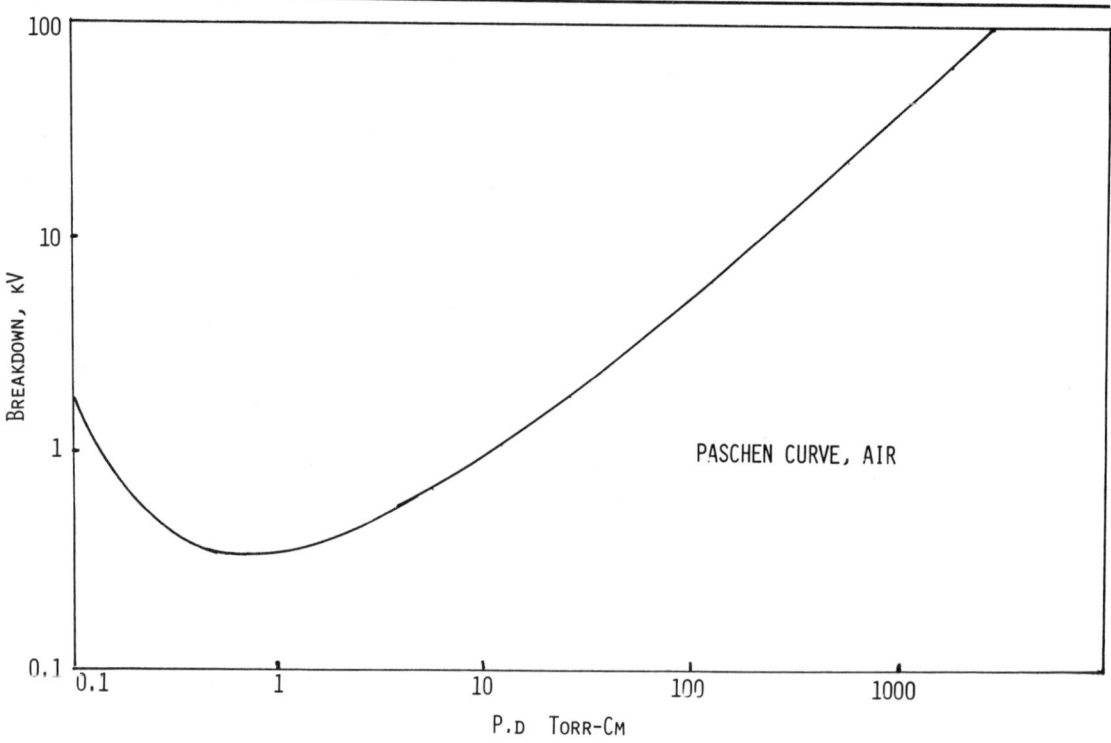

Fig. 6-7. Paschen curve, air.

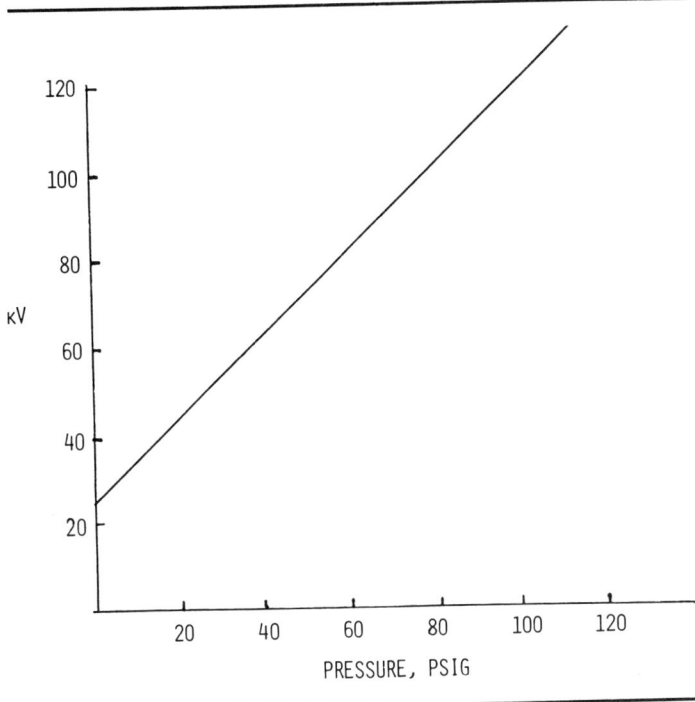

Fig. 6-8. 100 kV Maxwell gap self-break curve.

The equation is modified to:

$$E = 24.5\,p + 6.7\,\beta\,\frac{p^{1/2}}{R_{eff}^{1/2}} \tag{6-5}$$

where $R_{eff} = 0.115\,R$ for spheres, and $R_{eff} = 0.23\,R$ for cylinders.

Beta (β) is a term that takes geometry into account. Figure 6-9 shows the correction curve and, as an inset, the reason for the correction. It has been noted that the breakdown strength of air is a function of spacing. So is the FEF a function of spacing. These factors interact to produce the correction curve.

For this case, $\dfrac{S}{R} = \dfrac{0.754}{0.794} = 0.95$, $\beta = 0.9$, and $P = 1$. With this:

$$E = 24.5 \cdot 1 + 6.7 \cdot \frac{0.9}{\sqrt{0.115 \times 0.794}}$$

$$= 44.45\ \frac{kV}{centimeter}$$

180 Spark Gaps

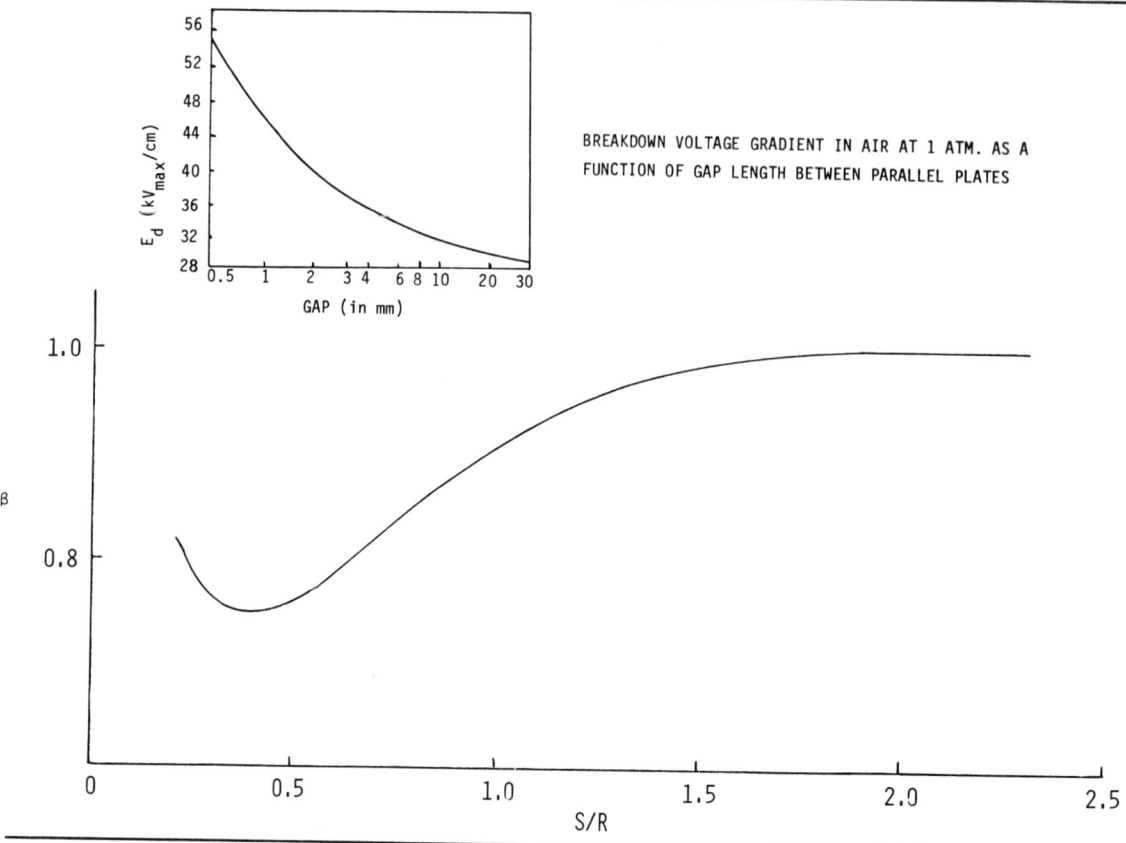

Fig. 6-9. Breakdown voltage gradient in air at 1 atmosphere as a function of gap length between parallel plates.

By the FEF, the effective spacing is:

$$S_{eff} = \frac{S}{FEF} = \frac{0.754}{1.327} = 0.568 \text{ centimeters} \tag{6-6}$$

This yields a self break of:

$$44.45 \text{ kV/centimeters} \times 0.568 \text{ centimeters} = 25.3 \text{ kV}$$

This is excellent agreement, better than is often observed by measurement.

6.1.3 Triggering

Assuming that the gap is designed as a closing switch, there are two basic methods of actuation: either overvolt the switch or reduce the holdoff capability of the dielectric. The former can be done directly as in the Marx bank, sharpening switches, etc., or by an electrical signal as in the field-distortion gap.

6.1 Gas Switches

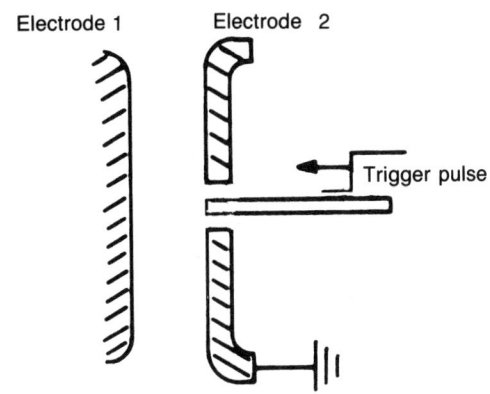

Electrode polarity and trigatron characteristics

Range	Jitter	Trigger Electrode	Opposite Electrode	Host Electrode	Mode
↑ wider ↑	↑ shorter ↑	+	−	+	A
		−	−	+	D
		−	+	−	B
		+	+	−	C

Fig. 6-10. Electrode polarity and trigatron characteristics (courtesy of EG&G, Inc., Electronic Components Division, Salem, Massachusetts).

The dielectric properties can be changed to induce breakdown by incident radiation, pressure drop, heating, E-beams, etc. In any event, closure has two parts: statistical and formative. These parts lead to delay and, usually, a jitter that is nominally approximately 10 percent of the delay.

Simple two-element switches are usually rather poor for closely timed operation. Two techniques are used to get around this, neglecting severely overvolted sharpening gaps. One is the *rope* switch, which takes advantage of statistics. A series connection of several two-element switches is used. With a normal distribution of delays, a few gaps will tend to break one or two standard deviations below the mean. These closures overstress the switches operating near the mean and severely overstress the statistically late units. Experimental results with ten 20 kV switches in a 200 kV rope gave a maximum jitter of 11 nanoseconds and a standard deviation of 3 nanoseconds. This was a factor of three better than a single 200 kV gap performance.

Radiation triggering is common in two-electrode gaps. In multiple-gap systems, an optical coupling path is often present, so that the first gap to fire illuminates the others. An external pulse, for example, a laser, can be used for the initial pulse. Because O_2 and N_2 require 990 A and 790 A photons for photo-ionization and windows are difficult for these wavelengths, most radiation sources are not external. Photoemission from Cu, Al, and Fe have thresholds of 2600 to 300 A. Windows with incoherent sources, such as mercury lamps, external sparks, and super-radiant N_2, have been used. As of 1968, some 161 papers had been published on the topic of laser triggering alone.

Use of laser triggering does not guarantee low jitter. If the laser pulse is short with respect to the gap delay, the usual 10 percent of closing time jitter is observed. V.S. Komelkov quotes a 50 Hz, four-gap system operating at 50 kV/gap with 0.1-nanosecond jitter employing laser triggering.

The Russians have also reported several E-beam triggered gaps, back to 1939. This reduced gap holdoff to about half the static value, and polarity effects were not significant. Gaps are typically operated at 70 to 80 percent of self-break, this should provide good triggering.

The most common illumination-type gap is the trigatron, a three-electrode gap. Field distortion is also present and contributes to breakdown. Figure 6-10 shows the geometry and relationships of polarity on the electrodes for various conditions. The mode column refers to the EG&G operating designations. It is to be noted that EG&G usually specifies Mode A except for a few low-voltage, low-energy units that exhibit long (100 to 1000 nanoseconds) delay times. The EG&G data is for fairly relaxed operation in terms of rise times and jitter obtainable with spark gaps.

An advantage of the trigatron is the relatively low voltage trigger requirement, particularly with respect to field distortion gaps. Ion Physics Corporation (IPC) has reported on such switches up to 10 mV with 200 kV triggers and a six-channel, 4.5 MV switch in the ARES electro-magnetic pulse (EMP) generator with a rise time of less than 10 nanoseconds. Erosion around the trigger pin is one of the problems associated with the trigatron. The trigger plasma tends to induce the main discharge into a region near the electrode.

Field-distortion gaps are often used where very low jitter is desired. The price appears in the trigger system requirements. The author is convinced that even a lightning stroke is probably both too slow and too low in voltage for proper triggering. Figure 6-11 is a schematic of such a gap. The midplane electrode can actually be located at a wide range of positions. The constraint is that it is biased to the potential that would exist at that plane in the two-electrode geometry. This is to avoid perturbing the static field distribution.

If it is in the midplane, the triggering sequence would go as shown in Fig. 6-12. The input trigger swings the midplane toward the potential of Electrode 1 (V1) and away from 2. The overvoltage breaks down the midplane-electrode 2 gap. The plasma then couples the midplane toward V2, so that now the midplane to electrode 1 gap overvolts and breaks. This cascade cannot be avoided in the midplane case.

When the trigger electrode is nonsymmetric, at approximately ⅓ or ⅔, Fig. 6-13 is obtained. In this case, the trigger is driven to V TRIG, equal to the voltage difference V1 − V2. For convenience, assume V2 = 0. Then the voltage between electrode 2 and the trigger is just V1, and the voltage between electrode 1 and the trigger is 2V1.

Because the spacings are 1:2, the electric field is the same in both cases and twice the initial value. This is the SOV (*simultaneous overvoltage*) mode and leads to very fast switching. The Maxwell 50 kV gap is of the latter sort, and a trigger dV/dt of 2 kV/nanoseconds is specified. Other sources quote 4 to 10 kV/nanoseconds as the rate

6.1 Gas Switches

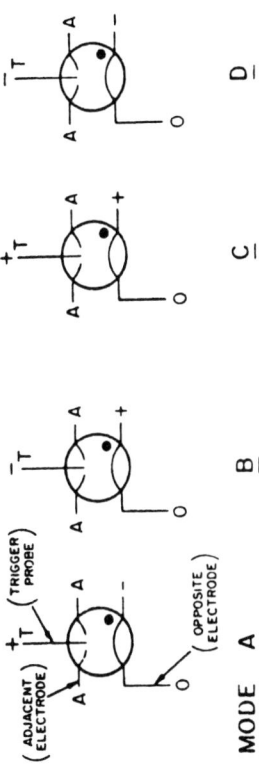

EG&G MODEL NO.	E-E RANGE, KV MIN/MAX (1,12)		SBV, KV	V_T MIN. TRIG kV (OPEN CIRCUIT)	MODE	RECOMMENDED EG&G XFMR	X DELAY TIME ns AT 70% SBV	AT 40% SBV	MAX SIMULTANEOUS DISCHARGE CONDITIONS (LIFE 5,000 - 20,000 SHOTS)	MAX REPETITIVE DISCHARGE CONDITIONS (LIFE 1 - 5 MILLION SHOTS)
GP-89	0.7	2.1	2.6	5	C	TR-1647	100	1000	25 JOULES	3 MILLI COULOMBS/SHOT
GP-90	1.3	3.4	4.2	5.5	C	TR-148A			5 kA PEAK	P/R = 18 WATTS/OHM
GP-91	4.4	10.	12.5	7	A,C	TR-1733			0.1 COULOMB	I_b = 35 mAdc
GP-92	8.0	20.	25.	7	A				E/R = 250 J/OHM	I_p = 8 Aac
GP-828	0.4	1.6	2.0	7	A,B	TR-1647	30	300	200 JOULES	4 MILLI COULOMBS/SHOT
GP-31B	2.0	6.0	7.5		A	TR-148A			7.5 kA PEAK	P/R = 32 WATTS/OHM
GP-20B	3.5	11.	14.	10		TR-1808			0.2 COULOMB	I_b = 60 mAdc
GP-46B	8.0	20.	25.			TR-1777			E/R = 750 J/OHM	I_p = 8 Aac
GP-85	2.0	6.0	8.0		A,B	TR-153	30	300	2000 JOULES	4 MILLI COULOMBS/SHOT
GP-86	6.0	15.	20.	20	A	TR-180B			25 kA PEAK	P/R = 50 WATTS/OHM
GP-87	10.	24.	30.			TR-1700			0.4 COULOMBS	I_b = 100 mAdc
GP-70	12.	36.	42(10)			TR-1777			E/R = 5000 J/OHM	I_p = 10 Aac
GP-30B	2.0	6.0	7.5		A,B	TR-153	30	300	2500 JOULES	10 MILLI COULOMBS/SHOT
GP-22B	6.0	15.	19.	20	A	TR-1700			50 kA PEAK	P/R = 115 WATTS/OHM
GP12B	10.	24.	30.			TR-1777			0.5 COULOMBS	I_b = 200 mAdc
GP-14B	12.	36.	42(10)						E/R = 12,5000 J/OHM	I_p = 15 Aac
GP-41B	12.	36.	42		A,B	TR-153	30	300	400 JOULES	
GP-32B	20.	48.	60(10)	20		TR-1700			50 kA PEAK	
GP-15B	25.	69.	86(10)			TR-1777			E/R = 100,000 J/OHM	
GP-74B	40.	100.	120(10)	20	A	TR-153	30	300		
						TR-1700				
						TR-1777				
GP-81B	40.	100.	120(11)							

Fig. 6-11. Gap mode designations (courtesy of EG&G, Inc., Electronic Components Division, Salem, Massachusetts).

184 Spark Gaps

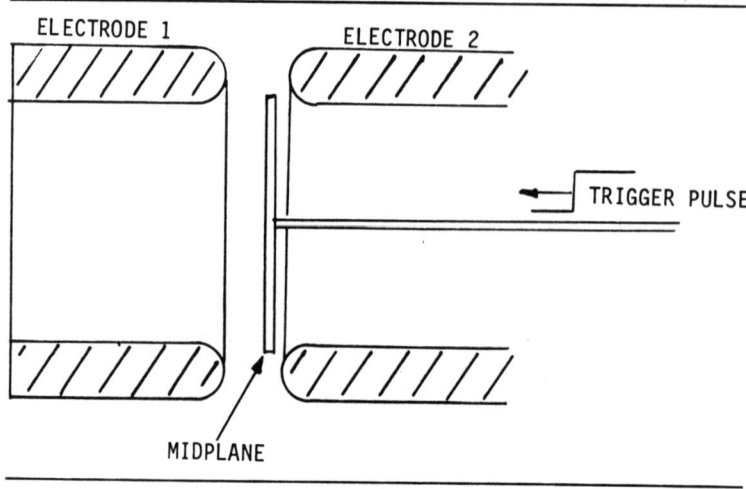

Fig. 6-12. *Field-distortion switch.*

of change of trigger voltage required for SOV operation. Note also that the trigger voltage is required to swing from:

$$\frac{V_1 - V_2}{3} \text{ to } -(V_1 - V_2) \text{ or } \frac{4}{3}(V_1 - V_2) \quad (6\text{-}7)$$

Hence, SOV operation has negative voltage gain!

6.1.4 Recovery

Spark gaps are not normally used as a high repetition rate switch. An exception, historically, would be the rotary gap. Pages 292–293 in Glasoe and Lebacqz present a summary of several such gaps. The text cites one that was designed for 10 MW at 1000 A and 1000 hour life. The jitter in such a system is usually not acceptable in modern systems, and other approaches are used.

The time to recovery is complex and dependant upon material, geometry, gas composition and pressure, arc length, and current. With respect to gases, recovery time increases in the following order: H_2, SF_6, CO_2, O_2, He, Air, N_2, Ar.

Geometry can be used to improve recovery time. Deionization takes place primarily at gas-to-structure contacts, so a series of metallic disk *doughnuts* has been introduced in some designs that also tends to grade the voltages and cool the gases in the gap. The time to recovery for an arc in nitrogen is shown in Fig. 6-14 for a particular gap geometry. Based on this, 1 kHz operation seems feasible. Heating becomes a greater problem than recovery at high repetition rates, and a blown system is normally employed. Even in low repetition rate systems, gap recovery limits the rate of reapplication of voltage. These limits must be taken into account in design of the charging circuit in order to avoid a latch-up problem.

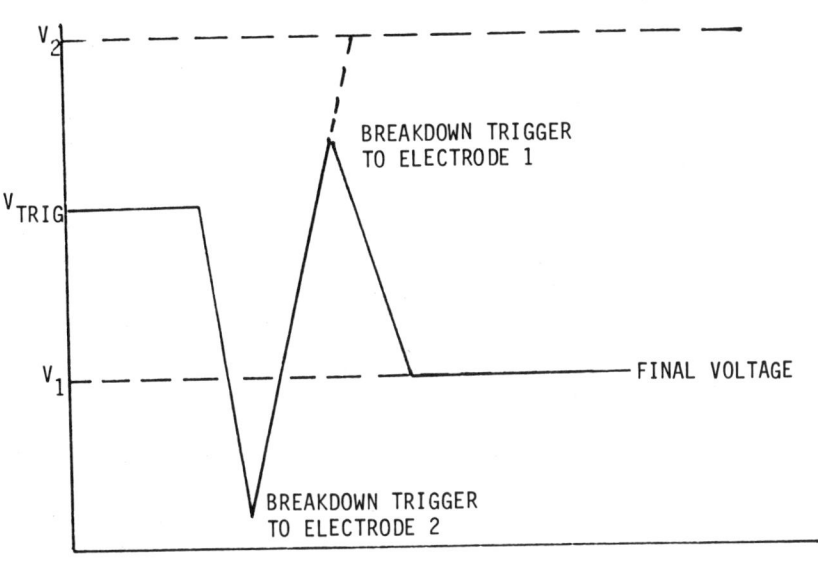
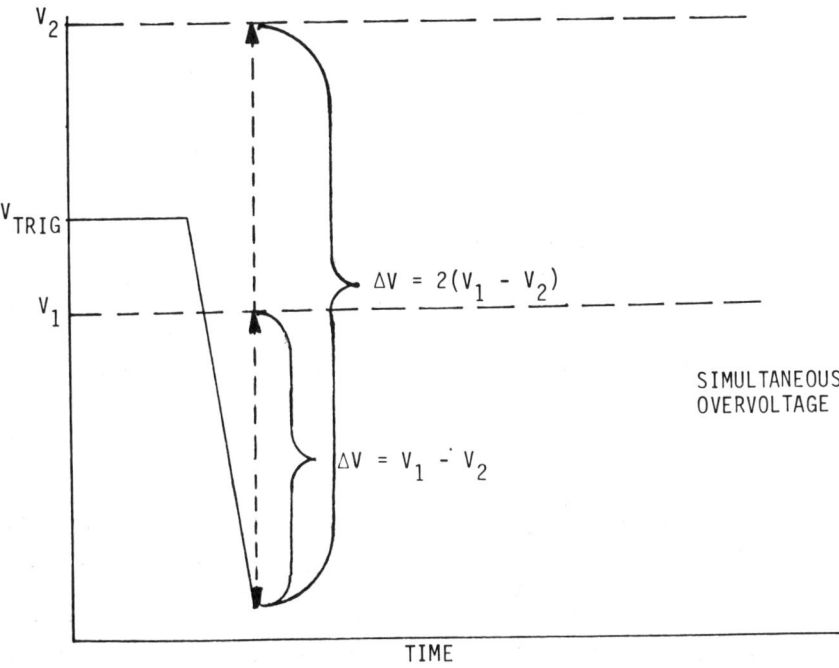

Fig. 6-13. *Triggering sequence for nonsymmetric midplane spark gap.*

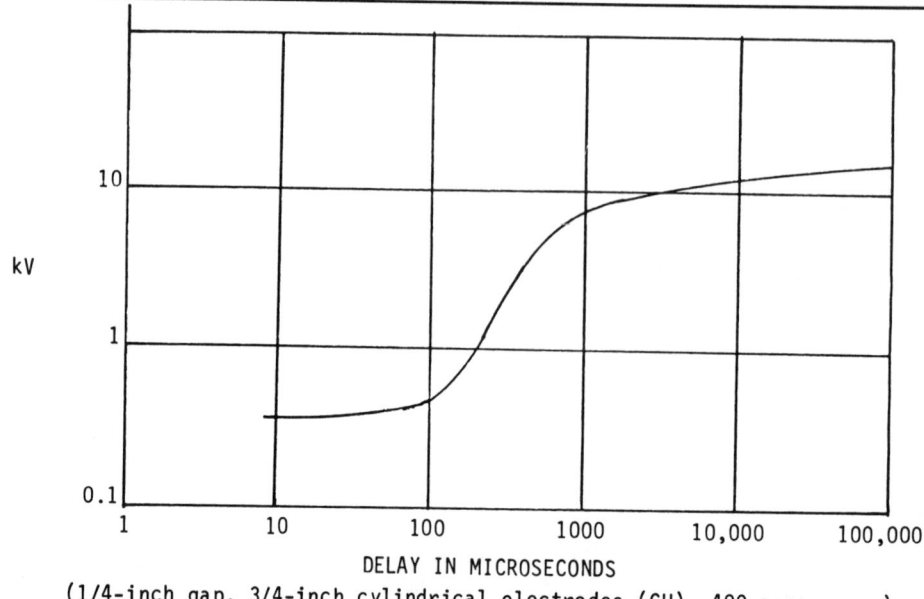

Fig. 6-14. *Arc recovery strength in N_2 at 1 atmosphere (courtesy of G.A. Farrall and J.D. Cobine, IEEE Trans. on Power Apparatus Systems, "Recovery Strength Measurements in Arcs from Atmospheric Pressure to Vacuum," PAS-86 No.8 © 1967 IEEE).*

6.1.5 Inductive and Resistive Considerations

The inductance and resistance associated with gap turn-on may produce major effects on pulse rise time and energy-transfer efficiency from the energy store to the load.

The inductance is treated only briefly because it is both difficult to calculate and usually small compared to other terms. One estimate is that it amounts to about 15 nH/centimeter for a 1-millimeter diameter channel. In the 100 kV Maxwell gap, the length was 0.754 centimeter, corresponding to 11.3 nH. Quoted value is 20 nH, and this includes geometric contributions of the structure. J.C. Martin points out that a wire fed from a disk has an inductance given by:

$$L = 2[\ln (\text{disk radius}) - \ln (\text{wire radius})] \qquad (6\text{-}8)$$

If the wire is small, the disk size and arc length are the major terms.

The resistive phase is considerably more complex to handle. Several formulas have been developed to describe this phase of the gap.

Sorensen and Ristic present a formula of the form:

$$R(t) = 2 \left(\frac{p^{1/2}}{E Z_o^{1/3} t} \right)^3 10^4 \qquad (6\text{-}9)$$

where p is nitrogen pressure in atmospheres, E is in MV/m, Z_o is in ohms, and t is in nanoseconds. Incidentally, Z_o, while well defined in this experiment as the characteristic impedance of the switched line, is not so easily specified in most cases. The equation differs from the most commonly used expression, but was derived for gaps of 2 to 20 mils.

Ray O'Rourke of Maxwell Laboratories has developed a formula for the resistive time as:

$$\tau_R = \frac{1^{1/3}}{Z^{1/3} E^{4/3}} \left(\frac{P}{P_o}\right)^{1/3} \quad (6\text{-}10)$$

where l is the arc length and P_o is the density of air at STP (standard temperature and pressure).

The most commonly used equation is from J.C. Martin and is:

$$\tau_R = \frac{88}{Z^{1/3} E^{4/3}} \left(\frac{P}{P_o}\right)^{1/2}$$

The units are defined by example. Application of the equation depends upon defining Z correctly. Also, you might ask, what is E? Does the FEF of the geometry enter in? The answer is no. The symbol E is the simple gradient, expressed in megavolts/meter obtained by dividing gap voltage by gap spacing. Returning to the 100 kV gap operating at STP so that self break is 25 kV, assume 20 kV operation. Thus:

$$E = \frac{20 \text{ kV}}{0.754 \text{ centimeter}} = 2.65 \quad \text{MV/meter} \quad (6\text{-}11)$$

and:

$$\tau_R = \frac{88}{3.667 \, Z^{1/3}} \frac{24}{Z^{1/3}} \quad \text{nanoseconds} \quad (6\text{-}11)$$

In a 50Ω system, τ_R is 6.5 nanoseconds. This fact points up the importance of operating at high E fields. If the above is repeated at 100 kV and 78 psig, the result is τ_R = 1.75 nanoseconds.

When the load is of very low impedance, difficulties increase. Even lumped-component PFNs can be designed to 5Ω or less impedance. The resistive phase time constants increase to 14 and 3.8 nanoseconds in the calculations above.

Consider Fig. 6-15 as a model of the switch system. If the time-varying resistor is neglected, the pulse rise time across the load goes simply as:

$$V_L = V(1 - e^{-t/\tau}) \quad (6\text{-}12)$$

where $\tau = L/Z$, the characteristic time constant.

When the resistive term is added, there are two ways of handling the problem. One, a la Martin, is simply:

$$\tau_{TOT} = \tau_l + \tau_R \quad (6\text{-}13)$$

Spark Gaps

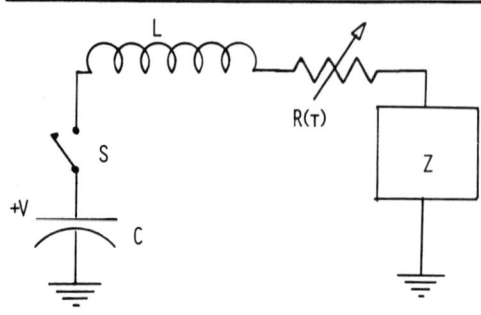

Fig. 6-15. Spark-gap model.

Another method falls back on the definition that:

$$\tau = \frac{L_R}{Z} \quad \text{so} \quad L_R = \tau Z \tag{6-14}$$

This converts the resistive term to an equivalent inductance that can be added to the geometric inductance. These should be added in quadrature so that:

$$L_{eff} = \sqrt{L_{geo}^2 + L_{res}^2} \tag{6-15}$$

leading, finally, to:

$$V_L = V(1 - e^{-t/\tau'}) \tag{6-16}$$

where:

$$\tau' = \frac{L_{eff}}{Z} \tag{6-17}$$

As an example, return to the 100 kV gap operating at 100 kV. With a 5Ω load and 20 nH geometric inductance:

$$\tau_R = 3.8 \text{ nanoseconds}$$
$$L_R = 3.8 \cdot 10^{-9} \cdot 5 = 19 \text{ nH}$$
$$L_{geo} = 20 \text{ nH}$$
$$L_{eff} = \sqrt{400 + 361} = 27.6 \text{ nH} \tag{6-15}$$
$$\tau' = \frac{27.6}{5} = 5.52 \text{ nanoseconds}$$

The expected 10 to 90 percent time would be:

$$2.2 \cdot 5.52 = 12 \text{ nanoseconds}$$

Heating of the gap occurs in two fashions. One is simply the product of gap current and the typical 150 V arc drop. This might be the major term in pulses of slow rise and long duration and is quite easy to estimate.

When the pulse is short, the situation becomes more complex. Here the energy loss during the resistive phase can become a significant term. This is especially true at high repetition rates or in some inversion generator circuits.

Because the voltage across the switch drops as (neglecting constant term):

$$V_S = V e^{-t/\tau_R} \quad (6\text{-}18)$$

It seems reasonable to assume the current rises as:

$$I_S = I(1 - e^{-t/\tau_R}) \quad (6\text{-}19)$$

The power is just V · I so that energy lost in the switch is:

$$U = \int VI = VI \int_0^\infty (e^{-t/\tau_R} - e^{-2t/R}) \, dt \quad (6\text{-}20)$$

$$U = \frac{VI \, \tau_R}{2} \quad (6\text{-}21)$$

This looks simple and might even be correct—if I can be determined. It is not limited by the switch, but by external elements as well. In general, it is a complex term composed of an exponential, a sine wave, and nonlinear components.

In the example considered:

$$V = 100 \text{ kV}$$
$$\tau' = 5.52 \text{ nanoseconds}$$
$$Z = 5 \text{ so } I = 20 \text{ kA and}$$
$$U = \frac{10^5 \times 2 \times 10^4 \times 5.5 \, 10^{-9}}{2} = 5.5 \text{ J}$$

This result is optimistic because τ' was based on only the switch 20 nH and did not include the capacitor and interconnect terms.

Airesearch, in an analogous calculation, assumed that I was approximated by a ramp of the form:

$$I_{(t)} = \frac{dI}{dt} \cdot t \quad (6\text{-}22)$$

using the maximum dI/dt expected, leading to:

$$U = \int_0^\infty \frac{dI}{dt} V e^{-t/\tau_R} t \, dt \qquad (6\text{-}23)$$

$$U = \frac{dI}{dt} V \tau_R^2 \text{ (not } \tau') \qquad (6\text{-}24)$$

As yet, good experimental data and model agreement are best described as deficient.

6.1.6 Erosion

The major life-limiting mechanism of spark gaps is electrode erosion. The EG&G data sheets specify the gap life in terms of coulomb transfer. This is because a fairly close relationship exists between charge transfer and metal removal. Even so, a factor of four is quoted for range-of-charge transfer. This leads to the conclusion that more than charge transfer is involved. Actually, two mechanisms are involved. One is from bombardment by charged particles. This is a heating effect and is very sensitive to current density and gas composition. When peak currents and rates of change of current are minimized, erosion rates of 10 micrograms per coulomb are typical for copper.

At very large currents or at large values of dI/dt, tree-shaped eruptions are observed that consist of plasma jets of electrode material, and erosion rates increase rapidly. The mechanism is primarily of dI/dt origin, related to the rate at which the current can spread and consequent levels of local heating.

A number of materials for electrodes have been tested for life properties. Steels, brass, and copper are not as good, in general, as molybdenum and tungsten. The best seem to be copper-tungsten alloys, such as Elkonite 10W3 (57 percent W; 43 percent Cu) or Elkonite 30W3 (66 percent W; 34 percent Cu).

It might be noted in passing that Glasoe and Lebacqz quote erosion rates of about 1 microgram/coulomb for tungsten in air with 10 to 20 times this rate in oxygen-free atmospheres.

6.2 LIQUID SWITCHES

A second type of dielectric used in gaps is a liquid. These are also fluids, capable of being flowed and also have self-healing properties between closures. The generally higher dielectric strength permits closer gap spacings with a lowering of switch inductance and rise time. Hydraulic forces arising from energy deposited in the arc channel can be troublesome. It is possible to build quite compact systems that include both switch and energy store or PFN in one closed system.

The dc self-breakdown of liquids is not as stable as for gases, due to entrained bubbles, impurities, and density gradients. Often an accessory system of filtering, deionizing, and pressurizing is needed. Even with these functions, liquid systems are usually employed only in pulse-charged systems.

One advantage of liquid gaps is the relative safety of high-pressure liquids with respect to gases. Also, the pressure required for the liquids are modest compared to the gas pressure employed in high-voltage, fast-closing gas dielectric gaps.

The spark channel in liquids is very narrow, which tends to produce enhanced electrode erosion. Particularly in cases where the dielectric is organic, carbon particles adhere to the anode so that the cathode erosion rate is higher.

The simple overvolted switch has been studied by several investigators. J.C. Martin reports jitter of 3 to 4 percent of the delay time with closure gradients of 400 kV/centimeter in oil and 300 kV/centimeter in water. These values were for microsecond pulse charge times. He also obtained multichannel switching in a 34-centimeter edge-plane geometry using CCl_4 at 400 kV. The byproducts of CCl_4 in this system were less than desirable.

A 200 kV multichannel water switch has also been reported. Rise time varied from 25 to 10 nanoseconds as the number of channels varied between 2 and 6. Because the load was only 1.6Ω, this is an extremely good closure performance.

A simpler, single-channel 3 MV gap using oil and a 5-centimeter spacing had a closure time of 40 nanoseconds. Lifetime of the switch was something over 100 pulses.

A rise-time relationship, developed by Physics International is:

$$\tau_R = \frac{5 \, p^{1/2}}{Z_0^{1/3} E^{4/3}} \tag{6-25}$$

which differs from the air gap relationships in that p is in terms of grams/cubic centimeter, because a liquid is involved. This relationship led to the choice of pressurized SF_6 instead of oil for a 5 MV switch.

Three-electrode field-distortion switches have been used in some applications. Water gaps to 3 MV and oil gaps to 5 MV are reported by Martin. These three electrode geometries have been attempted for repetitive operation with indifferent results. Dielectric purity is difficult to maintain. A rotary switch operating at 60 to 120 pps and 50 kV has been built that would operate for 2 to 6 hours with ±10-microsecond jitter.

To complete the spectrum of triggering techniques, both trigatron and laser radiation have been tried. Aksenov and others tested a 40 kV trigatron with water insulation. With the gap operating at 20 to 35 kV, a 22 kV trigger pulse was used. When the main and trigger pulse were applied simultaneously, the delay time was 60 to 80 microseconds. If the main pulse was delayed by about 80 microseconds with respect to the trigger gap breakdown, the switch delay was reduced to less than 10 microseconds.

Laser triggering calls for more powerful lasers than are needed for gases. The laser might induce breakdown in the liquid or at an electrode surface. A laser pre-ionized gap remains at low impedance for several hundred nanoseconds, so the trigger can precede the main charge pulse. Analysis of liquid gap behavior has been difficult due to lack of well-formulated theories of liquid breakdown.

6.3 VACUUM GAPS

The vacuum gap is unique in that, in principle, nothing is used for the dielectric. This avoids the problem of cooling the dielectric or of chemical decomposition in the arc. However it results in a very different sort of behavior, since switch closure must operate on a much different mechanism. One way or another, therefore, the vacuum must be spoiled by introduction of some charge carrier.

In the first place, a vacuum gap usually operates in the range of 10^{-5} to 10^{-7} Torr and some so-called vacuum gaps simply lie somewhere on the low-pressure side of the Paschen curve minimum. Vacuum gaps have a number of interesting properties:

1. Wide voltage range for a given geometry
2. High dielectric strength
3. Rapid recovery
4. High current capability
5. Low inductance
6. Quiet operation

As usual, there are associated problems. The two most significant are jitter and life.

Unlike most other gaps, voltage holdoff is not controlled by spacing, but by electrode surface. Since this varies from shot to shot, and with life, the self-break point is difficult to predict.

Performance of a vacuum gap can be impressive. J.M. Lafferty reports on a 150 kV, 100 kA design. The lifetime of the switch was limited to about a thousand shots due to erosion. The lifetime limit is usually not due to the insulators, even though a metal plasma is formed. The heating usually causes the insulators to form a vapor barrier at the surface, which prevents metallization. As a result, ceramic insulators are more likely to metallize than plastics or organics.

The outgassing due to the arc heating requires pumping to maintain vacuum. Pump oil can produce carbon deposits, which will cause internal tracking. As a result, either cold traps or oilless pumps are required. Best operation is with cathode triggering. The electrons freed are rapidly accelerated to bombard the anode. Anode triggering depends on ion bombardment of the cathode, a slower two-step process.

Plasma injection is used with lower pressure switches. There are not enough residual gas molecules to initiate an arc in such a gap, so the injected plasma meets this need. An injector, preferably in or near the cathode, is used and delays of 200 to 400 microseconds are typical.

Laser triggering has been employed, with a high-energy focused beam to generate clusters of ionized particles. Switching is polarity dependent. A negative (cathode) target gives about 1 microsecond rise time and a positive target about 100 microseconds. Cost and complexity have limited the use of such systems. Because the arc is in a vacuum, when the source is removed the plasma expands and cools rapidly. This excellent recovery characteristic is shown in Fig. 6-16.

Triggering is accomplished by somehow introducing some charge carriers into the vacuum space. These are readily accelerated producing ions by bombardment of the electrodes and result in the formation of a metallic plasma to close the gap.

One method is an internal spark, not too different from a trigatron. Most of the literature on this type of gap describes devices that operate in the 10^{-1} to 10^{-4} Torr range, not too different in pressure from a thyratron.

6.4 SOLID-DIELECTRIC SWITCHES

The remaining category is the solid-dielectric switch. Solid-dielectric switches offer the advantage of very low impedance and high current capability. These switches are

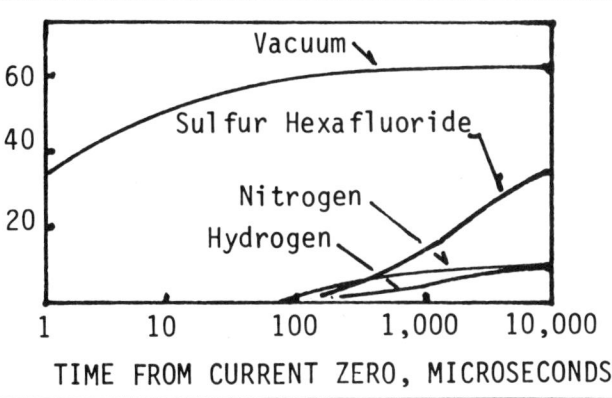

Fig. 6-16. Recovery strength for vacuum and gas. [2]

commonly used as dc voltages up to 100 kV, and operation has been reported at levels as high as 500 kV, pulse charged. A significant and obvious disadvantage to solid dielectric switches is their lack of self-healing and the necessity of replacing the dielectric after each event. Triggering of such switches is accomplished either by electromechanical puncturing for a microsecond jitter or by means of an exploding foil or wire situated between (and insulated from) the main electrodes to achieve modest jitter of tens of nanoseconds. In the latter case, not only the dielectric but also the trigger assembly, must be replaced after each shot.

The simplest form of solid-dielectric switch is one in which a sheet of polyethylene or mylar is pressed between a pair of electrodes and then overvolted. Rather erratic behavior can be expected from such a simple arrangement because the discharge path will usually be a flashover path, which defeats the whole purpose of the very narrow gaps and consequent low inductance inherent in the solid-dielectric switch. Although the flashover path could be increased sufficiently in principle to eliminate the flashover mode, it is more practical to overcome this behavior by purposely weakening the dielectric in one or more spots by stabbing it before insertion between the electrodes.

Figure 6-17 gives the self-break curves for three thicknesses of polyethylene that were partially punctured prior to insertion in the switch. The depressions, always near the positive electrode, were thus filled with air, which broke down on application of the high-voltage pulse, thereupon distorting the field in the polyethylene to breakdown. The air breaks down readily because its lower dielectric constant causes most of the initial stress to appear in the air pockets.

Stabbing can also be accomplished, though less precisely, in dc-charged switches. A deformation is caused by a remotely operated hammer or weight driving some sort of punch or tack into a thin copper or aluminum sheet electrode, eventually causing intrinsic breakdown in a very small volume. A solid dielectric switch belonging to this class can be a simple extension of the hammer-and-tack switch described above. In this switch, the deformation takes place between one electrode and a midplane trigger foil, or it might utilize one or more electrically or explosively triggered intermediate electrodes.

The simplest form of a three-electrode solid switch is shown in Fig. 6-18. An advantage of the midplane foil is the growth of numerous channels from the highly stressed

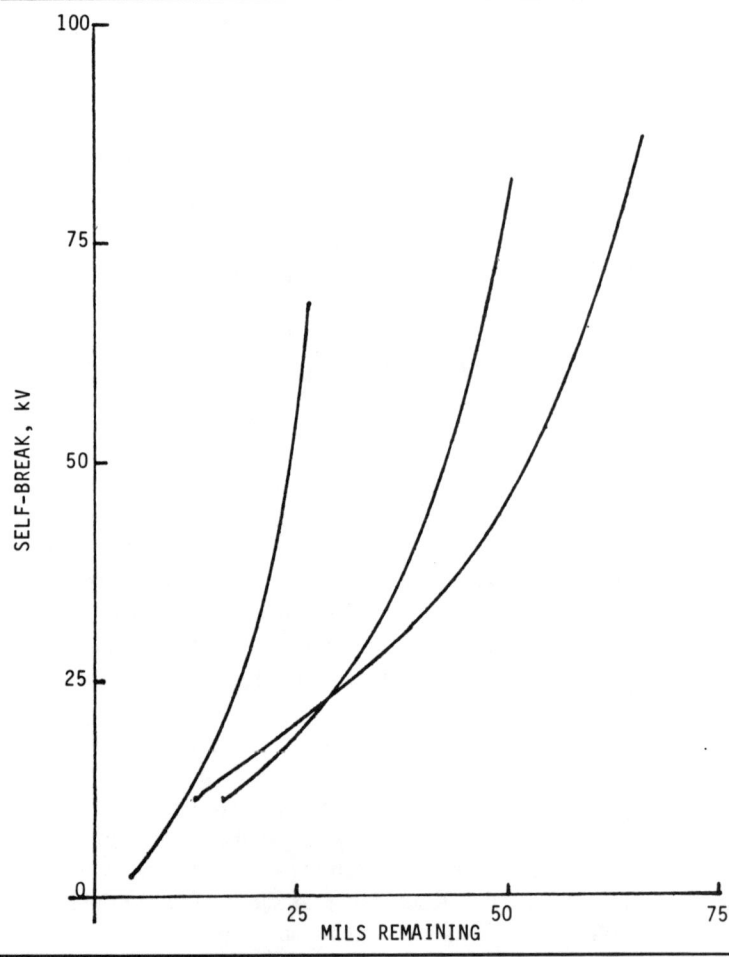

Fig. 6-17. Self-break curve for polyethylene switches.

edges of the grounded trigger foil. An experimental determination of the number of channels formed as a function of the main gap ratio implied satisfactory operation for $0.3 < q < 0.6$ where:

$q = y/(x + y)$
y = dielectric thickness between trigger and ground
x = dielectric thickness between trigger and $-V$

These tests were made at a constant 5 kV/mil for 4-mil and 6-mil mylar inserts. As the stress was increased in the range 3 to 6 kV/mil, the number of switch channels was observed to increase.

The same switch, triggered by a capacitively coupled 35 kV trigger signal rising in 20 nanoseconds, operated with a delay of 19.7 and jitter ± 10 nanoseconds when used to switch a 20-microsecond, 1Ω stripline charged to 50 kV.

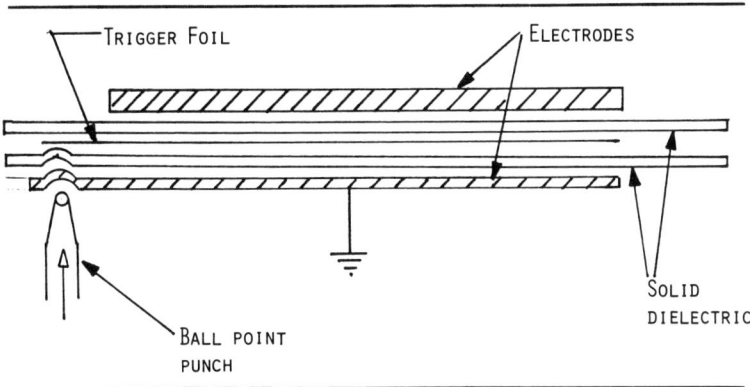

Fig. 6-18. Mechanical triggering method. [2]

A detailed account of solid dielectric switches reported between 1956 and 1966 is given by Komelkov.[2] Of the eight references cited, which describe triggering schemes such as exploding wires, explosive detonators, electrodynamic hammers and lasers, only one, attributed to Komelkov and Aretov in 1956, was of Soviet origin. Their switch utilized a ring of six explosive detonators to switch 1.4 MA at 40 kV through polyethylene insulation.

A high-power switch has been described in which the midplane foil (0.025-millimeter-thick copper) is separated from one main electrode by the principal dielectric, 0.5-millimeter-thick polyethylene, and from the other main electrode by a thin, prepunctured sheet of 0.075-millimeter-thick polystyrene. The puncture is a hole of $5 \cdot 10^{-4}$ square-millimeter area. Because breakdown of the switch is mainly attributed to explosive pressure generated in the vaporization of the trigger foil, a small puncture in the polystyrene can be considered more effective in concentrating the discharge on a small portion of the foil, thus facilitating vaporization. It is noted that the explosive pressure so generated is sufficient to puncture the polyethylene even when no voltage is impressed across the main insulation.

This switch was designed to satisfy inductance requirements $\cong 10^{-8}$ H and has closed successfully on pulses of up to 430 kA, passing 170 coulombs at 40 kV. Delay times in the microsecond range with jitters (standard deviation) of a few hundred nanoseconds were observed. Erosion of the main electrodes was distributed over the electrode faces by changing the position of the prepuncture in the trigger insulation. The triggering signal was directly coupled between the foil and main electrode from a capacitor discharge producing a 300 kHz signal of 100 kA peak at 10 to 20 kV.

Solid-dielectric switches can also be triggered by means of a focused high-brightness laser directed in a coaxial configuration as has been described for liquid and solid switches. This method has been demonstrated by Strickland, who applied dc voltages of 30 to 85 kV across gaps insulated with Lexan and Teflon 0.010 inch to 0.020 inch thick. Currents carried were 0.6 to 1.7 kA.

The particular breakdown phenomena that control the solid LTS are not established. The self-focusing of the laser beam due to nonlinearities in the dielectric as well as further energy deposition in the resultant plasma probably has some effect.

Strickland reported delay times as low as 2 nanoseconds with jitter of 2 nanoseconds. A transition point in delay was observed when the applied voltage was 40 kV across 10-mil Lexan. As this point, delay jumped discontinuously from 10 nanoseconds to the microsecond range. Multichannel switching was initiated by sandwiching a thin foil between two dielectric sheets. Laser-induced breakdown of the first gap caused an overvolting of the remaining gap, which discharged in numerous channels around the edges of the foil.

Relatively few materials have been widely used in solid dielectric switches. Of these, Mylar has the greatest dielectric strength and will therefore allow the shortest breakdown gap. For example, 10 mils of Mylar or 60 mils of polyethylene would be appropriate for a 150 kV switch. However, if low inductance is of prime importance, it should be noted that stabbed polyethylene can be made to break down in enough current-sharing channels to compensate for its increased thickness. In the event an automatic-feed system is desired for rapid replacement of spent dielectric, a flexible material such as polyethylene or Teflon is indicated. For applications requiring precise, machined thicknesses, clearly a rigid material such as Lexan or polystyrene is required. No mention has been found in the literature of Kapton, which has a dielectric strength about twice that of Mylar.

Solid-dielectric switches can be used wherever fast-rising, high currents at moderate voltages are required and sufficient time is available between pulses to replenish the dielectric and trigger foil (if used). The short gap spacings minimize inductance and have associated high field intensities, which reduce the resistive phase to a few nanoseconds. The resistive phase, t_R is given by J.C. Martin for solids of unit density (for example, polyethylene) by:

$$t_R = \frac{5}{Z^{1/3} E^{4/3}} \qquad (6\text{-}26)$$

where Z is the driving impedance in ohms and E is the applied field in MV/centimeter.

If precise synchronization with other prior events must be accomplished, an electrical or LTS triggering technique may be required. Synchronization with later events may be adequately achieved by means of a probe sampling the current rising in the switch. In the latter case, or when synchronization is unimportant, the simplicity of the overvolted stab switch or mechanically initiated switch can result in relatively substantial cost reduction.

6.5 REFERENCES

1. Burton, J.E. et al., "Multiple Channel Switching in Water Dielectric Pulse Generators."
2. Denholdm, A.S. et al., "Review of Dielectrics and Switching," Energy Sciences, Inc., Air Force Weapons Laboratory Report AFWL-TR-72-88, February 1973.

3. Dunbar, W.G., "High Voltage Design Guide for Airborne Equipment," Boeing Aerospace Company, AD-AC-29-268, AFAPL-TR-76-41, June 1976. See also W.G. Dunbar, "High Voltage Design Guide-Aircraft," AFWAL Aero Propulsion Laboratory Report AFWAL-TR-82-2057, Vol. IV, Jan. 1983.
4. Martin, J.C., "Nanosecond Pulse Techniques," Circuit and Electromagnetic System Design Notes (AWRE), Note 4, April 1970.
5. Martin, J.C., "DC Breakdown Voltages of Non-Uniform Gaps in Air," Dielectric Strength Notes (AWRE), Note 16, June 30, 1970.
6. Martin, J.C., "Multichannel Gaps," Switching Notes (AWRE), Note 10, March 5, 1970.
7. Neil, G.R. and R.S. Post, "Multichannel, High-Energy Railgap Switch," Rev. Sci. Inst. *49*, No. 3, March 1978.

Chapter 7

Thyratrons and Ignitrons

by
W.J. Sarjeant
(State University of New York at Buffalo)
and
D. Turnquist
(Impulse Engineering)

THE FIRST ITEM OF DISCUSSION IS THYRATRONS: THEIR USE, ADVANTAGES, AND DISADvantages. A *thyratron* can be defined as a ceramic or glass tube containing a heated cathode, a metallic anode, and one or more grids. It is filled with low-pressure hydrogen or deuterium (deuterium provides more hold-off voltage but a longer recovery time).

7.1 TRIODE AND TETRODE THYRATRONS

In the schematic of a tetrode thyratron (Fig. 7-1) notice that a baffle is located behind the grid to shield the grid from stray electrons emitted by the cathode. If there were no baffle, electrons from the cathode, emitted in a thermal distribution, would occasionally pass through the grid aperture into the anode region, causing avalanche multiplication and electrical breakdown of the gas in the tube. The baffle appears relatively larger in the schematic than it actually is. A hot cathode is positioned at the base of the thyratron. Independent of various construction geometries and materials, the cathode always must have a high thermal electron emissivity. A negative potential on the grid repulses the electron cloud and prevents the high-energy electrons from escaping to the anode and prefiring the tube.

The spacing of the anode to grid is derived from the Paschen breakdown curve for hydrogen. The electron collision frequency is lower on the left side of this curve, resulting in a small probability of a random breakdown in the device. The switch must have a short resistive-phase switching time (the time in which it collapses from a very high equivalent resistance to a very (low one), which must be balanced against recovery and forward hold-off requirements. This time is intimately related to the time it takes for the electron to accelerate from the cathode to the anode, which has a distance factor ($v_d \cong 10^7$ centimeters/second). The designer must use this closure time to specify a switching time for a device. Gas pressure and geometry are generally fixed to hold off the requisite voltage.

7.1 Triode and Tetrode Thyratrons 199

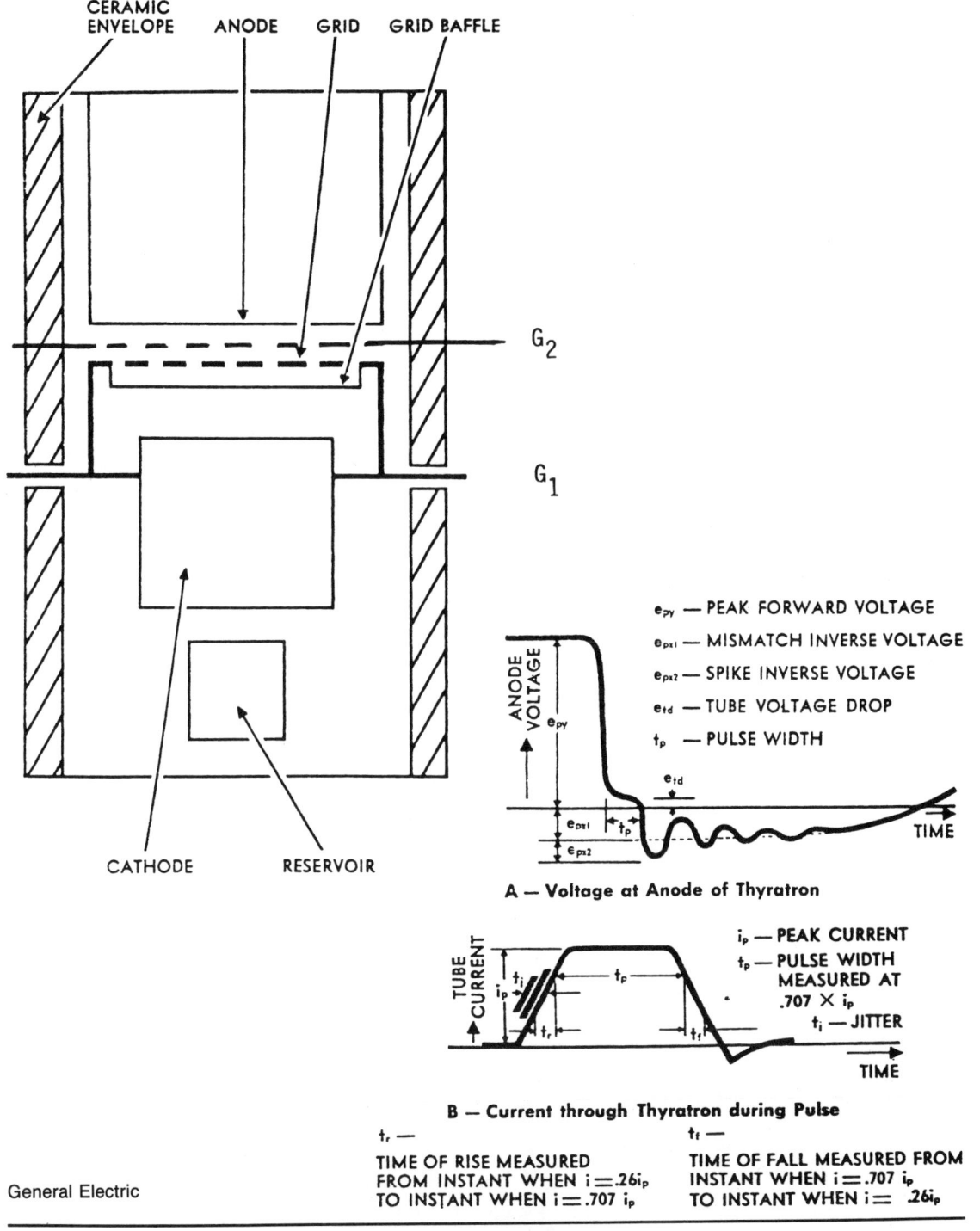

Fig. 7-1. Thratron geometry and pulse shapes. [1]

Extremely low-inductance tubes, constructed with small baffles, might be developed in the foreseeable future. There can be a large negative bias potential applied to the grid (if the device is pulse charged), allowing higher forward hold-off. The ratings are exceeded in experimental systems by more than a factor of two. An extremely deep potential well can be produced in the grid region using this technique. The well would permit none of the high-velocity end of the electron distribution to pass through the grid structure and on to the anode. Passing of electron distribution to the anode could cause breakdown.

A tetrode thyratron contains a second grid (Fig. 7-1). If there is a negative potential on this grid when the tube turns on, the electron concentration must build up from the initial electron number density to the space/charge limit ($\cong 10^6$/cubic centimeters), as $N_o e^{\alpha t}$. If N_o could be significantly increased by a very powerful trigger pulse or a large lower-grid forward bias, growth rates and switching times could be much faster. A positive potential on the lower grid generates an intense discharge (100 mA); a high negative bias on the upper or control grid (depending on the baffling, -100 to -1500 V) in conjunction with high-power triggering has given tube 1/e times of 2 nanoseconds or less.

When the tube has conducted and the current has fallen to zero, the electrons and ions should recombine to make the device electrically neutral and ready for another application of high voltage. If the anode goes suddenly negative, the positive ions remaining in the tube are drawn to it. If it moves strongly negative, avalanche ionization can occur because of the rapid acceleration of the positive ions into the neutral hydrogen. The tube will then arc back to a point on the anode, damaging the tube. In a triode tube, the arc tends to remain at the same point on the anode at high repetition rates and burns through the tip of the vanes of the cathode. In contrast, in a tetrode tube there is always a forward bias on the lower grid in the tube; the anode arcs, but the main discharge remains in a glow mode as it turns on. The forward-bias glow is reestablished very rapidly. When the tube arcs back, the anode still suffers some superheating at the arc contact point, but the cathode area is in the glow mode (this arrangement eliminates cathode superheating). Figure 7-1 schematically illustrates a true tetrode thyratron tube, drawn with the type of grid structure just described. There is no disadvantage in using tetrodes instead of triode tubes, except a slight price differential. Because the advantages in lifetime and switching speed always favor the tetrode thyratron, this type is preferable to triode tubes. There is a tetrode thyratron equivalent for every triode thyratron, although there are some internal geometrical variations among manufacturers.

At the bottom of the tube in Fig. 7-1 is depicted the titanium hydride reservoir of hydrogen. When the reservoir is heated, an equilibrium vapor pressure of hydrogen (or deuterium) is established. As the hydrogen is gradually absorbed by the ceramic and electrodes, a reservoir of the gas is required for replenishment. Tubes reach end of life when all the reservoir hydrogen is dissipated, and the gas pressure thus decreases, increasing the hold-off and precluding triggering.

Figure 7-1 also shows the voltage fall parameters. When the tube turns off, there is generally an inverse voltage. If this inverse voltage is excessive, the tube arcs back. The right waveform sketch defines most of the various pulse parameters. Note that rise and fall times can be specified in a number of ways, with the time from the 30 to 70 percent amplitude points being one of these. The parameter of most concern is the rate

7.2 GLASS THYRATRON CONSTRUCTION

Figure 7-2 shows a cross-sectional view of the construction of a glass triode thyratron. By very careful control of the dI/dt, a tetrode version (CX1159) of this type of tube was run at 30 times its P_{pk} rating for a considerable period of time. The only geometrical difference between glass and ceramic thyratrons is that the glass types have a long stem, making the heat conduction from the anode quite difficult. The grid is totally

of change of current through the system, dI/dt. The values P_B, which is given by the product of $e_{py} \cdot I_{pk} \cdot prf$, has been experimentally determined to be a poor indicator of true performance limitations of the switch.

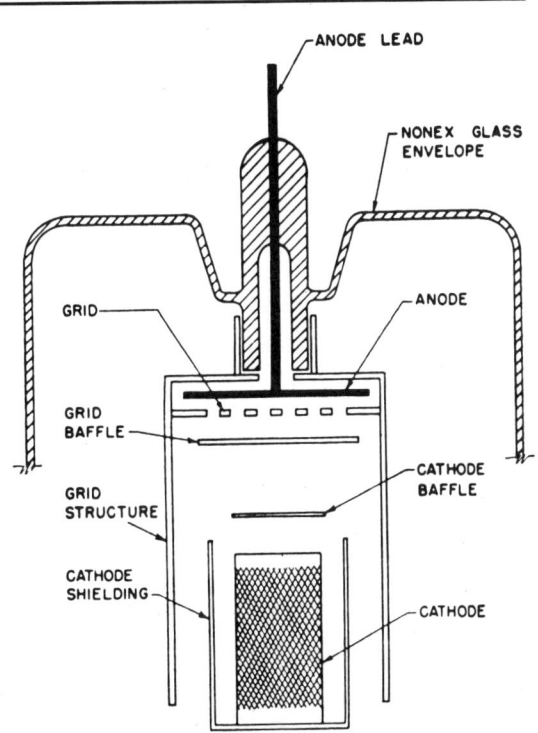

Anode–	Molybdenum, Copper
Grid–	Copper, Molybdenum
Cathode–	BaO, SrO, CaO, Coating on Tungsten
	Barium Aluminate Impregnated Tungsten
Gas–	Hydrogen or Deterium at approx. .5 Torr
Reservoir–	Hydride material, Titanium, Tantalum, etc.

Fig. 7-2. Basic thyratron configuration and materials summary (courtesy of S. Goldberg and J. Rothstein, "Hydrogen Thyratrons," Advance in Electronics and Electron Physics Vol. 14, Academic Press, 1961).

enclosed to prevent long surface discharges and arc-backs. Figure 7-2 gives the various components of the tube and a listing of typical materials for the structural elements. The construction of the anode is the most important factor. Despite some advantages, glass tubes cannot tolerate the high thermal loading necessary for high dI/dt circuits as the result of the long thermal path through the anode seal.

A typical radar modulator has several microhenries of inductance in the discharge loop. Older radar modulators often used large anode inductors to limit dI/dt to $\cong 1000$ A/microsecond. Glass tubes perform well in continuous duty if the dI/dt is low; if the dI/dt is high, the discharge can arc along the length of the metal stem, causing the tube to leak and fail.

Figure 7-3 indicates different types of cathode structures. In relatively low dI/dt circuits with multimicrosecond pulses, where the entire cathode is being used, there can be large peak current allowed (>2 kA/square centimeter). There is a propagation velocity limit ($\cong 10^7$ centimeters/second) of the glow into the cathode vane structure for a very fast pulse; thus, if the pulse is of a short duration, the discharge propagation can extend only a short distance. The cathode then can be made relatively short for short-pulse systems ($\cong 100$ nanoseconds) to improve thermal transfer capabilities and reduce inductance.

7.3 CERAMIC THYRATRON CONSTRUCTION

At the bottom of Fig. 7-3 is a sketch, drawn to a 1:1 scale, of the very first commercial ceramic hydrogen thyratron tube with all components labeled. The device, with a 25 kV hold-off, is obviously very small. The trigger pulse is applied to the grid, and the discharge rises through the grid baffle structure to the anode. The baffle illustrated is punctured; in other configurations, the discharge must bypass the baffle. The effective grid area is that essentially geometrically closed off by the baffle. Relations to determine device scaling limits for desired dI/dt and peak currents have been well-established in recent years.

7.4 THYRATRON SWITCHING IN PULSE-FORMING NETWORK TYPES OF HPE

7.4.1 Discharge Circuit

Figure 7-4 shows a typical HPE circuit, comprising a thyratron, a trigger generator that applies a positive pulse to the thyratron, a resonant charging network, a pulse-forming network, and a load. When the tube is triggered with a positive voltage pulse, the pulse starts the breakdown in the grid-cathode space, increasing the electron density to the level where anode-grid avalanche breakdown commences. Current increases through the tube until it is limited by the external inductance in the discharge loop of the circuit. (The circuits that have been used are limited by the external inductance.) In this application area, the perversity of the situation becomes acute; in fast-discharge lasers the rate of current rise is ultimately limited by the intrinsic inductance of the switch tube itself, the PFN capacitors, and the HPE-load electrical interface. The lowest inductance tube currently available (HY-5313) is only $\cong 10$ nH total inductance.

When the anode voltage falls in a triode tube, a remarkably powerful grid spike rises to a large fraction of the anode potential for 3 to 4 nanoseconds, and often damages the

7.4 Thyratron Switching in Pulse-Forming Network Types of HPE

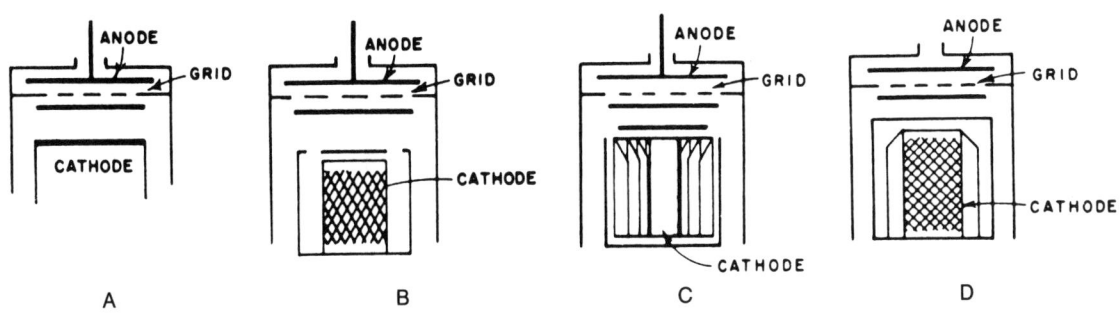

A   Plane parallel thyratron structures with different cathode arrangements.

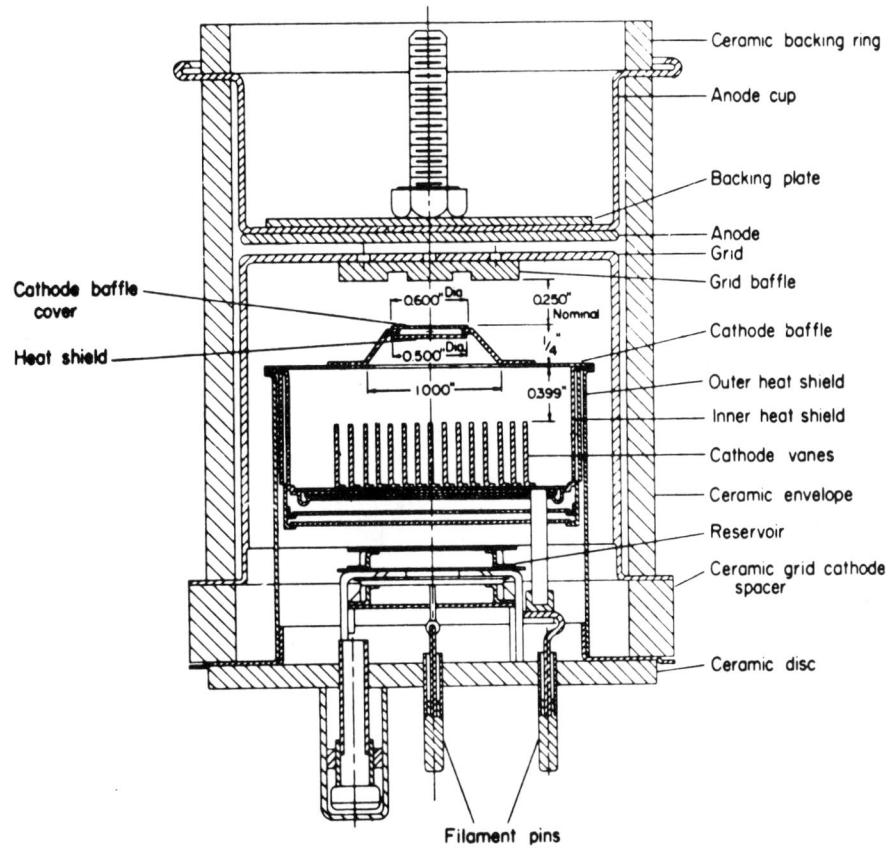

B   Hydrogen thyratron tube type 7322/1802.

Fig. 7-3. First commercial ceramic hydrogen-thyratron tube (courtesy of S. Goldberg and J. Rothstein, "Hydrogen Thyratrons," Advance in Electronics and Electron Physics Vol. 14, Academic Press, 1961).

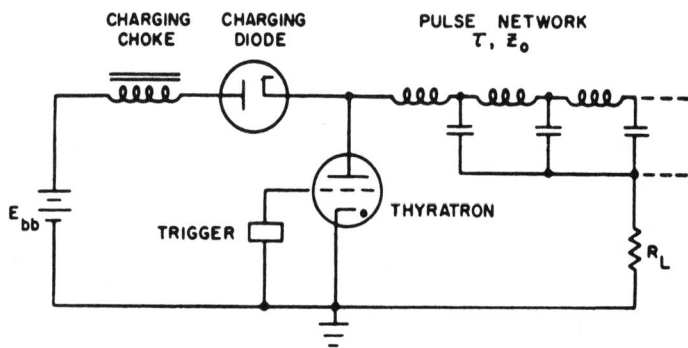

A Hydrogen thyratron pulse generating circuit.

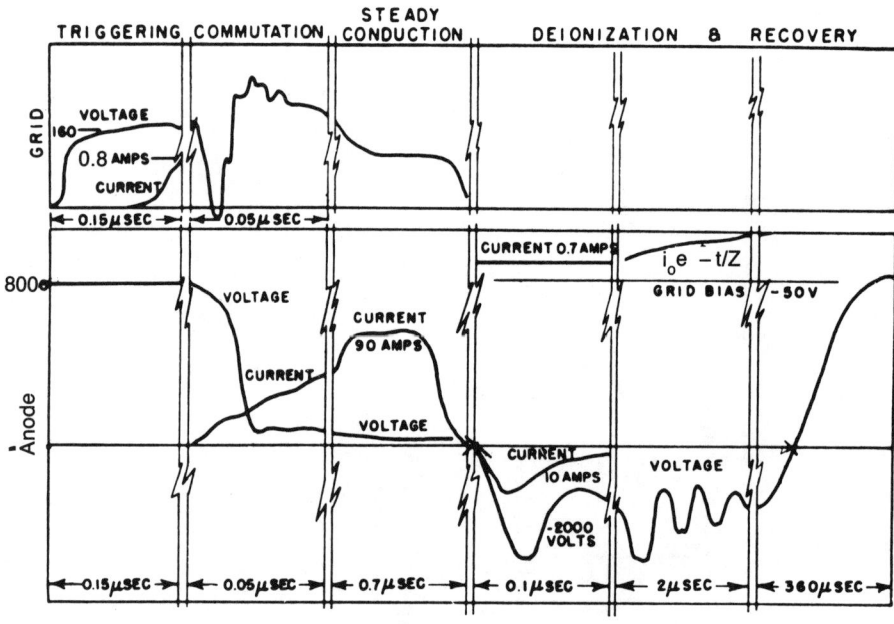

B Typical events in the pulsed operation of a hydrogen thyratron.

Fig. 7-4. *Typical HPE circuit and thyratron pulse shapes (courtesy of S. Goldberg and J. Rothstein, "Hydrogen Thyratrons,"* Advance in Electronics and Electron Physics *Vol. 14, Academic Press, 1961).*

trigger generator. A tetrode tube, because of its forward bias on the auxiliary grid, circumvents this with a constant impedance in the auxiliary grid region on the order of a few hundred ohms. A tetrode tube can even be driven with a solid-state trigger generator. When the current turns off, a small positive ion clean-up current of about 5 percent of I_{pk} remains. Recharging can begin after full recovery (Fig. 7-4B). The oscillations at the bottom of the figure are caused by the stray capacity of the tube resonating with the lead inductance; the oscillations can be dampened with an inductor/resistor network in the anode of the tube (an option not suitable for very low inductance excimer laser drivers). Recovery time is typically a few microseconds for small tubes. If the plasma is cut off between the cathode and the anode, recovery time is reduced somewhat. If a very large negative voltage is suddenly applied to the tetrode grid, the electrons will be repelled back to the cathode, and recovery time can be significantly shortened, particularly in large tubes (from 250 microseconds down to 20 microseconds or less). When the tube turns off, a pulse-transformer trigger generator places a very large inverse spike onto the grid, making the recovery time very short. If the negative anode voltage is excessive, an inverse diode across the tube becomes necessary.

7.4.2 Thyratron Breakdown and Recovery

Figure 7-5 shows how a tube actually does break down. The initial field distribution from cathode to anode is parabolic. Positive charge is injected into the grid region, and space charge penetrates toward the anode potential past the grid baffle. The bottom sketch indicates that in a space/charge limit with no trigger, baffles force thermal electrons to follow a particular path along which they cannot see the anode potential. A long path length and thus a long static pulse breakdown time (several hours) controls the prefire rate. To reduce the inductance problem in high dI/dt tubes, an open-grid baffle structure allows the largest permissible discharge cross-sectional area with a reduced hold-off time. An electron still cannot pass through directly with the cathode baffle there. The precise baffle configuration depends on the application of the tube, especially on the peak current per pulse the tube must transfer. If there is a great deal of current per pulse, the grid geometry causes a series of discharge columns. The magnetic fields exert a pressure on each glow column, and the round geometry eventually causes the current flow to pinch itself off at $\cong 10$ kA/ square centimeter.

7.4.3 Charging Techniques

The resonant-charging circuit with a diode (Fig. 7-4A) has the charging waveforms illustrated in Fig. 7-6. Whatever voltage is left on the PFN at the starting point, the voltage on the PFN rings up to a peak of $2E_{DC} + E_{inverse}$ designated $2E_{BB} + e_{pxi}$ in the figure. As the voltage starts to decay the diode becomes reverse-biased and decouples the PFN from the charging supply. The PFN is then discharged, and afterwards, during the recovery time, the thyratron suddenly turns off. As the repetition rates increase to case C in the figure, a net dc current builds up through the choke (linear charging) with severe heating of the choke. With the presently available solid-state charging circuits, the fast recovery times possible preclude the need to operate in this mode. For example, in a 1000 Hz system there is a millisecond interpulse period. Designs in large systems are routinely executed for 200 microsecond resonant charge times. High repetition rate switch thyratrons have been built that have demonstrated a reliable level of performance in high peak power applications with many kilowatts of average power at repetition rates in excess of 20 kHz.

7.4.4 Triggering

Thyratrons are triggered in the ways shown in Fig. 7-7. The third class of trigger (Type C) is preferable. If a thyratron is triggered with the first circuit, when that tube

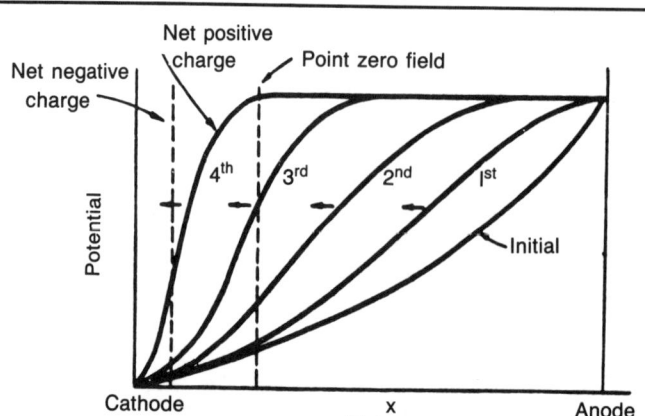

A Successive steps in the breakdown of a gas diode containing a thermionic cathode showing the development and measurement of the plasma front.

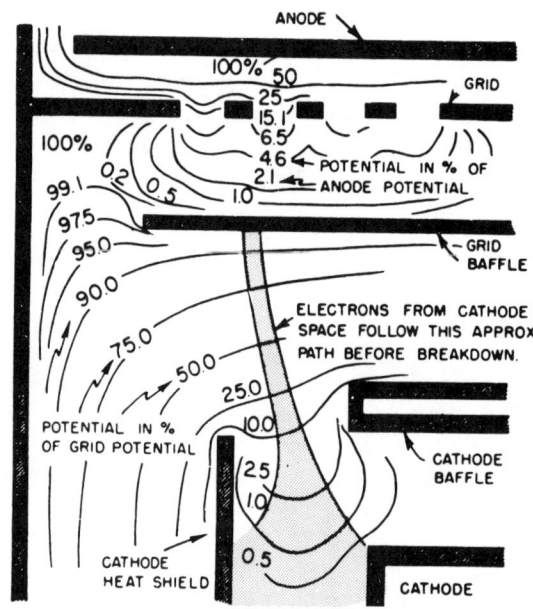

B Equipotential lines in the grid-cathode space and in the space between grid baffle and anode of 4C35. Lines are identified in percentage of grid voltage (in grid space) and in percentage of anode voltage (in anode space).

Fig. 7-5. *Thyratron breakdown time history (courtesy of S. Goldberg and J. Rothstein, "Hydrogen Thyratrons,"* Advance in Electronics and Electron Physics *Vol. 14, Academic Press, 1961).*

is turned off during the recovery phase, the grid-cathode impedance is determined by the resistor value. That in turn determines the recombination time in the grid-cathode region, because all the recombination current passes through that resistor. Circuit A operates well at low repetition rates and is very inexpensive.

7.4 Thyratron Switching in Pulse-Forming Network Types of HPE

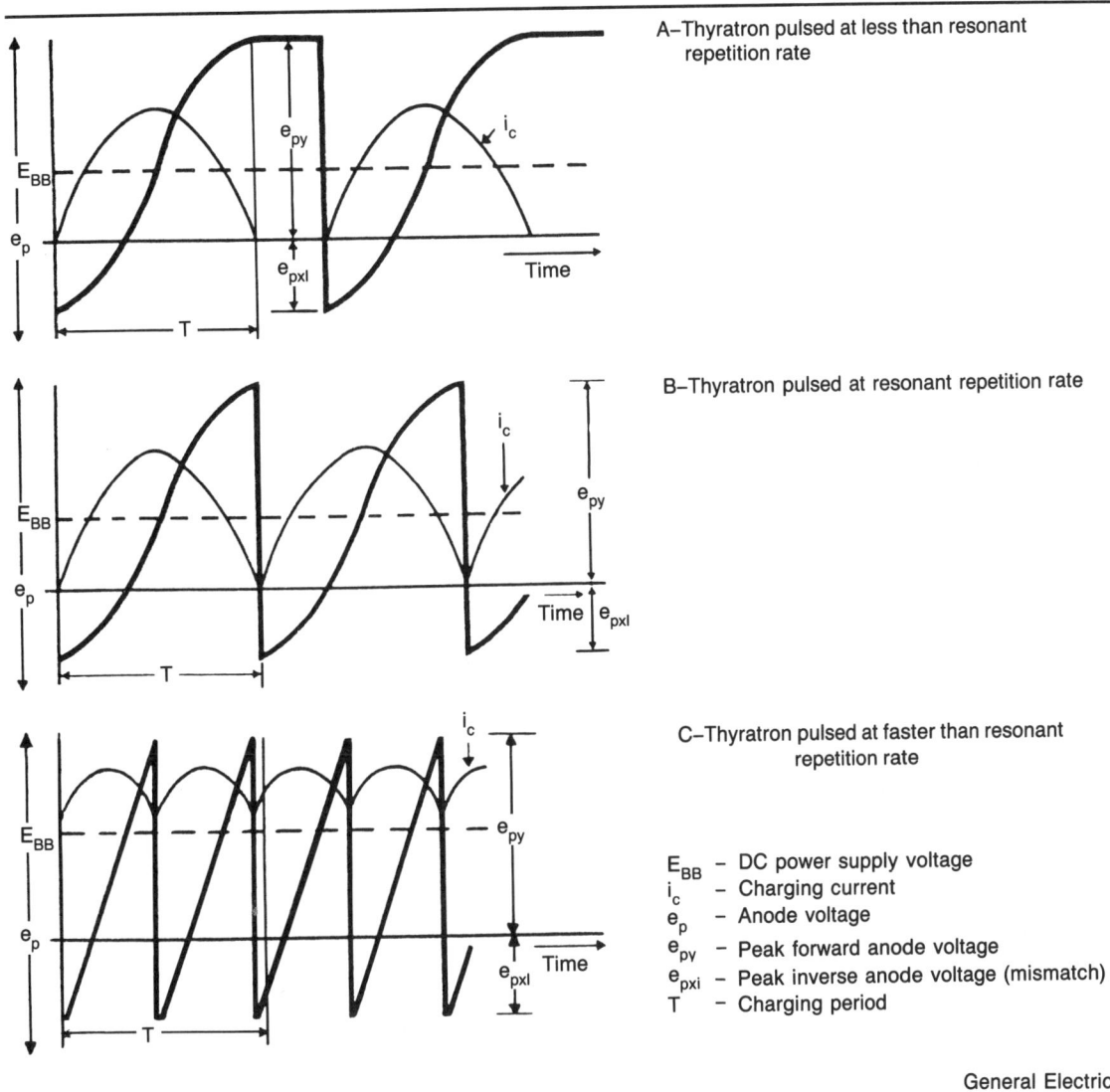

Fig. 7-6. Wave shapes of charging voltage and current. [1]

In the second circuit, B, a negative bias is applied to reduce recovery time. A 50Ω resistor provides fast recovery, unfortunately requiring considerable power to be dissipated in it.

The third circuit pictured is the preferred one. It can be shown that with this network (or a somewhat modified version), when the thyratron turns off, the collapse of the trigger load current gives rise to a very large negative spike on the grid, which reduces the tube recovery to a few microseconds (for megawatt average power tubes, this becomes $\cong 20$ microseconds just because of their larger size). The isolating resistor usually

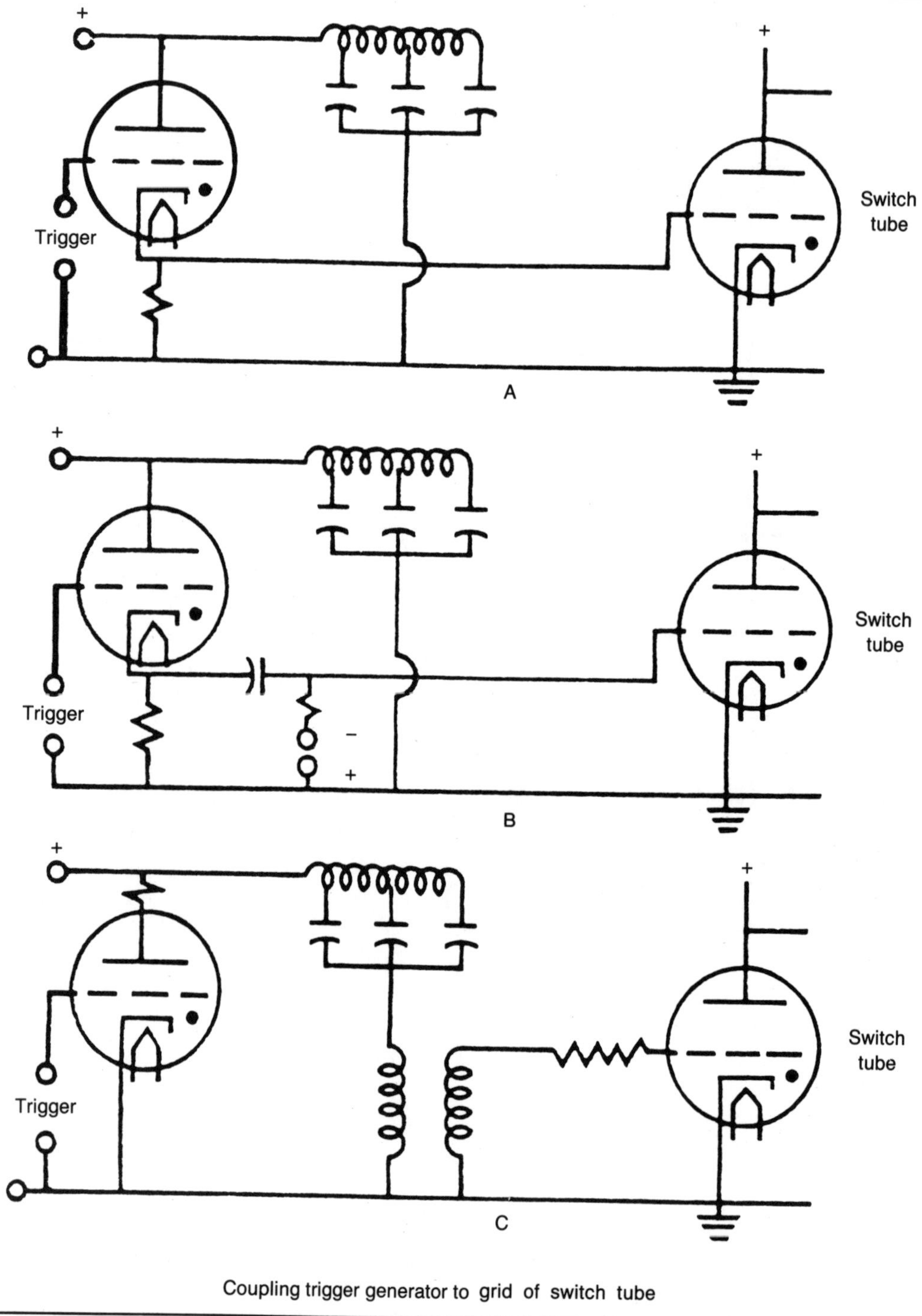

208   Thyratrons and Ignitrons

Coupling trigger generator to grid of switch tube

Fig. 7-7. Trigger-pulse shapes and pulse generators. [1]

7.4 Thyratron Switching in Pulse-Forming Network Types of HPE

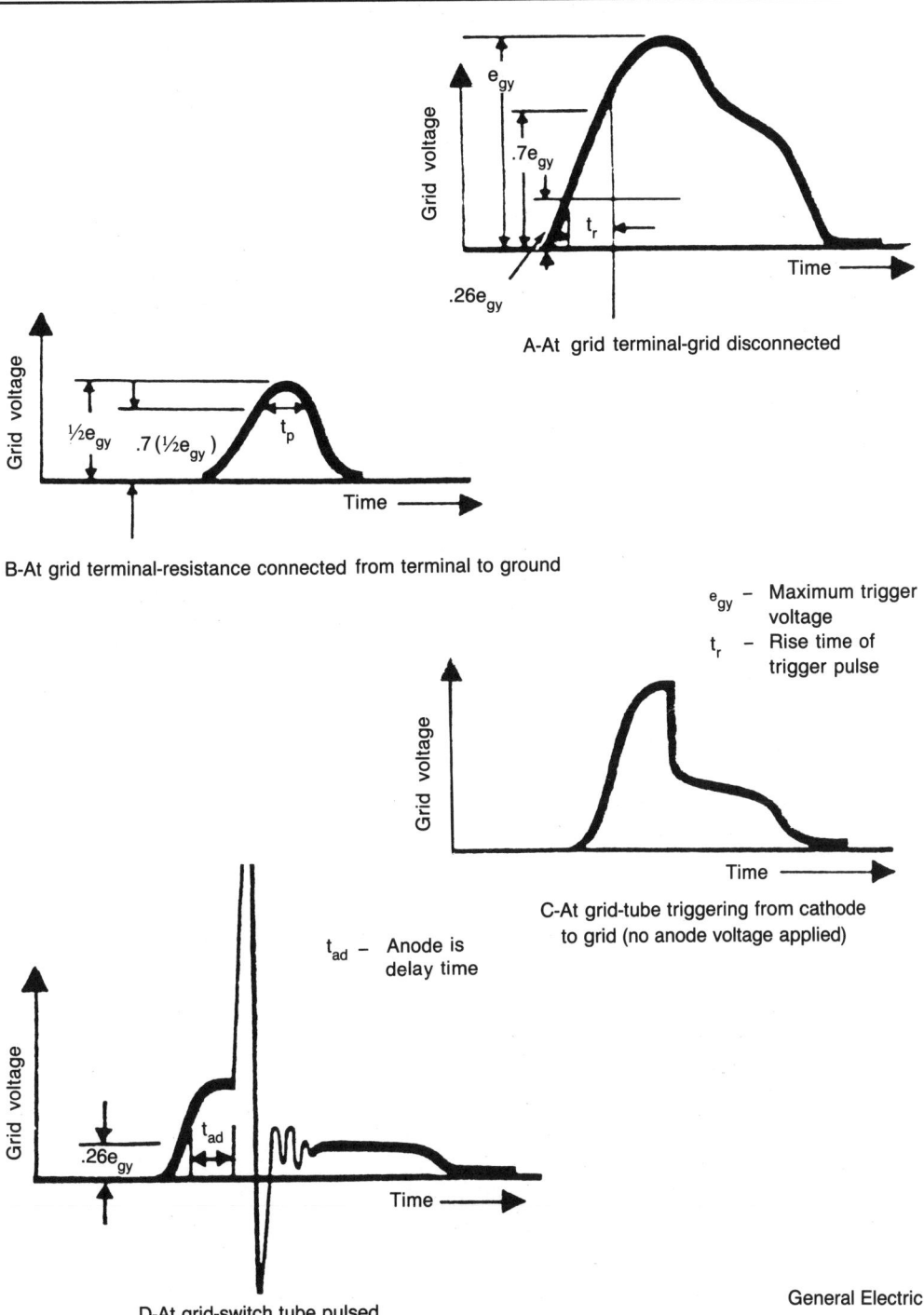

Fig. 7-7. Trigger-pulse shapes and pulse generators (continued). [1]

is a 10Ω, 10 W wire-wound resistor. Despiking elements are unnecessary in this circuit if a tetrode tube is used. For triodes, a small spark gap from grid to ground is advisable.

The transformer is available commercially (for example, EG&G, Inc.), and is designed to give a 1 to 4 kV trigger pulse into a 50Ω load. The trigger pulse width should be longer than the HPE discharge time to ensure that the tube remains on without interruption during the switching interval. In a current reversal mode, it is possible for the tube to turn off and then on, increasing anode heating, which reduces its lifetime as well as giving rise to the possibility of damaging internal arcing on restrike.

A graph of the grid waveforms for the triode tube is at the bottom of Fig. 7-7. The peak amplitude of the grid spike is between 30 and 80 percent of e_{py}. There is a delay time from application of the trigger pulse to when the grid breaks down. This can be made small by using a high dV/dt trigger pulse (same argument as in triggering a spark gap) and by using a tetrode geometry with auxiliary-grid forward bias to reduce the electron density growth time.

The most efficient means of triggering high dI/dt thyratrons is with transformers; the least efficient way to trigger high dI/dt spark gaps is with transformers.

7.4.5 Typical Application—Klystron Load

Figure 7-8 is a simplified diagram of a typical PCS using a PFN energy storage. For high-pulse fidelity in a pulse transformer, the turns ratio can rarely exceed 7:1. The load shown is a klystron with an impedance characteristic given by:

$$I_L = kV_L^{3/2} \tag{7-1}$$

The design operating point is normally selected with a reflection coefficient of slightly less than one resulting in a load impedance somewhat smaller than the PFN impedance (the voltage reverses on the PFN at the end of the PFN discharge time, preferably by about 5 percent or 10 kV, whichever is smaller). The load impedance being slightly less than the PFN impedance means that when the tube turns off, a small negative voltage remains during the recovery time, as shown in Fig. 7-9. Behavior with $R_L > Z_n$ is illustrated in Fig. 7-10; the curve clearly shows the long forward conduction period. After conduction, the resonant charging network begins recharging the PFN. A negative aspect of thyratrons is that the energy deposited into the tube can be substantial during the initial portion of the recovery period, for 0.1 to 10 microseconds while the tube is carrying current. Positive ion heating can account for 70 percent of the heating in the thyratron, causing internal sputter damage, and can drive hydrogen into the electrodes. This spike inverse voltage can be minimized by putting a snubber network on the open end of the transmission line or PFN, as shown in Fig. 7-11. The values R and C, empirically determined, are on the open-circuited end of the line. With that network installed, the inverse energy dissipation is reduced by a factor of two. Such a critical damping network is also highly recommended for any trigger circuit.

7.4 Thyratron Switching in Pulse-Forming Network Types of HPE

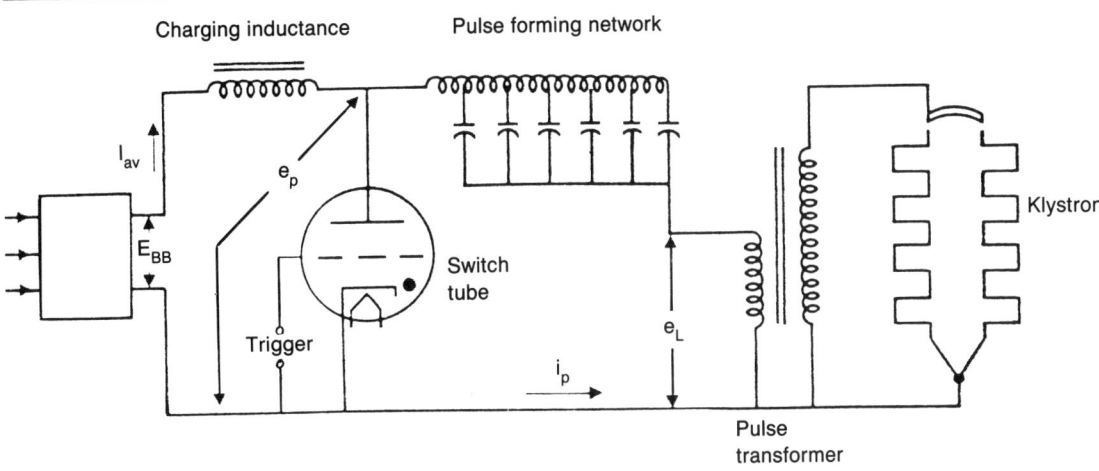

Simplified diagram of system using line type modulation

Discharge Circuit

During the charging cycle, the pulse-forming network appears as a large capacitance receiving energy from the power supply. During the discharge, the network is the source of energy. The closing of the switch causes the charged network to discharge. A constant voltage appears at the network terminals, while a voltage wave traverses the network and reflects backward to the terminals. During this time, a current flows in the discharge circuit. The magnitude of this current, neglecting losses, is:

$$i_p = \frac{e_{py}}{Z_o + R_L}$$

where i_p is pulse current (amperes)
e_{py} is voltage on network prior to closing switch (volts)
Z_o is network characteristic impedance (ohms)
R_L is load impedance (ohms)

For an ideal pulse-forming network, the network impedance appears as a pure resistance:

$$Z_o = \sqrt{\frac{L_n}{C_n}}$$

where L_n is network inductance (henrys)
C_n is network capacitance (farads)

However, since the network contains no actual resistive elements, no power is actually dissipated.

The current pulse is sustained in the discharge circuit until the reflected voltage pulse reaches the network terminals. This time interval, known as the pulse width, is:

$$t_p = \frac{\sqrt{L_n C_n}}{2C_n Z_o}$$

where t_p is pulse width (seconds)

General Electric

Fig. 7-8. Typical PCS using a PFN energy storage. [1]

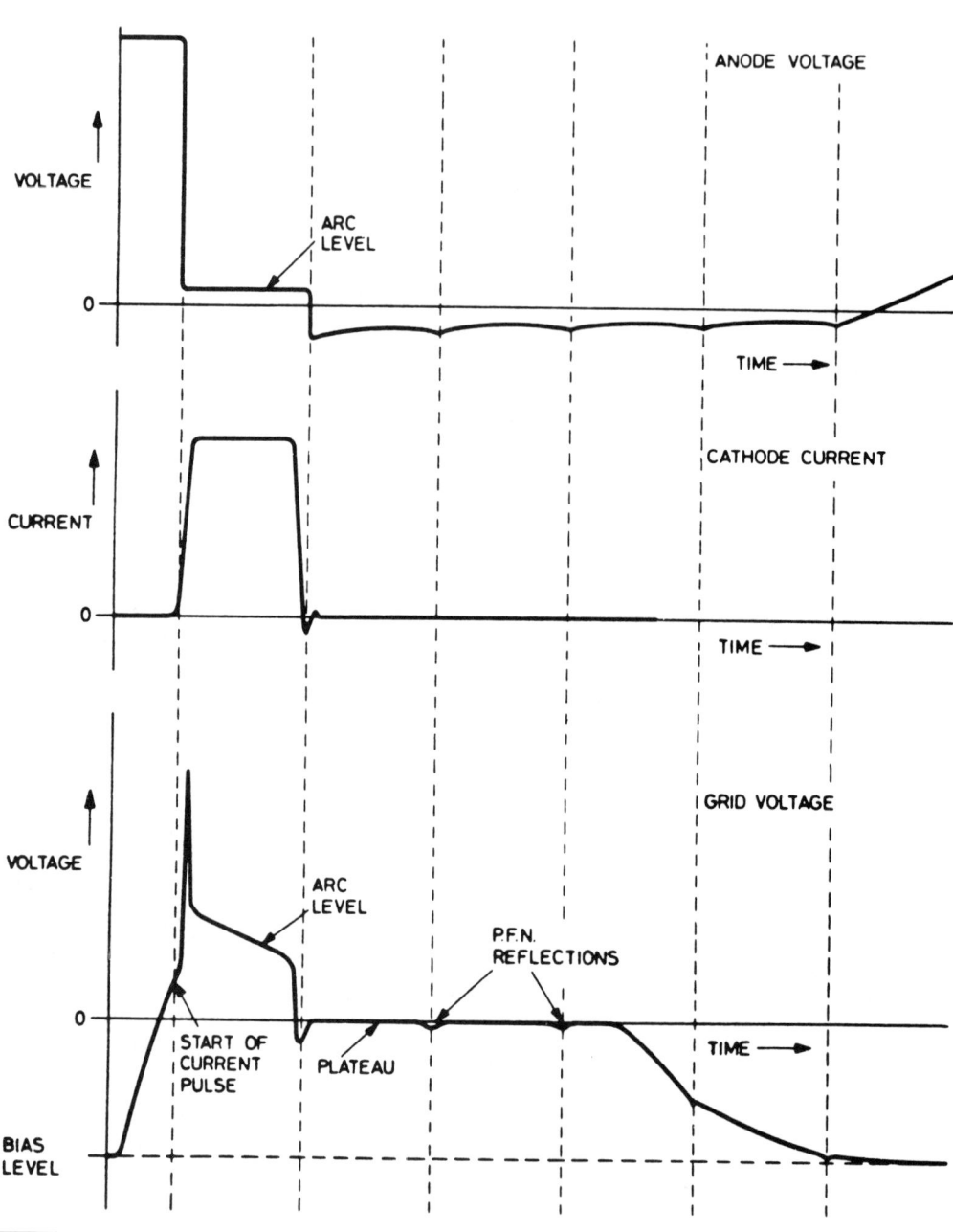

Fig. 7-9. Thyratron recovery for R_L less than Z_N (courtesy of EEV Inc., Hydrogen Thyratrons—Product Data).

7.4 Thyratron Switching in Pulse-Forming Network Types of HPE

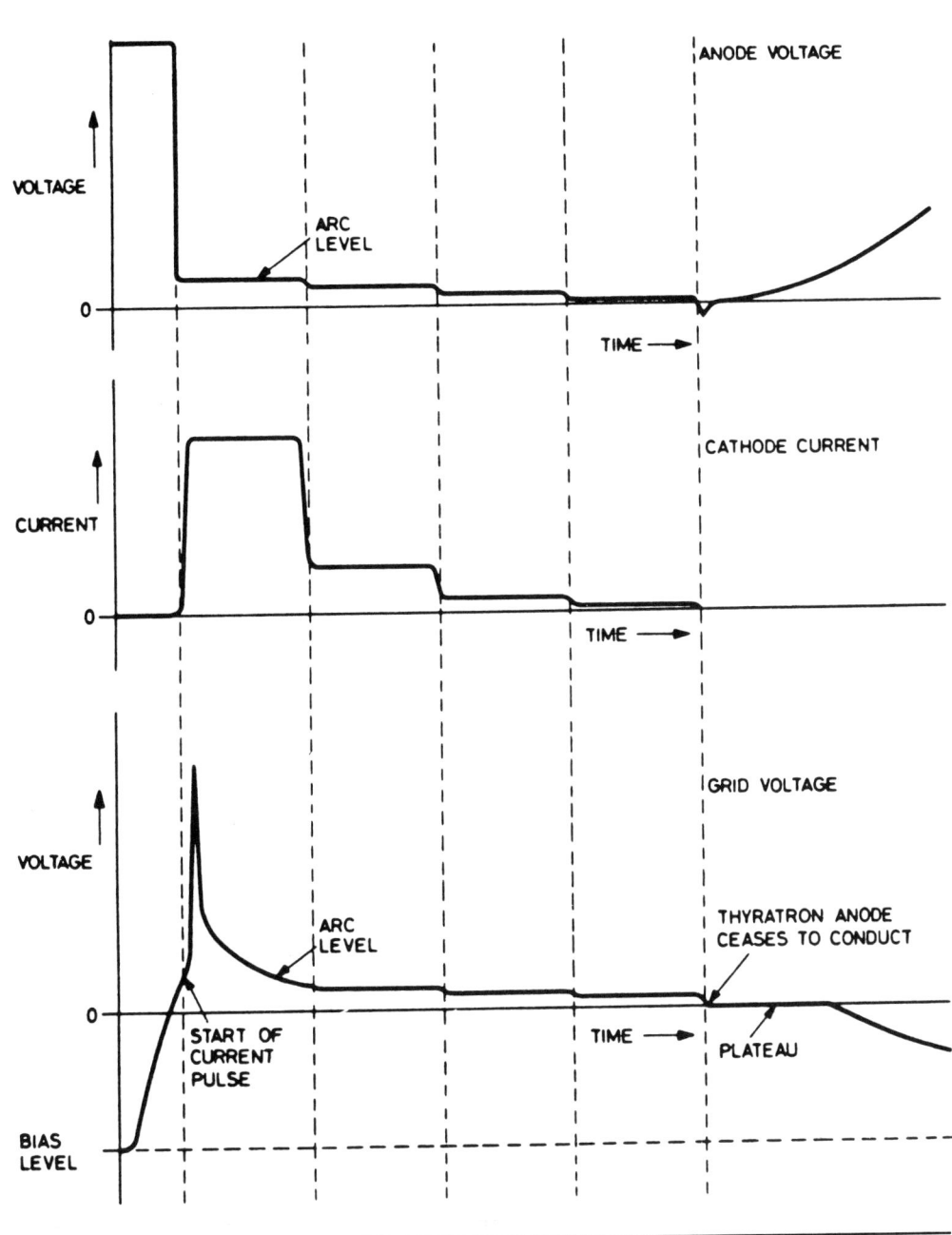

Fig. 7-10. Thyratron recovery for R_L greater than Z_N (courtesy of EEV Inc., Hydrogen Thyratrons—Product Data).

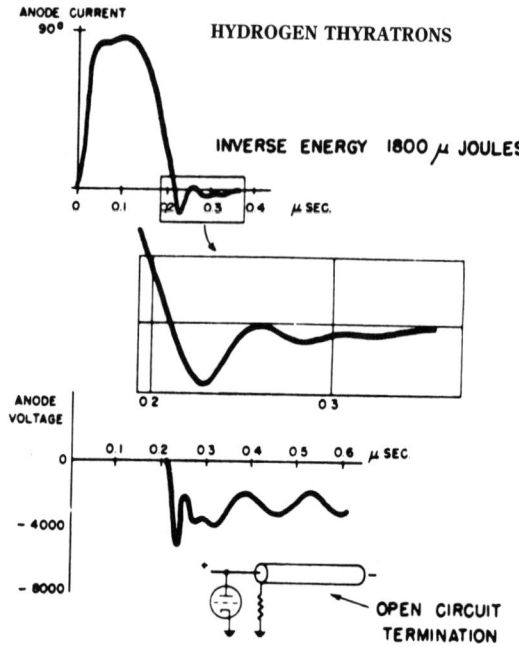

Measurement of spike inverse voltage and inverse current with a pulse network as normally used, that is, open circuited at the far end.

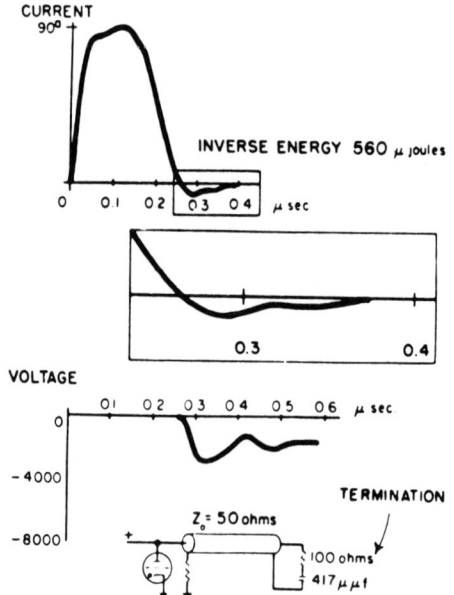

Same as above except that the pulse network has been terminated at its far end as shown.

Fig. 7-11. Damping of inverse-spike voltage with an RC snubber network (courtesy of S. Goldberg and J. Rothstein, "Hydrogen Thyratrons," Advance in Electronics and Electron Physics *Vol. 14, Academic Press, 1961).*

7.5 OVERVIEW OF COMMERCIAL AND DEVELOPMENTAL THYRATRON CHARACTERISTICS

7.5.1 Present Peak-Voltage and Peak-Current Limits

Burkes' Switching Report presents an overview of the tubes available up to quite recently, including developmental types. A 5 kA, 250 kV tube can be purchased. All of the ratings in Fig. 7-12 are for long-current (multimicrosecond) operation in pulse-modulator service. The tube on the right is the low-inductance one being developed for LANL for 20 to 100 kA, 50 kV short-pulse operation. The one at the 250 kV point is the tube used in the very large HPE developed at Fort Monmouth; the balance of points denote all the other commercial devices available. A somewhat more recent profile, which is a little different and more encouraging than this, will be presented later. Although Fig. 7-12 implies that 500 kA cannot be achieved at 250 kV, this is now recognized to be potentially possible, with development of larger device structures undertaken.

7.5.2 Pulse-Modulator Tubes

Figure 7-13 presents a tabulation of the characteristics of several pulse modulator tubes currently on the market, primarily for radar modulator service. These are standard tubes and meet military specifications. Most standard tubes last about 2000 hours at full average power, but some high-reliability types for weather radar applications are available that last much longer (> 15,000 hours). These are all pulse-modulator tubes

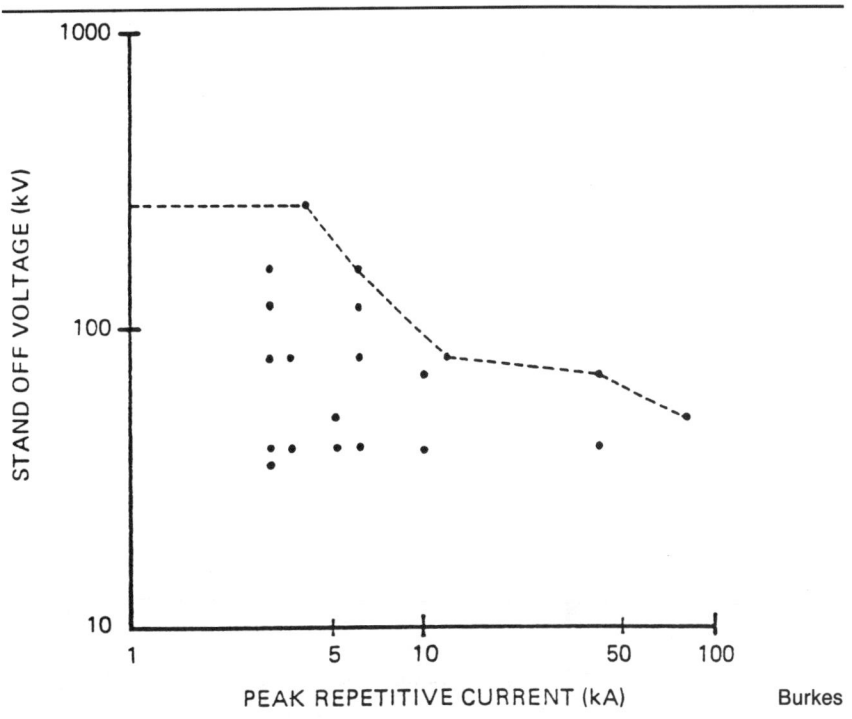

Fig. 7-12. Peak current-voltage profile of available thyratrons. [4]

JAN Type	7621	7322	8614	8479
Maximum Ratings				
Peak Anode Voltage, epy (kv)	8	25	40	50
Peak Anode Current, ib (a)	90	1000	4000	5000
Average Anode Current, ib (Adc)	0.1	2	7	8
RMS Anode Current, Ip (Aac)	2	36	125	200
Average Power Switched (kW)	0.4	25	140	200
Pb x 10^9 (epy x prr x ib)	2.7	20	160	400
Current Rate of Rise, di/dt (ka/µs)	1	5	5	10
Anode Delay Time, tad (µs)	0.4	0.5	1	0.3
Jitter, tj (ns)	10	5	10	2
Filament Power, Pf + Pres (W) Heater + Reservoir	22	176	288	484
Weight (lbs)	0.13	2	6.5	45
Seated Height (in.)	1.6	4.3	5.5	16
Diameter (in.)	1.15	3.3	4.5	9.5

Fig. 7-13. Characteristics of conventional pulse-modulator thyratrons (courtesy of EG&G, Inc, Electronic Components Division, Salem, Massachusetts).

and are generally identical, regardless of the source. They have relatively modest hold-off voltages, peak currents, and average currents.

7.5.3. Tube-Heating Factors

Consider the heating factor, P_b. The calculated P_b is often well below the tabulated limit when the tube is destroyed in high-peak current, short-pulse applications. The answer to this problem is to neglect P_b when calculating whether a tube is likely to operate safely in a given application. EG&G has determined that the correct scaling parameter, the Π_b factor, must be calculated from:

$$\Pi_b = \frac{dI}{dt} \cdot e_{py} \cdot \mathrm{prf} \tag{7-2}$$

EG&G made these calculations after examining numerous different tubes and dozens of applications. The results are consistent for all instances. At $1 \cdot 10^{17}$ the anode is warm and requires some air cooling. At $3 \cdot 10^{17}$, oil cooling is necessary, and the anode is showing a dull red color. At $200 \cdot 10^{17}$, high-pressure water cooling and large anode structures are required. (All these represent continuous-duty figures.) The switch will operate for many thousands of hours with a dull red anode. To exceed $200 \cdot 10^{17}$, the problem of the current distribution must be solved, and anode hot spots must be eliminated. A resistive anode medium with high thermal conductivity would be a potential solution to this problem. Years ago, in the design of one of the first TEA lasers, a cathode and a number of separate emitters were individually inductively ballasted. A thyratron tube that looks similar to this, with annular rings for the electrodes and rods protruding from a collector plate for the anode elements, all going to a second collector plate at the top, is conceivable. It could be vapor cooled, but it would be a difficult structure to align. Another possibility is a tube with a radial electron flow geometry, with cathodes in the middle, and with anodes on the side in layers. Calculations show the tube inductance could be reduced to about 1 nH this way.

7.5.4 Summary of Areas of Applications of the Hydrogen Thyratron

Figure 7-14 is a summary of different applications. The small nitrogen lasers operate at modest voltages and powers but require a short switching time and tube structures optimized for short (10 to 50 nanoseconds) pulse durations. The pulsed gas excimer *uv* lasers require 1 to 2 orders of magnitude higher peak currents to be switched at very high rates of rise of current. For large systems, the RMS anode currents increase significantly as desired average electrical average powers switched per tube approach 200 kW. Combining this thermal heat load with a low-inductance structure of long lifetime demands significant technology-base development. The adiabatic-mode thyratron switches operate for a few minutes at average powers of up to 1 MW and then are allowed to cool down over a long period of time. The extension of this tube geometry to multimegawatt average power levels, up to 500 kV e_{py}, would result in a dramatic engineering design capability for long lifetime, high-reliability power-conditioning systems applicable to both laser fusion and military applications.

	Small Lasers		Pulsed Gas UV Lasers	Isotope Separation (Proposed)		Adiabatic Mode	
	Target Desig	Nd Laser		LLL	LASL	Maps 250	Maps 40
Anode Voltage, epy (kv)	20-30	4	15-35	15-30	20-30	250	40
Peak Anode Current, ib (ka)	0.4-2	0.015	20-50	0.1-4	10-20	20	40
Average Anode Current, Ib (Adc)	0.005	0.015	0.001-0.1	1-10	≤10	2	50
RMS Anode Current, Ip (Aac)	<3	0.5	5-100	10-200	200-300	200	1400
Pulse Width, tj (μs)	0.2-0.5	0.1	<0.1	~1	~1	2	10-20
Repetition Rate, prr (kHz)	0.015	10	0-10	10-50	~1	0.05	≤0.5
di/dt (ka/μs)	2	1.5	<1000	~100	60-100	30-60	20
Average Power (kW)	0.05-0.1		0.001-1	20-100	50-200	500	1000
Major Limits or Problems	No heater power is a requirement		Switch inductance; Switch resistance	High prr; Recovery requirement; Unknown dissipation; Life	High rms current; High peak current; Life	High voltage; High peak current	No kickouts allowed; High peak current; High rms current; Heater power undesirable

Fig. 7-14. Summary of hydrogen thyratron applications (courtesy of EG&G, Inc., Electronic Components Division, Salem, Massachusetts).

7.5 Overview of Commercial and Developmental Thyratron Characteristics

7.5.5 Developmental Thyratrons

This is a profile sketch of some of the switch tubes developed over the past few years. The MAPS-40 tube has been run at modest repetition rates (<120 Hz), and Figs. 7-15 and 7-16 show what it looks like. It is a large tube (almost 18 inches high), costs $15,000, and is heavy. The cathode has a very high ratio (near unity) of emission area to unused area. It is estimated that this tube size could be designed to switch 3000 C in a single shot if it were properly triggered and baffled. The heater power is roughly several kilowatts; with additional water cooling, the tube is capable of operating at at least 0.8 MW continuous average power throughput.

Fig. 7-15. Photograph of the MAPS-40 megawatt thyratron (courtesy of EG&G, Inc., Electronic Components Division, Salem, Massachusetts).

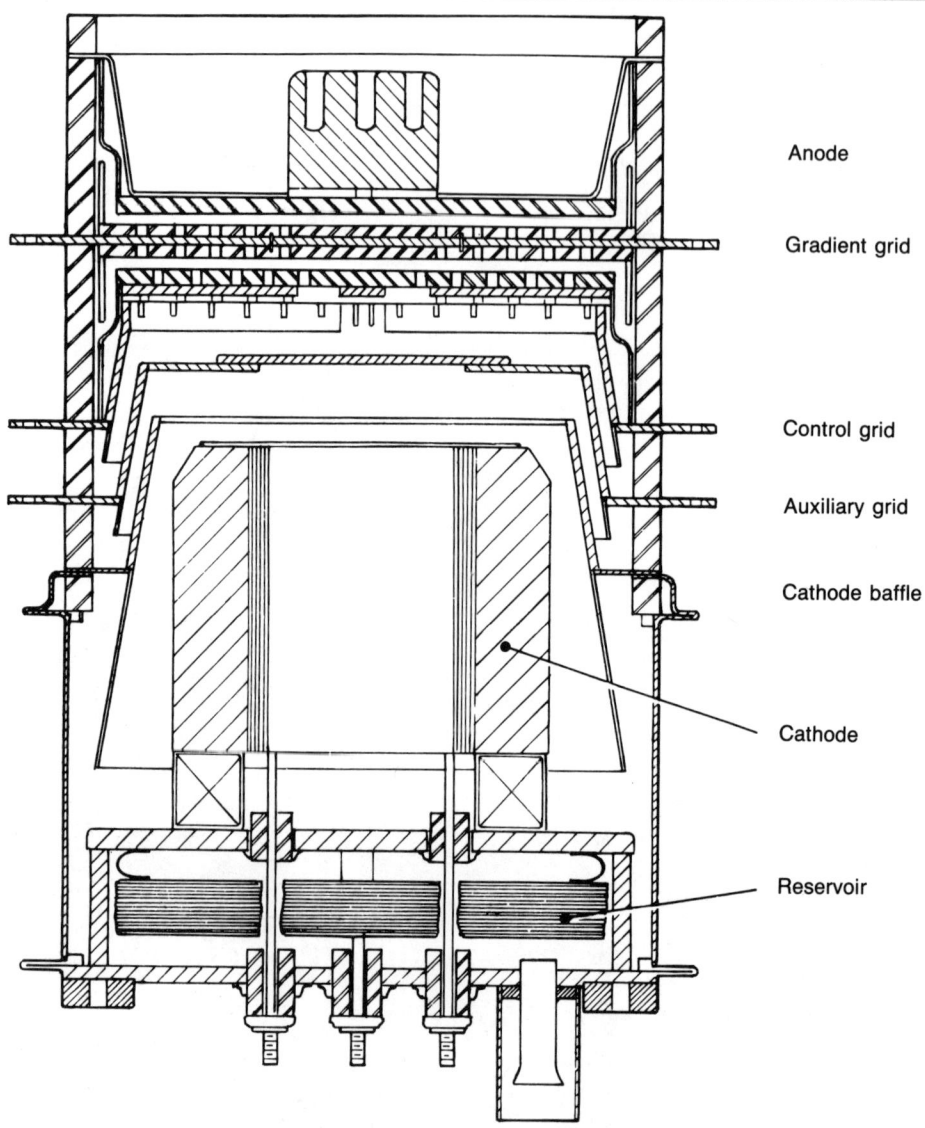

Fig. 7-16. Cross-sectional view of the MAPS-40 megawatt thyratron (courtesy of EG&G, Inc., Electronic Components Division, Salem, Massachusetts).

In continuous-duty experimental use only, the heater can practically be turned off once these tubes are started. If it is necessary to operate a tube several times over any of its ratings for a short period of time in an experimental apparatus, the cathode is brought up to temperature, the HPE is turned on at a modest repetition rate, heat loading power is increased to maximum for some 30 to 40 seconds, and then the heater is turned down. This procedure is allowable because the cathode designs used today are based on military applications, which demand a short heat-up time. A lot of unneces-

7.5 Overview of Commercial and Developmental Thyratron Characteristics 221

sary power thus flows into the heater. This power can be reduced by a factor of at least 10, at the expense of warm-up time, with heat-conserving cathode baffle geometries. All the side baffles shown in Fig. 7-16 are placed to prevent the sputtering of metal onto the ceramic walls of the tube.

The rather large distance between the control grid and the cathode in Fig. 7-16 reduces triggering requirements, but the impedance during the triggering time is relatively high (50 to 100Ω). It might be possible to build the same type tube with shorter grid-cathode spacings. The inductance of a foreshortened coaxial design could be quite low and the 10 kV or so needed to trigger the tube is not difficult to acquire today. For extremely low inductance applications, thyratrons can be mounted in a coaxial housing, grounded at the cathode baffle flange, as illustrated for the HY-5311 shown in its housing in Fig. 7-17. This creates an incomparable 30 to 40 nH switch that can operate in continuous duty service up to $\cong$ 100 kW average power.

It is pertinent to review the performance achieved for the MAPS-40. If the average current and voltage are multiplied, the result is 900 kW of average power at 120 pulses/second (Fig. 7-18). No other switch tube has ever done that. This demonstrates the potential average power capabilities for future, very fast switch-tube devices. The MAPS-40 is a rather fast device. Its inductance is 50 to 60 nH, and in this application it turns on in a few nanoseconds (the current rise time was circuit-inductance limited to $\cong$ 1 microsecond). Future applications might require a powerful switch tube for pulse

Fig. 7-17. Tetrode thyratron in a commercial, low-inductance housing (courtesy of EG&G, Inc., Electronic Components Division, Salem, Massachusetts).

Parameter (Units)	Specification Objectives: Rating	Specification Objectives: Operation (1)	Representative Performance: Full Power Test	Special Tests (1)	Special Tests (2)	Special Tests (3)
epy (kV)	40	44	44	40	36	50
ib (kA)	40	44	44	75	36	50
egy (kV)	1.5 to 4.0	—	2	—	—	—
tp (μs)	—	10	10	—	—	—
prr (Hz)	500	125	125	—	77	50
Ib (A dc)	50	50	50	—	20	24
Ip. (kA ac)	1.48	1.48	1.48	—	0.85	1.1
Pb (10^9 va/s)	400	242	242	—	—	—
dik/dt (kA/μs)	20	20	20/40	75	36	50
td (μs)	—	0.2	<0.2	—	—	—
Δtad (μs)	—	0.1	<0.1			
tj (μs)	0.02	—	<0.02			
Ef (Vac)	15±1.5	—	—			
Eres (Vac)	15±1.5	—	—			
If (A ac)	70	—	66			
Ires (A ac)	40	—	40			
tk (sec)	900	—	1200			
Life (pulses)	—	5×10^6	*	Highest Peak Current	Continuous Operation	Highest Voltage

*0.3×10^6 pulses achieved to date without discernible change in performance.

Fig. 7-18. Performance profile of the MAPS-40 megawatt thyratron (courtesy of EG&G, Inc., Electronic Components Division, Salem, Massachusetts).

durations of <100 nanoseconds. This present data base is important because it provides the baseline information for thermal heat loading, cathode capacities, and grid-heating structure that occur in a long pulse mode, which are directly applicable to short pulse length scaling of present devices.

This tube has run up to 36 kA in a 20 microsecond pulse at 120 Hz continuously (2), and might run for several thousand hours under these conditions. The peak current and peak voltage were reduced slightly from the adiabatic case (1) in that test. It is possible to operate up to 50 kV at lower dI/dt (keeping Π_b conserved), although some kick-outs may be anticipated over the long 20 microsecond pulse duration. The tube performance for short (<1 microsecond) pulses remains to be determined.

The heat-load calculations show that the cathode structure can operate at >0.8 MW continuous duty so that the cathode has an enormous capacity for average power. If a short-pulse device is required, the cathode height can be greatly reduced, while retaining all the other ratings. There is no need to maintain the original height because of the propagation time of the glow discharge into the vane structure.

In some of the early experiments, the molybdenum anode was drilled through by arcs at the ceramic-metal interface. This happened because the current was not symmetrically directed into the tube; it was taken into one side. The arc went up the opposite side of the ceramic and melted a hole in the anode. The point is that symmetrical current feeds are necessary in large systems to prevent asymmetric magnetic fields from forcing the discharge against the side of the tube.

Figure 7-19 shows a schematic view of the HY-4105, the 100 kV tube developed and operated at 80 A peak for 30 seconds for a total charge transfer of 2000 C. It has five sections, and a control grid and auxiliary grid above the cathode structure. This is a special cathode structure to allow it to sustain a current of 80 A for 30 seconds and recover to hold off 100 kV on the anode; the cathode would essentially be at ground potential. The gradient grids were differentially biased at 20 percent of the anode voltage. This was done by resistive and capacitive grading. A gradient grid denotes that at some point between the anode and cathode, there is an equipotential plane, which is perforated to allow the discharge to pass. It is fixed at a voltage level that tracks a fraction of the anode voltage. Imagine a tube that does not have horizontal equipotential lines at all points between the anode and the control grid. Inserting one gradient grid after another forces equipotential lines in this region. That provides a lower degree of field enhancement, so a higher overall anode-cathode stress can be supported for a given geometry. Without the gradient grids, there are few equipotential lines, and field emission occurs causing prefires. A tube with gradient grids that will hold off 100 kV can be built 2 ft. high in air or $\cong 10$ inches high in oil.

The drift-type grid cavities are essentially a stack of tubes in series with long drift spaces between sections. Almost all the high-voltage tubes have been built that way, allowing a long insulation length between sections to protect from flashover. A short, high creep-strength insulator (20 to 30 kV per inch) can be built like the MAPS-40, using the planar gradient-grid structure. A dramatic reduction in height is thereby possible. With the continuing development of new technologies that can control inside surface tracking, considerably higher stresses (approximately 50 kV per inch) might become possible in oil.

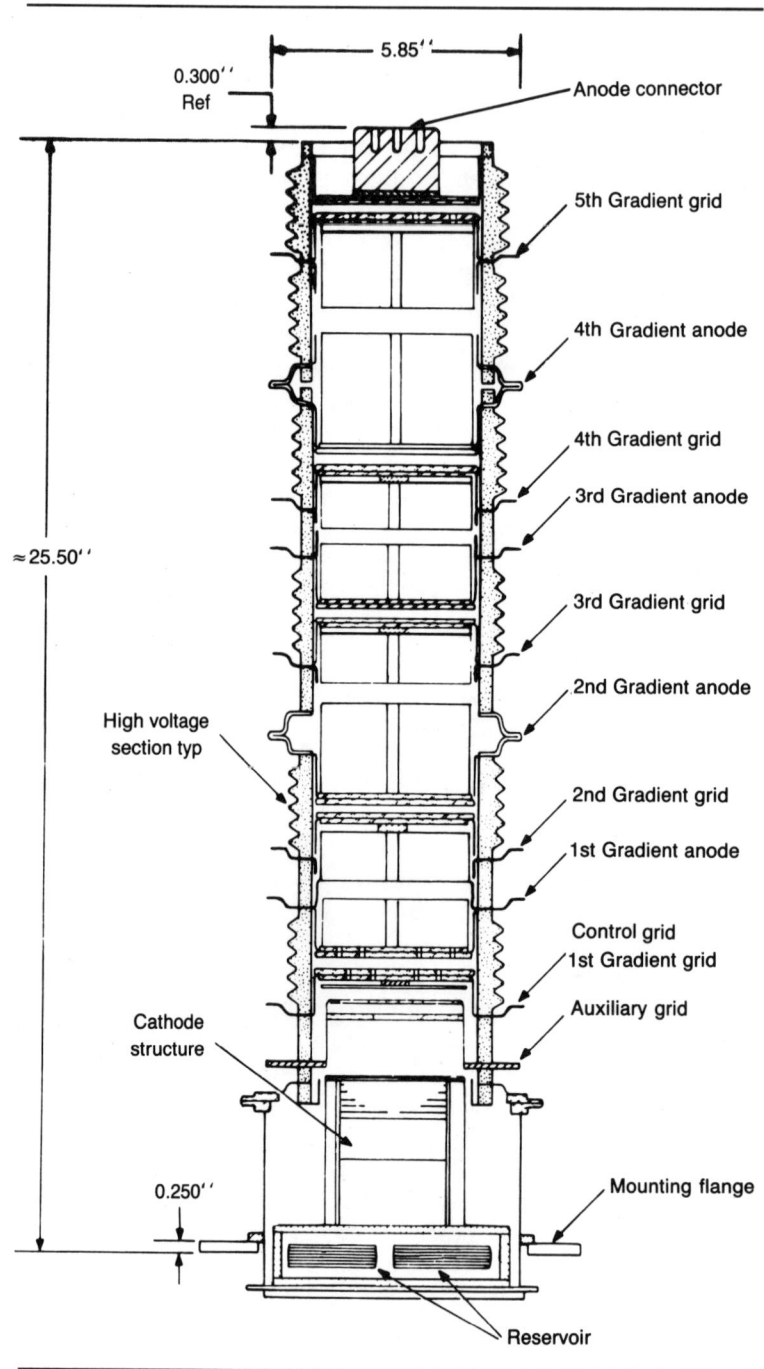

Fig. 7-19. Type HY-4105 long-pulse thyratron operating at 80A for 30 seconds with a 100 kV hold-off and total charge transfer of 2000 C (courtesy of EG&G, Inc., Electronic Components Division, Salem, Massachusetts).

PROGRESS IN THYRATRON CAPABILITY

Single-Stage
- 10^{12} a/s 45 kV

Multi-Stage
- Anode fall time reduced from 40 ns to 10 ns
- Isolation between stages from 0.40 to 0.95
- 40 nH, 150 kV tube built and in evaluation
- 11 stage tube built, with good triggering
- Operation at 10^7 a/s/v demonstrated
- Feasibility of 250 kV, 40 ka, 10 ns, 10 kHz, thyratron is established

OPERATING CONDITIONS FOR NEW THYRATRONS WITH EXTENDED RATINGS

TUBE TYPE	MAP 40 HY-7	CX-1171	MAPS 250	HY-4105	HY-5313
Voltage holdoff (kv)	40	80	215	100	45
Peak current (ka)	40	3	14	80	20
Pulse width (us)	10	2	5	2×10^7	0.06
di/dt (a/s)	4×10^{10}	9×10^{10}	2×10^{10}	—	10^{12}
Average current (Adc)	50	<0.25	3	20	—
Average power (kW)	1000 (burst) 10:1	— (also double ended)	325 (burst)	1000 (burst) 10:1	—

Fig. 7-20. Summary of ongoing development of submicrosecond pulse duration, high rate of rise thyrations.

Figure 7-20 is a summary of ongoing research on 50 to 250-kV, 20 to 100 nanosecond pulse width, burst-mode and continuous-duty devices at megamp/microsecond current rates of rise. The technology involved in the tubes discussed above is being applied to this problem. There has been a significant increase in repetition rates (>1 kHz) required in many short-pulse applications, so switch recovery times must be minimized. The dI/dt problem is almost entirely in the anode area. The cathodes are currently feasible, but the anode structures require further development. The cathode performance is space-charge emission limited. The current that can be extracted from a cathode for the time of the discharge is a function of thermal emissivity and total cathode area. That problem is being studied for short-pulse applications, and much higher peak currents may be possible. Anode and grid-baffle thermal damage are two of the factors that may limit switch performance at these short pulse lengths.

7.6 DEVELOPMENT OF HIGH dI/dt THYRATRONS

Some years ago, high-current pulse triggering of tetrode tubes was studied with the intent of increasing dI/dt, as, indeed, was observed (up to 400 kA/microsecond). A heavy current pulse was put in the auxiliary grid. A very large current pulse into the trigger grid somewhat later to test the forward bias effect. Suddenly, the current switch time decreased. The scenario is: the current builds up as $N_o e^{\alpha t}$ and is inductance limited—injecting the intense trigger pulse increases N_o and reduces the switching time. This effect is the basis for current fast-switch tube development.

7.6.1 Developmental Devices and Characteristics

Figure 7-21 shows the HY-3004 tetrode used in the first dI/dt experiments in a LANL program. It has an auxiliary grid, a grid cup, and a grid baffle. One is above the next, so there is no direct view from anode to cathode. Figure 7-22 illustrates the performance initially achieved in a coaxial test assembly using low-impedance coaxial cables, where dI/dt was in excess of 500 kA/microseconds.

Figure 7-17 illustrated a tetrode, large-cathode tube now available in the integral low-inductance housing shown. Finally it is possible to buy a thyratron that is engineered at the factory to work in its own low-inductance housing. The connected inductance is 30 nH or less. It will switch 20,000 A in a short pulse, and can be pulse charged to $\cong 50$ kV. The assembly is meant for liquid cooling. For sealed operation, the inside is filled with Dow Corning DC-200 fluid, and there are two connections (the auxiliary forward-biased grid and the trigger grid), with the cooling being achieved through the fins. The tube can be resonantly charged up to 35 to 40 kV and pulse charged up to 50 kV.

Figure 7-22 shows that when there is a large inductance in the switch-load interface (B), the current rises very slowly and the voltage falls very quickly. The circuit performance is inductance limited. If, on the other hand, the loop inductance is made very low, the current pulse looks like (A). Then the resistive phase time of the tube is the switching time-limiting factor, and the voltage fall time is quite long. The integral of the product of the voltage fall time and current rise time gives the joule heating in the tube structure. A new developmental tube has operated pulse charged to 50 kV, and demonstrated a single-shot dI/dt in excess of 1 MA/microsecond.

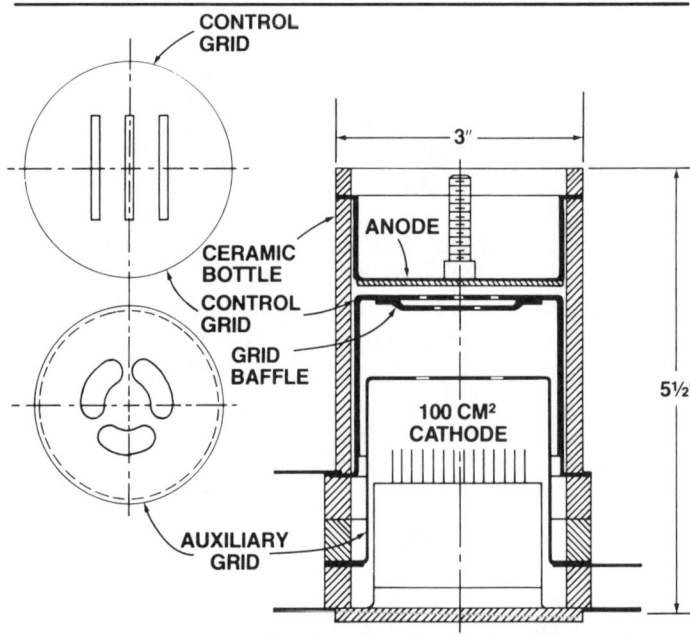

Fig. 7-21. Cross-section of the experimental low-inductance HY-3004 tetrode thyratron used in the first high DI/dt experiments at EG&G (courtesy of EG&G, Inc., Electronic Components Division, Salem, Massachusetts).

7.6 Development of High dI/dt Thyratrons

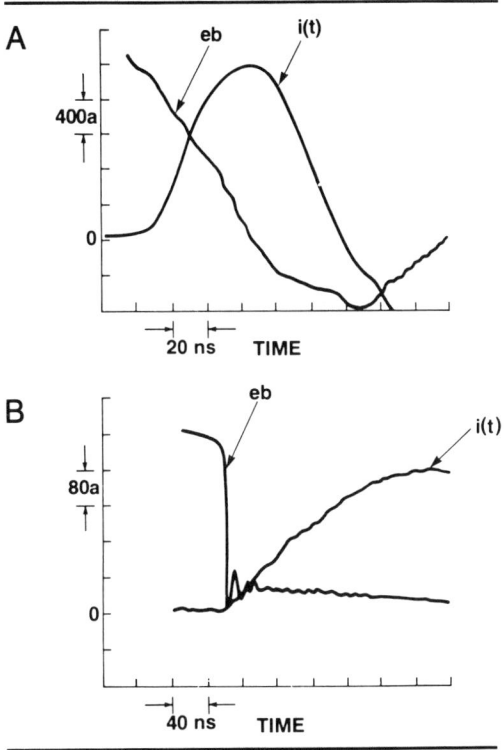

Fig. 7-22. Initial performance of the low-inductance HY-3004 tetrode thyratron (courtesy of EG&G, Inc., Electronic Components Division, Salem, Massachusetts). [5]

7.6.2 Generalized Switch Resistive Phase Fall Time Relation

Figure 7-23 is a calculation of the resistive phase fall time versus the pressure for many of the known switch devices. It shows actual performance. The LMPV is the liquid metal plasma value, and the resistive-phase fall time is in microseconds. A vacuum spark gap spans 250 to 500 nanoseconds. The cross-field closing switch (CFCS) is somewhat faster (30 to 50 nanoseconds), and the hydrogen thyratron covers 10 to 50 nanoseconds. The krytron is much maligned, but is a very good single-shot device, never intended for high repetition rates, with a resistive phase fall time under 2 nanoseconds. Spark gaps approach 0.5 to 1.5 nanosecond resistive-phase fall times, but can also be single-channel inductance limited in rise time. This is the first time this type of calculation has examined all types of glow and arc switches. The study was carried out by Bob Caristi at EG&G.

It is clear that other alternatives to create a fast switch system must be explored using conventional kilohertz switching elements. There are several other techniques to hold off electric fields with ferrites and such, which can aid in reducing apparent switching times by a factor of two to four. At very high repetition rates, the energies deposited into the ferrites may cause excessive heating and fracture.

Figure 7-24 is a plot of dI/dt per volt against device gas pressure. This is in units of amperes/second/volt for the switch element. If the inductance of the circuit is 1 nH,

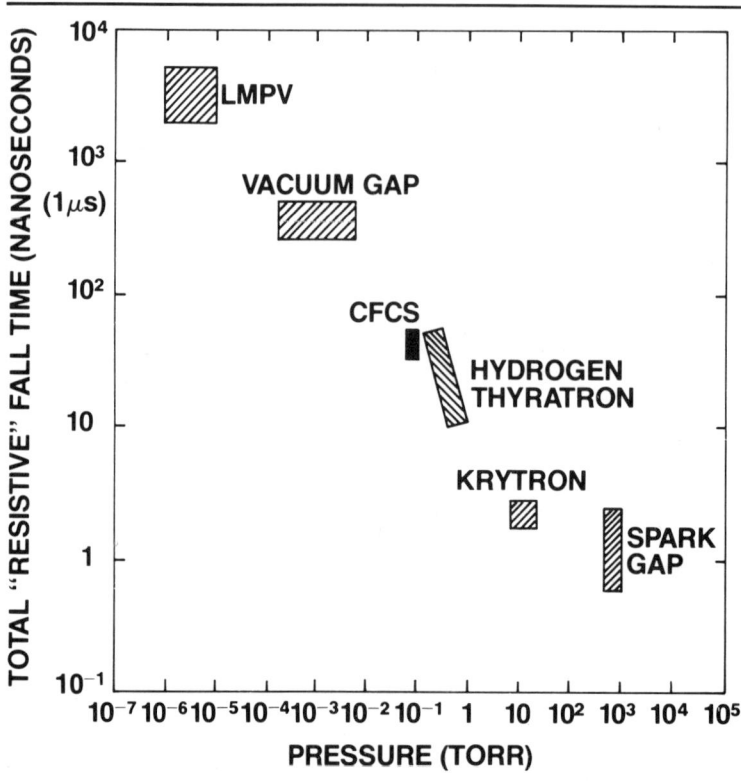

Fig. 7-23. Calculated resistive phase fall time against pressure for many of the known switch devices.

a certain switching performance can be achieved for a given device. The spark gap is far to the right. The resistive phase limitation of the device is to the left of the line, and the circuit inductance limitation is to the right. For example, if the loop inductance is 100 nH, it is impossible to achieve 10 MA/microseconds. This is a very useful graph.

7.6.3 Summary of Developmental Thyratron Parameters

Figure 7-25 is a summary of available specialty thyratrons and some projections in scaling up devices now under development. This is an up-to-date representation of switch parameters, and it gives performance criteria in realistic units for repetitive pulse widths up to 20 microseconds. The range in dI/dt and the rated inverse voltages that the tubes can sustain are listed self-consistently. For example, the HY-5313, is capable of 800 kA/microsecond, at 50 kV and 10 nH tube inductance. This is the latest tube produced in this family, and the total loop inductance in its housing is 20 nH. The device will switch 20 kA at 50 kW of average power, and it is available today. If it is desired to replace spark gaps in a megajoule impulse generator operating at 10 pulses/second, and the system has to last some 10^{10} shots, the table shows where the cathode technology can be scaled to. All the projected numbers here are reasonable estimates worked out with the EG&G staff. The large cathode tube is fairly fast with $\cong 100$ nH inductance, and

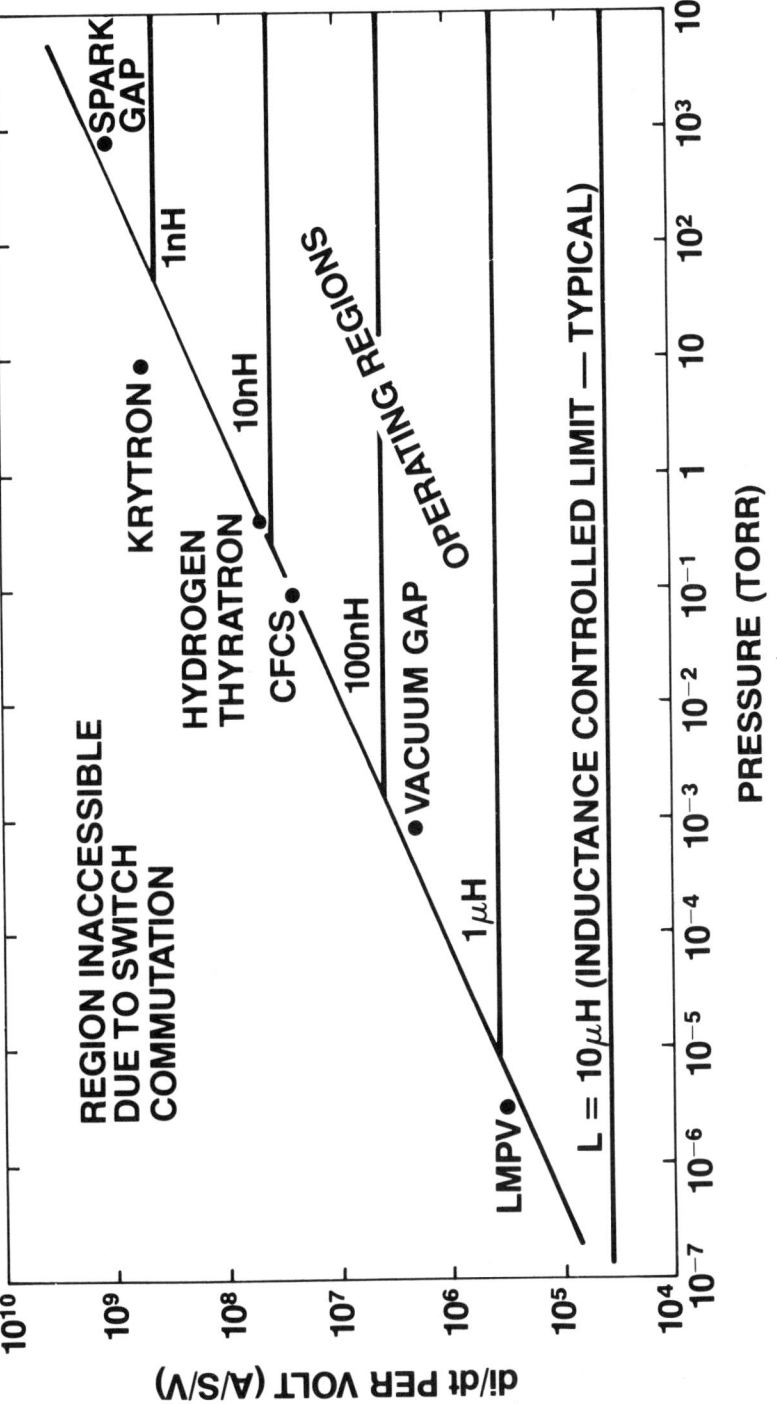

Fig. 7-24. Limiting dI/dt per volt as a function of switch commutation time and system inductance (courtesy of EG&G, Inc., Electronic Components Division, Salem, Massachusetts). [5]

Tetrode Thyratron Characteristics

	Standard Tetrodes	New Low-Inductance Design Resonantly Charged	New Low-Inductance Design Pulse Charged	Large Cathode Tube	Developmental Devices — New Very Low-Inductance Concept – Pulse Charged	Multiple Gradient Grid Tube	New Large Cathode Gradient Grid Tube	Long Pulse Tube
Representative EG&G tubes	HY-5	HY-53	HY-5313	HY-7		HY-5505		HY-4105
Resistive fall time to I/e, ns	4	3	2	4	<1	4	4	4
Inductance, nH	50	15	10	50	<2	60	<100	100
Voltage, epy, kV	1-70	1-35	1-50	1-40	1-100	1-250	>250	1-250
Peak current, Ip, kA, for pulse widths:								
(a) ≤100 ns	20	20	20	200	200	100	300	
(b) 1–20 μs	5	6	6	80	80	20	100	
(c) ≥20 s	8	8	8	20	20	15	100	80 A
Average power, kW	50	50	50	500 1000 (burst)	2000	2 500 (burst)	50	2000 Coulombs (once every 5 min)
RMS current	125	125	125	1600	3000	1000	3000	80
Average current	8	8	8	20 50 (burst)	20	15	100	80
di/dt, kA/μs	5	500	800	150	>2000	10	500	NA

NOTE:

Rated peak inverse voltage kV, for pulse widths: ≤100 ns: 25 kV
1–20 μs: 15 kV
20 μs–20 s: 1 kV

Fig. 7-25. *Characteristics of developmental tetrode thyratrons (courtesy of EG&G, Inc., Electronic Components Division, Salem, Massachusetts).*

its lifetime is likely to be in excess of 2000 high-voltage on-hours. This table also considers future possibilities. The very low inductance tube is an annular type of structure that is projected to switch in a 1/e fall time of 0.3 nanosecond. All of these average power numbers are calculated for thermal loadings of cathodes and anodes demonstrated in current developmental tubes and all are self-consistent.

This is a summary of where the thyratron-switch tube world is and the direction in which it could develop. It looks sensible up to >250 kV, the limiting number that arises when producing a short thyratron for resonant charged, long-pulse types of applications. All of these are large tubes. Very small tubes are also imaginable. If an ultrasmall thyratron was desired, calculations say that a 1/e time of 100 picoseconds is feasible. Such a tube could be an interesting replacement for krytrons and vacuum tubes as very fast switches in many of Pockel Cell circuits.

7.7 OTHER THYRATRON APPLICATIONS

Other applications for thyratrons are depicted in Fig. 7-26. In very large systems, a thyratron can be efficiently used for an inverse diode (A). In design studies for 0.2 to 1 MW average power modulators, it is as efficient to use a thyratron as it is a triggered charging diode. Once that decision is made, a thyratron is the obvious choice for (A) because there is already a common heater and cathode. Thyratrons have often been used as high-powered end-of-line (EOL) clippers. (Glasoe and Lebacqz contains more information on that. The EOL clipper basically prevents the voltage on the PFN from reversing.) Circuit (B) is a useful circuit that is rarely used today. It is a clipper to control the peak pulse voltage and return the excess energy to the input capacitor. Circuit (C) is the Blumlein HPE. Blumleins operating up to 213,000 V were built a number of years ago using the MAPS 250, and they are currently in use. So there are actually few technological problems other than component development foreseen at the 250 kV level. Technological challenges certainly arise at the higher voltage levels.

7.8 IGNITRONS

Ignitrons are generally most reliably used as single-shot switches of multicoulombs of charge. Basically, (Fig. 7-27) an ignitron is a pool of mercury in a low-pressure envelope with an insulated anode structure. An abnormal glow discharge is generated around the igniter, producing some electrons, which are accelerated toward the anode. This action causes lanche ionization and gives rise to a glow discharge that can carry up to several hundred coulombs of charge. Ignitrons are not generally in use today because they have been replaced by solid-state technology for many commercial applications. The device on the right side of Fig. 7-27 is a gradient-grid ignitron created to handle the Stanford linear accelerator, (SLAC) switching applications. It has the same argument in its favor as the gradient-grid thyratron. It was designed to operate at 360 pulses/second, 50,000 W average power, but it failed to meet these specifications primarily because of anode heat management and tube recovery difficulties. There is a long heat path out of the anode—the same problem as with glass thyratrons. A ceramic ignitron structure may be beneficial, but development in the area has been minimal.

The unique aspect of an ignitron is that it can transfer more charge than any other glow-discharge device known, reliably and for a long discharge duration with virtually

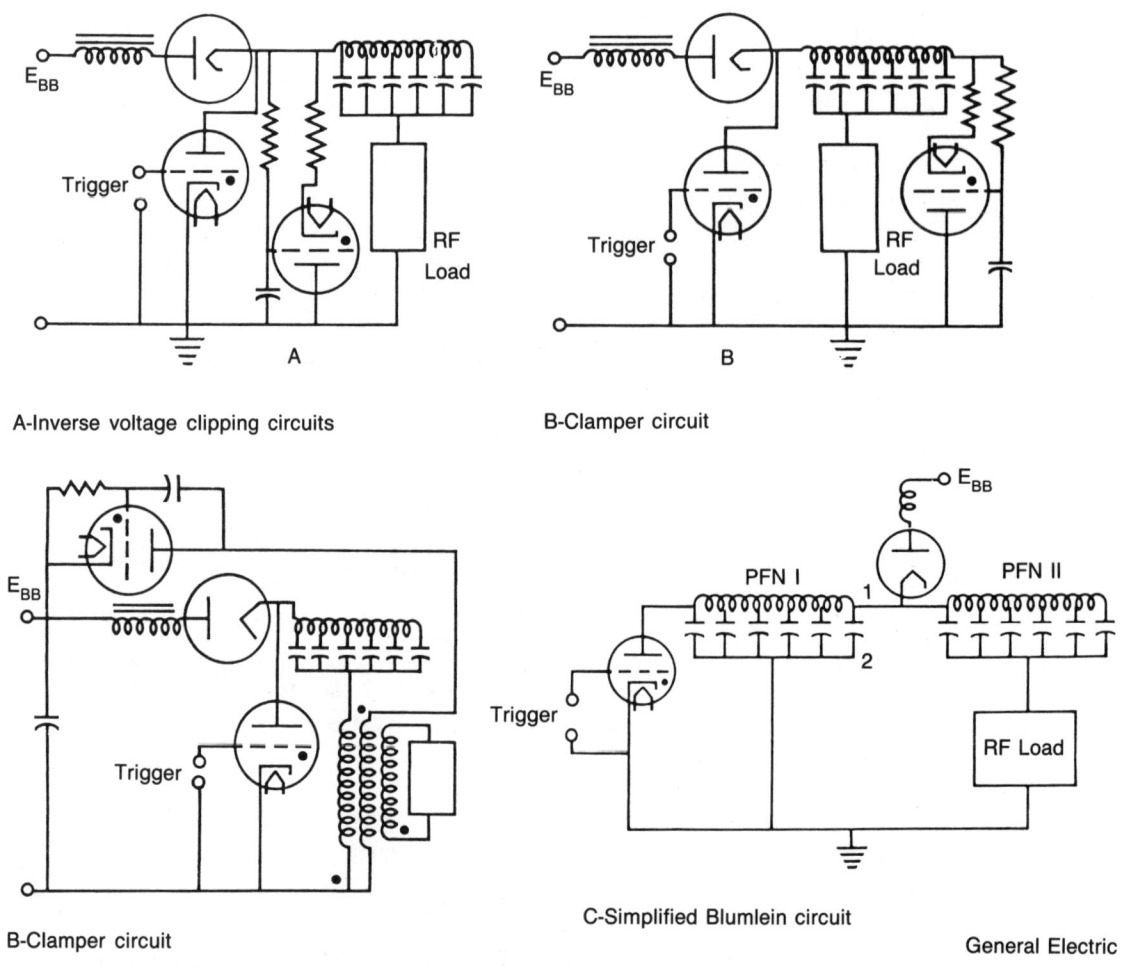

Fig. 7-26. *Other applications of thyratrons.*

no internal damage, except the gradual erosion and contamination of the graphite igniter. This igniter is usually a small piece of graphite extending into the mercury pool. A small arc is generated by the trigger voltage pulse over the surface of the graphite. The proper way to use an ignitron in a high repetition rate circuit is to pulse charge it. Another reason to use the device is its low cost.

7.8.1 Characteristics of Standard Ignitrons

Several standard ignitrons are shown in Fig. 7-28. The 37248 has been run up to 250 Hz and held off up to 60 kV in experimental systems. Its molybdenum anode permits reversal of the current through it (a perfect bidirectional switch). This is done safely through a secondary element called a holding electrode. A large capacitor is discharged into the holding electrode at the time of triggering the ignitor, creating a plasma that

7.8 Ignitrons 233

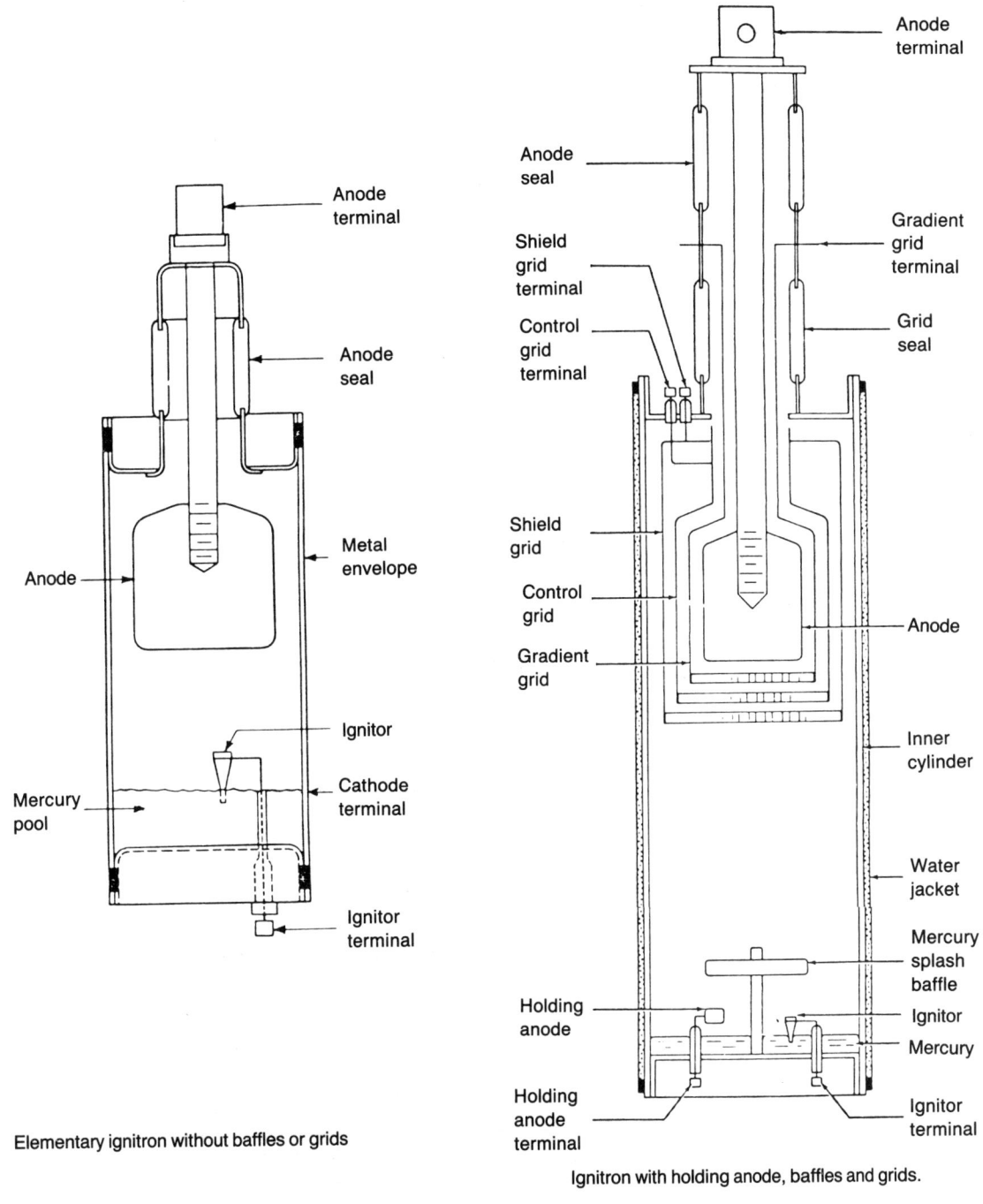

Fig. 7-27. Cross sectional views of the basic types of ignitrons. [11]

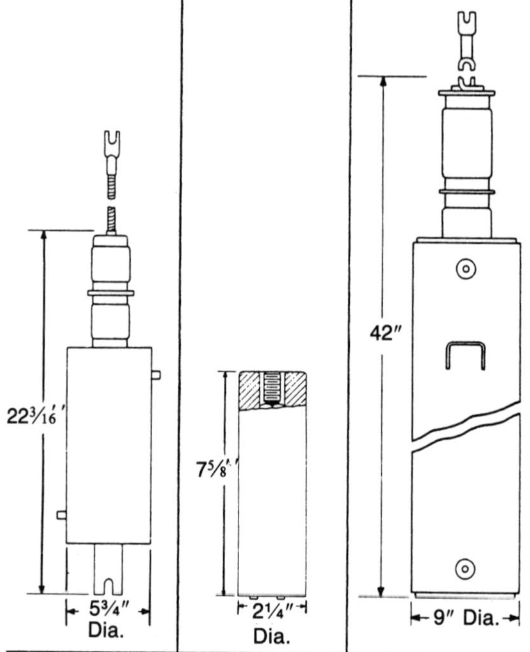

234 Thyratrons and Ignitrons

	GL-5630	GL-37248	GL-6228
Anode Voltage (kilovolts)	35	50	50
Peak current (kiloamperes)	20	25	30
Total charge (coulombs)		15	
Ionization time (microseconds)	0.8	0.5	0.8
Discharge Rate (pulse/minute) typical		2	
Anode Material	Graphite	Molybdenum	Graphite
Cooling	water jacket	clamp	water jacket
Rigid length (inches)	22-3/16	7-5/8	42
Diameter (inches)	5-3/4	2-1/4	9

Fig. 7-28. Representative commercial ignitrons made by Richardson Electronics. [11]

7.8 Ignitrons   235

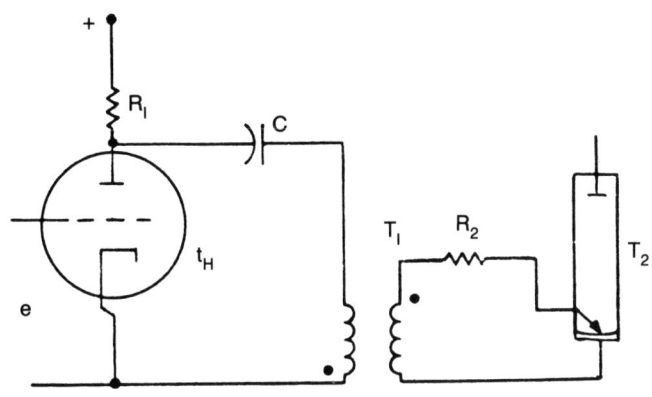

e   – Signal voltage
T_H – Thyratron
C   – Capacitor
R_1 – Charging resistor
R_2 – Discharge limiting resistor
T_1 – Pulse transformer
T_2 – Ignitron

A-Fundamental ignitor firing circuit

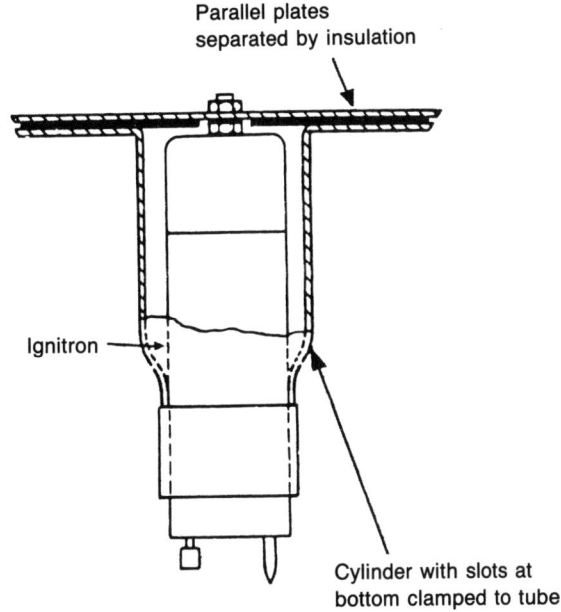

B-Ignitron in coaxial mounting

Richardson Electronics

Fig. 7-29. Low-inductance housing for the GL-37248 type of ignitron and a fundamental igniter firing circuit. [11]

remains on during the main discharge time. When the current reverses, the cathode becomes the anode, and the anode becomes the cathode. Graphite anode ignitrons are all designed for use only in critically dampened discharge circuitry where the current does not reverse. Severe anode damage will occur if the current reverses.

A very simple and practical low-inductance housing for this tube is illustrated in Fig. 7-29. The anode must be warmed and the cathode cooled to operate at maximum voltage and high repetition rates. The temperature control is necessary to control the hold-off and minimize arc-backs. In the fundamental igniter trigger shown, a thyratron discharges a small capacitor into a pulse transformer, which triggers the ignitron (through the resistor into the igniter).

7.8.2 Triggering Techniques

At high repetition rates, the trigger circuit needs a carefully shaped pulse, as is shown in Fig. 7-30. In this particular circuit a 1:1 pulse transformer is used. The storage capacitor, C_1, is charged to 4,000 V (0.1 to 1 μF; low inductance and ESR) and discharged into T_2, which triggers the ignitron through a shaping network. This network comprises a noninductive 10Ω resistor, in series with a 200 pF capacitance across the saturating inductor of 10 μH unsaturated inductance. With this network, an igniter pulse is produced that effectively goes up toward 6 kV if the capacitor is charged to 4 kV. As the igniter current grows, the inductance decreases to maintain an intense, sustaining current in the igniter area. This network reduces the delay time (to <100 nanoseconds) and jitter (to <30 nanoseconds) considerably and operates the same way as the pre-ionization in a TEA laser. The saturating ferrite gives a steep current gradient to produce the ionization density in the cathode area of the tube quickly. Another tactic is to pretrigger the holding electrode with a fast-pulse circuit, giving a double-discharge type of ignition. These devices might be reasonably considered for continuous use up to 100 Hz repetition rates, and they are also very efficient. They are probably no use whatsoever in driving excimer lasers, but they might be of interest in some types of large fusion lasers.

7.8.3 Stacking Ignitrons

The most reliable way of ignitron stacking is shown in Fig. 7-31, using a trigger transformer and two shaping networks to trigger each ignitron separately. The difficulty with the (A) circuit is that only the two capacitors provide triggering energy for the upper ignitron. In fast circuits, these must be small to prevent the upper tube from being

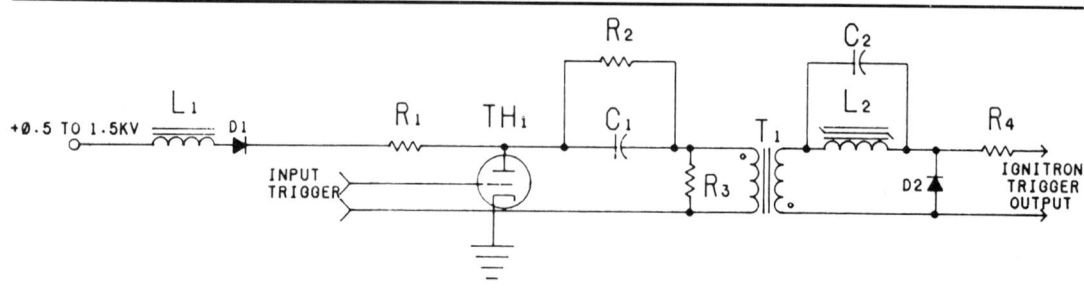

Fig. 7-30. Trigger circuits for high repetition rate, low-jitter firing of ignitrons.

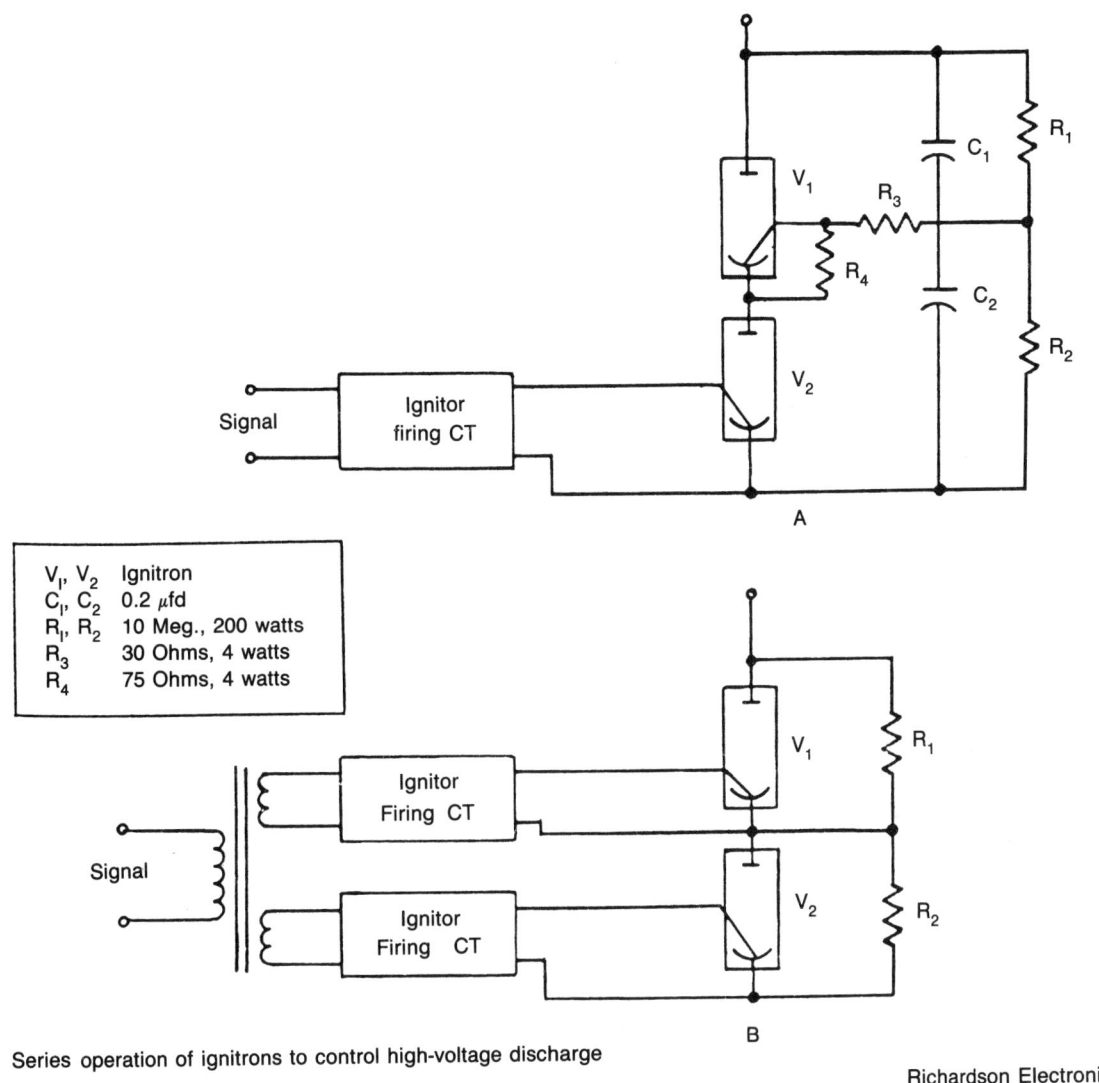

Fig. 7-31. Series operations of ignitrons to achieve reliable switching of voltages in excess of the single-tube rating. [11]

very heavily triggered and damaging the igniter. A considerable amount of high-frequency ringing, which causes increased anode dissipation in the device, can also be found in this circuit.

7.9 REFERENCES

1. "Hydrogen Thyratrons—Theory and Application," General Electric Technical Literature.

2. K.J. Germeshauser, "Pulsed Generators," (G.N. Glasoe and J.V. Lebacqz, eds.), Vol 5, pp. 335-354. Radiation Laboratory Series, McGraw-Hill, New York, 1948.
3. S. Goldberg and J. Rothstein, "Hydrogen Thyratrons," Academic Press, 1961. pp. 207-264. See also: "Research Study on Hydrogen Thyratrons," Vol. 1 (1956) by S.T. Martin and S. Goldberg; Vol. II (1956) by S. Goldberg; and Vol. III (1957) by S. Goldberg and D.F. Riley, Edgerton, Germeshauser and Grier, Boston.
4. T.R. Burkes, "A Critical Analysis and Assessment of High-Power Switches," Naval Surface Weapons Center Report No. NP30/78, September, 1978.
5. D. Turnquist, R. Caristi, S. Friedman, S. Mertz, R. Plante, and N. Reinhardt, "New Hydrogen Thyratrons for Advanced High-Power Switching," Conf. Record of the 2nd IEEE International Pulsed Power Conference, June 12-14, 1979.
6. B.R. Gray, "Evaluation of State-of-the-Art Hydrogen Thyratrons at Extended Ratings," Conf. Record of 11th Modulator Symposium, September 1973.
7. J.J. Hamilton, S. Mertz, R. Plante, D. Turnquist, N. Reinhardt, J. Creedon and J. McGowan, "Development of a Forty-Kilovolt Megawatt Average Power Thyratron (MAPS-40)," paper presented at the 13th Pulse Power Modulator Symposium, June 1978.
8. S. Friedman, S. Goldberg, J. Hamilton, S. Mertz, R. Plante, and D. Turnquist, "Multigigawatt Hydrogen Thyratrons with Nanosecond Rise Times," paper presented at the 13th Pulse Power Modulator Symposium, June 1978.
9. EG&G Technical Literature.
10. EEV Inc. Technical Literature.
11. General Electric Ignition Technical Literature, now licensed to and manufactured by Richard Electronics, Ltd. Franklin Park, Illinois.

Chapter 8

Charging Systems

by
W. Nunnally
(University of Texas at Arlington)

PULSE POWER SYSTEMS ARE USED TO OBTAIN MULTIPLICATION OF THE AVAILABLE CONtinuous (ac) power. Power multiplication occurs when an amount of energy, E, is accumulated over a long period of time, T_A, in an intermediate energy store, usually a capacitor or pulse-forming network (PFN), and then discharged in a much shorter period of time, T_D. If the power during accumulation and discharge is respectively constant, P_A, and P_D, the power multiplication is:

$$\frac{P_D}{P_A} = \frac{E/T_D}{E/T_A} = \frac{T_A}{T_D} \qquad (8\text{-}1)$$

Charging systems are used to charge or accumulate energy in one or more intermediate energy stores. This chapter examines dc and ac charging systems, which use nonresonant and resonant circuits to transfer charge from a dc power supply to the intermediate energy store or from one intermediate energy store to another.

8.1 THE DC CHARGING SYSTEMS

In dc systems, the transfer of charge manifested by voltages and currents is governed by RC and L/R time constants. Circuits that responed to these time constants are illustrated in Fig. 8-1. The transients in RC and LR circuits require three to four time constants to reach steady state or complete charge transfer effectively. Recognizing the system type or basic components aids in determing system response.

The rectifier transformer portion of dc power supplies is discussed in detail in Chapter 1. However, the ac power available and various ac transformer connections are an important part of the power available and various ac transformer connections are an important part of the power supply and thus of charging systems. Four methods of con-

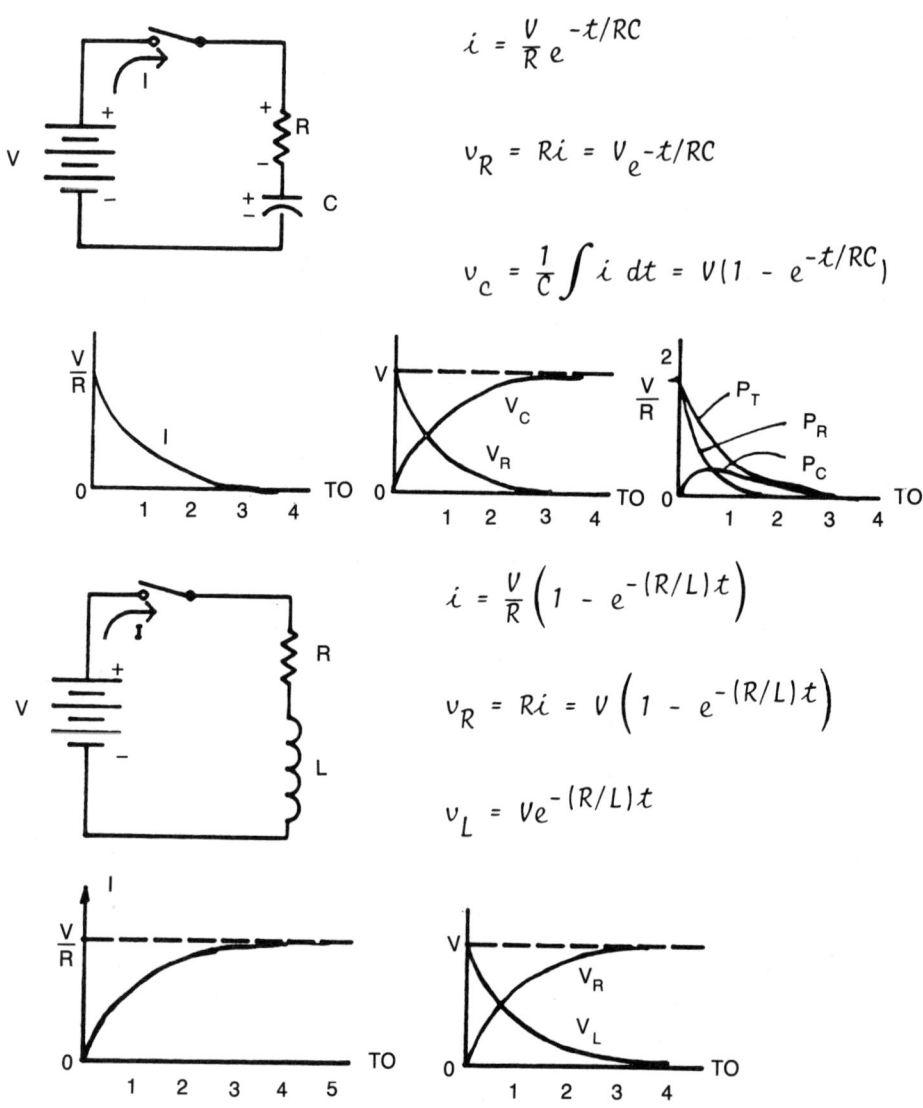

Fig. 8-1. Basic circuit time constants.

necting ac delta and wye transformer windings in three-phase systems are illustrated in Fig. 8-2. The basic advantages and disadvantages of the various primary-secondary transformer winding connection combinations are listed in Table 8-1. Combinations of these windings are also used to give lower—higher frequency ripple—as mentioned in Chapter 1.

Principally, wyes operate with high voltages low currents, and it has a neutral connection. The center point or neutral of the wye can be grounded (Fig. 8-2), and the system can be grounded. In fact, if the neutral is not used, third harmonics occurs.

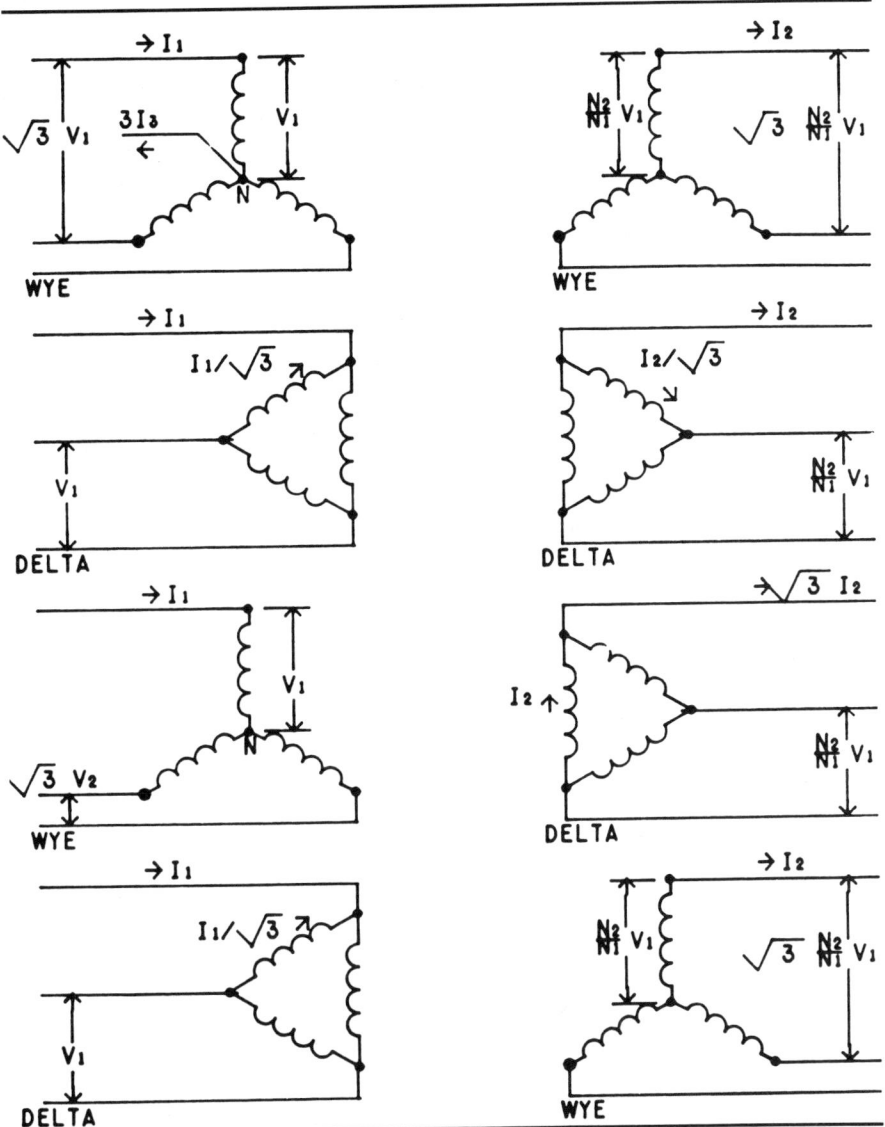

Fig. 8-2. Three-phase transformer connections.

The delta-delta configuration, on the other hand, operates with generally higher currents and lower voltages, and it remains balanced, even with unbalanced loads.

8.1.1 Charging Circuits

A delta primary and a wye secondary are usually used so that the advantages of a delta on one side cancel the disadvantages of a wye on the other side, and vice versa. In a high-voltage supply, a delta primary-wye secondary transformer will operate with lower ac line-phase voltages and higher secondary-output voltages with only moderate voltages to ground.

Table 8-1. Transformer Primary-Secondary Connections.

A. Wye-Wye Characteristics
1. Can be used with or without neutral connection
2. Neutral connections, primary, and secondary allow independent loads.
3. Relatively lower currents
4. Higher voltages
5. Large third harmonic components must flow to source through primary neutral to have sinusoidal line voltages.
6. Unbalanced loads lead to unbalanced voltages when neutral not used in primary. (Difference currents must have equalizing path.)

B. Delta-Delta Characteristics
1. Does not have or require a neutral
2. Voltages are low or moderate.
3. Relatively large currents required.
4. Windings must be nearly similar to prevent large circulating currents.
5. With unbalanced loads, voltages tend to remain balanced.

C. Delta-Wye or Wye-Delta Characteristics
1. Usually used instead of delta-delta or wye-wye because separate disadvantages and advantages tend to compensate
2. The primary line voltages do not have a third harmonic component and are sinusoidal.
3. Unbalanced loads do not require unbalanced source.
4. Loading is essentially independent of winding uniformity.
5. Delta-wye usually used for high-voltage power supply.

Power supplies are sized by relating the ac input power to the required dc power through power supply efficiency. For a single-phase system, the input power P_I is determined by:

$$P_I = Il_{RMS} \cdot El_{RMS} = I_{line_{RMS}} \cdot E_{line_{RMS}} \qquad (8\text{-}2)$$

and for a three phase (wye or delta), the input power is:

$$P_I = I_{line_{RMS}} \cdot E_{line_{RMS}} \cdot \sqrt{3} \qquad (8\text{-}3)$$

The output power P_0 for a dc power supply is:

$$P_0 = I_{dc_{MAX}} \cdot V_{dc_{MAX}} \qquad (8\text{-}4)$$

However, it is possible to operate dc supplies, for short periods and low duty factors, up to twice their rated current. Thus, for some applications, it might be possible to purchase only one-half the continuous power supply required. As mentioned in Chapter 1, rectifiers are usually specified for 2.5 times the rated output current and voltage.

For a resonant ac charge sytem, the output power can be determined by:

$$P_0 = I_{dc_{AVG}} \cdot V_{dc_{MAX}} \qquad (8\text{-}5)$$

The input power and the output power are related by using the power supply efficiently. Most power supplies have an efficiency of approximately 80 percent. The inefficiency is due to transformer core loss, copper winding loss, and rectifier loss. Thus, if the dc power requirements are known, the input ac power P_I requirements can be determined from:

$$P_I = 1.2 \cdot P_0: \quad \text{dc charge}$$

$$P_I = 1.2 \cdot 1.15 \cdot P_0: \quad \text{ac charge} \qquad (8\text{-}6)$$

NOTE: For a resonant ac charge system, the output power is multiplied by an additional factor of 1.15.

8.1.2 The dc Power Supply Control Methods

The various charging systems are distinguished by the methods and circuits used to regulate, control, or limit the transfer of charge from the ac line through the dc power supply to the intermediate energy store (usually a capacitor of PFN).

For the ideal power supply (Fig. 8-3), the input power is equal to the output with a voltage and current level shift due to the transformer or:

$$P_{dc} = P_{ac}$$

$$I_{dc} = I_{ac_{RMS}} \cdot \frac{N1}{N2}$$

$$V_{dc} = V_{ac_{RMS}} \cdot \frac{N2}{N1} \qquad (8\text{-}7)$$

Several types of dc charge systems can be discussed using the ideal power supply model.

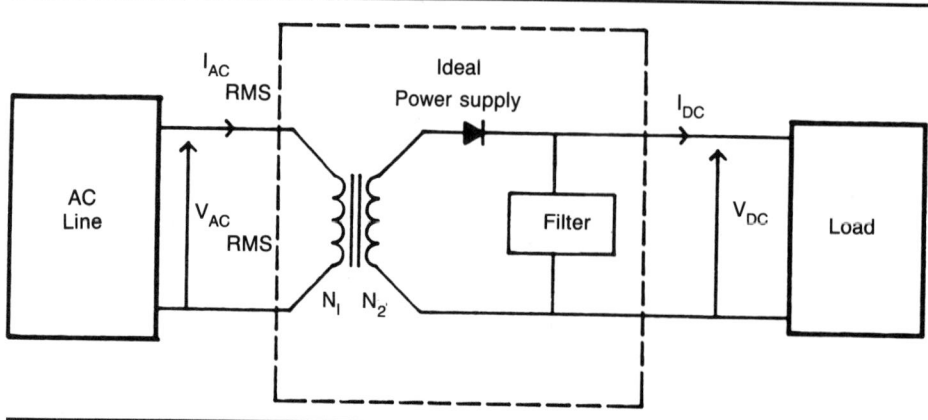

Fig. 8-3. Ideal power supply.

8.1.3 Impedance Limited Charge System

One method of controlling the transfer of charge to the intermediate energy store is to limit the rate of charge transfer or current. This is usually a prudent requirement because real-world power supplies have finite current ratings.

The maximum power supply current when charging a capacitor, C, from an ideal supply (Fig. 8-4) is determined by a resistance, R, in series with the power supply output when the ac contactor is closed (t = 0), or:

$$I_{dc \text{ MAX}} = \frac{V_{dc \text{ MAX}}}{R} \tag{8-8}$$

Similarly, the power supply current can be limited by a reactance in series with the power supply input (Fig. 8-5). The power supply load impedance when the contactor

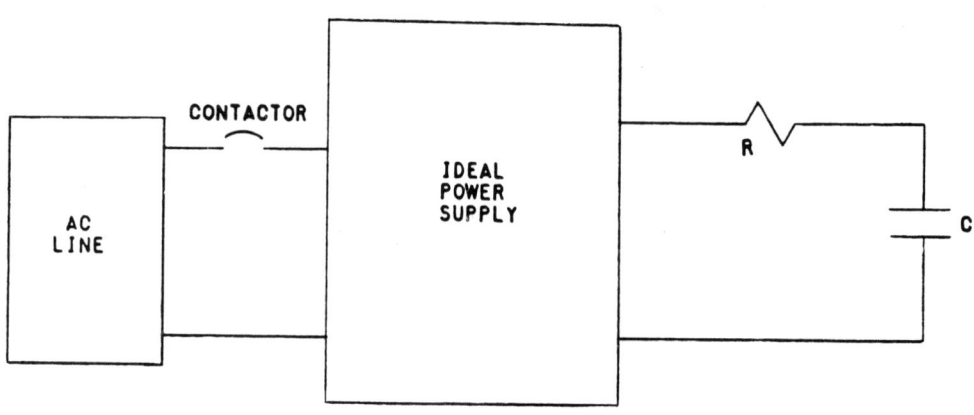

Fig. 8-4. Resistance charging system.

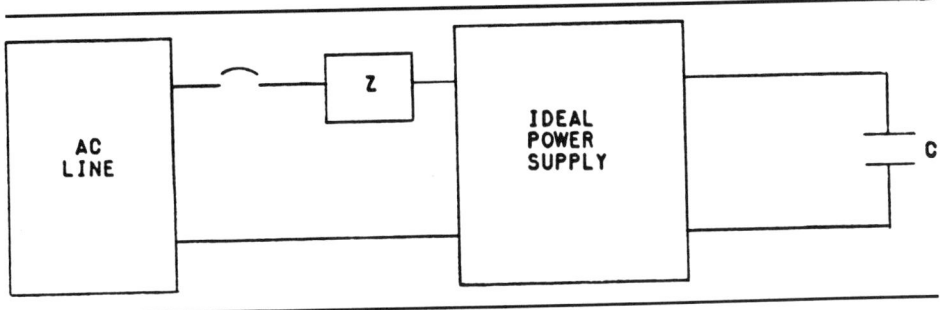

Fig. 8-5. Reactance charging system.

is closed (t = 0) initially looks like a short circuit (R = 0). Thus, the initial power supply input current maximum is limited to:

$$I_{ac\ peak} = \frac{V_{ac\ peak}}{|Z|} = \frac{V_{dc\ max} \sqrt{2}}{|Z| \left(\frac{N2}{N1}\right)^2} \quad (8\text{-}9)$$

The type of Z can be inductive, capacitive, or resistive. Generally, capacitors of the desired impedance and power rating (voltage, current) are easier to locate than inductors or resistors. Only resistors are usually used in the dc supply output, and usually a combination of the two above methods is used.

8.1.4 Charging Efficiency

A digression would be helpful at this point to examine system charging efficiency or the efficiency of transferring energy from the power supply to the intermediate store. The power supply must deliver a total energy E_p during the charge time T_C, or:

$$E_p = \int_0^{T_C} i_C(t) \cdot v_C(t)\, dt \quad (8\text{-}10)$$

where $i_C(t)$ is the charge current flowing into the capacitor and $v_C(t)$ is the capacitor voltage. The capacitor current and voltage for a series resistor R are given by:

$$i_C(t) = \frac{V_{dcM}}{R} e^{-t/RC}$$

$$v_C(t) = V_{dcM}(1 - e^{-t/RC}) \quad (8\text{-}11)$$

The capacitor energy E_C, at full charge, which occurs at $t \cong 3RC$, is:

$$E_C = \frac{1}{2} C \cdot V_{dcM}^2 \quad (8\text{-}12)$$

246 Charging Systems

The charging or charge transfer efficiency is given by

$$\eta_C = \frac{E_C}{E_p} \tag{8-13}$$

which is at most 50 percent for a series resistance, and $t \geq 3RC$.

8.1.5 Voltage Ramp Charge System

One of the problems with a resistance/reactance limited charge system is the capacitor voltage approaches the desired value asymptotically. When charging a capacitor to large voltages, it is advantageous to reach the desired voltage level and discharge the energy store as soon as possible to avoid probabilistic breakdown problems and reduced capacitor lifetime. The capacitor voltage increases at a slower rate as the charging current decreases; this is due to the reduced difference between the constant power supply voltge and the charging capacitor voltage. Thus, to maintain the charging rate, the power supply voltage must be increased as the capacitor voltage increases. Increasing the power supply voltage linearly, as with the motorized variac (Fig. 8-6), will provide a constant charging current and a linear capacitor voltage increase. However, the voltage ramp charge system is susceptable to load faults (capacitor short) when the power supply voltage is maximum, which results in a large current surge from the power supply that must be accommodated.

8.1.6 Monocyclic Constant Current Charge System

A passive method of providing a constant charging current regardless of load impedance (short or open) is the monocyclic network (Fig. 8-7). The LC network is designed to be resonant at the ac line frequency, ω_0. In this case, the current into the power supply, I_1, is given by:

$$I_1 = \frac{E_1}{\left| \omega L \right|} \tag{8-14}$$

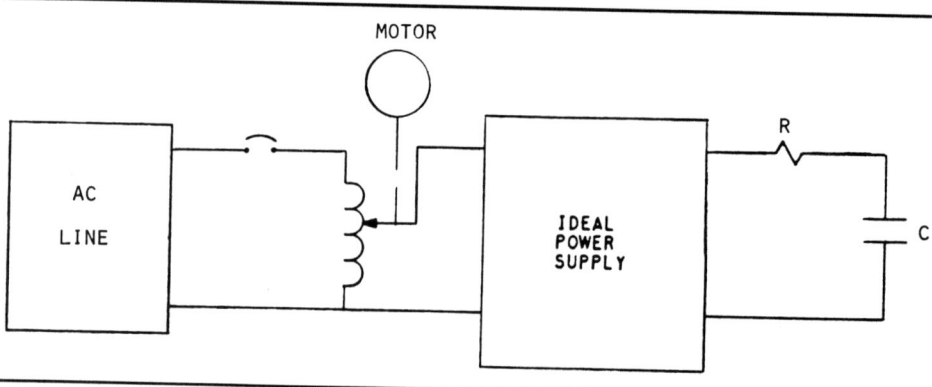

Fig. 8-6. Voltage-ramp charging system.

8.1 The dc Charging Systems 247

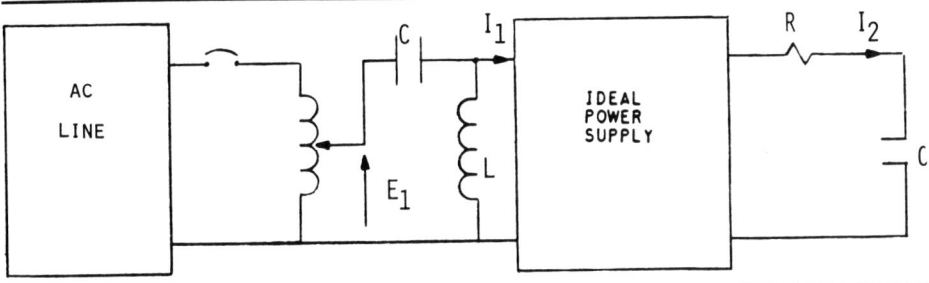

Fig. 8-7. Single-phase monocyclic network.

The ideal power supply charging current, I_2, is given by:

$$I_2 = \frac{N1}{N2} \cdot I_1 = \frac{N1}{N2} \cdot \frac{E_1}{|\omega|} \tag{8-15}$$

The variac in Fig. 8-7 is used to set E_1 and thus the constant charging current amplitude.

The constant current required to charge a capacitance C to voltage V_o in time T is given by:

$$I_2 = \frac{C \cdot V_o}{T} \tag{8-16}$$

which can be related to the input voltage:

$$E_1 = \frac{N2 \cdot |\omega L| \cdot C \cdot V_o}{N1 \cdot T} \tag{8-17}$$

The current surge rise time seen by the ac line when closing the contactor of Fig. 8-7 can be determined using the equivalent circuit of Fig. 8-8. The impedance of a capacitor charged by a constant current ideal power supply through a resistor is given by:

$$Z = \left(\frac{N1}{N2}\right)^2 \cdot \left(R + \frac{t}{C}\right) \tag{8-18}$$

which increases linearly with time, t.

The term t/C is small compared to R near $t = 0$ such that the initial impedance seen by the ac line and the monocyclic network is:

$$Z = \left(\frac{N1}{N2}\right)^2 \cdot R \tag{8-19}$$

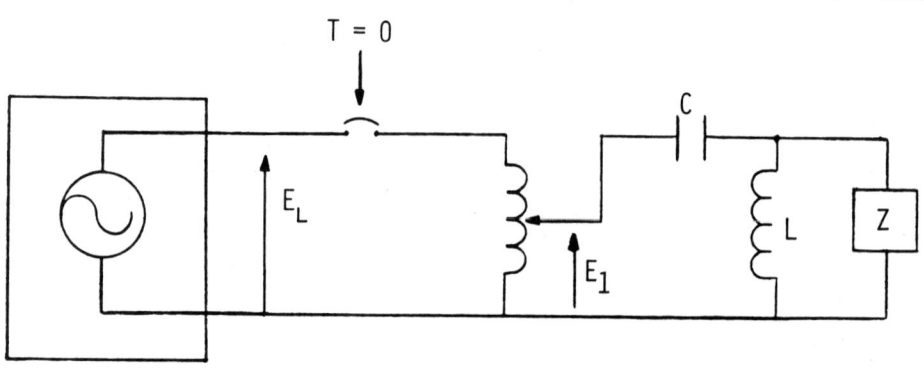

Fig. 8-8. Ac line current surge.

The ac line current surge occurs in a time given by:

$$t_s = 3\left[2R \cdot \left(\frac{N1}{N2}\right)^2 \cdot C\right] \quad (8\text{-}20)$$

which is usually much shorter than the capacitor charge time, T.

8.1.7 Constant Current Charging Efficiency

The energy delivered by the power supply is given by:

$$E_p = \int_0^T V_{ps} \cdot i_{ps}\, dt$$

$$= \int_0^T I_{ps}^2 \cdot (R + t/C)\, dt$$

$$= I_{ps}^2 \cdot \left(RT + \frac{T^2}{2C}\right) \quad (8\text{-}21)$$

and

$$E_C = \frac{C \cdot V_o^2}{2} = \frac{I_{ps}^2 \cdot T^2}{2C} \quad (8\text{-}22)$$

such that constant current charging efficiency given by:

$$\eta = \frac{E_p}{E_C} = \left(1 + \frac{2RC}{T}\right)^{-1} \quad (8\text{-}23)$$

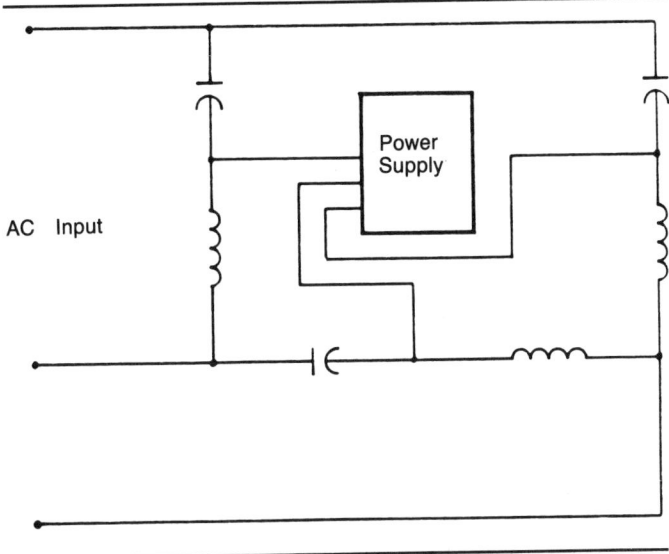

Fig. 8-9. Three-phase monocyclic.

is very good. A three-phase monocyclic hookup is shown in Fig. 8-9. In single-phase systems, the power supply can be connected across either the resonant L or C. However, the power supply does not experience high-frequency transients associated with contact closure when connected across the capacitance.

The monocyclic contant current charge system will deliver a constant current into any load. Thus, load short circuits will not damage the power supply with overcurrents. On the other hand, a load open circuit will produce extremely high voltages as the system tries to force the constant current into the high, open-circuit impedance. Usually a high-voltage insulation failure shorts the power supply or the load to present a low-impedance load to the system. For this reason, safety spark gaps across the power supply output are used to ensure that the output voltage reaches only a predetermined maximum value. The monocyclic system uses a resistor in series with the power supply output to isolate the power supply from transients. The value of R also determines the ac line source current envelope rise time.

8.1.8 Constant Power Charging System

A constant power charging system (Fig. 8-10) uses electronics to limit the power going to the load. Constant power charging systems monitor the power going into the load and modify the pulse width so that during any given period only a finite amount of power is delivered. The supply output voltage is initially ramped on to limit turn-on surges. Constant power charging systems use power more efficiently as shown in Fig. 8-11. Because the power is controlled with electronics, the R in the secondary can be very small. Constant power charging systems are not more efficient, but they use power more effectively.

The power delivered during a charging cycle by various power supplies is shown in Fig. 8-11. In an RC-charge system, the power is highest when the system is turned

250   **Charging Systems**

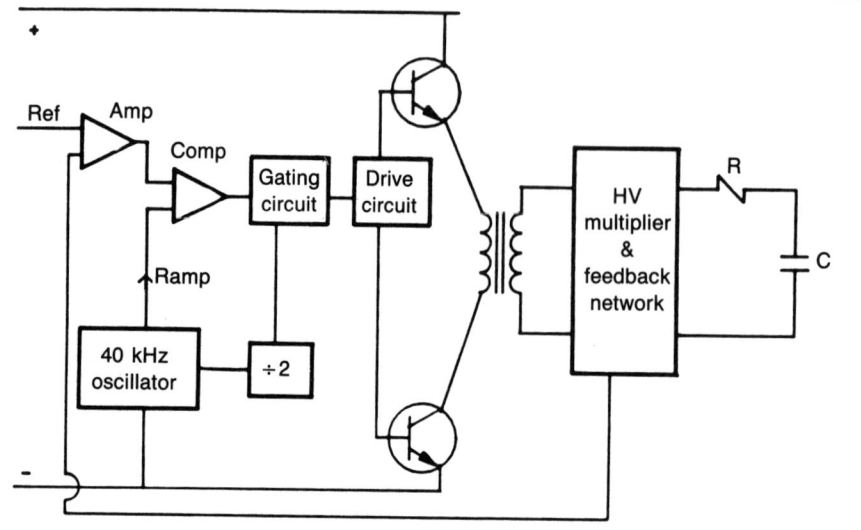

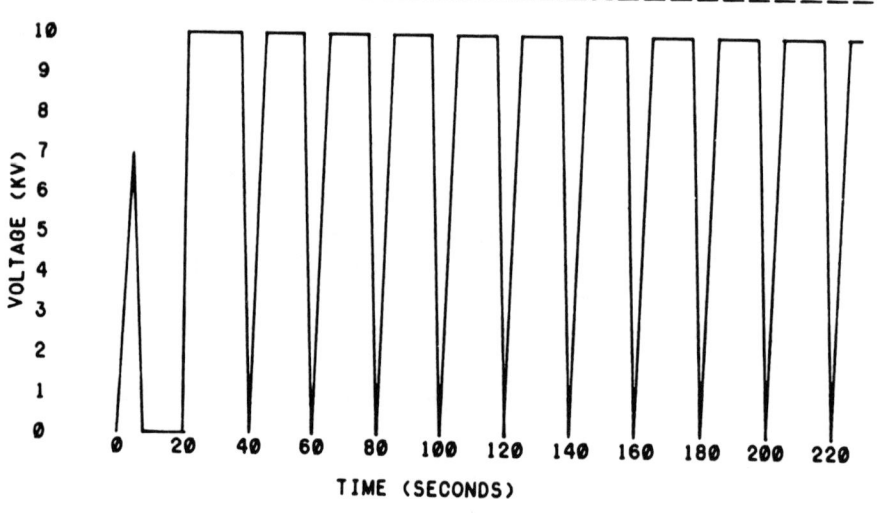

Fig. 8-10. Constant power charging circuit using a switching power supply (inverter).

on. A constant dc voltage is multiplied by an exponentially decaying current, so the output power delivered drops. In a constant-current charging supply, the power begins at some value multiplied by IR and increase linearly because the output voltage increases linearly with time. Constant power systems use the same power continually to charge a capacitor, and thus a lower-cost-peak-power supply can be used. The ac line transient generated when turning off the power supply must also be evaluated. Minimum power is being delivered to the load in a resistance/reactance charge system as the voltage

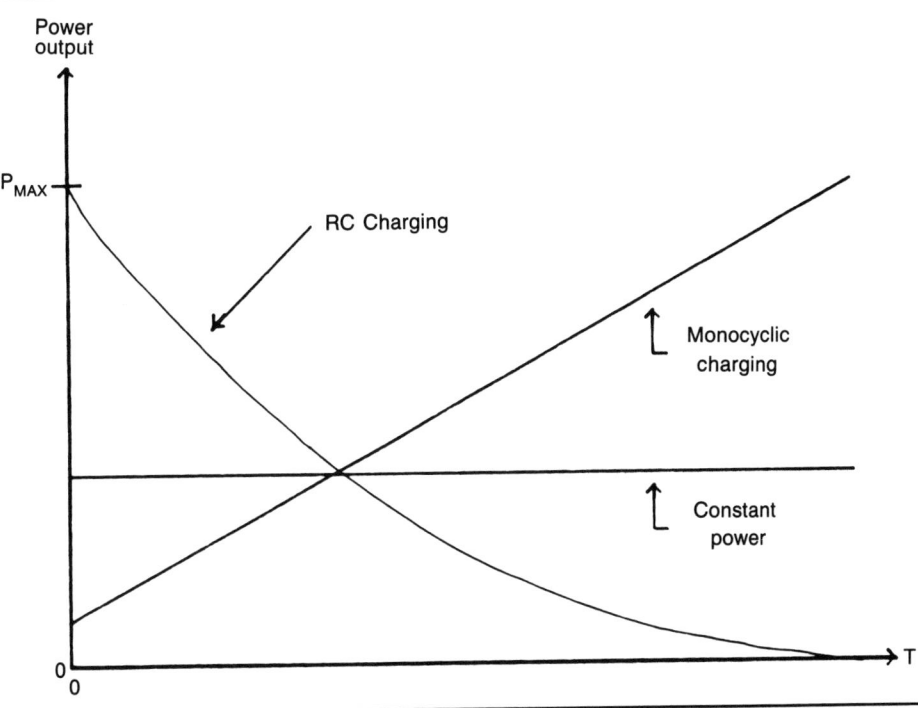

Fig. 8-11. *Maximum power required versus system.*

reaches the desired level, and it is very easy to open the contactors. In the monocyclic constant current charging systems, maximum power is being delivered to the load when the desired voltage is reached; opening the breakers is difficult and can generate an ac line transient.

8.1.9 Phase-Control Charging Systems

Phase-control systems regulate the charging of an energy store by only passing a fraction of each ac line half cycle through to the power supply. The fraction of each cycle passed can be related to the phase angle of the ac waveforms (Fig. 8-12). The switch is triggered at a fraction of the ac cycle determining the firing angle. The amount of charge transferred to the capacitor can be monitored electronically and regulated by varying the firing angle. Similarly, load faults can be detected, and the power supply disconnected from the ac line by removing the trigger signal from the switches.

Phase-control systems are commercially available. There are several features available that assist in charging systems; two of them are current limit and soft start. The current limit uses electronics to monitor the output current going through the ac line and holds it to a certain level by forbidding the advancement of the phase angle past a certain point. Using a current limiting phase control unit, a constant charge rate can be obtained as shown in Fig. 8-13. Whenever current is pumped into a capacitor, the maximum current occurs at $t = 0$. The soft-start feature brings up the power to the

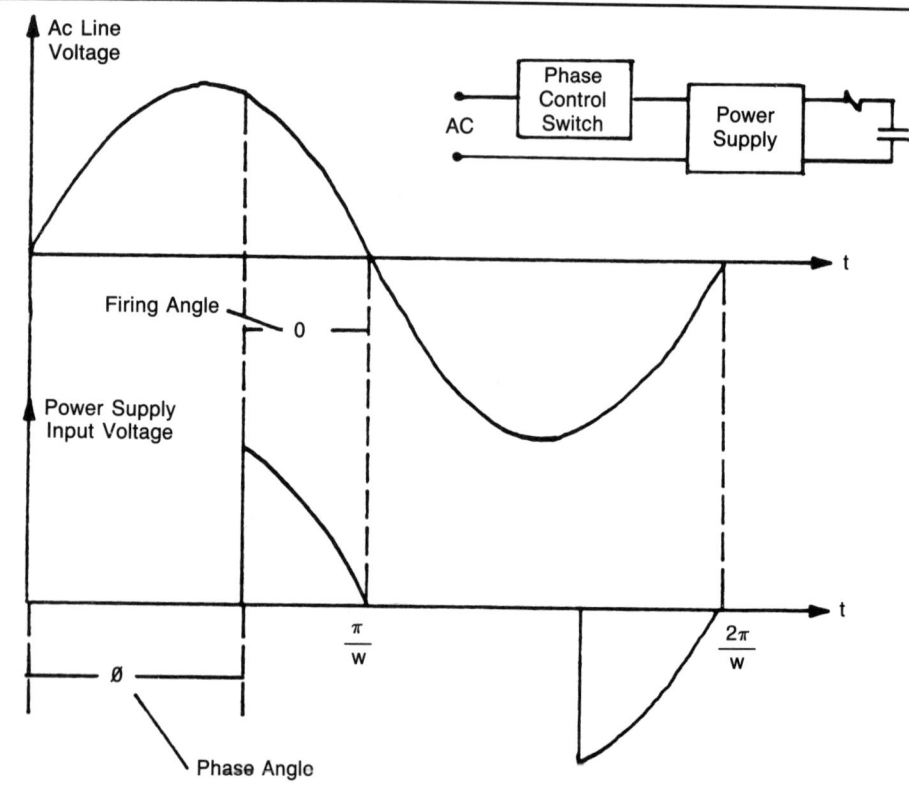

Fig. 8-12. *Phase angle waveforms.*

switch drive system slowly to prevent the switches, usually SCRs, from slamming on. This action reduces initial current surges. The charge rate is thus constant and can be limited to the maximum current rating of the power supply. The sharp turn-on and turn-off transients are a disadvantage of the phase-control system. The transients cause extra losses in the power supply transformer and diodes that must be considered. Figure 8-14(A) illustrates a hybrid SCR three-phase system, using one SCR and one reverse diode with no ground wire. For full control, six SCRs must be used in a three-phase system as shown in Fig. 8-14(B). These systems should be bought, not built. Before buying one, determine system transients to allow the manufacturer to incorporate a proper snubber in the design. Figure 8-15 illustrates typical hybrid and full-control waveforms.

8.1.10 Saturable Reactor-Charging System

Saturable reactor (SR) charging systems are essentially phase-control systems. The unique and most advantageous characteristics are the use of saturable reactor (magnetic amplifier) switches, which do not wear out, are not subject to main-system transients, and operate at high temperatures (up to 150 °C depending on components chosen). The saturable reactor switching system changes state from a large impedance to a finite impedance in series with the power supply each half-cycle.

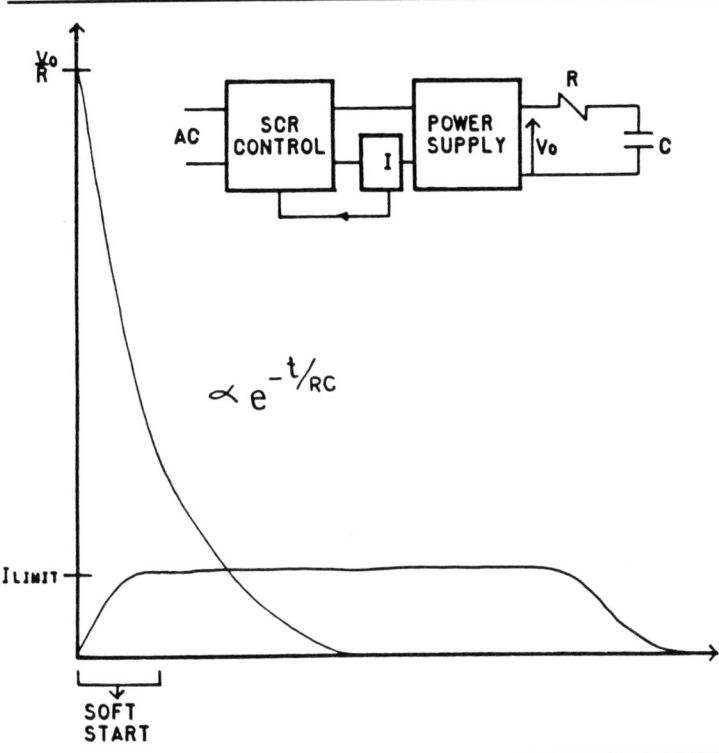

Fig. 8-13. *Constant-current charging using SCR-current limit.*

The operation of a single-phase saturable reactor is illustrated in Fig. 8-16. An SR consists of three windings: A and B series windings and a control winding arranged on a three-leg magnetic circuit (core).

In the off state, the control winding current is zero ($I_C = 0$), and no control flux is present in the core ($\phi C = 0$). The core is designed to remain unsaturated during the ac line voltage cycle in the absence of control flux ($\phi A = \phi B < \phi$ saturation). The series inductive reactance of the unsaturated windings, A and B, is large compared to the total power supply impedance, such that only a small fraction (as low as 6 percent) of the line voltage appears across the power supply input (Fig. 8-17). To prevent leak charging, a short must be placed across the capacitor until the unit is to be charged.

An increase in the dc control current produces a control flux in the core, which adds to the flux in one winding, B, and subtracts from the flux in the other winding, A. As the ac cycle current increases, the saturation flux exceeded in core leg B but not in core leg A, or:

$$\phi_A = \phi_{A_{ac}} - \phi_C < \phi_S \tag{8-24}$$

and

$$\phi_B = \phi_{B_{ac}} + \phi_C > \phi_S \tag{8-25}$$

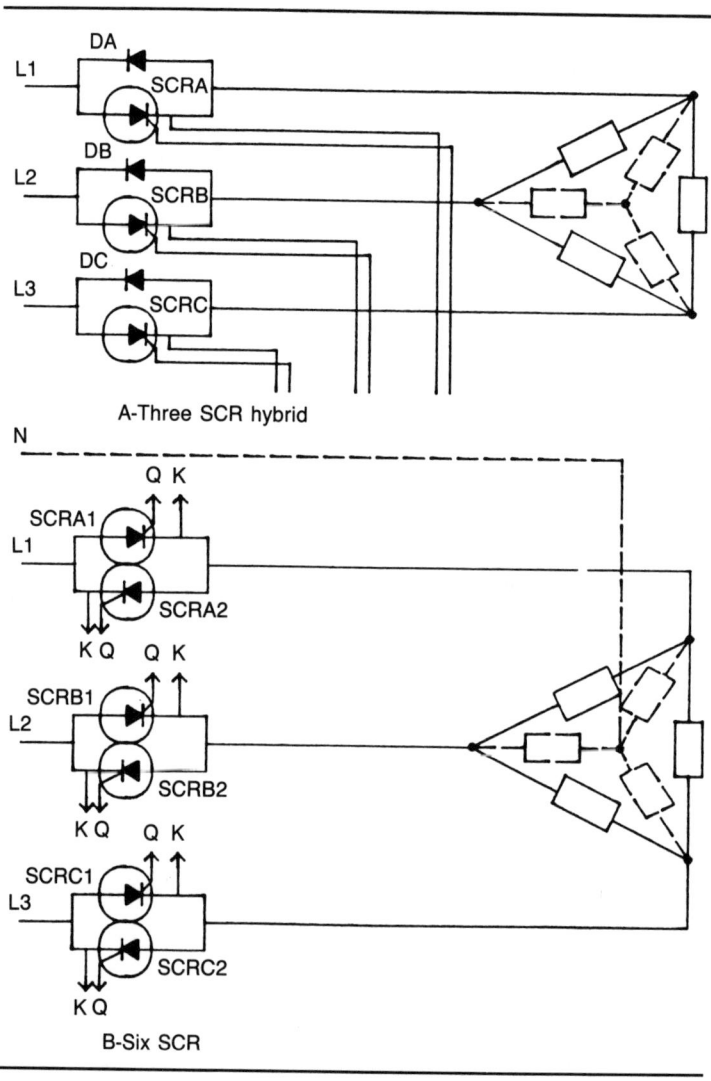

Fig. 8-14. Phase-control charge system.

Until either winding A or B saturates, the transformer coupling from both windings to the control winding is equal and opposite. The control circuit output impedance is thus not transformer coupled into the ac circuit. As one winding saturates, the impedance and the corresponding voltage across that winding drops to a low value, and the same voltage then appears across the load (power supply). Simultaneously, the transformer coupling from the saturated winding decreases, leaving a net transformer coupling through the unsaturated winding. This action places the transformed control power supply impedance in series with the power supply (Fig. 8-18). As shown in Fig. 8-18, the output impedance of the control dc power supply through transformer action thus determines the maximum current fed to the power supply. If the control supply ac output impedance

8.1 The dc Charging Systems

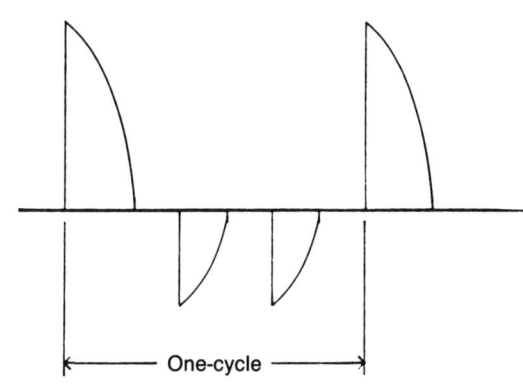

A Six SCR full-control output waveforms.

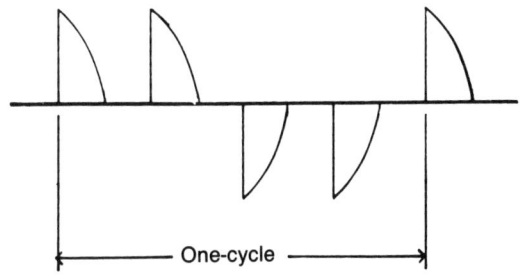

B Three SCR hybrid output waveforms.

Fig. 8-15. SCR phase-control output waveforms.

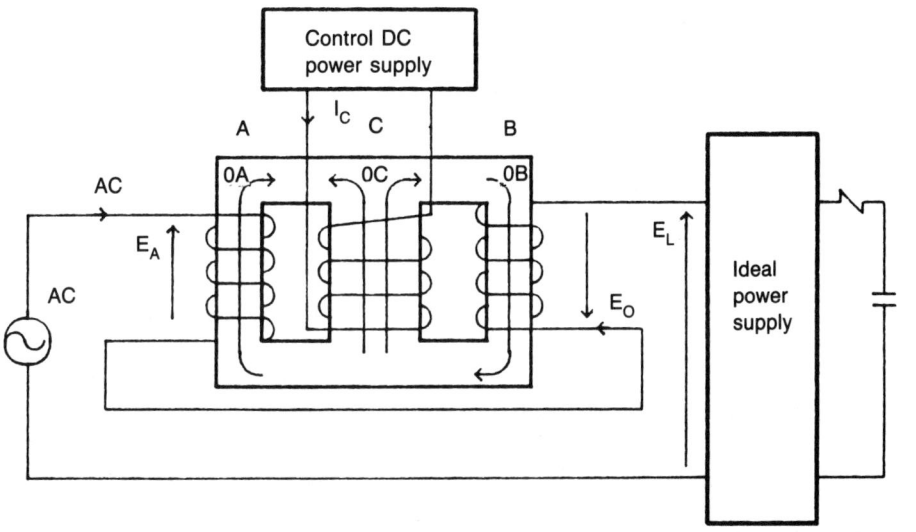

Fig. 8-16. Single-phase saturable reactor control.

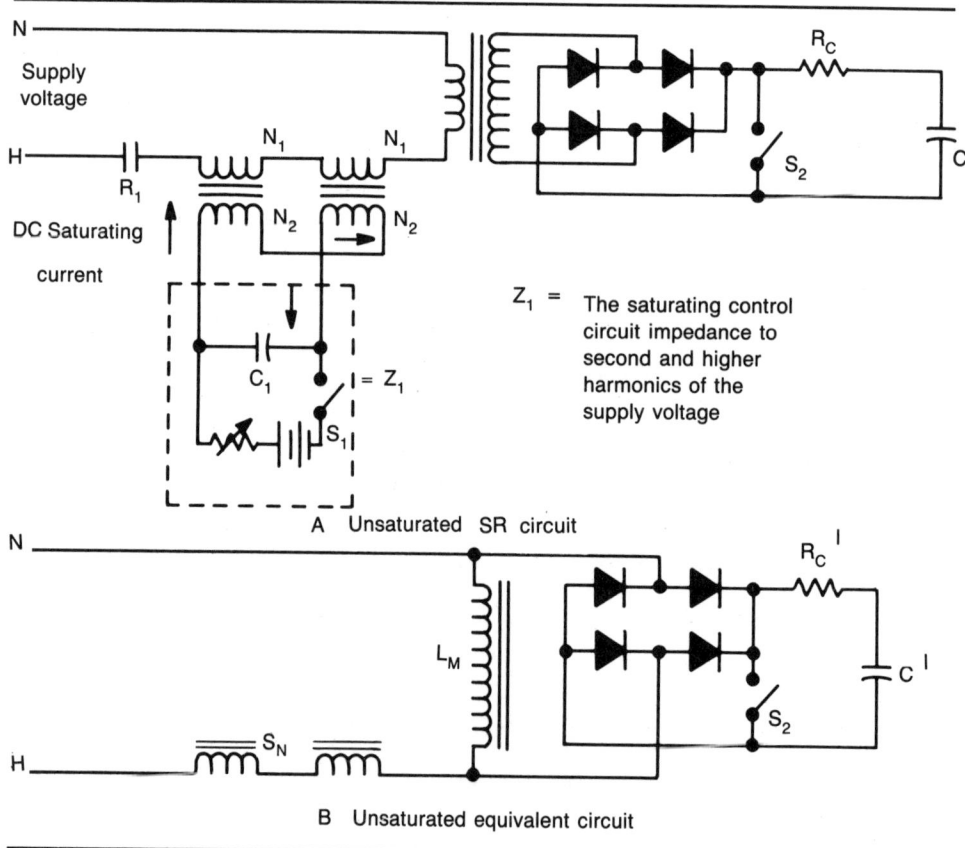

Fig. 8-17. *Saturable reactor OFF condition.*

is low (large capacitor), essentially all of the supply voltage (up to 94 percent) can be applied to the power supply input. If the ac control impedance is large, the voltage applied to the power supply is limited through the ratio of the transformed control circuit impedance and the reflected power supply impedance. In the latter case, the power supply output is essentially a square wave with a duration equal to the time either winding is saturated.

As the first ac half-cycle comes to an end, the saturated winding unsaturates to again isolate the power supply from the ac line. On the opposite half-cycle, the opposite winding saturates at the same voltage level (flux level) with similar results in the power supply output.

The current rise time constant is determined by the inductance and resistance of the control winding and power supply impedance if the control dc supply is used to turn on the supply. A simple feedback circuit monitoring the power supply current can be used to modify the SR control winding current to maintain a constant current charge system. SR systems can thus be used as constant current systems up to about 80 percent of their rated output.

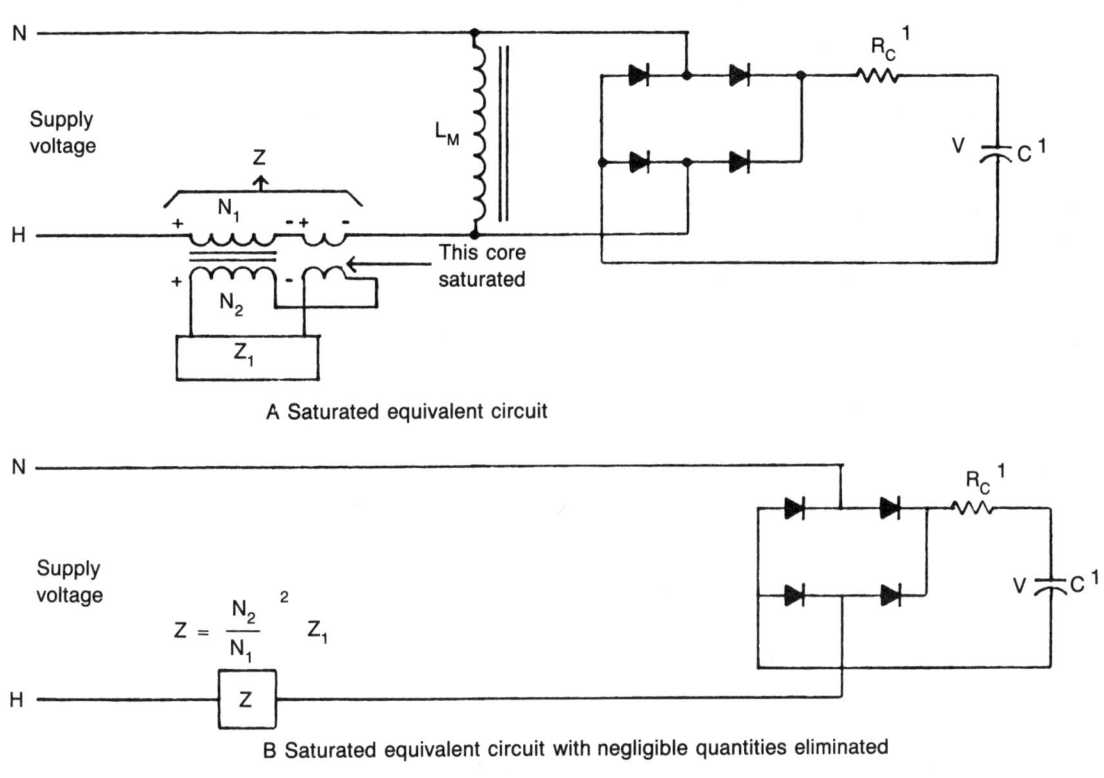

Fig. 8-18. Saturable reactor ON condition.

8.2 THE AC CHARGING SYSTEMS

In ac or resonant charging systems, the transfer of charge is governed by $(LC)^{1/2}$ time constants. The basic resonant charge circuit is shown in Fig. 8-19. Charge is transferred from the output of a low-impedance power supply or large filter capacitor, C_1, through a charging inductor with inductance L and total circuit resistance, R, to an intermediate energy storage capacitance or PFN with total capacitance, C_2. If the power supply output voltage (initial filter capacitor voltage) is V_o and switch S1 is closed at $t = 0$, the resulting current is given by:

$$i(t) = \frac{V_o}{\omega L} \exp\left(-\frac{R}{2L}\right) \sin \omega t \qquad (8\text{-}26)$$

where:

$$\omega = \left[\frac{C_1 + C_2}{LC_1C_2} - \left(\frac{R}{2L}\right)^2\right]^{1/2} \qquad (8\text{-}27)$$

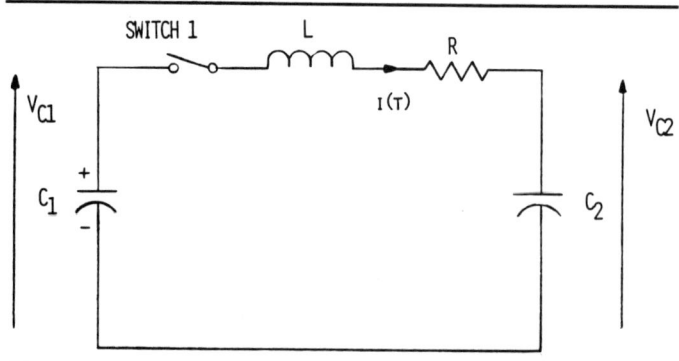

Fig. 8-19. Basic resonant charge circuit.

The capacitance voltages are given by:

$$v_{C_1}(t) = V_o \left\{ 1 - \frac{C_2}{C_1 + C_2} \left[1 - \frac{\exp\left(-\frac{R}{2L}t\right)}{\omega} \left(\frac{R}{2L}\sin\omega t + \omega\cos\omega t\right) \right] \right\} \quad (8\text{-}28)$$

and

$$v_{C_2}(t) = \frac{V_o C_1}{C_1 + C_2} \left[1 - \exp\frac{\left(-\frac{R}{2L}t\right)}{\omega} \left(\frac{R}{2L}\sin\omega t + \cos\omega t\right) \right] \quad (8\text{-}29)$$

For most resonant charge systems, the resistance is minimized to maximize voltage transfer such that:

$$\frac{R}{2L} << \left(\frac{1}{LC}\right)^{1/2} \cong \omega \quad (8\text{-}30)$$

The equations for current and capacitor voltages are simplified to:

$$i(t) = V_o \left[\frac{C_1 C_2}{(C_1 + C_2)L}\right]^{1/2} \exp\left(-\frac{R}{2L}t\right) \sin\omega t \quad (8\text{-}31)$$

$$v_{C_1}(t) = V_o \left[1 - \frac{C_2}{C_1 + C_2}(1 - \cos\omega t)\exp\left(-\frac{R}{2L}t\right) \right] \quad (8\text{-}32)$$

$$v_{C_2}(t) = V_o \left(\frac{C_1}{C_1 + C_2}\right)(1 - \cos\omega t)\exp\left(-\frac{R}{2L}t\right) \quad (8\text{-}33)$$

The maximum voltage on capacitor C_2 occurs at the half-period $T/2$ of the resonant oscillation given by:

$$\frac{T}{2} = \pi \left(L \frac{C_1 C_2}{C_1 + C_2} \right)^{1/2} = \frac{\pi}{\omega} \qquad (8\text{-}34)$$

and has a value of:

$$V_{C_2 \text{max}} = \frac{V_o C_1 \cdot 2}{C_1 + C_2} \cdot \exp\left(-\frac{R}{2L} \frac{\pi}{\omega} \right) \qquad (8\text{-}35)$$

The charging efficiency or the ratio of the energy delivered by the power supply finally stored in C_2 at time $T/2$ is given by:

$$\eta_C = \frac{\int_0^{\pi/W} v_{c2}(t)\, i(t)\, dt}{\frac{1}{2} C_2 V_{C2\,\text{max}}^2} \qquad (8\text{-}36)$$

or:

$$\eta_C \cong \left(1 - \frac{\pi}{4Q} \right), \quad Q = \frac{\omega L}{R} \qquad (8\text{-}37)$$

Note from Eq. (8-35) that it is possible to charge C_2 to almost twice the power supply voltage if $C_1 \gg C_2$ or nearly complete energy transfer is possible if $C_1 = C_2$. The maximum energy and voltage transfer ratios are shown in Fig. 8-20, and the voltage

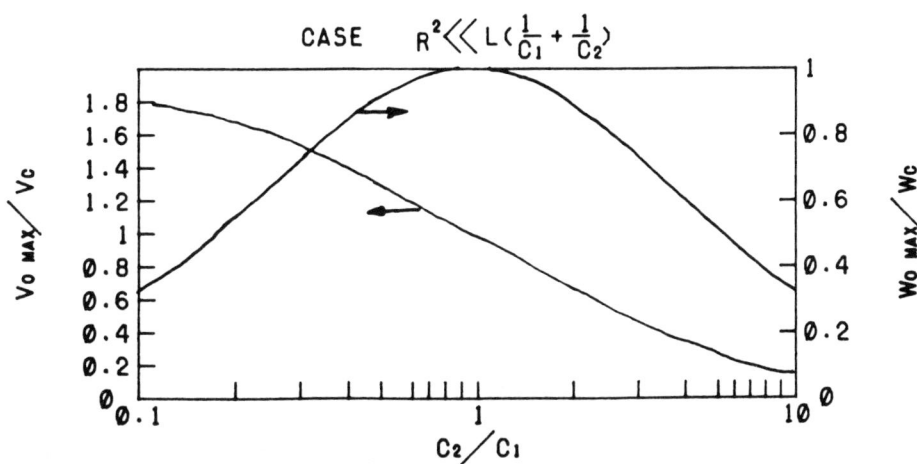

Fig. 8-20. Calculated resonant-charge efficiency and voltage transfer.

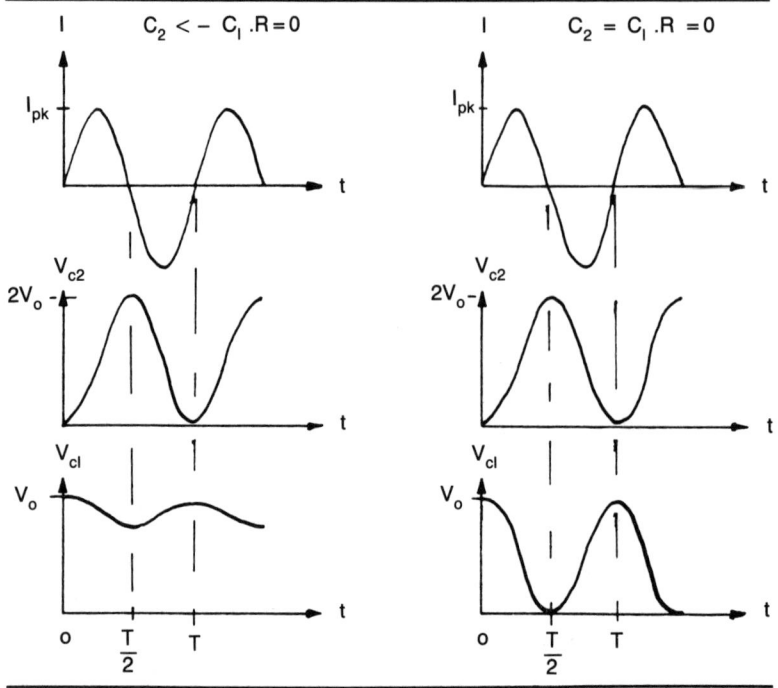

Fig. 8-21. Resonant circuit waveforms.

and current waveforms for the two cases $C_1 = C_2$ and $C_1 >> C_2$ are shown in Fig. 8-21. The peak current during charge occurs at $T/4 = \pi/2\,\omega$, and is given by:

$$I_{pk} = V_o \left[\frac{C_1 C_2}{(C_1 + C_2)L} \right]^{1/2} \cdot \exp\left(-\frac{R}{L} \cdot \frac{\pi}{\omega}\right) \tag{8-38}$$

The standard charge system also contains a series-charging diode shown in the simplified circuit of Fig. 8-22 to prevent the charge from returning to C_1. A *command charge system* is a a resonant charge system that utilizes a switch element, thyratron, SCR, hard tube, etc., as S1, which closes to permit irregular or charge on command and then opens at current zero. Figure 8-22 also illustrates a basic load and output switch S2, which is closed to discharge the intermediate energy store C_2 into the load. In a resonant charge system, the intermediate energy store would automatically recharge when S2 is opened. Typical waveforms for the circuit of Fig. 8-22 are shown in Fig. 8-23.

Again, a small digression is necessary to define current quantities that are relevant to load and component ratings. The average of the current i(t) is defined by:

$$I_{AVG} = \frac{1}{t} \int_0^{T_p} i(t)\, dt \tag{8-39}$$

8.2 The ac Charging Systems

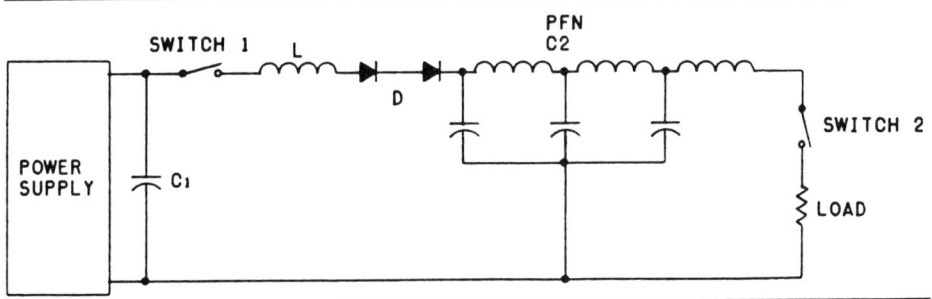

Fig. 8-22. Command or resonant-charging system.

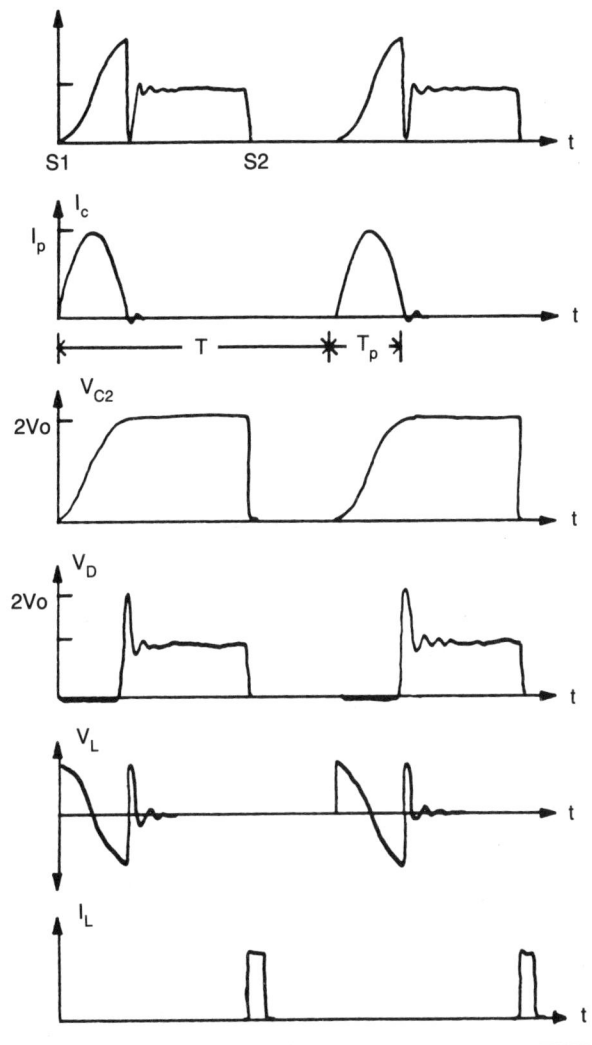

Fig. 8-23. Resonant-charge waveforms.

where T is the repetition period, and T_p is the duration of the current $i(t)$.

Another important quantity is the RMS current that is used to determine power dissipation of components and is given in general by:

$$I_{RMS}^2 = \frac{1}{t} \int_0^{T_p} i(t)^2 dt \qquad (8\text{-}40)$$

For a sinusoidal half-cycle $i(t)$ with half period, T_p and peak current I_{pk} and interpulse period T as shown in Fig. 8-24A, the average and RMS currents are given by Fig. 8-24A and then:

$$I_{AVG} = \frac{2I_{pk}}{\pi} \left(\frac{T_p}{2T}\right) \qquad (8\text{-}41)$$

$$I_{RMS} = I_{pk} \left(\frac{T_p}{4T}\right)^{1/2} \qquad (8\text{-}42)$$

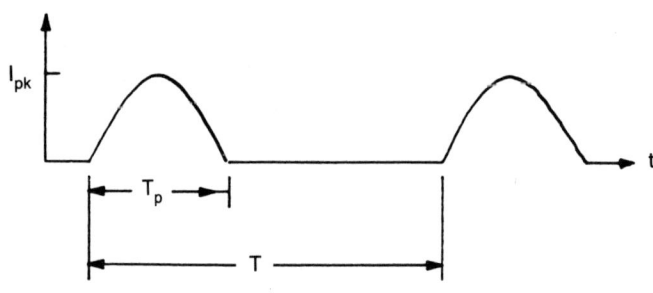

A Sinusoidal repetitive pulse.

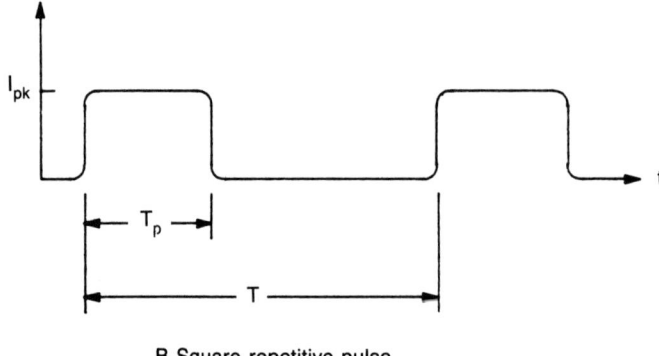

B Square repetitive pulse.

Fig. 8-24. Average and RMS waveforms.

Similarly, for a square current pulse with amplitude I_{pk}, duration T_p and interpulse spacing T as shown in Fig. 8-24B. The average and RMS currents are given by:

$$I_{AVG} = I_{pk} \left(\frac{T_p}{T}\right) \qquad (8\text{-}43)$$

and:

$$I_{RMS} = I_{pk} \left(\frac{T_p}{T}\right)^{1/2} \qquad (8\text{-}44)$$

The pulse repetition rate is determined by the interpulse spacing T or:

$$\text{PRF} = \frac{1}{T} \qquad (8\text{-}45)$$

The individual components of the basic resonant/command charge system can be specified using the average and RMS current values.

8.2.1 Charging Inductors

A basic circuit model of charging inductor is shown in Fig. 8-25, where R_c represents core eddy current losses, and R_w represents the inductor winding resistance. The interwinding capacitances are represented by C_{ww} and winding-to-core

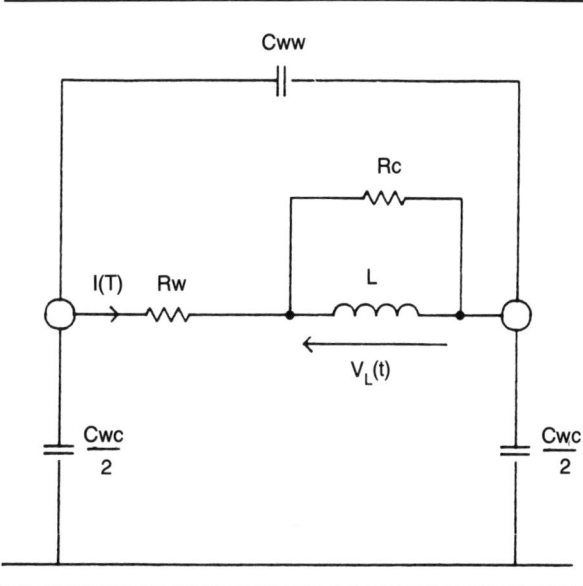

Fig. 8-25. Charging choke equivalent circuit.

capacitances by C_{wc}. The inductance of the charging choke is determined from the simple circuit of Fig. 8-19, or:

$$L = \left(\frac{1}{C_1} + \frac{1}{C_2}\right) \cdot \left(\frac{T_c}{\pi}\right)^2 \qquad (8\text{-}46)$$

where T_c is the half period of the resonant charge oscillation or the charge time. The RMS is used to size the inductor wire in order to insure acceptable copper losses.

The energy dissipated in the inductor windings per charge pulse is given by:

$$E_{pw} = \int_0^{T_c} i^2(t) \cdot R_w \, dt \qquad (8\text{-}47)$$

$$= T \cdot I_{RMS}^2 \cdot R_w$$

The energy lost to eddy currents per pulse is given by:

$$E_{pc} = \int_0^{T_c} \frac{V_L^2}{R_c} \, dt \qquad (8\text{-}48)$$

which for a sinusoidal half-cycle is equal to

$$E_{pc} = \frac{\omega^2 \cdot L^2}{2R_c} \cdot I_{RMS}^2 \qquad (8\text{-}49)$$

When specifying a charging choke, the following requirements and information are used.

a. Inductance $L(+6, -2\ \%)$, linear
b. Peak current I_{pk}
c. Charge time T_c
d. Repetition rate $PRR = \frac{1}{T}$
e. RMS current $I_{pk} \cdot \left(\frac{T_c}{2T}\right)^{1/2}$
f. Peak voltage $\omega L = \frac{2\pi L}{T_c}$
g. Core potential   This is, core to be grounded, etc.
h. Capacitance screen location and circuit connection

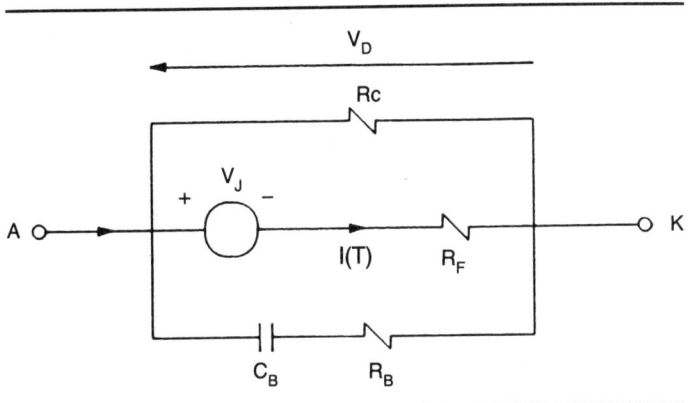

Fig. 8-26. Charging diode equivalent circuit.

8.2.2 Charging Diode

A simple model of a forward-biased charging diode or diode stack is shown in Fig. 8-26. The instantaneous forward power deposited in the diode is given by:

$$P_F = V_j\, i(t) + R_F\, i^2(t) \tag{8-50}$$

The value R_F represents the diode forward resistance where V_j is the semiconductor junction voltage drop. Thus, the energy deposited in the diode unit per pulse is:

$$E_{pd} = \int_0^T \left(V_j\, i^2(t) + R_F\, i^2(t) \right) dt \tag{8-51}$$

or

$$E_{pd} = T\, V_j\, I_{AVG} + T R_F\, I_{RMS}^2 \tag{8-52}$$

Without biasing resistors, R_B can represent series diode leakage impedance, R_j, and C_B can represent series combination of the individual diode junction and stray capacitance, C_j. The resistance R_B and the capacitance C_B can also represent the series combination of dc biasing resistors and ac biasing capacitors placed across each diode in a stack. Charging diodes are specified with the following parameters and recommendations:

Peak reverse voltage	$4V_o$—safety factor of 2
Average current	$2I_{AVG}$—safety factor of 2
RMS current	I_{RMS}
Recovery time	$\leq T_c/10$
Turn-on time	$(1/e)\, T_p/10$
Repetition rate	PRR = $1/t$

Bias capacitors   nC_B ≥ 10 C_j (each diode)

Bias resistors $\frac{R_B}{n} \le \frac{R_R}{10}$ (each diode)

Number of series diodes   n

8.2.3 Transient Suppression Circuit Design

A more complicated command charge system is shown in Fig. 8-27. In this section, the effect of stray and component capacitances will be added to the circuit model. The previous section designed the major circuit components that determine the first-order circuit voltages and currents. The stray capacitances produce unwanted and potentially damaging transient currents and voltages during Switch S1 closing and opening, during diode commutation, and during load Switch S2 closing and opening. Equivalent circuits will be used to determine effective methods of controlling the transient oscillations by adding additional snubber RC networks.

The objective of designing snubber networks is to control the unwanted oscillation by adding a reactive component that dominates the value of the natural or stray component and then adding a dissipative component to make circuit Q approximately one or two. Thus the equivalent circuit of the transient network is very important in determining the correct snubber system.

8.2.4 Charging Switch Snubber

The command charge switch of Fig. 8-27, when opened at the end of a charging cycle, experiences a large voltage transient. The transient is due to residual current flowing in the stray inductance L_S and the stray capacitance C_{S1}, as shown in the equivalent circuit of Fig. 8-27A. The snubber capacitor, C_S, is chosen much larger than the switch stray capacitance C_{SS}. The equivalent circuit is then reduced to Fig. 8-28B when $C_1 \gg C_S, C_{SS}, C_{S1}$ and $C_S \gg C_{S1}$. The snubber resistance, R_S, is chosen using:

$$R_S = \left(\frac{L_S}{C_{S1}}\right)^{1/2} \cdot \frac{1}{\left(Q^2 + \frac{1}{4}\right)^{1/2}} \tag{8-53}$$

to damp the resultant circuit of Fig. 8-28C to a Q of two. Note that the stray capacitance C_{S1} must be known to accurately design the snubber circuit. Ideally, C_S could be chosen such that $C_S \gg C_{S1}, C_{SS}$ or $C_S' = C_S$. However, at higher voltages, the required value and voltage rating of C_S can become very expensive. Thus a compromise value of C_S is chosen. As a rule of thumb, most snubber capacitances are on the order of 1 nF.

8.2.5 Charging Inductor Snubber

The charging inductor also experiences transients due to residual current flow when either the switch S_1 of Fig. 8-27 opens or the diode stack commutates. The equivalent circuit of the changing inductor transient circuit is shown in Fig. 8-29A. First, the stray

8.2 The ac Charging Systems 267

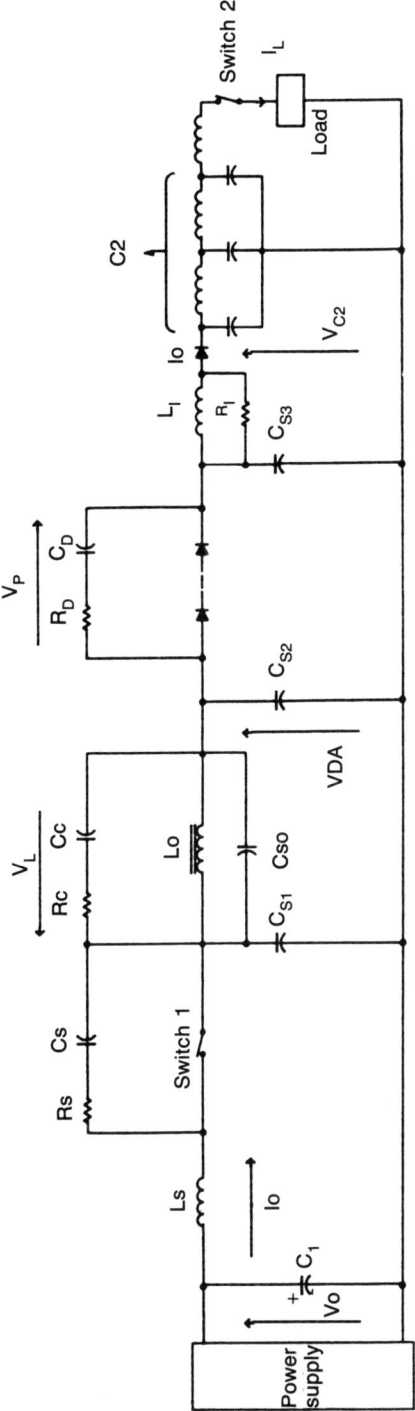

Fig. 8-27. Charging schematic for snubber design.

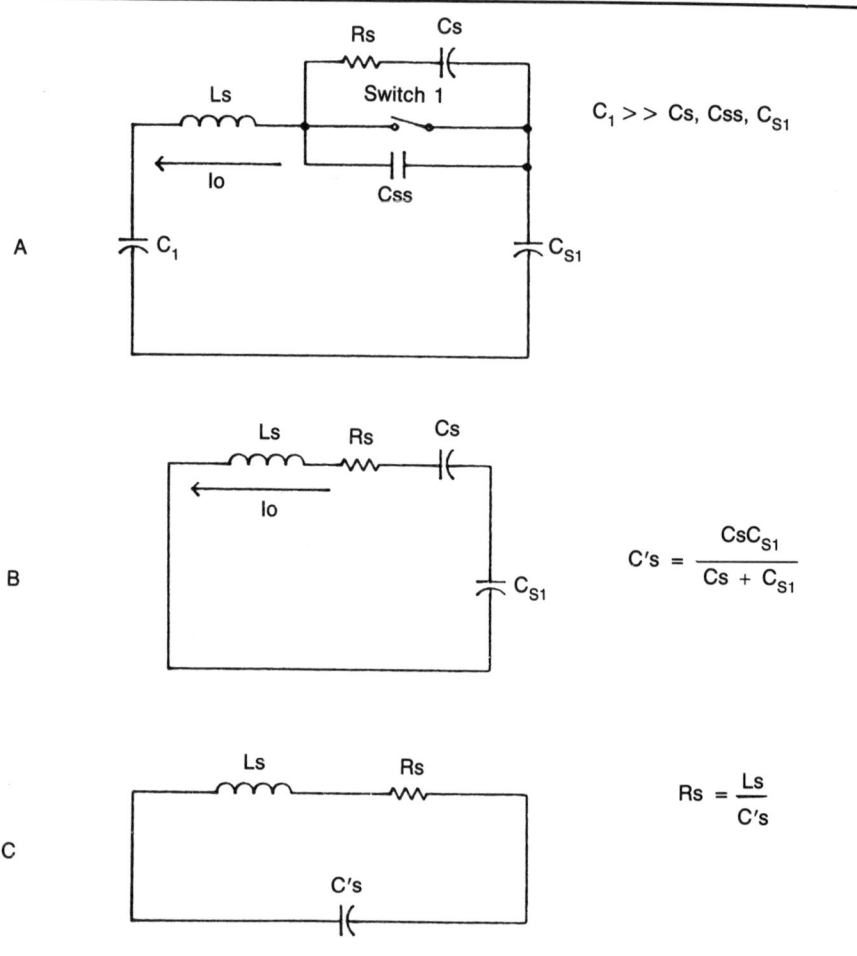

Fig. 8-28. Switch snubber.

capacitances are consolidated in C_S' as shown in Fig. 8-29B. Then C_C is chosen to swamp C_S' or $C_C \gg C_S'$, which yields the circuit of Fig. 8-29C. Now R_C is chosen to provide a Q of 2 or:

$$R_C = \left(\frac{L_C}{C_C}\right)^{1/2} \cdot \frac{1}{\left(Q^2 + \frac{1}{4}\right)^{1/2}} \tag{8-54}$$

8.2.6 Diode Snubber

When switch S_1 of Fig. 8-27 is closed or when the diode stack commutes, the diode stack experiences large, high-frequency transient voltages up to twice the operating voltage. In order to prevent the transient reverse voltages, a snubber circuit is placed across the diode stack as illustrated in Fig. 8-30. An equivalent circuit is taken from Fig.

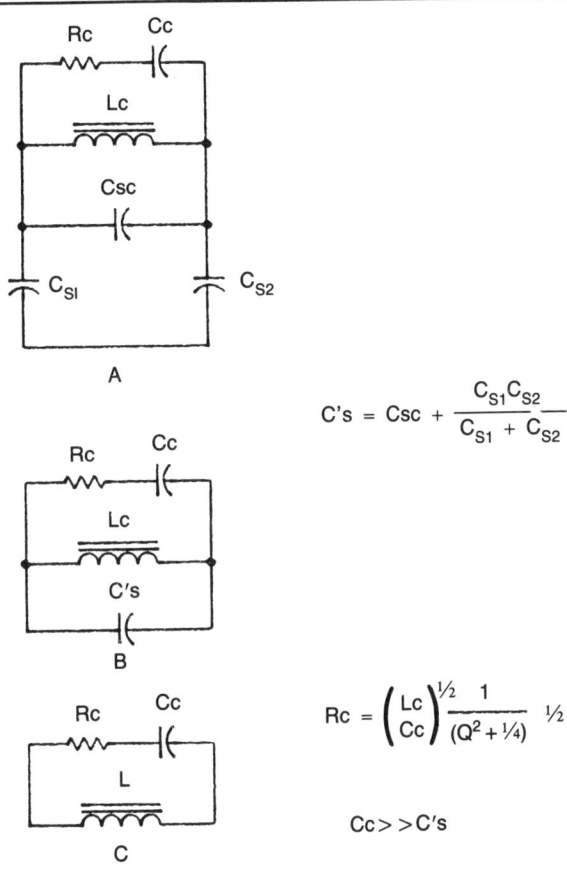

Fig. 8-29. Charging inductor snubber equivalent circuits.

8-27 as shown in Fig. 8-30A. Then the stray capacitances are consolidated in Fig. 30B. The snubber capacitance, C_D, is then chosen to swamp the effect of C_{S23} or $C_D >> C_{S23}$. Then R_D is chosen to provide a time constant greater than the diode reverse recovery time or:

$$R_D C_D > T_{RR} \qquad (8\text{-}55)$$

Most diodes have reverse recovery times on the order of 2 microseconds and some as short as 100 nanoseconds. Many commercial 60 Hz diode stacks have $R_D C_D$ snubbers with time constants on the order of 1 microsecond. The diode snubber also affects the forward voltage across the diode stack during turn on. For very high frequency circuits at high current, the diode turn on times are on the order of 0.5 to 1 microsecond. Without an appropriately designed snubber network, large forward voltages appear across the diode during turn on, which can produce large anode dissipation.

270 Charging Systems

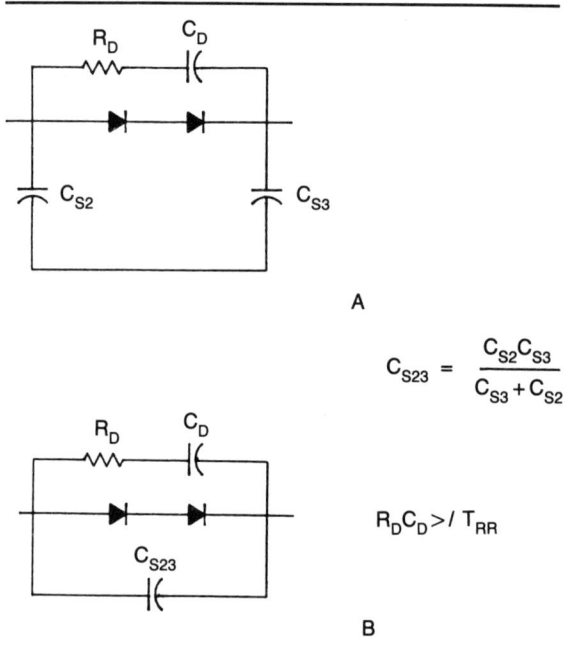

Fig. 8-30. Diode stack snubber equivalent circuits.

8.2.7 Isolation Network

In order to isolate the diode stack from the switching transients of S_2 in Fig. 8-27, a snubber/isolation network is provided to decouple the charge network from the output circuit during pulse output. The parallel LR network shown in Fig. 8-27 (R_I, L_I) is used to maintain a constant current through the charging circuit during the time S_2 is closed. Ideally the constant current value is zero. The circuit of Fig. 8-27, after adding the previously designed snubbers, can be reduced to the circuit of Fig. 8-31A. After consolidating the resistors and capacitors and neglecting the other inductances, the equivalent circuit of Fig. 8-31B results. When S_2 is closed ($t = 0$) the current in the circuit of Fig. 8-31B is given by:

$$i(t) = \frac{V_o}{R_{LOAD} + R' + R_I} \cdot e^{-t/2R_I e'} \cdot \left(1 + \frac{t}{2R_I C'}\right) \quad (8\text{-}56)$$

which has a slope of zero at $t = 0$. Thus the $L_I R_I$ network minimizes the rate of change of current in the charging circuit for an e-folding time of $R_I C'$. Usually $R_{LOAD} + R'$, especially R', are much larger than R_I, such that the $L_I R_I$ network only affects the cur-

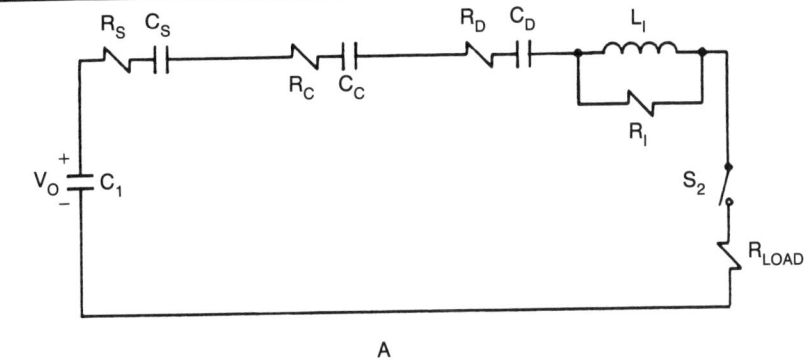

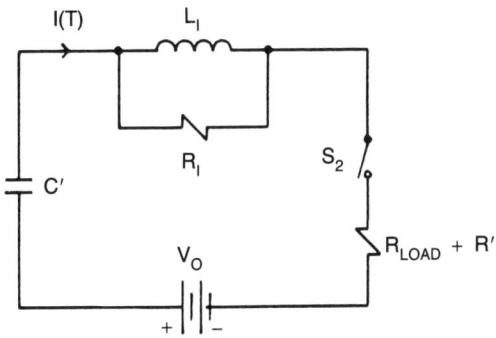

Fig. 8-31. Isolation network equivalent circuits.

rent slope and thus the voltage seen by the charging circuit while S_2 is closed. The value of L_I and R_I are determined by:

$$L_I = R_I^2 C' \tag{8-57}$$

The value of $R_I C$ should be such that:

$$R_I C' = 10 T_p \tag{8-58}$$

where T_p is the output pulse length.

8.2.8 Diode-Surge Curcuit and Ratings

Charging diodes are usually specified by average and RMS current ratings. However, the diode surge ratings are important if system malfunction is expected (Murphy's probability $\cong 1$). Figure 8-32 illustrates the circuit for the failure mode in which S_2 and S_1 remain closed and the charging diode experiences a surge consisting of the discharge of the power supply filter, C_1, plus the power supply short circuit current until the ac breaker opens. The equivalent circuit of Fig. 8-32B indicates the two components of the diode surge current. The diode surge parameters should be defined by two parameters. The peak surge current should be designed to pass the initial current peak, and the diode I^2t rating should be able to handle the entire current pulse of Fig. 8-32C. Note that this value of I^2t depends on the opening time of the ac breaker.

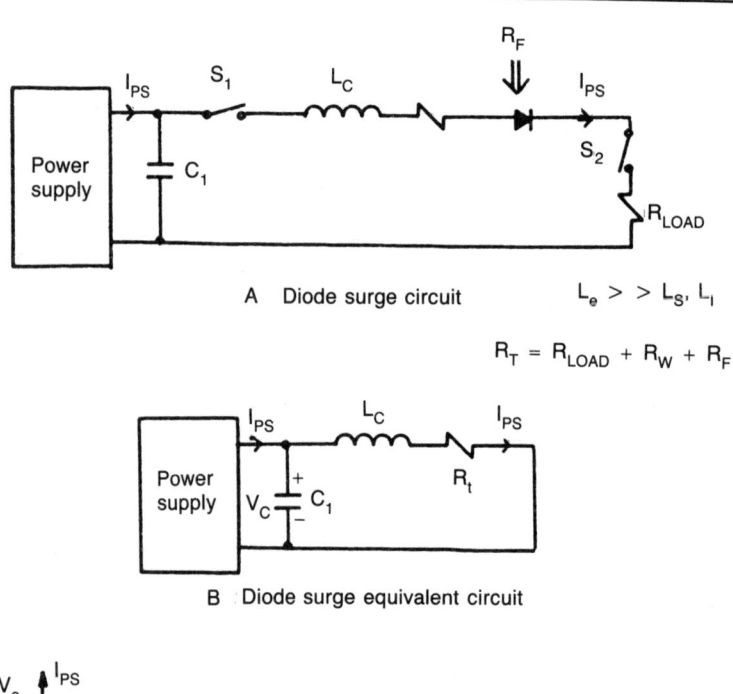

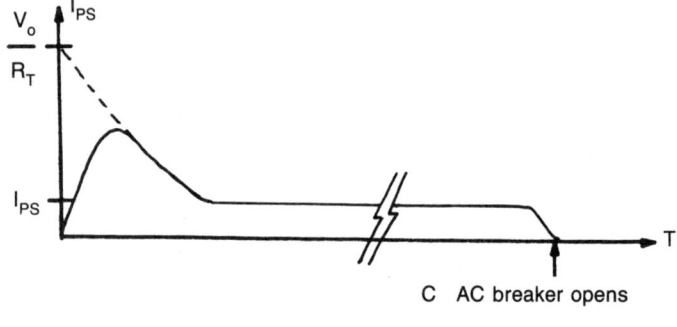

Fig. 8-32. Diode surge circuits and waveform.

8.2.9 Alternate Charging Systems

Several other circuits for charging the intermediate energy store or for regulating the final voltage of the intermediate energy store are shown in Fig. 8-33. The circuit of Fig. 8-33A illustrates a two-step, transformer charging method. This circuit is usually used when fast charging of C_2 is desired upon command of switch 1. The circuits of Fig. 8-33B and C are used to control the voltage on the intermediate energy store represented by the PFN. In the circuit of Fig. 8-33B a resistance is switched into the charging transformer secondary when the voltage on the PFN has reached a desired value during the resonant charge cycle. This effectively places a resistance in series with the resonant

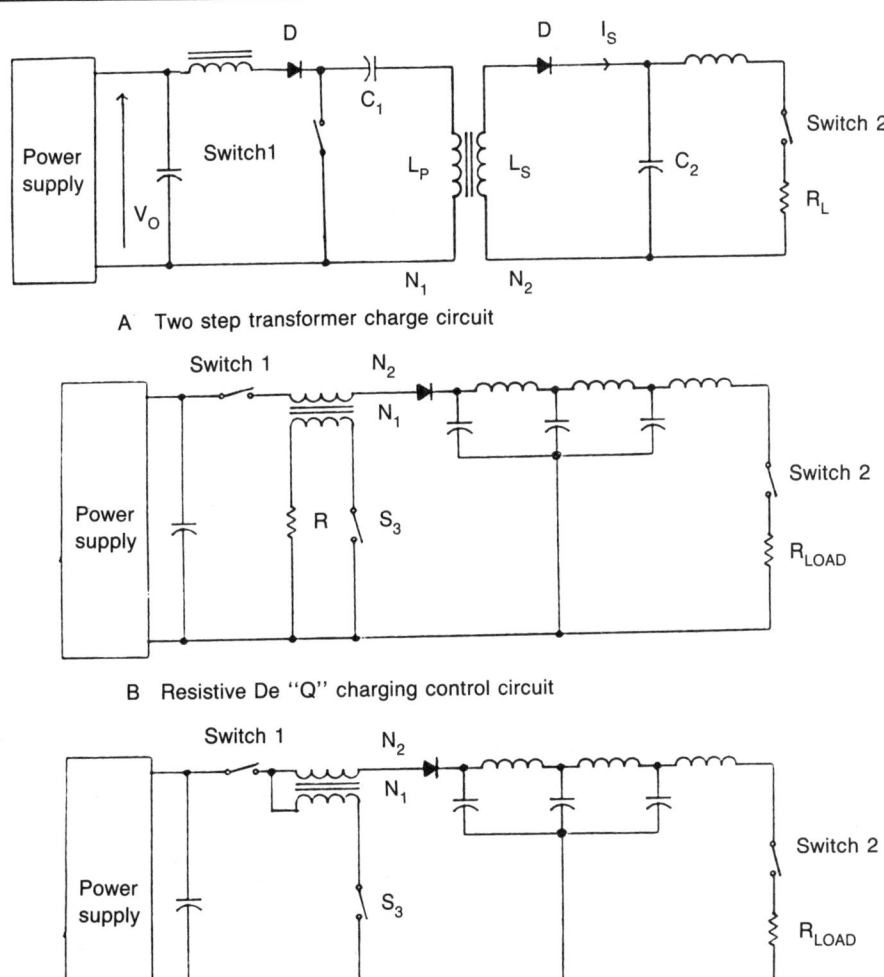

Fig. 8-33. Alternate charging systems.

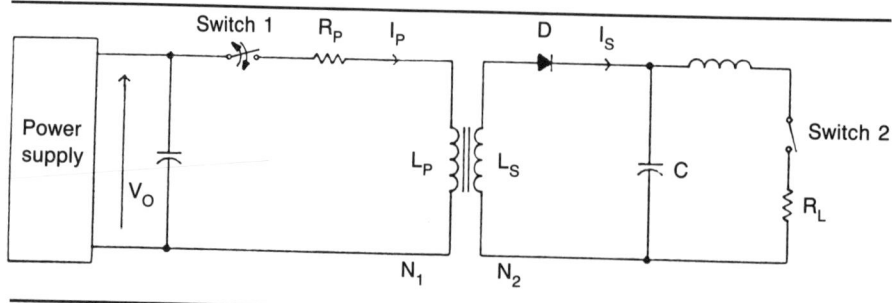

Fig. 8-34. Flyback charging circuit.

charge circuit and spoils the circuit quality factor (Q), thus the name "DeQ" circuit. The circuit of Fig. 8-33C is a reactive DeQ circuit. At the time the PFN reaches the desired voltage, switch S_3 is closed. The charging current and thus the inductor primary flux is not zero. The flux is removed from the charging inductor by way of the secondary winding, and the energy is returned to the low-impedance power supply filter capacitor.

Another method used to command charge the intermediate energy store is the flyback system shown in Fig. 8-34. This system requires opening and closing a switch in the primary of the transformer. This method is also a two-stage process. First, the required energy or some quantum of the required energy is stored in the primary inductance of the transformer. Then the primary switch is opened, the inductor flux collapses and reverses the polarity of the voltage on the transformer secondary. This action in turn, forward biases the secondary diode and transfers the inductively stored energy to the secondary capacitance. This system has the advantage that only the energy stored in the transformer is available to a secondary fault.

8.3 REFERENCES

1. T.M. Sprague, "Capacitor Bank Charging System," Los Alamos Scientific Laboratory Technical Note, 1965.
2. W. Willis, "Constant Current (Monocyclic) Charging Circuit," Los Alamos Scientific Laboratory Technical Note, 1973.
3. B. Hayworth and D. Warrilow, "Constant Power Charging Supplies," "Electro-Optical Systems Design," 19, p. 42.
4. R.W. McMillian, "How to Pick the Best Power Supply Type for Capacitor Charging in a Pulsed Laser," "Laser Focus," February 1977, p. 62.
5. H.K. Jennings, "Charging Large Capacitor Banks in Thermonuclear Research," "Electrical Engineering," June 1961.
6. J.L. Brewster, el al., "Design Studies for Ultra-fast, Low-Impedance High-Peak-Power Pulsed Systems," Field Emission Corporation, Air Force Weapons Laboratory Report AFWL-TR-65-21, November 1965.
7. P. Mlynar and G.C. Seaborn, "High Voltage Silicon Rectifier Designer's Handbook," Westinghouse, 1963.
8. R. Murray, Editor, "Silicon Controlled Rectifier Designer's Handbook," Westinghouse, 1964.

9. A. Schofield, "Saturable Reactor Charging Systems," Los Alamos Scientific Laboratory Technical Note, 1979.
10. G.N. Glasoe and J.V. Lebacqz, Editors, "Pulse Generators," Dover Publications, Inc., NY, 1965.
11. G.J. Thaler and M.L. Wilcox, "Electric Machines: Dynamics and Steady State," John Wiley & Sons, Inc. New York, New York 1966.

Chapter 9

High-Voltage Air Core Pulse Transformers*

by
Gerald J. Rohwein
SANDIA NATIONAL LABORATORIES
Albuquerque, NM 87185

AIR CORE PULSE TRANSFORMERS ARE TYPICALLY FOUND IN APPLICATIONS INVOLVING very high peak power levels (up to 10^{11} W) or ultrahigh RF frequencies. They differ from the more common iron or ferrite types in that no magnetic core materials are used to channel the magnetic flux through the winding to assure flux linkage. With air-core transformers, flux linkage is strongly dependent upon the physical proximity of all turns of one winding with respect to all turns of the other winding. Consequently, the coupling coefficients of air-core transformers tend to be lower than for magnetic core transformers, particularly with high-gain transformers that have thick windings. However, because air-core transformers do not have magnetic cores, they are not limited in current-handling capacity by saturation of the magnetic materials or frequency limited by the composition of the core.

The lack of core saturation and frequency limits are the most significant advantages of air core pulse transformers. They are, therefore, well suited for use with large, high-current primary power sources such as parallel capacitor banks which often operate at peak current levels of over a megampere and to such applications as high-voltage triggering where the discharge frequency may be several megahertz.

9.1 DISCUSSION

There are many devices that could be included in a discussion of air core pulse transformers. This chapter will cover only the general types designed for high-voltage pulse generation and energy transfer applications. Special emphasis is given to pulse-charging systems that operate up to the multimegavolt range.

---

*This work was supported by the U.S. Department of Energy, under Contract DE-AC04-76-DP00789.

9.1.1 Types of Air Core Transformers

There are two basic types of high-voltage air core pulse transformers that can be operated in the megavolt range. The first and most common is the single-layer helical wound transformer (Fig. 9-1). The second is the spiral-strip type (Fig. 9-2). These transformers differ from each other primarily in the configuration of the secondary windings, which, as is shown in this book, accounts for a significant difference in their resistance to insulation failure from fast pulses. The primary windings for either type, however, whether single or multiple turn, can be designed in a variety of ways without affecting electrical breakdown characteristics of the transformers. For reasons of high-voltage isolation, the low-voltage primary winding is typically placed outside the secondary in either type transformer so that the high voltage output of the secondary can be led out through the center of the assembly.

With helical transformers, high-voltage standoff between the primary and secondary windings is provided by an insulated space between the windings. The space can be uniform or tapered in the longitudinal direction. When tapered, the insulation thickness

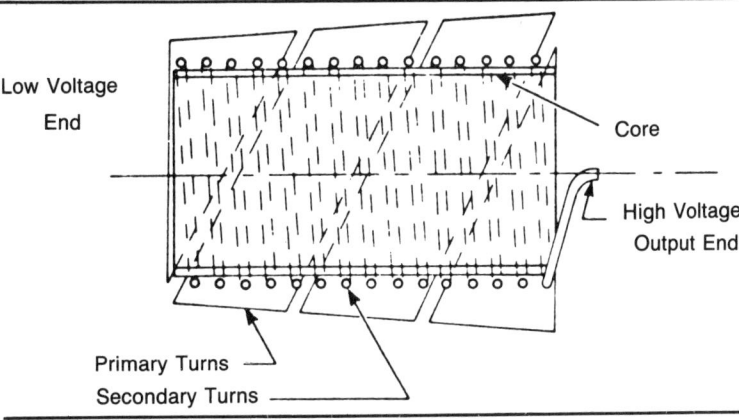

Fig. 9-1. Basic helical-wire transformer.

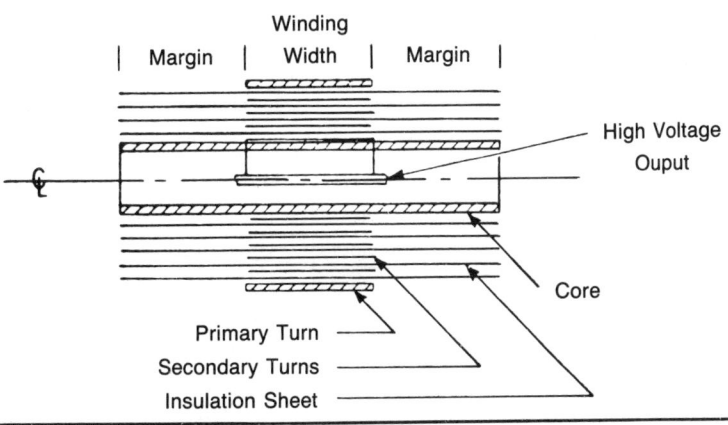

Fig. 9-2. Basic spiral-strip transformer.

(usually oil) increases with the voltage along the length of the coils. With spiral-strip transformers, voltage standoff is largely a function of the radial thickness of the secondary winding because the turns directly overlay each other. The winding stack, therefore, has a pure radial voltage gradient between the high-voltage inner turns and low-voltage outer turns. Like helical type transformers, the open volume in spiral strip transformers is usually insulated with oil. However, because of the high winding density, spiral-strip windings must ordinarily be vacuum impregnated to displace air from the secondary winding.

9.1.2 Common Problems

Two problems are common to both types of transformers. One problem is electrical breakdown turn-to-turn or between windings. The other problem is partial shorting from eddy currents induced in voltage-grading devices and structural components that are present in high-voltage transformers. Turn-to-turn breakdown is common with helical transformers used in charging systems for pulse-forming lines (PFL). This problem arises from fast-rising voltage transients (usually less than 10 nanoseconds) generated by the discharge of the PFL that are fed into the output of a direct coupled transformer. The turns of a helical winding are inductively, and transit time isolated from each other. The capacitance components between turns and from each turn to ground (Fig. 9-3) are not sufficient to grade fast-rising transients. Consequently, a voltage pulse approaching the full amplitude of the transient can momentarily appear across the final turns of the secondary and cause breakdown.

9.1.3 Voltage-Grading Techniques

This problem was corrected in the FRIZZ[1] transformer design by adding a capacitive voltage grading dish across the output turns of the transformer as shown in Fig. 9-4. Spiral-strip transformers are inherently less prone to breakdown from fast voltage transients because the interturn capacitance components are all directly in series to ground (Fig. 9-5). They are comparatively large because of the large surface area and close spacing between turns of the spiral strip. Consequently, a fast voltage pulse is capacitively graded through the thickness of the winding. The principal weakness of simple spiral strip windings, however, is their tendency to break down from the thin edges of the strip. An arc breakdown typically originates from the edge of one of the final turns,

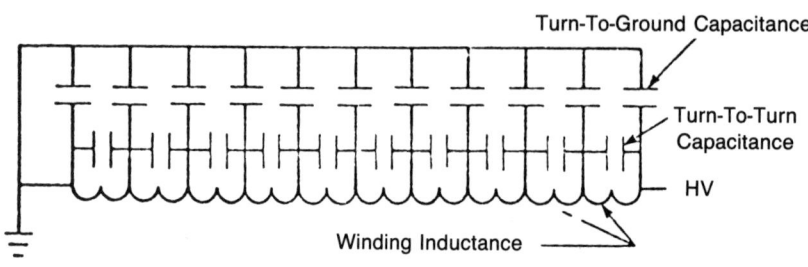

Fig. 9-3. Equivalent circuit along the length of helical wound transformer.

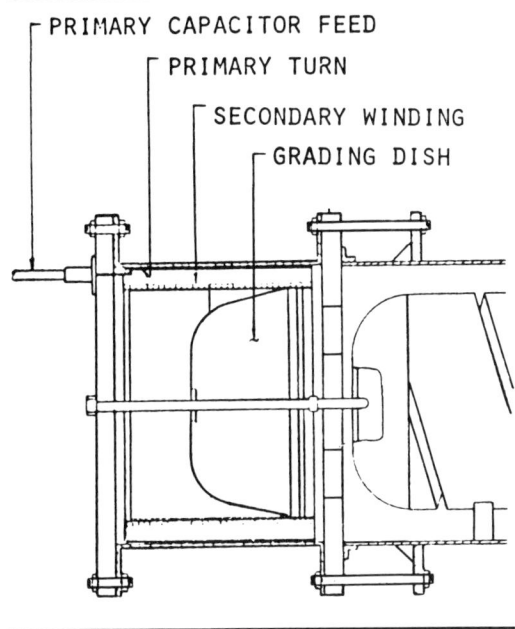

Fig. 9-4. FRIZZ, helical transformer.

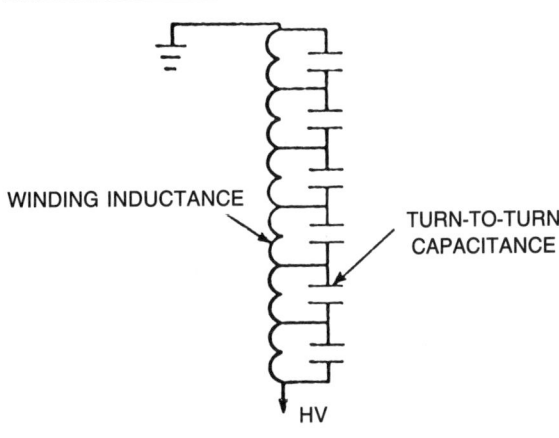

Fig. 9-5. Equivalent circuit through thickness of spiral-strip winding.

flashes across the insulation margin, and closes with the low-voltage primary. Such breakdowns practically always damage the insulating film in the margin and leave a heavy carbon path through the oil.

These edge breakdowns result from highly enhanced electric fields associated with equipotentials that emerge from between the high-voltage turns and bend sharply around the edges of the turns (Fig. 9-6). The high fields can be reduced and edge breakdown eliminated by placing a coaxial shield across the margins to shape the equipotentials into

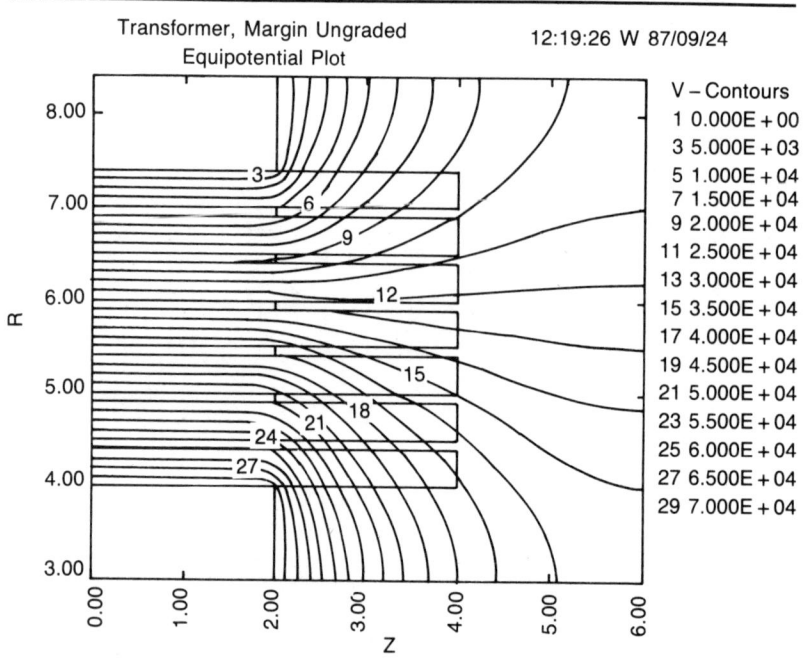

Fig. 9-6. *Equipotential profile at the edge of unshielded and upgraded spiral winding.*

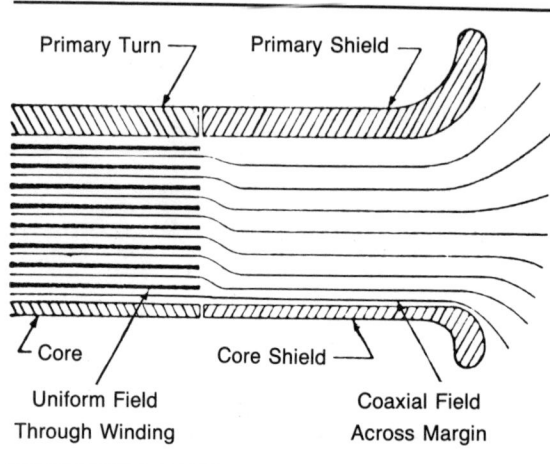

Fig. 9-7. *Coaxial shield across spiral-strip transformer margin.*

a coaxial distribution that is nearly parallel with the uniform field through the thickness of the winding (Fig. 9-7).

Adding the voltage-grading structures to either type of transformer can lead to the second major difficulty with air core transformers: eddy-current shorting of the transformer by the grading structures. Although these structures can be slotted, eddy currents can be induced as illustrated in Fig. 9-8, a diagram of eddy currents in cylindrical

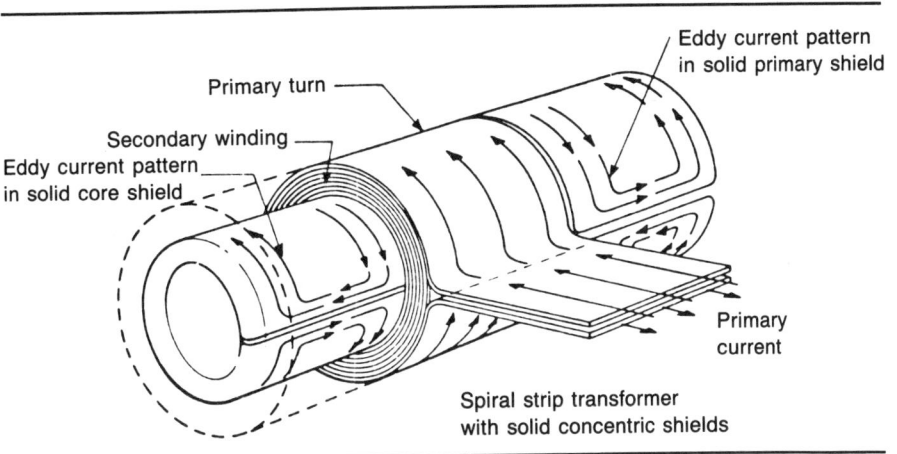

Fig. 9-8. *Eddy current pattern in solid concentric shields.*

shields around the margins of a spiral-strip transformer. The effect of eddy-current shorting is to decrease the gain and energy transfer efficiency of the transformer by an amount roughly proportional to the magnitude of the eddy currents. In one instance, cylindrical shielding reduced the gain of a spiral strip transformer by 58 percent[2] compared to an equivalent unshielded transformer.

The problem of electrostatic shielding and grading structure design must be concerned with making these devices transparent to magnetic fields. One grading technique that has been employed successfully with transformers used in single-shot service is to fill the open volume of the transformers with a resistive liquid, such as water or a solution of water and copper sulfate.[3,4] In these cases, voltages are resistively graded, but the resistivity of the solution is high enough so that no current of any magnitude can be induced in the solution. Resistive solution grading is adequate for single-shot transformers but less satisfactory for repetitive-pulse applications because of resistive losses in the grading solution.

The most successful method developed for grading spiral-strip transformers[5] is concentric ring cage shielding across the margins as shown in Fig. 9-9. Magnetic fields diffuse freely through the ring cage without inducing eddy currents in the elements, yet the proper electric field distribution is maintained in the margins. Each ring, of course, must have at least one circumferential gap to prevent current from flowing in the hoop direction of the rings. Spiral-strip transformers shielded in this manner have been operated successfully in PFL charging applications to 3 MV with over 90 percent energy transfer efficiency from the primary to secondary capacitor.[5,9] Figure 9-10 is a diagram of a ring-shielded transformer that operates with high efficiency in the multimegavolt range.

9.2 TRANSFORMER CIRCUIT ANALYSIS

Pulse-transformer circuits often utilize capacitor banks as the primary energy store. One reason for using capacitor banks is that capacitors are capable of developing higher peak power levels than any other electrical device. The transformers in these circuits

282 High-Voltage Air Core Pulse Transformers

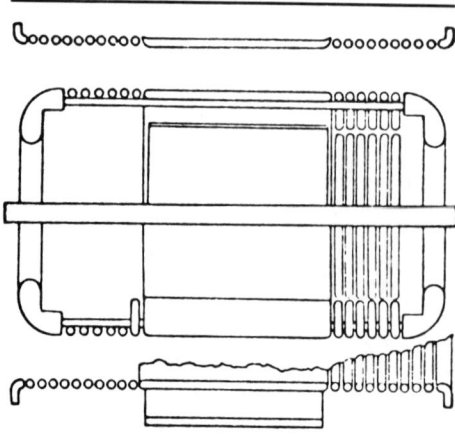

Fig. 9-9. Ring cage shield for spiral-strip transformer.

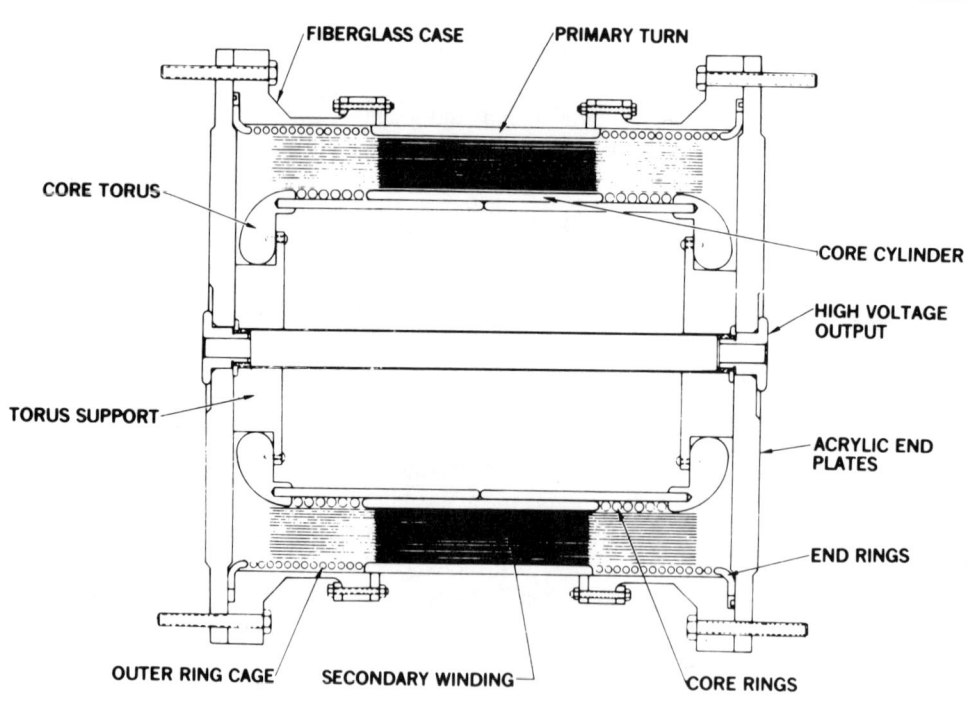

Fig. 9-10. Three megavolt transformer.

must, therefore, be capable of delivering the high primary power pulse to a load on secondary at higher voltage or higher current with a predetermined transformation ratio.

9.2.1 Ideal Relationships

For transformers with perfect coupling, no winding resistance, and negligible interturn capacitance (Fig. 9-11), a simple set of relations can be derived that relate the primary and secondary current (I_P, I_S) and voltage (V_P, V_S) to the primary and secondary turns (N_P, N_S) of the transformer:

$$I_P N_P = I_S N_S \tag{9-1}$$

$$\frac{V_P}{V_S} = \frac{N_P}{N_S} \tag{9-2}$$

The primary and secondary impedances (Z_P, Z_S) are:

$$Z_p = \frac{V_p}{I_p} \tag{9-3}$$

$$Z_S = \frac{V_S}{I_S} \tag{9-4}$$

from which

$$\frac{Z_p}{Z_S} = \left(\frac{N_p}{N_S}\right)^2 \tag{9-5}$$

From these ideal relationships, it can be seen that a transformer can be used to change current, voltage, or impedance. However, because no transformer is perfect and often both the primary and secondary sections of the circuit have resonant characteristics, the analysis of such circuits is somewhat more complex. A rigorous treatment of various pulse transformers circuits by W.H. Bostick can be found in Glasoe and Lebacqz.[6]

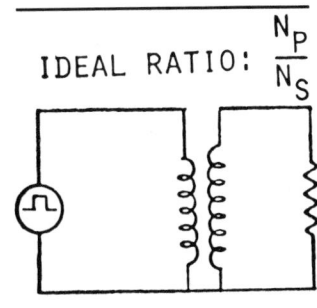

Fig. 9-11. Simple transformer circuit with perfect coupling.

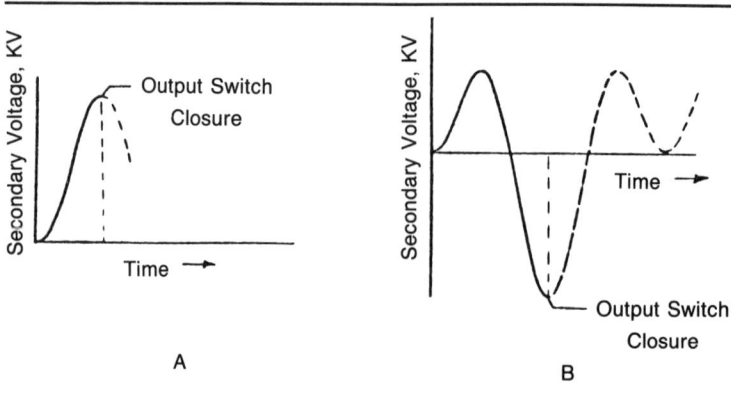

Fig. 9-12. (A) First-swing charge cycle. (B) Dual-resonance charge cycle.

Of interest are transformer circuits where low voltage capacitor banks are used for charging high-voltage, pulse-forming transmission lines by means of a voltage step-up transformer. Such charging circuits can be operated in matched- or off-resonance modes,[7] but they are generally matched as nearly as possible to maximize energy transfer efficiency. In the matched frequency mode—that is, with the open-circuit frequencies of primary and secondary sections of the circuit equal ($L_1C_1 = L_2C_2$), two cases are of practical interest. The first case is swing charging, where maximum secondary voltage is reached on the first excursion (Fig. 9-12A). The second case is dual-resonance charging, where maximum voltage occurs on the second or reverse voltage excursion of the secondary (Fig. 9-12B). The secondary voltage (V_2) as a function of time for a matched frequency circuit with damping and the transformer coefficient of coupling (K) less than one is given by:

$$V_2(t) = \frac{V_0}{2} \sqrt{\frac{L_2}{L_1}} \; e^{-t/T} \left[\cos \frac{\omega t}{\sqrt{1-K}} - \cos \frac{\omega}{\sqrt{1+K}} \right]$$

where

$$T = \frac{4L_1L_2}{R_2L_1 + R_1L_2} (1 - K^2) \text{- damping time constant}$$

V_0 = Charge voltage on primary capacitor

L_1 = the inductance of the transformer primary circuit

L_2 = the inductance of the transformer secondary circuit

ω = the radian frequency of primary and secondary circuits

$$K = \frac{M}{\sqrt{L_1 L_2}}$$

M = the mutual inductance.

The time to the first and second voltage peaks is approximately:

$$\tau_1 = \frac{\pi \omega}{2} \text{ first peak} \qquad (9\text{-}7)$$

$$\tau_2 = \frac{3}{2} \pi \omega \text{ second peak} \qquad (9\text{-}8)$$

With Equation 9-5, it can be shown that the second or reverse voltage peak exceeds the first for all values of $K < 0.8$.[7] This relationship implies that energy transfer efficiency decreases for first-swing charging as the coupling coefficient decreases, but it increases with lower coupling at the second peak.

9.2.2 Dual-Resonance Charging

Transfer efficiency at the second peak increases to a theoretical maximum of 100 percent when $K = 0.6$, which is the most common coupling condition for the so-called dual resonance charge cycle. This and other discrete coupling coefficients that satisfy requirements for 100 percent energy transfer are given by:[8]

$$K = \frac{2n - 1}{2n^2 - 2n + 1} \qquad (9\text{-}9)$$

for n = 1, 2, 3, 4, ... and K = 1, 0.6, 0.385, 0.28, ...

If $K = 1$, the secondary voltage maximum is reached on the first voltage excursion. However, with air-core transformers, this operating condition is generally unrealistic because of the practical difficulty in constructing high-voltage, high-gain transformers with coupling coefficients greater than about 0.85. Within the range 0.6 to 0.85, it is possible to produce a wide variety of transformer designs. For this reason, dual-resonance transformers having the lower coupling coefficients of 0.385, 0.28 which reach maximum voltage on the third or fourth voltage excursions, are of little practical interest.

9.2.3 Circuit Tuning

In a dual-resonance transformer circuit operating with $K = 0.6$, it is desirable to use a transformer that by itself has a coupling coefficient greater than 0.6.[5] This permits exact resonant matching with the primary and secondary capacitors by adding small tuning inductors to both sections of the circuit (Fig. 9-13). The values for primary and secondary tuning inductances (L_{t-1} and L_{t-2}) can easily be found if the primary and secondary capacitances (C_1, and C_2) and the inductances (L_P, L_S and M) of the trans-

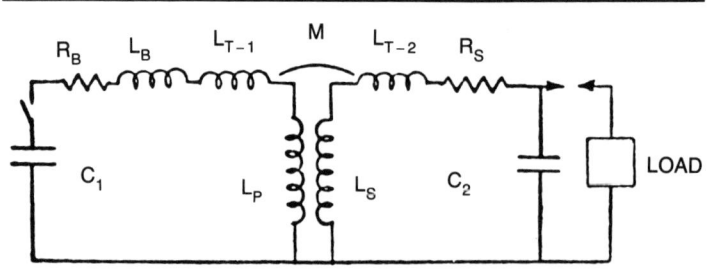

Fig. 9-13. Transformer circuit with tuning inductors.

former are known. From the relation for coupling coefficient, set $K = 0.6$:

$$\frac{M}{\sqrt{L_1 L_2}} = 0.6 \tag{9-10}$$

The values of total primary and secondary circuit inductance, L_1 and L_2, become:

$$L_1 = \frac{1}{L_2}\left(\frac{M}{0.6}\right)^2 \tag{9-11}$$

$$L_2 = \frac{1}{L_1}\left(\frac{M}{0.6}\right)^2 \tag{9-12}$$

Because $L_1 C_1 = L_2 C_2$, substitution gives:

$$L_1 = \frac{M}{0.6}\sqrt{\frac{C_2}{C_1}} \tag{9-13}$$

and

$$L_2 = \frac{M}{0.6}\sqrt{\frac{C_1}{C_2}} \tag{9-14}$$

and the turning inductances are:

$$L_{t-1} = L_1 - (L_p + L_b)$$

$$L_{t-2} = L_2 - L_s$$

where L_b is inductance of the primary capacitor bank.

From the foregoing equations, it can be seen that any transformer with a coupling coefficient greater than 0.6 can be adapted to a dual-resonance circuit. As previously indicated, unless the circuit coupling is close to 1.0, it is more advantageous from the standpoint of transfer efficiency to operate in the dual-resonance mode with $K = 0.6$. With repetitive-pulse systems, high transfer efficiency is essential to prevent excessive energy dissipation in the system components from residual energy ringing through the system after discharge of the secondary capacitor. When properly tuned, a dual-resonance charging system will operate with an overall efficiency between 90 and 95 percent, including switch losses.[5,9]

9.3 TRANSFORMER INSULATION

Proper application of insulating materials is an essential consideration in the design of high-voltage pulse transformers. As a minimum, it is necessary to know the effective dielectric strength of the materials and the voltage stresses that they will be subjected to in service. In reality, these two aspects are a complex of many factors, the details of which are beyond the scope of this book. It is useful, however, to summarize some of the general characteristics and modes of failure of dielectric materials.

9.3.1 Liquid and Solid Insulation Breakdown

For pulse transformers, the types of materials most commonly used are liquids and film or cast solids. These materials are usually used in combination. Failure in one invariably leads to failure of both. The initiating mechanism can be prompt in nature—such as bulk breakdown from over stressing—or gradual as a result of deterioration of dielectric strength from partial discharges. The process of dielectric deterioration from partial discharges includes gas productions in the liquids that can lead to breakdown from gaseous inclusions and erosion of solids in contract breakdown with the liquids.

Three characteristics common to liquid and solid dielectrics used in pulsed voltage applications are: (1) time dependent breakdown strength in the approximate range of 10^{-6} to 10^{-9} seconds, (2) breakdown strength that decreases with a weak dependence on increased stressed area in the range of a few square centimeters to several thousand

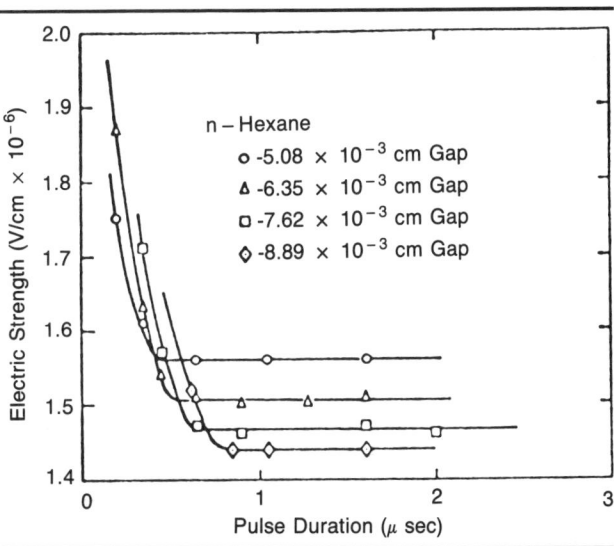

Fig. 9-14. Breakdown strength of n-hexane as function of electrode gap and pulse duration. (Courtesy of Ion Physics Corporation, Record of the "High-Voltage Technology Seminar," Sept. 29-30, 1969, "Liquid Insulation.")

square centimeters, and (3) breakdown strength which decreases with increasing insulator thickness below 1 centimeter. Data also exist that show the dielectric strength of liquids, such as transformer oil and n-Hexane, decrease by a factor of three with increases in pulse voltage rate of rise from 10^2 to 10^4 kV/centimeter-microsecond.[13] Figures 9-14 and 9-15 illustrate pulse time and electrode gap dependence on breakdown for several oils.[10,11] Figure 9-16 is a plot of dielectric strength as a function of thickness for Mylar polyester film.[12] Although the trends for liquids and solids are similar, the actual

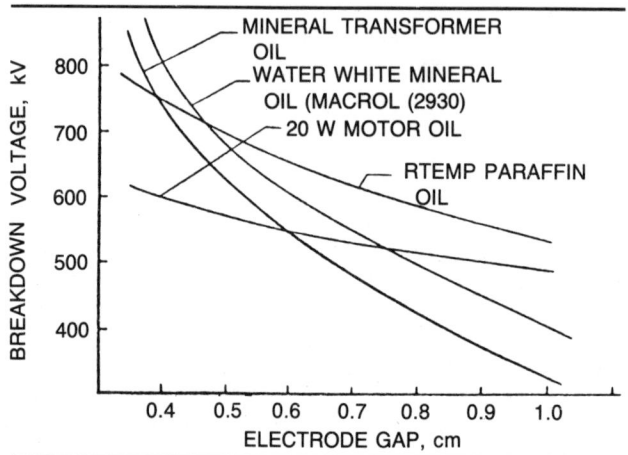

Fig. 9-15. Impulse breakdown strength of various oils as a function of electrode gap with a 1.5 microsecond dual-resonance pulse shape and 10 square centimeter electrodes (courtesy of Sandia National Laboratories).

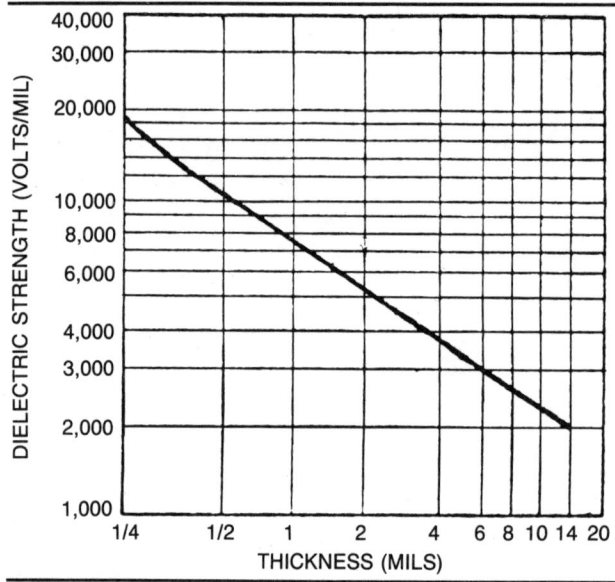

Fig. 9-16. Dielectric strength of Mylar polyester film as a function of thickness. (Courtesy E.I. Duport, Co.)

strength levels for given pulse times and thicknesses can be very much different. With few exceptions, liquid dielectrics tend to be weaker than solids, particularly solid films.

9.3.2 Time-Dependent Breakdown Model

Various theoretical models have been developed to explain dielectric breakdown phenomena. Some consider liquids and solids to be homogeneous dielectrics, and others ascribe breakdown to the presence of various impurities or faults in the media. For practical purposes, any theory that considers liquids or solids totally homogenous can have only limited application. However, the effects of impurities and faults are inclined to neglect the fact that eventually it is the insulator itself that breaks down.[14] Thus, there is no single theory that is generally accepted for all cases.

A model for time-dependent breakdown of solids by Kuffel and Abdullah[15] is illustrated in Fig. 9-17. Breakdown strength that decreases as a function of time is described as resulting from four different initiating mechanisms, which operate in different time regimes. The highest strength a material can have (its intrinsic strength) is obtained experimentally under the best of conditions when all extraneous effects are removed so that breakdown is dependent on only the material and temperature. Intrinsic breakdown occurs in times of 10^{-8} seconds or less at stress levels in excess of 10^6 V/centimeter. Intrinsic breakdown is postulated to be electronic in nature and assumed to occur when electrons in the dielectric gain enough energy from the applied field to cross the forbidden energy gap from the valency to the conduction band. The criterion condition is formulated by solving an equation for energy balance between the gain of energy by conduction electrons from the applied field and their loss to the lattice.

The second and next lower initiating mechanism is streamer breakdown that, with electrodes embedded in the material to avoid external effects, occurs from an electron entering the conduction band of the insulator at the cathode and drifting toward the anode under the influence of the field. The electron gains energy between collisions and loses energy on collisions. On occasions when the path is long enough to exceed the lattice ionizing energy, an additional electron is produced by collision. The process, if repeated, can lead to the formation of an electron avalanche which, if a critical size is

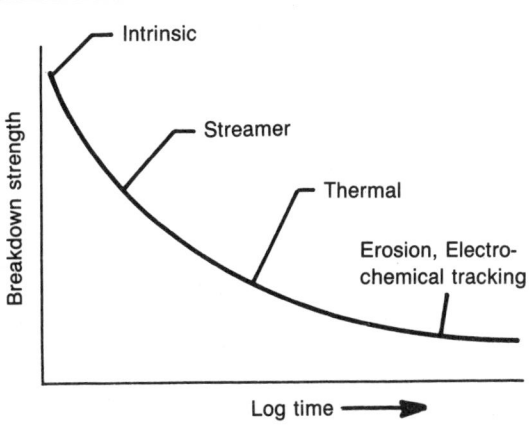

Fig. 9-17. Solid dielectric breakdown model. (Reprinted with permission from E. Kuffel and M. Abdullah, "High-Voltage Engineering," © 1970 Pergamon Press, Inc.)

exceeded according to Seltz,[16] will result in a streamer being formed. The fields at the streamer tip, estimated to be approximately 10^7 V/centimeter, exceed the intrinsic electric strength; complete breakdown ensues. The concept is similar to the streamer theory (developed by Loeb and Meek)[16] for gas breakdown and collisional ionization, leading to electron avalanches and streamer formation in liquids (by Goodwin and MacFadyen).[18]

Thermal breakdown of a solid is associated with a decrease in electric strength with increasing temperature. The temperature rise in a solid might occur as a result of heat transferred from an external source or direct dissipation in the solid from electric stressing. In one instance, when a dc field is applied to a dielectric at room temperature, the conduction current is generally very low, but it increases rapidly with increasing temperature associated with heat generated by the current. If the heat is not conducted away, the solid can go into a thermal runaway condition where the electric stress exceeds the dielectric strength of the material at the elevated temperature. For ac voltages, heat is generated inside the dielectric as a result of dipole relaxation. The losses associated with ac fields are usually much greater than dc fields because the loss in relaxation phenomena such as dipolar motions depend upon the rate of change of the field. Consequently, thermal breakdown strength is generally lower for ac fields and decreases with increasing frequency.

Erosion breakdown, as the name implies, involves long-term degradation of a solid from arcs or from partial discharges generated within or at the boundaries of the solid and an electrode surface or another insulating medium. Repeated surface discharges across a solid insulator surface, for example, can gradually remove enough material along preferential paths until breakdown occurs. Cavities, voids, or particle inclusions that have lower breakdown strengths than the solid dielectric result in corona generation and eventual failure from within the volume of the material.

9.3.3 Breakdown from Voids

The problem of breakdown from gas-filled voids is aggravated by the fact that the gas usually has a lower dielectric constant than the solid, and the electric field in the void is correspondingly increased according to:[19]

$$E_b = \frac{3\epsilon_1 E_o}{2\epsilon_1 + 1} \qquad (9\text{-}15)$$

where E_o is the field in the solid in the absence of the void and ϵ_1 is the permittivity of the solid. The same relation applies to bubbles in a liquid.

In addition to the foregoing failure modes, solids (particularly soft ones) can be subject to electromechanical failure. This type of failure is the result of electrostatic pressure produced by the electric field. A two-dimensional mechanical stress is thereby produced in the insulator with compression through the thickness and tension in the lateral direction. If the stresses exceed the tensile or compressive strength of the material, it will fail mechanically.

Liquid dielectrics exhibit the same general characteristics as solids with respect to failure modes and dielectric strength dependence on temperature, pulse time, stressed area, and thickness in the short gap ranges (< 1 centimeter). They differ from solids in their sensitivity to particle and chemical contamination, absorbed gases and increased strength at high hydrostatic pressures.

9.3.4 Martin's Breakdown Formula

Probably the most widely used relationship for impulse breakdown of liquid dielectrics is the empirical formula developed by J.C. Martin[20] for gaps greater than 1 centimeter:

$$E t^{1/3} A^{0.1} = K \qquad (9\text{-}16)$$

where

$E =$ field strength in MV/centimeter
$t =$ effective pulse time, the time the voltage is greater than 0.63 of its maximum value
$A =$ electrode area in square centimeters
$K =$ 0.5 for transformer oil

The expression has been experimentally verified for areas up to 10^6 square centimeters and pulse times from less than 20 nanoseconds to at least 1.0 microseconds. For small electrode spacings a gap dependence of $d^{-1/4}$ in the expression gives a better fit to breakdown data:[21]

$$E\, t^{1/3}\, A^{0.1}\, d^{-1/4} = K \qquad (9\text{-}17)$$

Martin's liquid breakdown formula applies specifically to bulk breakdown conditions with comparatively long time intervals (at least minutes) between shots so that the effects of any disturbances in the liquid such as partial discharges or streamers that do not completely cross the electrode gap have time to heal.

9.3.5 Corona-Inception Voltage

Gas generated by partial discharges and streamers in the captive oil volume of a transformer cannot generally be tolerated because of shot-to-shot buildup. The transformer designer must therefore be as concerned with the corona inception voltage (CIV) as with dielectric strength. Corona or partial discharge damage in transformers has been a well-known problem in ac power transformers and low-voltage pulse transformers for over 50 years. It has only recently received attention as a major problem with high-voltage pulse transformers. The primary reason is that high-voltage pulse transformers were practically always used in single-shot applications where mild corona discharges had little or no effect over a life of five to ten thousand shots. However, with repetitive pulse systems which must have life expectancies of at least 10^9 shots, corona damage has become a dominant consideration in pulse transformer design. Figure 9-18 is a photograph of damage to Mylar film produced by repetitive pulsing with partial discharges for several

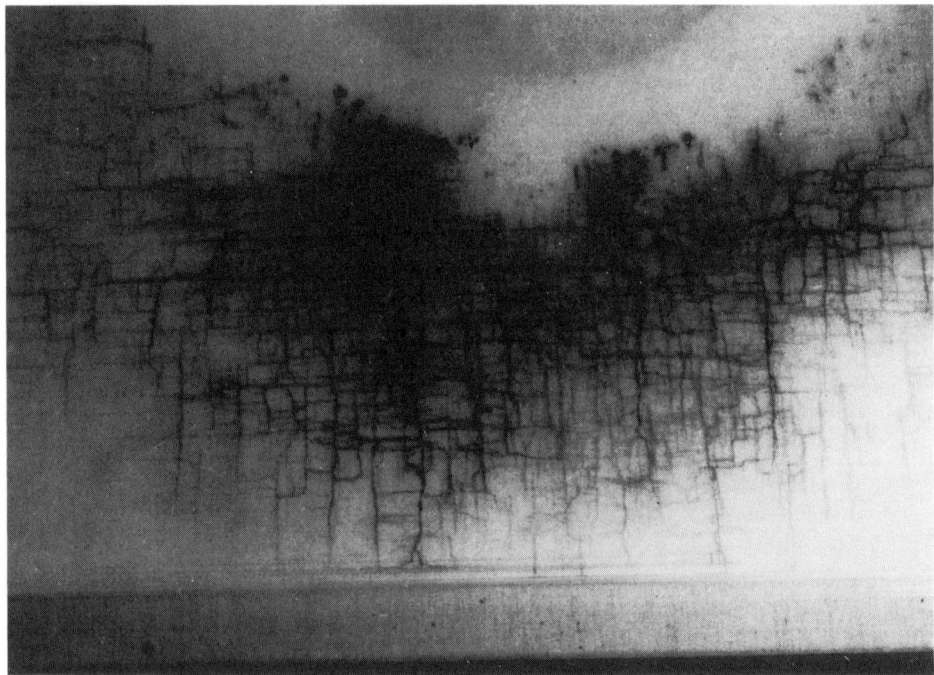

Fig. 9-18. Damage to Mylar film from partial discharges, typical polymer behavior.

thousand shots. This type of failure is common in spiral-strip transformers when partial discharges occur in the oil adjacent to the film surface.

Clark[22] gives a brief description of the products of corona in mineral insulating oils. He shows that the effects of ionization associated with corona bear a strong resemblance to the effects of heat cracking insofar as the products which are formed. The primary gaseous products include:

Hydrogen	H_2
Methane	CH_4
Carbon Monoxide	CO
Carbon Dioxide	CO_2
Ethane	C_2H_6
Ethylene	C_2H_4
Acetylene	C_2H_2

In addition to the gaseous products, carbon particles and unsaturated liquid hydrocarbons are formed. The liquids have poor oxidation stability and lower dielectric strength. Thus, corona discharges over a long period result in the buildup of low-strength liquids that eventually render the oil useless as a dielectric.

It is well known that moderate carbon particle, dust and moisture contamination in many insulating oils does not decrease the dielectric strength appreciably under fast pulse conditions in the range of nanoseconds to at least 10 microseconds. These contaminates do, however, lower the CIV, which is not as sensitive to pulse time.[23] With silicone oils,

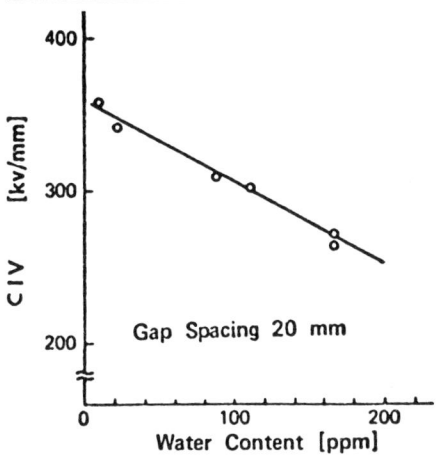

Fig. 9-19. Relation between water content and value of CIV in silicone liquid. (Reproduced with permission, Kuwahara et al, "Partial Discharge Characteristics of Silicone Liquids," IEEE Transactions on Electrical Insulation, Vol. EI-11, No. 3, September 1976 © 1976 IEEE.)

Fig. 9-20. Mylar-copper disk laminate test.[23]

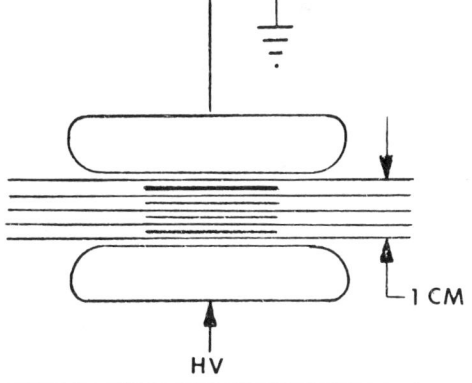

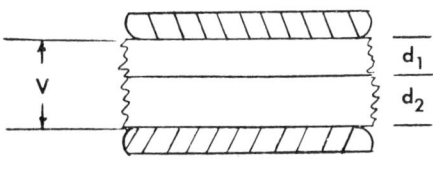

Fig. 9-21. Voltage stresses in a two-dielectric insulation.

for example, a reduction of nearly 30 percent has been measured between dry oil samples and ones with a moisture content of 170 parts per million[24] (Fig. 9-19). The CIV reduction is linear over that range and fits the equation:

$$\text{CIV} = -0.55X + 3610 \text{ kV/centimeter} \qquad (9\text{-}18)$$

X = water content, parts per million

Other experiments with oil-impregnated Mylar-copper disk laminates sandwiched between flat plate electrodes (Fig. 9-20) showed similar results with respect to partial discharge effects. A variety of oils were pulse tested in both barrel-fresh condition and in the dry, degassed condition. In some cases, breakdown of the laminate from repetitive partial discharges differed by as much as 30 percent. Table 9-1 summarizes the oils tested and the nominal breakdown levels.

Table 9-1. Onset of Partial Discharges in Various Oils Tested in Laminate Structures.

	DRY DEGASSED CONDITION kV/CENTIMETER	BARREL-FRESH CONDITION kV/CENTIMETERS
Mineral Oil, Inhibited	325	250
X-ray Oil (Marcol 2930)	590	510
Silicone Oil (Dow Corning 200)	360	305
MIPB (Monsanto 1238)	620	410
DAA (HISOL SAS 10 E)	460	390
Paraffin Oil (RTEmp)	650	380

Stresses are mean values and are intended to illustrate comparative properties of the oils in two conditions.

9.3.6 Electric Stress in a Two-Dielectric System

To determine the actual stress in the oil and Mylar film, it is necessary to compute the stress distribution in the two-dielectric systems (Fig. 9-21) according to the relation:

$$E_1 = \frac{V_o}{d_1 + d_2\,(\epsilon_1/\epsilon_2)} \text{ kV/centimeter} \qquad (9\text{-}19)$$

$$E_2 = \frac{V_o}{d_1\,(\epsilon_2/\epsilon_1) + d_2} \qquad (9\text{-}20)$$

where

V_o = the total voltage across the electrodes, k_V
E_1 and E_2 are the voltage stresses in materials 1 and 2
ϵ_1 and ϵ_2 are the dielectric constants of materials 1 and 2
d_1 and d_2 are the thicknesses of materials 1 and 2

Considerable caution must be exercised in interpreting dielectric strength and partial discharge data because of differences in test conditions. Factors already mentioned such as pulse time, sample thickness, and area can give results that differ by as much as an order of magnitude. This can be seen by comparing the CIV for silicone and mineral oil reported by Kuwahara et al.[23] with that obtained in the laminate tests. In the first case, measurements were made using closely spaced needle-plane electrodes with applied voltages between 15 kV and 20 kV. This resulted in a CIV of 3367 kV/centimeter for silicone oil and 3518 kV/centimeter for mineral oil. With the laminates, which were 1-centimeter thick and tested above 300 kV, the onset of partial discharges was observed at 360 kV/centimeter for silicone oil and 324 kV/centimeter for mineral oil with both oils in the dry degassed condition. These two sets of results emphasize the necessity of searching for data or conducting tests as near as possible to the actual service conditions that dielectrics will experience. Lacking such specific data, most common insulating liquids such as mineral oils, paraffin-base oils (that is, common motor oils), and silicone oils can be repetitively stressed to at least 150 kV/centimeter without encountering bulk breakdown or partial discharge problems. If the oils are carefully filtered to remove particle matter above 1 micron and degassed before use, the margin of safety will be increased by at least 50 percent. When the highest possible performance must be obtained from an insulation system, there is no alternative to conducting tests as near as possible to anticipated service conditions.

9.4 REFERENCES

1. T.H. Martin, "FRIZZ—A High Voltage Impulse Tester," SC-RR-71, 0341, June 1971.
2. G.J. Rohwein, "Design of Pulse Transformers for PFL Charging," Proc. of 2nd IEEE Int'l. Pulsed Power Conf., Lubbock, TX, June 1979.
3. D. Finkelstein, P. Goldberg, and J. Schuchatuwitz, "High-Voltage Impulse System," Rev. Sci. Instr., Vol. 37, No. 2, February 1962.
4. J.C. Martin, P.D. Champney, and D.H. Hammer, "Notes on the Construction Methods of A Martin High Voltage Pulse Transformer," School of Elec. Engr., Cornell University, June 1967.
5. G.J. Rohwein, "A 3 MV Transformer for PFL Pulse Charging," IEEE Trans. on Nucl. Sci., Vol. NS26, No. 3, June 1979.
6. G.N. Glasoe and J.V. Lebacqz, "Pulse Generators," (Dover Publications, Inc., NY, 1965).
7. E. Cook, L. Reginato, "Off Resonance Transformer Charging for 250 kV Water Blumlein," Proc. of 13th Pulsed Power Modulator Symposium, Buffalo, NY, June 1978.
8. C.R.J. Hoffmann, "A Tesla Transformer High-Voltage Generator," Rev. Sci. Instr., Vol. 46, No. 1, June 1975.
9. M.T. Buttram and G.J. Rohwein, "Operation of a 300 kV, 100 Hz, 30 KW Average Power Pulser," Proc. of 13th Pulsed Power Modulator Symposium, Buffalo, NY, June 1978.
10. F. Tse, W. Bell, M. Mulcahy and P. Bolin, Liquid Insulation, H.V. Tech. Seminar, Ion Physics Corp., Boston, MA, September 1969.
11. G.J. Rohwein, Unpublished Oil Breakdown Data, Sandia Laboratories, Albuquerque, NM, January 1980.
12. E.I. DuPont DeNemours & Co., Mfg. Data on Electrical Properties of Mylar Polyester Film, February 1968.

13. K.C. Kao and J.P.C. McMath, "Time-Dependent Pressure Effect in Liquid Dielectrics," IEEE Trans. on Elec. Insulation, Vol. E1-5, No. 3. September 1970.
14. L.L. Alston, Editor, "High-Voltage Technology," (Oxford Univ. Press, 1968).
15. E. Kuffel and M. Abdullah, "High-Voltage Engineering," (Pergamon Press, 1970).
16. F. Seitz, Phys. Rev. 73, 833, 1949.
17. L.B. Loeb and J.M. Meek, "The Mechanism of Electric Spark," (Oxford Univ. Press, 1940).
18. D.W. Goodwin and K.A. MacFadyen, Proc. Phys. Soc., London, B 66 85, 815 (1953).
19. J.D. Jackson, "Classical Electrodynamics," (J. Wiley and Sons, New York, London, 1962).
20. J.C. Martin, "Comparison of Breakdown Voltages for Various Liquids Under One Set of Conditions," AWRE Report, 1965.
21. I.D. Smith, Pulse Breakdown of Transformer Oil, AWRE Report, 1965.
22. F.M. Clark, "Insulating Materials for Design and Engineering Practice," (J. Wiley and Sons, Inc., New York, London, 1962).
23. G.J. Rohwein, Unpublished Results of Oil-Laminate Tests, Sandia Laboratories, 1979.
24. H. Kuwahara, K. Tsuruta, H. Munemura, T. Ishil and H. Shiomi, "Partial Discharge Characteristics of Silicone Liquids," Trans. IEEE Power Engineering Society, January 1975.

Chapter 10

Measurement Techniques

by
W.L. Willis*

THIS CHAPTER IS RESTRICTED TO MEASUREMENTS OF VOLTAGE AND CURRENT. ALSO, although the measurement themselves should be as quantitative as possible, the discussion is rather nonquantitative. Emphasis is on types of instruments, how they may be used, and the inherent advantages and limitations of a given technique.

A great deal of information can be obtained from good, clean voltage and current data. Power and impedance are obviously inherent if the proper time relationships are preserved. Often an associated physical event that is difficult to determine, such as a time-varying load characteristic or the time of light emission, can be evaluated from the V-I data.

The scarcity of text is rather surprising. "High-Voltage Engineering" has no treatment of current measurements, and Frungels' "High Speed Pulse Technology" devotes only nine pages to the subject.

The lack of active high-voltage devices, such as 50 kV operational amplifiers restricts measurement devices to passive elements, primarily R and C.

10.1 VOLTAGE MEASUREMENTS

There are several well-developed techniques for voltage measurements. These include:

>Spark gaps
>Electrostatic meters
>capacitive dividers

---

*Written while at the Los Alamos National Laboratory. Currently at the Northrop Research and Technology Center.

Resistive dividers
Mixed RC dividers
Electro-optic effect

10.1.1 Spark Gaps

A great deal of work is documented that relates breakdown voltage to gap geometry and such related parameters as gas dielectric and pressure. The most commonly considered geometries are spherical, point-to-plane, and uniform field.

The limitations are obvious with respect to an engineering instrument. Probably the greatest limitation is the one-point nature of the indication. Neglecting use as a protective device, spark gap measurements seem most appropriate for a laboratory environment. There are some advantages. One is that the use of large spheres to avoid corona will permit measurements to very high voltages. Also, neither electrode must be at ground potential so that voltages of large differential can be addressed. If one electrode is at ground potential, it becomes possible to vary the gap until breakdown occurs without undue complexity. One side point is that the literature below is based on arc breakdown. The gap current might be limited by an external ballast, but there are several regimes of breakdown. To ensure accuracy, enough current must be permitted to flow for an arc to occur.

The Americans and Europeans differ in their choice of preferred geometry. Europeans seem to prefer sphere-to-sphere geometries, but American investigators have a preference for point-to-plane gaps. The Europeans occasionally make disparaging remarks about the American preference, but the point-plane has advantages in terms of size, capacitance, etc. Figures 10-1 and 10-2 are tabular data taken from Frungel and from Kuffel and Abdullah, respectively. In Fig. 10-1, at a 0.1-centimeter gap between 2-centimeter spheres, a breakdown voltage of 4.65 kV has been obtained. In Fig. 10-2, the same configuration is quoted at 4.7 kV. The agreement between investigators is quite good.

It is worth noting (Fig. 10-1) that whether or not one sphere is grounded makes a difference. Also, polarity can make a difference, particularly with the point-plane geometry. The data is usually taken with a source of short wavelength illumination on the gap to reduce the statistical scatter. This is particularly important for impulse measurements. The tables of Figs. 10-1 and 10-2 extend into the megavolt range but are really only applicable from dc to about 25 kHz. Frungel states that with 1-centimeter gaps, 31 kV impulses as short as 100 nanoseconds can be measured to 1-2 percent. A 1-millimeter gap will measure to 10^{-10} second.

Other effects of significance on gap breakdown voltage include nearby grounds and humidity. A rule of thumb for proximity effects is that three to five sphere diameters of separation are required. Humidity effects are complex. For a fixed geometry, humidity increases hold-off strength. Figure 10-3 from Kuffel shows this effect. Figure 10-4, also from Kuffel, shows how the increase in hold-off varies with gap spacing at fixed humidity for several sphere sizes. Errors of 1 to 5 percent are easily possible just due to weather effects.

When the breakdown voltage versus spacing for spheres is examined (Figs. 10-1 and 10-2), a nonlinear relationship is noted. This is avoided in the point-to-plane gap,

10.1 Voltage Measurements

CALIBRATION VOLTAGES FOR SPHERE SPARK GAPS

		D = 2 cm					D = 5 cm			
		1 Sphere Grounded		Symmetr. Potential			1 Sphere Grounded		Symmetr. Potential	
s	s/D	$\mathfrak{E}_0$	U_{max}	$\mathfrak{E}_0$	U_{max}	s/D	$\mathfrak{E}_0$	U_{max}	$\mathfrak{E}_0$	U_{max}
(cm)		(kv/cm)	(kv)	(kv/cm)	(kv)		(kv/cm)	(kv)	(kv/cm)	(kv)
0.1	0.05	48.12	4.65	48.12	4.65	0.02	47.17	4.65	47.17	4.65
0.2	0.10	43.40	8.13	43.12	8.08	0.04	41.44	8.03	41.44	8.03
0.3	0.15	42.17	11.44	41.44	11.29	0.06	39.16	11.28	39.16	11.28
0.4	0.20	41.82	14.54	40.97	14.41	0.08	37.98	14.40	37.85	14.35
0.5	0.25	42.22	17.60	40.95	17.45	0.10	37.33	17.47	37.03	17.33
0.6	0.30	43.20	20.28	41.28	20.48	0.12	36.94	20.44	36.53	20.21
0.7	0.35	44.28	23.59	41.61	23.39	0.14	36.69	23.36	36.23	23.16
0.8	0.40	45.34	26.30	41.93	26.15	0.16	36.59	26.25	36.11	26.03
0.9	0.45	46.43	28.90	42.17	28.73	0.18	36.52	29.08	36.06	28.83
1.0	0.50	47.32	31.20	42.41	31.20	0.20	36.69	31.90	36.03	31.70
1.1	0.55	48.21	33.35	42.64	33.50	0.22	36.92	34.71	36.07	34.45
1.2	0.60	48.96	35.28	42.82	35.68	0.24	37.15	37.45	36.16	37.22
1.3	0.65	49.66	36.98	43.00	37.76	0.26	37.45	40.22	36.25	39.93
1.4	0.70	50.35	37.50	43.16	38.31	0.28	37.76	42.97	36.32	42.57
1.5	0.75	50.96	40.03	43.34	41.70	0.30	38.05	45.54	36.43	45.20
1.6	0.80	51.57	41.33	43.47	43.47	0.32	38.33	48.04	36.52	47.70
1.7	0.85	52.13	42.71	43.62	45.20	0.34	38.64	50.48	36.62	50.22
1.8	0.90	52.65	43.84	43.77	46.83	0.36	38.90	52.80	36.71	52.73
1.9	0.95	53.11	44.85	43.87	48.20	0.38	39.17	55.00	36.80	55.10
2.0	1.00	53.52	45.80	43.96	49.70	0.40	39.46	57.25	36.88	57.48
2.5	1.25	55.59	49.98	44.36	55.72	0.50	40.76	67.20	37.23	68.50
3.0	1.50	57.15	52.70	44.61	60.48	0.60	42.02	75.60	33.52	78.20
4.0	2.00	59.70	56.87	44.88	67.08	0.80	44.12	88.63	37.96	94.87
5.0	—	—	—	—	—	1.0	45.71	97.75	38.23	108.0
6.0	—	—	—	—	—	1.2	47.04	104.8	38.41	118.5
7.0	—	—	—	—	—	1.4	48.14	110.2	38.55	127.1
8.0	—	—	—	—	—	1.6	49.03	114.3	38.64	134.3
9.0	—	—	—	—	—	1.8	49.90	117.8	38.70	139.9
10.0	—	—	—	—	—	2.0	50.64	120.7	38.75	144.8
12.0	—	—	—	—	—	—	—	—	—	—
14.0	—	—	—	—	—	—	—	—	—	—
15.0	—	—	—	—	—	—	—	—	—	—
16.0	—	—	—	—	—	—	—	—	—	—
18.0	—	—	—	—	—	—	—	—	—	—
20.0	—	—	—	—	—	—	—	—	—	—

^a Initial voltages and breakdown field strength of equal spheres in air with one grounded pole and symmetrical voltage distribution on the electrodes. Valid for b = 760 mm Hg and t = 20° (rel. air density δ = b × 293/760(273 + t) = 1.0).

Fig. 10-1. Calibration voltages for sphere spark gaps (courtesy of F.B.A. Frungel, "High Speed Pulse Technology," Academic Press, 1965).

Flashover voltages (50% values in impulse tests) for **alternating** voltages, for **direct** voltages of either polarity, and for full **negative** standard impulses and impulses with longer tails: one sphere earthed

Sphere gap spacing, cm	Kilovolts peak at 20°C : 1013 millibars											
	Sphere diameter, cm											
	2	5	6.25	10	12.5	15	25	50	75	100	150	200
0.05	2.8											
0.10	4.7											
0.15	6.4											
0.20	8.0	8.0										
0.25	9.6	9.6										
0.30	11.2	11.2										
0.40	14.4	14.3	14.2									
0.50	17.4	17.4	17.2	16.8	16.8	16.8						
0.60	20.4	20.4	20.2	19.9	19.9	19.9						
0.70	23.2	23.4	23.2	23.0	23.0	23.0						
0.80	25.8	26.3	26.2	26.0	26.0	26.0						
0.90	28.3	29.2	29.1	28.9	28.9	28.9						

This table is not valid for the measurement of impulse voltages below about 10 kV.
The figures in brackets, which are for spacings of more than $0.5D$, are of doubtful accuracy.

Fig. 10-2. Flashover voltages for alternative voltages (courtesy of F.B.A. Frungel, "High Speed Pulse Technology," Academic Press, 1965).

which is one reason why this method is popular in the U.S. However, the accuracy is not as good as is obtained with the spherical gap, and the usual practice of using a needle is not optimal. If the angle formed by the point (total angle, not angle from axis) is plotted against breakdown voltage, the curve in Fig. 10-5, from Frungel, results. The shape of the point is not important over a wide range, provided it is not too small.

Linearity with gap and polarity effects are shown in Fig. 10-6, also from Frungel. The polarity effect can be useful. An example is an X-ray tube, which uses this effect to achieve half-wave rectification as well as to permit measurement of the voltage on the nonconducting half cycle.

The third type of gap is the uniform field. Figure 10-7, from Kuffel, relates breakdown to gap as measured by several investigators. Two points stand out. One is the good agreement between investigators. The other is that when this figure is compared to Figs. 10-1 or 10-2, the breakdown strength is very close to the spherical case. The trouble of producing Rogowski electrodes is hardly worth the effort for the small difference.

10.1 Voltage Measurements

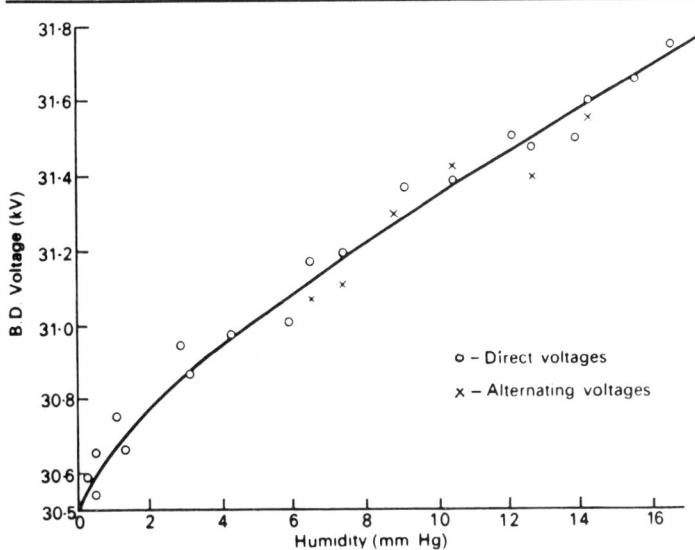

Comparison of the influence of humidity with direct and alternating voltages for 1·0-cm gap between 25-cm diameter spheres.

Fig. 10-3. Humidity effects on hold-off strength (reprinted with permission from E. Kuffel and W. Zaengl, "High Voltage Engineering," 1965, Pergamon Books, Ltd).

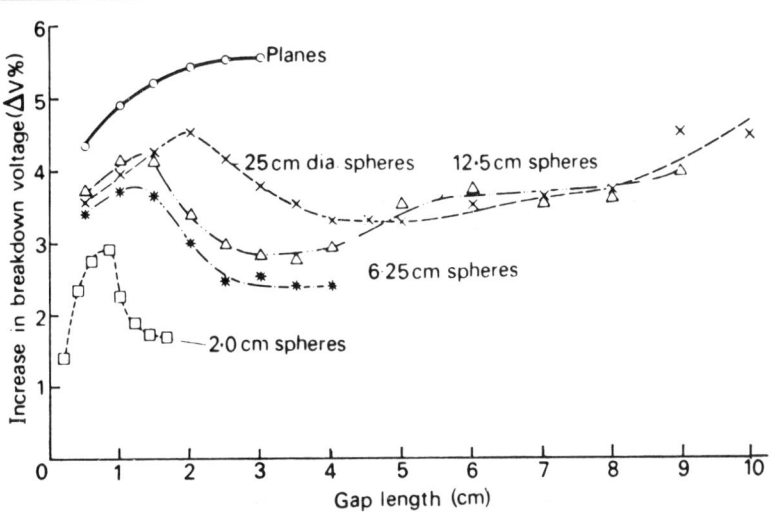

Increase in breakdown voltage for different gap lengths when humidity was changed from 0 to 17 mm Hg.

Fig. 10-4. Variation of hold-off with gap spacing (reprinted with permission from E. Kuffel and W. Zaengl, "High Voltage Engineering," 1965, Pergamon Books, Ltd).

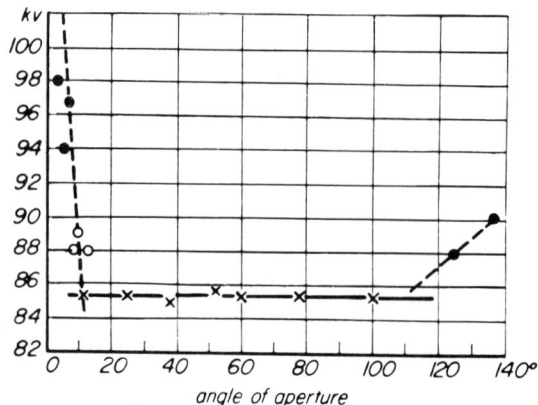

Effective spark voltage between two points, depending on the aperture angle at the sparking distance of 25 cm, by Weicker. Spark emission connected with glide-brush formation, glowing, and incomplete brush formation, respectively.

Fig. 10-5. Plot of point angle against breakdown voltage (courtesy of F.B.A. Frungel, "High Speed Pulse Technology," Academic Press, 1965).

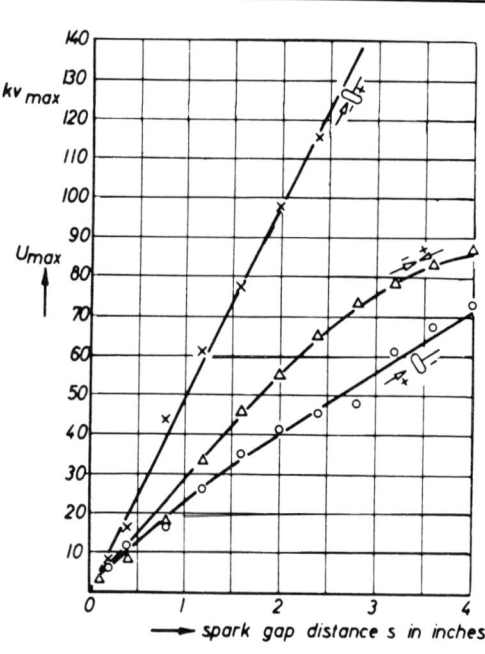

Fig. 10-6. Spark distance by + and - polarities of the point-plane (courtesy of F.B.A. Frungel, "High Speed Pulse Technology," Academic Press, 1965).

Breakdown Voltages of Uniform-field Gaps in Air at 20°C and 760 mm Hg

Gap length	Breakdown voltage as measured by					Breakdown voltage of sphere gap
	Schumann	Ritz	Holzer	Fisher	Bruce	
cm	kV	kV	kV	kV	kV	kV
0·1	4·5	4·54	—	—	—	4·6
0·2	8·0	7·90	—	—	7·56	8·0
0·3	11·3	11·02	—	—	10·60	11·1
0·4	14·4	14·01	—	14·1	13·54	14·1
0·5	17·4	17·00	—	16·9	16·41	17·0
0·6	20·3	19·90	—	19·6	19·24	19·8
0·7	23·2	22·80	—	22·3	22·04	22·7
0·8	26·1	25·70	—	24·9	24·81	25·8
0·9	28·9	28·50	—	27·5	27·57	28·6
1·0	31·7	31·35	31·66	30·3	30·30	31·0
2·0	59·6	58·7	61·20	55·7	57·04	58·0
3·0	87·0	85·8	86·94	—	83·19	85·0
4·0	114·0	112·0	113·04	—	109·0	112·0
5·0	140·0	138·5	137·8	—	134·7	137·0
6·0	166·2	163·8	163·44	—	160·2	164·0
7·0	191·8	189·9	187·74	—	185·6	190·0
8·0	216·8	215·0	212·88	—	211·0	215·0
9·0	241·2	240·0	237·78	—	236·3	240·0
10·0	266·0	265·0	263·0	—	261·4	265·0
11·0	290·4	290·0	288·2	—	286·6	288·0
12·0	—	315·0	313·2	—	311·6	312·0
13·0	—	—	338·1	—	—	336·0
14·0	—	—	363·2	—	—	362·0
15·0	—	—	387·7	—	—	388·0
16·0	—	—	412·6	—	—	412·0

Fig. 10-7. Breakdown voltages of uniform field gaps in air (courtesy of F.B.A. Frungel, "High Speed Pulse Technology," Academic Press, 1965).

10.1.2 Electrostatic Voltmeters

When two objects are charged with the same polarity, a repulsive force exists between them. This follows from Coulomb's Law:

$$F = \frac{1}{4\pi\epsilon_o} \frac{Q_1 Q_2}{d^2} \qquad (10\text{-}1)$$

This effect has been implemented in several ways to produce electrostatic voltmeters. These devices are marked by very high input impedances. A low voltage (120 V) Sensitive Research Instrument Corporation unit was measured at $4.37 \times 10^{16} \Omega$ and a capacitance of:

$$C = (15.1 + 154.7\,\theta) \times 10^{-12} \qquad (10\text{-}2)$$

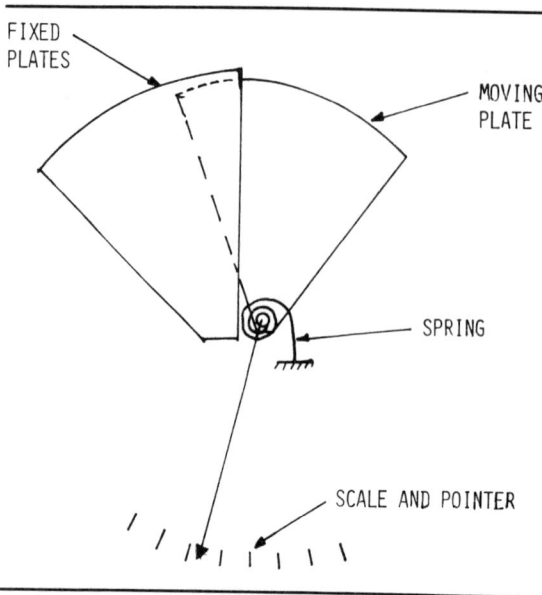

Fig. 10-8. Geometry of voltmeter.

where θ is the relationship between voltage indication and meter angle in radians. Figure 10-8 is a sketch of the geometry employed. The electrostatic forces cause a movable set of capacitor plates to move into a fixed set of plates against a spring force.

Figure 10-9 is a design that employs a linear rather than a rotational motion of the plates. The hemispherical structure provides shielding and a uniform field for the moving disk. In this case, the force is given by:

$$F = \frac{d^2}{2825\ l^2}\ V^2\ \text{gram-weight} \qquad (10\text{-}3)$$

Where d is the disk diameter in centimeters, l is the gap length in centimeters, and V is the applied potential in kilovolts. A capacitive divider and guard rings are included to prevent flashover. The small motions are detected by reflecting a light beam off the disk. A feedback system that provides an electrical force to prevent motion has also been designed. Figure 10-10 shows this feedback system. A detector measures the light beam deflection and feeds back a corrective force on the solenoid. This also provides an electrical output signal. Both these systems quote 0.1 percent accuracy.

Because of the V^2 dependence of the forces, the output is nonlinear. Linearizing mechanisms are possible but not usual. One consequence of the V^2 dependence is lack of polarity sensitivity. Also, ac waveforms will work, and a true RMS reading results. The upper frequency limit for high voltage units is 20 to 25 kHz and is due to heating effects. The capacitive reactance and high-voltage combination result in excessive current flow at high frequencies.

Safety precautions are required with these meters because they can be left charged due to the high internal resistance. Only a fraction of a joule should be present, but in-

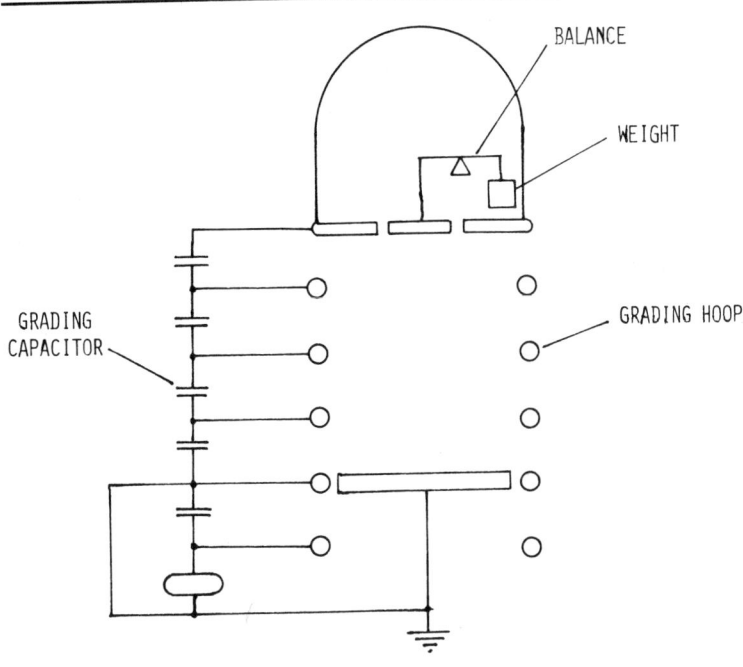

Fig. 10-9. Linear design electrostatic voltmeter.

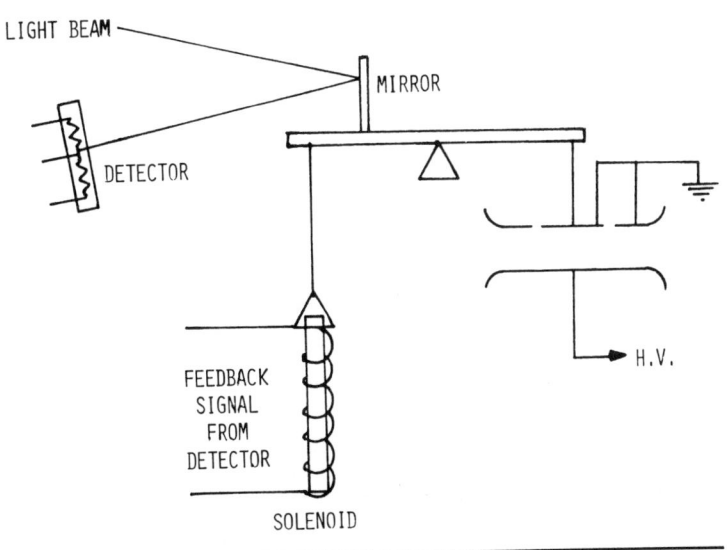

Fig. 10-10. Feedback system.

306 Measurement Techniques

jury due to muscular reflex can occur. Under some conditions, the meters can also explode.

10.1.3 Capacitive Dividers

Capacitive dividers were mentioned above with respect to voltage grading. Capacitance can be used in several ways as a voltage-measuring device. One way it can be used is as a series reactance with an ammeter. Frequency would be a necessary parameter for calibration. This is not a usual technique. A slightly more complex circuit, as shown in Fig. 10-11, makes a good peak-reading meter. Assuming a sinusoidal voltage:

$$v = V_o \sin 2\pi \text{feet} \qquad (10\text{-}4)$$

and that current in a capacitor is given by:

$$i = C\frac{dv}{dt} = 2\text{-}\pi f V_o \cos 2\pi f \, dt \qquad (10\text{-}5)$$

the charge through a rectifier, per cycle, is:

$$\int i \, dt = C \int dv = 2\pi f \, C \, V_o \int_{\frac{3}{4}f}^{\frac{1}{4}f} \cos 2\pi f \, dt = 2 \, C \, V_o \qquad (10\text{-}6)$$

The charge per second is $2CV_o f$ for equal $+$ and $-$ voltage peaks.

A true capacitive divider has both advantages and disadvantages. It is usually made up of a small, high-voltage capacitor in series with a larger, low voltage capacitor. It can reproduce fast events with small distortion and requires very little energy from the circuit. In practice, a number of snares become apparent. First, some inductance will be present in the divider and in the high-voltage connection. Self-resonance usually appears by 10 MHz (less on a poorly made or connected divider). Second, most recording instruments do not work well in the presence of high-voltage transients. This means that a long cable with associated capacitance might be required. This inevitably affects calibration. Third, stray capacitances affect the low-voltage capacitor.

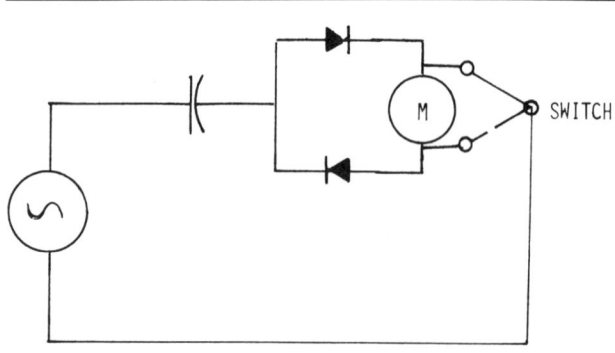

Fig. 10-11. Peak-reading meter circuit.

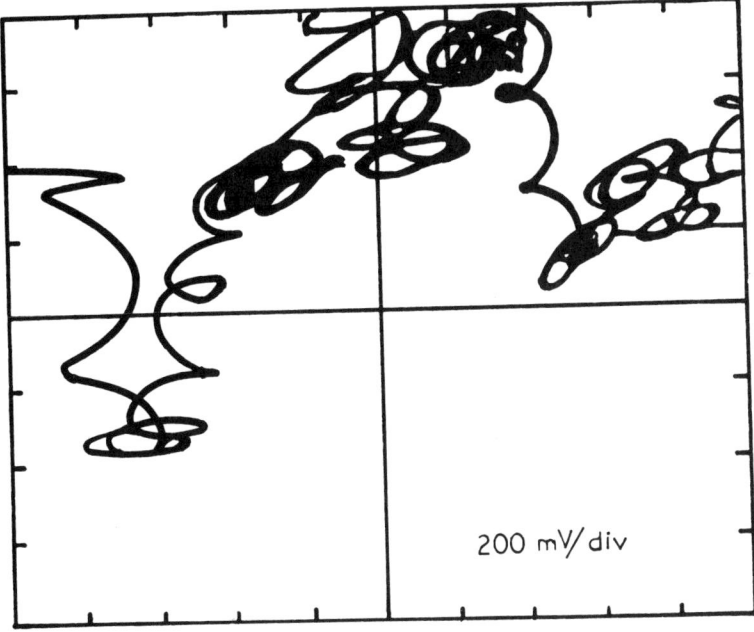

Fig. 10-12. Effect of high-voltage transient on oscilloscope.

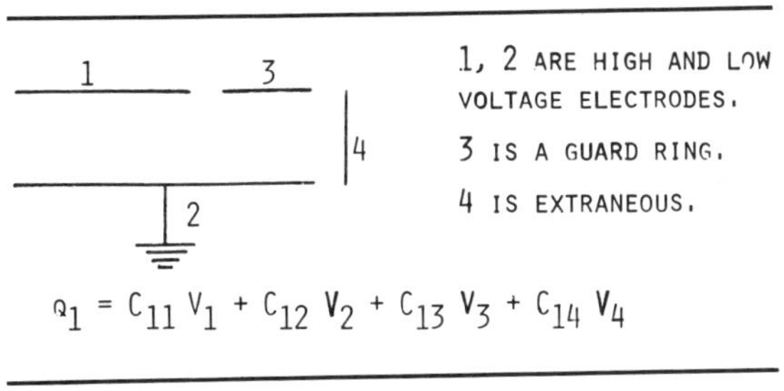

Fig. 10-13. Stray high-voltage transient effects.

Figure 10-12 shows the effect of the problem on an oscilloscope produced by proximity to a large current transient. Figure 10-13 shows a schematic representation of the problem. The consequence is that accurate results often require an in-place calibration.

A remaining difficulty is associated with connection to a recording instrument. These have an input resistance of some kind so that in the dc case the high-voltage capacitor soon charges to the applied voltage, and the output signal vanishes. In pulse measurements, the effect appears as a level shifting so that ground-referenced measurements are difficult. Often, as in fast-pulse work, it is desired to terminate cables in 50 or 75 Ω to prevent reflections. This produces obvious problems.

10.1.4 Resistive Dividers

As with capacitive dividers, measurements are possible using a large value series resistor and an ammeter. This is typical in high-voltage power supplies where response time is not critical. The only problem usually associated with such simple meters is heating of the metering resistor.

Victoreen makes a line of high-voltage resistors that are convenient for spark-gap bias circuits. One line, the MOX-5, ranges from 50 kΩ to 5000 MΩ with 12.5 W and 37.5 kV ratings. At 40 percent of rated wattage, the temperature rise is 60 C. The temperature coefficient is 500 ppm so that a 3 percent error is already present. The message is to consider heating effects in high voltage resistive divider circuits.

In ac or pulse measurements, a number of problems arise, related to:

> Residual inductance of resistors
> Stray capacitance
> Impedance drop in connecting high-voltage lead
> Impedance drop in ground return
> Stray oscillations

Probably the biggest problem is associated with stray capacitances. Consider the distributed circuit shown in Fig. 10-14(A). The LaPlace Transform of this circuit can be shown to be:

$$V = \frac{\sinh\sqrt{[RC_1P/(1 + RC_2P)]n}}{\sinh\sqrt{[RC_1P/(1 + RC_2P)N]}} V \qquad (10\text{-}7)$$

If $C_g = NC_1$, $C_c = C_2/N$, and $R_o = NR$, the step response for several cases is as shown in Fig. 10-14(B). Note that it is C_g, the parallel set of capacitances from resistors to ground, that really degrades the rise time. The C_2 terms tend to help, if anything. If the divider is potted, or oil immersed, the C_1 terms increase rapidly.

There is a technique that, at a price, almost eliminates this problem. Figure 10-15 shows the solution, which is a concentric cylinder of copper sulfate solution around the resistor column. This drives the normally grounded end of the strays to the same potential as the divider at any given point, preventing charge flow into the C_1 term. The solution works best with about 3 kΩ resistance, and the loading is the price mentioned. The product is about 15 nanoseconds, but structure can be seen on leading edges of 10 to 12 nanoseconds rise time.

10.1.5 Mixed Dividers

There is a class of dividers best described as mixed, in which RC combinations are employed to obtain a wide range of frequency response. The Tektronix 6015 is a commercial example where a combination of fixed and variable resistors, and stray and variable capacitors, can be set for excellent dc and ac (to 75 MHz) response. This corresponds to a step response of about 5 nanoseconds.

A detailed analysis is not attempted here, because several such efforts have already been performed. A few general comments are in order. First, the general problem to be solved is that of Fig. 10-14A. For very fast rise times, the capacitances associated with a mixed divider represent a low impedance. The paper "The Measurement of Short-

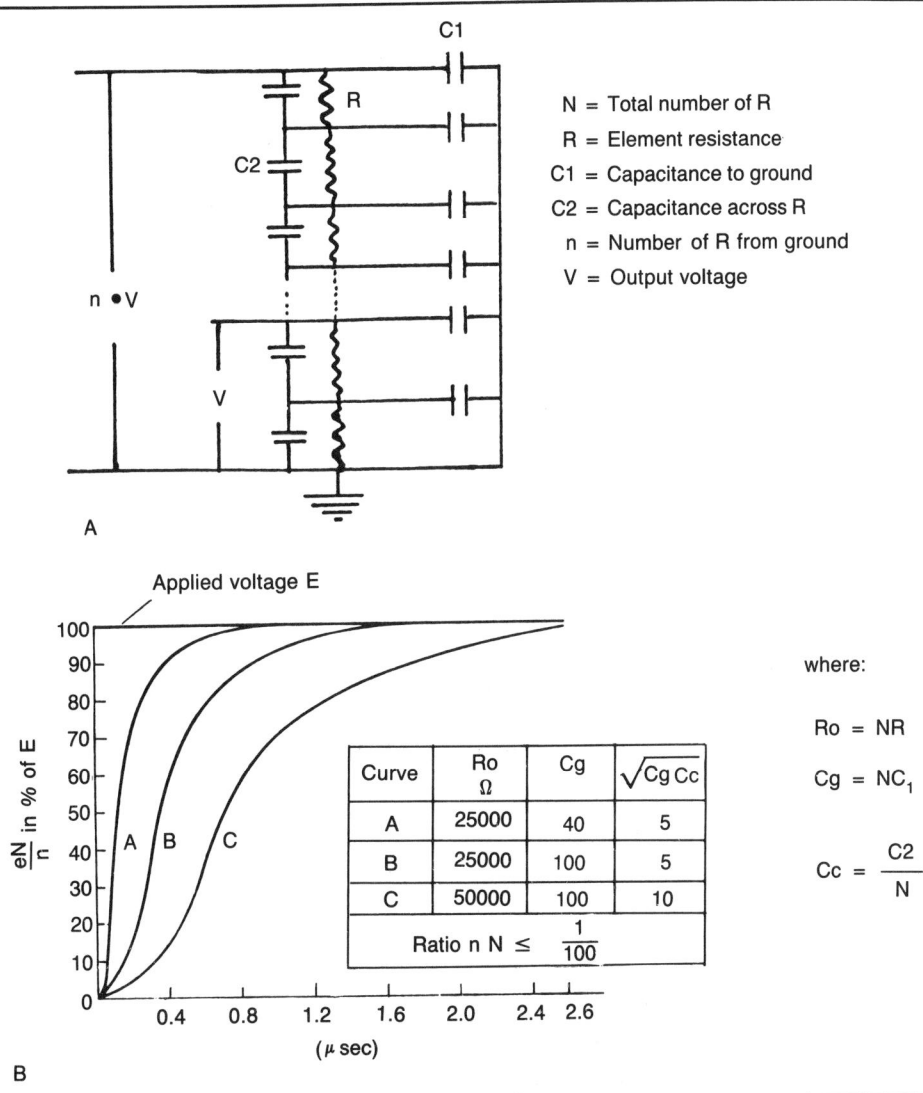

Fig. 10-14. (A) Distributed circuit. (B) Step responses (Reprinted with permission from E. Kuffel and W. Zaengl, "High Voltage Engineering," 1965, Pergamon Books, Ltd).

Duration Impulse voltages" by Creed and Collins describes an attempt to use $10,210\Omega$ as the input. There are six conclusions drawn, of which two are very relevant.[4]

"A high impedance divider can be satisfactorily made for the measurement of steeply rising impulses, but a measuring system cannot have a high impedance for the first few nanoseconds. Steep front tests can therefore be performed with a high-impedance system, but for measuring rates of voltage collapse across a gap, only a low-impedance system is possible."

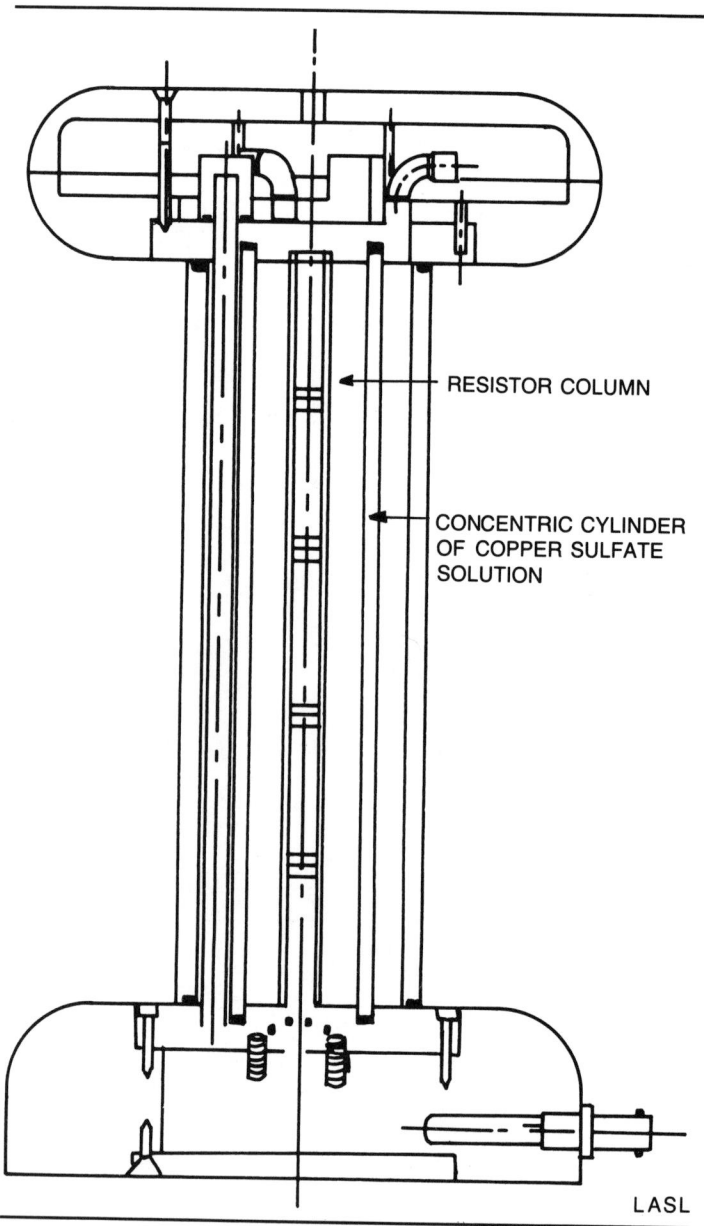

Fig. 10-15. Concentric cylinder of copper sulfate around resistor column.

This refers to the fact that spherical spark gaps were used as the standard for the measurements. The second point is:[4]

"When making very-short-time measurements with an impulse-voltage divider, it is important to consider all leads from the transmission-line concept; their surge impedance and length must be considered when studying the circuit."

10.1 Voltage Measurements

All mixed dividers will have some input capacitance. The inductance of connections will be self resonant somewhere, typically in the 10 to 20 MHz region. This topic is addressed in the paper by Creed, Kawamura, and Newi.[5] They describe an impulse measuring system in terms of five components: a divider, a recording oscillograph, a coaxial cable connecting the two, a lead to connect the divider to the high voltage terminal, and a lead or ground plane that connects the base of the divider to the grounded side of the test object. The latter two are often taken for granted.

10.1.6 Electro-Optic Effect

A somewhat exotic device for voltage measurement employs the Kerr cell. This is simply a capacitor filled with (usually) nitrobenzene. Most materials transmit polarized light with little or no effect. Nitrobenzene is one of the few materials that have indices of refraction that depend upon electric fields present. The effect is to rotate polarized light into another axis. This is expressed by:

$$n_\| - n_\perp = \lambda B E^2 \tag{10-8}$$

where λ is the wavelength of light, B the Kerr coefficient, E the electric field, and n the index of refraction. The resulting phase shift is given by:

$$\phi = 2 \int_0^L B E^2 \, dl \tag{10-9}$$

In practice, the Kerr cell is often used as an electro-optic shutter. The layout is sketched in Fig. 10-16. When the voltage V is applied, the Kerr cell rotates the polarized light 90°, and the crossed polarizer will transmit. This can be turned around with the detected light as an indication of the applied voltage. The output signal is unusual but can be deconvolved. The Bureau of Standards uses this method and quotes better than 5-percent accuracy on 300 kV pulses.

The light transmission is given by:

$$\frac{I}{I_m} = \sin^2 \left[\frac{\pi \, (V/Vm)^2}{2} \right] \tag{10-10}$$

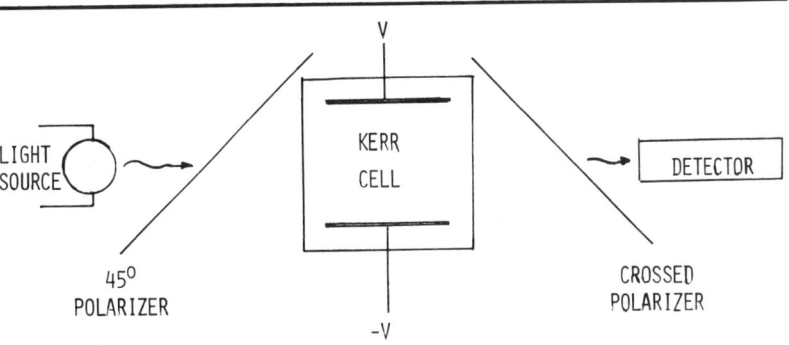

Fig. 10-16. Schematic of Kerr cell.

312 Measurement Techniques

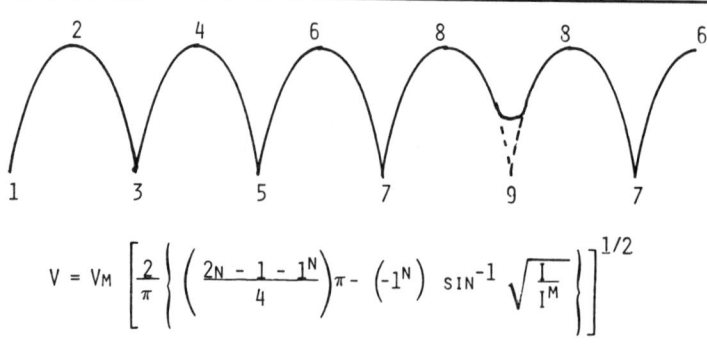

Fig. 10-17. Waveform of light transmission.

Note that V can be greater than V_m, limited usually by external flashover, so that the phase rotation is continuously increasing with voltage in a nonlinear manner. A wave form will look somewhat like Fig. 10-17. An expression that will track the voltage is given on the figure. The wavelength dependence is avoided in a measuring system by using monochromatic light, typically a HeNe laser.

The Kerr effect is very fast, and the measuring limit is either the detector or the self-resonance of the Kerr cell capacitance and the connection inductance. Advantages of the system are the lack of ground reference and optical coupling. A good reference is IEEE Transactions, "Instrumentation and Measurements," Dec. 1975, Vol. IM-24, No. 4.

10.2 CURRENT MEASUREMENTS

There are two basic current measurement techniques available. Either charge flow is measured directly, or the magnetic field is detected and the current is inferred in some manner. The second method is more flexible, particularly in high-voltage circuits, but the problem of detecting a representative portion of the field can be troublesome. Very little attention is given to simple ammeter and dc measurements. The only point to emphasize is that the ground return is a perfectly valid point in which to install an ammeter. This avoids floating-deck problems, protective covers, etc. The only significant consideration is that of multiple grounds, commonly due to the third or ground wire in the ac line.

Direct measurement of pulsed current is usually taken across a resistor, often called a CVR or *current viewing resistor*. Just as there is an inductance associated with a capacitance in voltage dividers, so there is an inductance included in practical CVRs. The $L(dI/dt)$ term introduces an error analogous to the resonance limit in voltage dividers.

Frungel describes the current viewing resistor of Fig. 10-18. He used 0.2 millimeter bronze disks for the parts labeled 2 and obtained 10^{-4} Ω with an associated 10^{-9} H. Part 1 is a set of hairpin-like CVRs. The geometry is neat, yet his discussion leads to a 35 kHz capability. The real world is not that grim. T&M Research Products, Inc. in Albuquerque, New Mexico, makes a variety of CVRs with frequency responses to 1000 MHz, corresponding to 0.36-nanosecond rise times. Construction is based on transmission line principles to achieve low impedance and inductance. The resistive element is a thin con-

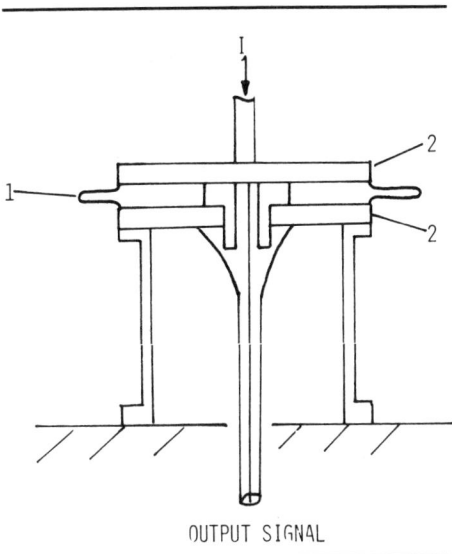

Fig. 10-18. Current-viewing resistor (reprinted with permission from E. Kuffel and W. Zaengl, "High Voltage Engineering," 1965, Permagon Books, Ltd).

stantin foil so that skin-depth effects are avoided, even at high frequencies. The temperature coefficient is very low.

The quantitative, low-impedance, low-distortion characteristics of such CVRs are very attractive. The drawback is that one side goes to ground through the measuring system, oscilloscope, etc. Thus the CVR is only applicable when the system permits grounding at the point of measurement.

When grounding poses a problem, magnetic field detection techniques are employed. The simplest, conceptually, is the Rogowski coil. This depends upon Lenz's Law: "When the flux linking a closed circuit is changing, the flux set up by the induced current is in such a direction as to tend to prevent the change in the flux linkage."

Note the emphasis on change. Rogowski coils inherently measure a term proportional to dI/dt, and integration is necessary to obtain a signal representing current. Figure 10-19 considers the relationship between the coil and the field to be measured. In Fig. 10-19(A), it is apparent that the single-turn loop around the wire lies in the plane of the magnetic field and will not produce a useful signal. Figure 10-19(B) shows the solution when the Rogowski surrounds the conductor. It should be emphasized that only the principle is easy where Rogowskis are concerned. In practice they can be quite difficult, particularly at high frequencies. A thorough study of the literature is recommended before attempting to design and build a Rogowski for quantitative work.

Another similar device is the current transformer. In the simplest case, with close coupling, insignificant core losses, and winding reactances large with respect to resistances, the currents go as the turns ratio:

314 Measurement Techniques

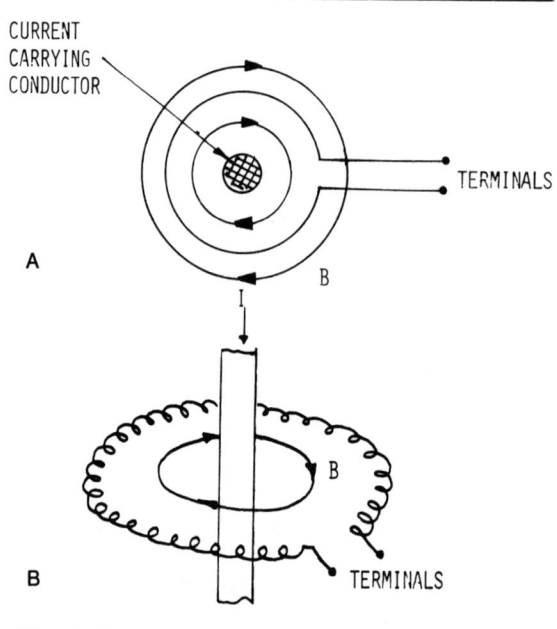

Fig. 10-19. (A) Loop in plane of field. (B) Rogowski pickup.

$$\frac{i_p}{i_s} = \frac{N_s}{N_p} \qquad (10\text{-}11)$$

where p and s refer to primary and secondary. Thus:

$$i_s = i_p \frac{N_p}{N_s} \qquad (10\text{-}12)$$

and if i_s flows through a load resistor, R_L:

$$V_{out} = i_p R_L \frac{N_p}{N_s} = K\, i_p \qquad (10\text{-}13)$$

Also, under ideal conditions, the impedance reflected into the primary is:

$$R_{REF} = \left(\frac{N_p}{N_s}\right)^2 (R_L + R_s) \qquad (10\text{-}14)$$

where R_s is the secondary winding resistance. Usually, N_p is one and N_s is quite large so that R_{REF} is very small.

Several manufacturers supply these transformers, including Pearson, Stangenes, Ion Physics, and English Electric Valve. The units produced by the latter are designed for thyratron measurements, and the current to be measured must be centered in the opening.

Not all grounding problems are avoided. The outer casing is at ground so that adequate insulation between the transformer and the current-carrying conductor is required. Also, the signal ground is at the point where the coaxial cable goes to a screen room or measuring instrument so that some noise pickup is often encountered.

Most of these units contain an iron core, which results in an annoying perturbation of the signal. In pulse measurements, the dc component of the pulse produces a spurious undershoot on the trailing edge.

Actually, because these are transformers, they respond to the rate of change of current. An internal integration is performed to yield a signal representing the current, and this is normally presented with a 50Ω output impedance to match cables. In fast pulse work the cable is usually terminated in 50Ω at the recording end as well. This halves the sensitivity so that a 0.1 V/A device becomes a 0.05 V/A device. Be sure to note the limits on lower as well as upper frequency response and the limits on charge per pulse in choosing the transformer for a given application.

Hall effect probes are also used to infer currents from the associated magnetic fields. Figure 10-20 shows the geometry involved. A charge carrier moving in the crystal is deflected by a simple:

$$F_Y = qv_x B_z \qquad (10\text{-}15)$$

where the subscripts are the axes of a rectangular coordinate system, and all terms are vectors. Charge accumulation at the surfaces produces a balancing electric field force so that:

$$\frac{V_{HALL}}{b} q = qvB \qquad (10\text{-}16)$$

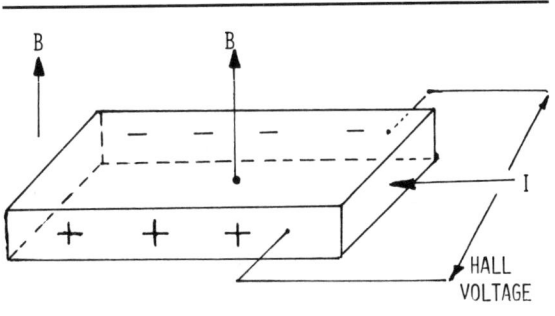

Fig. 10-20. Hall effect.

316 Measurement Techniques

and v is related to current by:

$$i = gnvbd \qquad (10\text{-}17)$$

where

n = carrier density
q = magnitude of electronic charge
v = carrier velocity
b = sample width
d = sample thickness

By manipulation:

$$V_{HALL} = \frac{iB}{qnd} = R_H \frac{iB}{d} \qquad (10\text{-}18)$$

If the Hall constant, R_H, is known and the current is preset, then the Hall voltage varies with the B field. In effect, it behaves somewhat like a Rogowski. The charge carrier mobility sets an upper limit but probes with thin sections are quoted up to several megahertz.

An analog of the Kerr effect also exists, which is Townsend rotation. A similar rotation of polarized light can be induced in a crystal with the magnetic field inducing the rotation. The Japanese have measured current in high-voltage power lines with such a system.

10.3 REFERENCES

1. F.B.A. Frungel, "High Speed Pulse Technology," Vols. 1–3, Academic Press, New York, 1965.
2. E. Kuffel and W. Zaengl, "High Voltage Engineering," Pergamon Press, New York, 1965.
3. LASL Dwg. No. 33Y180069.
4. F.C. Creed and M.M.C. Collins, "The Measurement of Short-Duration Impulse Voltages," IEEE Trans. Communication and Electronics, Vol. 82, pp. 621–630, November 1963.
5. F.C. Creed, T. Kawamura, and G. Newi, "Step Response of Measuring Systems for High-Impulse Voltages," IEEE Trans. on Power Apparatus and Systems, Vol. PAS-86, pp. 1408–1420, November 1967.

Chapter 11

Particular Applications

by
R.R. Butcher
XMR, Inc.

THIS CHAPTER DISCUSSES PARTICULAR APPLICATIONS. THE FIRST POINT OF DISCUSSION is the magnetron load. The voltage versus current characteristic of the load is shown in Fig. 11-1. The voltage increases to the point where the device breaks down; then it goes into a nearly constant glow voltage region as the current increases. When you drive the current hard enough, you get into a region that is called *abnormal glow*, and then it eventually will go into an arc at which point the voltage goes to a low value as the current increases. This is typical of a number of devices. Most gas loads exhibit at least similar behavior. The data presented have been taken on a krypton fluoride fast-discharge laser.[1]

11.1 LASER LOADS

11.1.1 Discharge Circuit

The schematic diagram for the laser circuit (Fig. 11-2) consists of a two-stage Marx bank charged to 30 kV, made up of 0.15 μF Maxwell capacitors and using Maxwell 50 kV spark gaps as switches. This Marx bank is connected to the cable pulse forming networks through an inductance, denoted by L_d. The PFN consists of 16 lengths of LANL Type 40/100 cable. This cable has a little over 30Ω (per cable) characteristic impedance, resulting in about 1.9Ω characteristic impedance for the pulse-forming line. The physical length of the line is 20 feet, which produces a 38-nanosecond, one-way transit time. Note that the laser load has an inductance, L. The voltage was measured across that inductance and the electrodes. The current, I, which goes through the laser, was monitored with a current viewing resistor (CVR).

11.1.2 Circuit Operation

When the spark gaps are triggered, the voltage rings up on the cable PFN through L_d (Fig. 11-3). For purposes of analysis, because the voltage rise time at the laser is

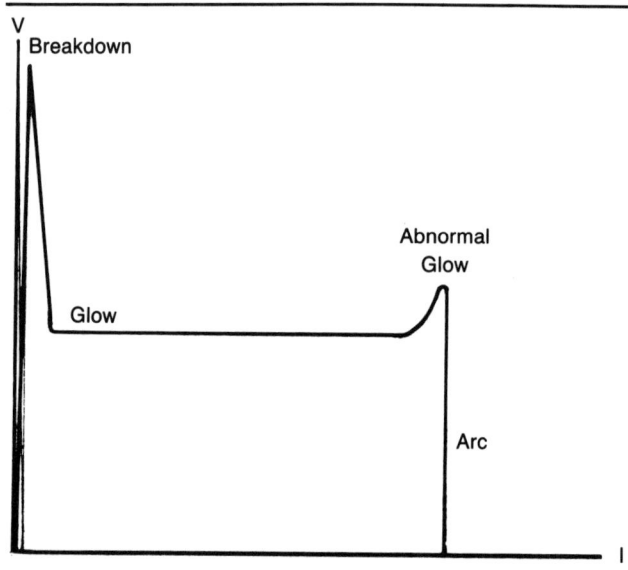

Fig. 11-1. Magnetron load characteristics.

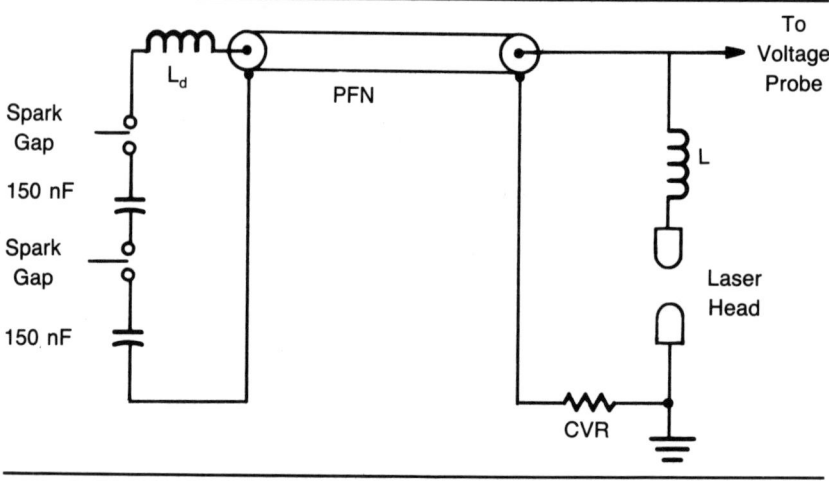

Fig. 11-2. Laser-system schematic.

somewhat longer than the transit time of the cables, the PFN is treated as a lumped 20 nF capacitor while it is charging. Treating the Marx as a single 75 nF capacitor charged to 60 kV results in the simplified circuit shown in Fig. 11-3. This is a series-resonant circuit as far as charging purposes go. Neglecting any resistive losses, the voltage on the cables can be approximated by the equation:

$$V_c(t) = V_m (1 - \cos \omega t) \tag{11-1}$$

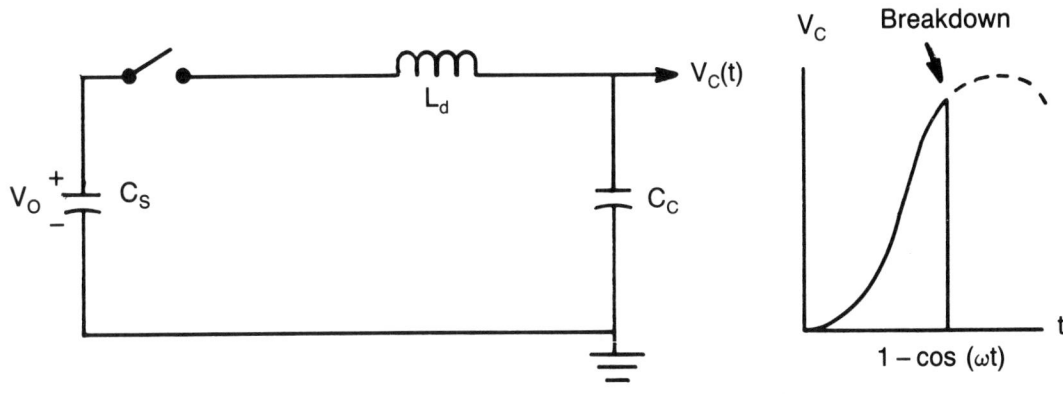

Fig. 11-3. Simplified schematic and charging waveform.

where the ringing frequency is calculated from:

$$\omega = \frac{1}{(L_d C_{eq})^{1/2}} \quad (11\text{-}2)$$

where L_d is the inductance of the driver, and the equivalent series capacitance is expressed by:

$$C_{eq} = \frac{C_c C_s}{C_s + C_c} \quad (11\text{-}3)$$

where C_s is the capacitance of the storage capacitor (or equivalent series capacitance in the case of a Marx generator), and C_c is the cable capacitance. Because the charge divides between series capacitors, the peak voltage is expressed by the relationship:

$$V_m = V_o \left(\frac{C_s}{C_s + C_c} \right) \quad (11\text{-}4)$$

where V_o is the initial voltage on the storage capacitor, or nV_o in the case of an n-stage Marx generator. It is worth noting that if $C_s \gg C_c$, the ringing frequency is determined by primarily C_c and L_d, which is true for the circuit of Fig. 11-3. Also, if the voltage is allowed to ring to its full peak value ($\omega t = \pi$), the voltage will nearly double. The time rate of change of voltage can be found by taking the derivative of Equation 11-1 and is expressed as:

$$\frac{dV}{dt} = V_m \omega \sin \omega t \quad (11\text{-}5)$$

which is useful for calculations involving this parameter.

11.1.3 Energy Transfer

Another point worth considering is the energy transfer efficiency:

$$\text{Eff} = \frac{E_c}{E_s} \qquad (11\text{-}6)$$

where the energies on the capacitors are found from:

$$E_c = \frac{1}{2} C_c V_c^2 \qquad (11\text{-}7)$$

$$E_s = \frac{1}{2} C_s V_o^2 \qquad (11\text{-}8)$$

and the voltage on the cable capacitor (assuming $\omega t = \pi$) is found from Equations 11-3 and 11-4:

$$V_c = \frac{2 V_o C_s}{(C_c + C_s)} \qquad (11\text{-}9)$$

The resulting energy transfer efficiency is calculated from Equations 11-6, 11-7, 11-8, and 11-9 as:

$$\text{Eff} = \frac{4 C_c C_s}{(C_s + C_c)^2} \qquad (11\text{-}10)$$

Solving for maximum transfer efficiency:

$$0 = \frac{d\text{Eff}}{dC_c} = 4 C_s \frac{d}{dC_c} \left[\frac{C_c}{(C_s + C_c)^2} \right] \qquad (11\text{-}11)$$

$$0 = \frac{4 C_s \left(C_s^2 - C_c^2 \right)}{(C_s + C_c)^4} \qquad (11\text{-}12)$$

either $C_s = 0$ or $C_s = C_c$ will satisfy this equation. Obviously the latter condition will result in the maximum efficiency, which from Equation 11-10 becomes:

$$\text{Eff} = \frac{4 C_c^2}{(2 C_c)^2} = 1 \qquad (11\text{-}13)$$

Therefore, the driver can be designed for maximum voltage gain or maximum efficiency, but not both.

11.1.4 Laser Electrical Characteristics

The laser is usually operated at a voltage, V_o, high enough to reach the breakdown threshold of the gas well before the peak of the charging voltage, that is, $\omega t < \pi$. In this case, some energy is delivered to the laser gas from the driver directly, because the PFN is being charged by the driver while being discharged by the gas load. This usually raises the laser output somewhat but results in extra heat being put into the laser gas after the useful work is done. The extra heat reduces overall efficiency.

Figure 11-4 is a typical voltage and current trace taken at the laser end of the cables. Instead of the voltage coming up with a 1-cosine function as expected, it is more rounded off. Again, the effect depends on the number of cables, the length of the cables, and resistive losses. For the particular conditions on this laser, the voltage charged up to about 32 kV, which was the breakdown voltage for the particular gas mix and electrode spacings used. The voltage then started to drop as the current started to come up, and the voltage might have been going towards some sort of a plateau or a glow voltage at about the time the current peaked. The vertical line drawn near 1.3×10^{-7}

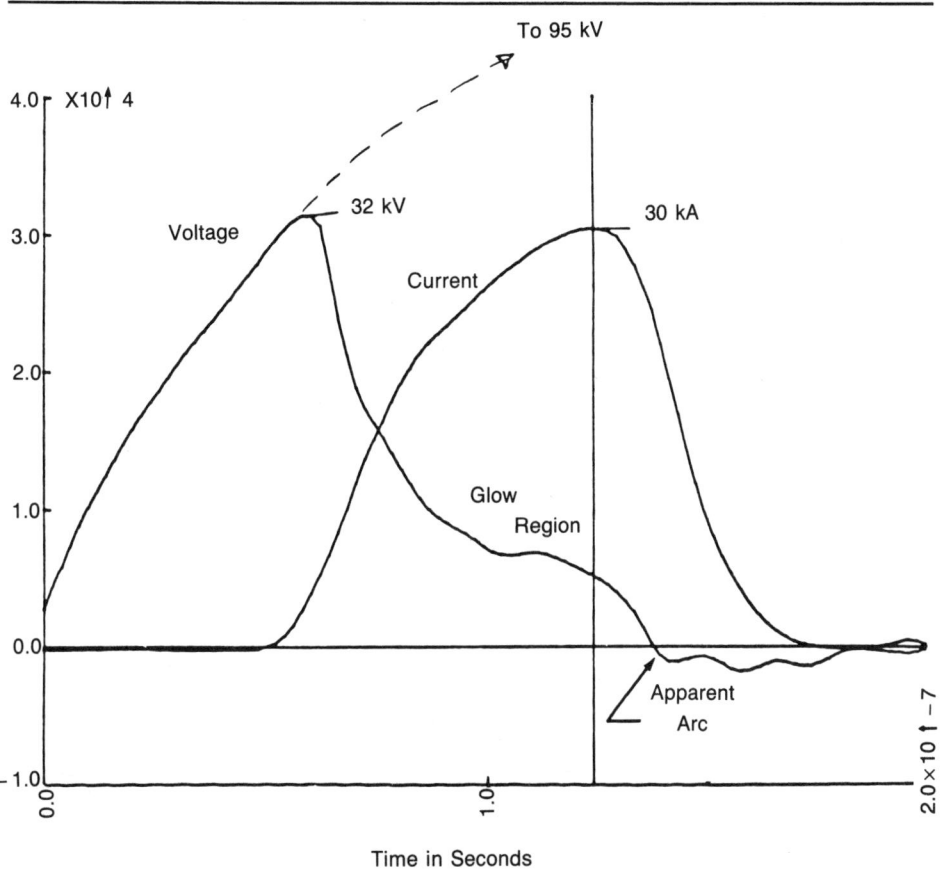

Fig. 11-4. Voltage and current waveforms.

322 Particular Applications

seconds indicates the time when the current reached its peak value. Shortly after the peak current, it appears that the discharge went into an arc, because there is a very small voltage appearing across the laser during the balance of the current pulse. This might not be the case, however, as is discussed later.

Now consider some other waveforms that were derived from the voltage and current. The voltage and current were the only waveforms that were actually measured. The rest of the data presented are derived from these voltage and current waveforms.

There are a number of factors that can be calculated from the voltage and current waveforms. One of them, for example, is the dI/dt (Fig. 11-5). This particular laser has a dI/dt of approximately 10^{12} A/seconds. This is used in other derivations. The charge that is flowing through the circuit (Fig. 11-6) is the time integral of the current, and in this particular case, there is a little over 2 mC that passed through the laser during the time of interest. All of these variables were calculated from the scope photographs of the voltage and current. A desktop calculator was used to digitize the waveforms and then calculate additional parameters. Another useful parameter is the instantaneous power pulse (Fig. 11-7). Note that this is the *apparent power* because it is the product of the voltage and current as measured. It is between 200 and 300 MW peak power, and it is about 70 or 80 nanoseconds in width. There is a little negative power at about 150 nanoseconds. The negative power occurs when the voltage went negative while the current was positive. It is power being returned to the circuit from the laser-head inductance. The negative power near 30 nanoseconds is due to digitizer drift.

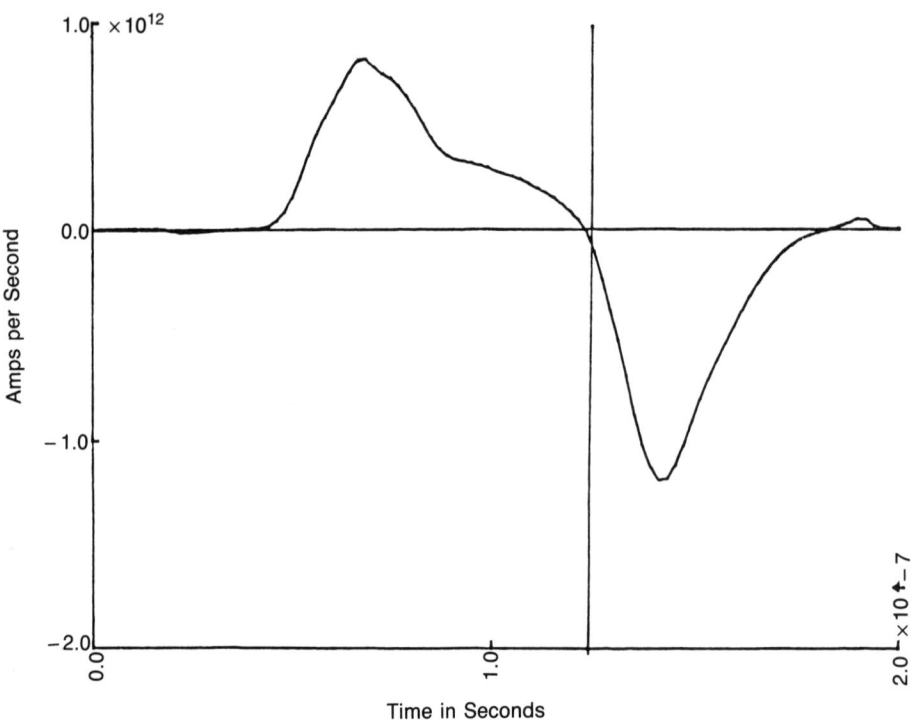

Fig. 11-5. dI/dt calculated from current.

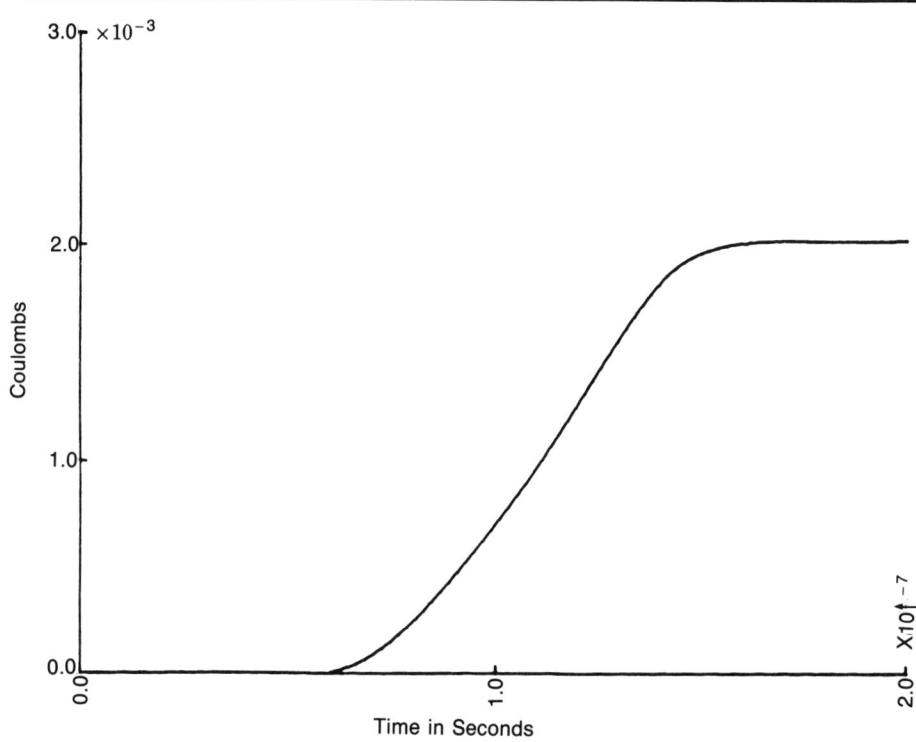

Fig. 11-6. Charge calculated from current.

The apparent energy deposited in the load (or at least going into the load whether it stays there or not) is found from the time integral of the apparent power. In the curve in Fig. 11-8, the energy is slightly less than 15 joules. Notice there is a small amount that was returned at around 150 nanoseconds.

11.1.5 Inductance Considerations

It would be useful to know the laser-head inductance. It is possible to calculate the inductance based on the geometry, which is a rather complicated geometry. There are other ways to check the value. One approach is to start with the apparent power, which is voltage times current. However, the voltage has two components. There is an IR drop, the resistive component, and there is the $L\,dI/dt$ term, which is the inductive component of the voltage. The apparent power is really the sum of those two times the current:

$$W = \left(IR + L\,\frac{dI}{dt}\right) I \qquad (11\text{-}14)$$

324   **Particular Applications**

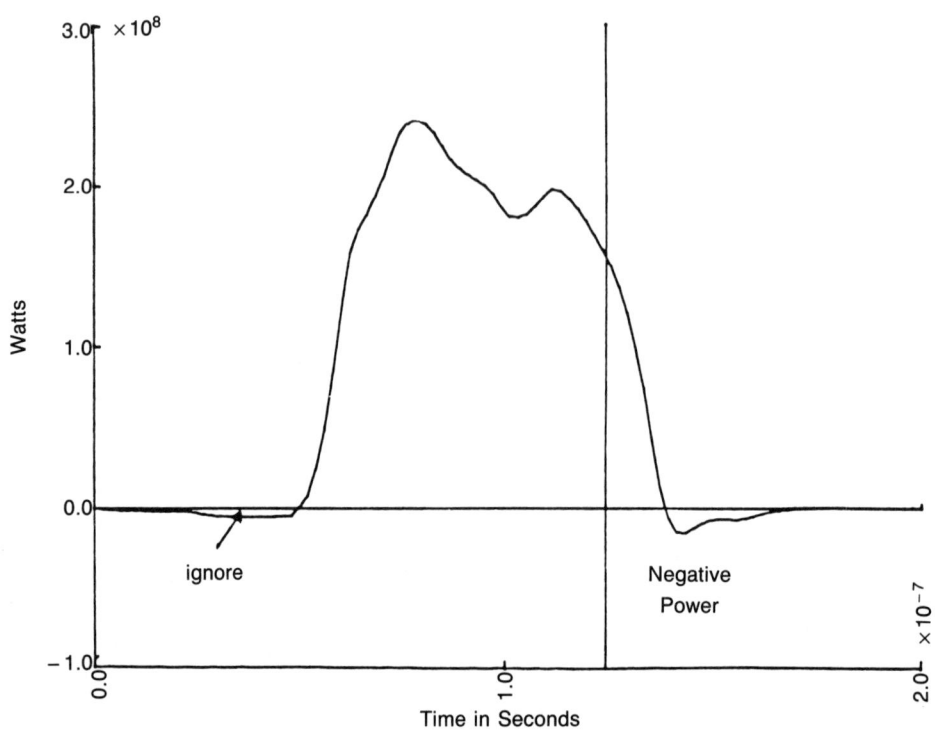

Fig. 11-7. Apparent instantaneous power in laser.

The time derivative of the power is expressed:

$$d\frac{W}{dt} = 2IR\frac{dI}{dt} + L\frac{dI^2}{dt} + LI\frac{d^2I}{dt^2} \tag{11-15}$$

Evaluating Equations 11-15 at a point where dI/dt is zero, that is, at the peak of the current pulse, results in two of the terms drop out leaving:

$$\frac{dW}{dt} = LI\frac{d^2I}{dt^2} \tag{11-16}$$

This can be rearranged to find the inductance as:

$$L = \frac{dW}{dt} \bigg/ \left(I\frac{d^2I}{dt^2}\right) \tag{11-17}$$

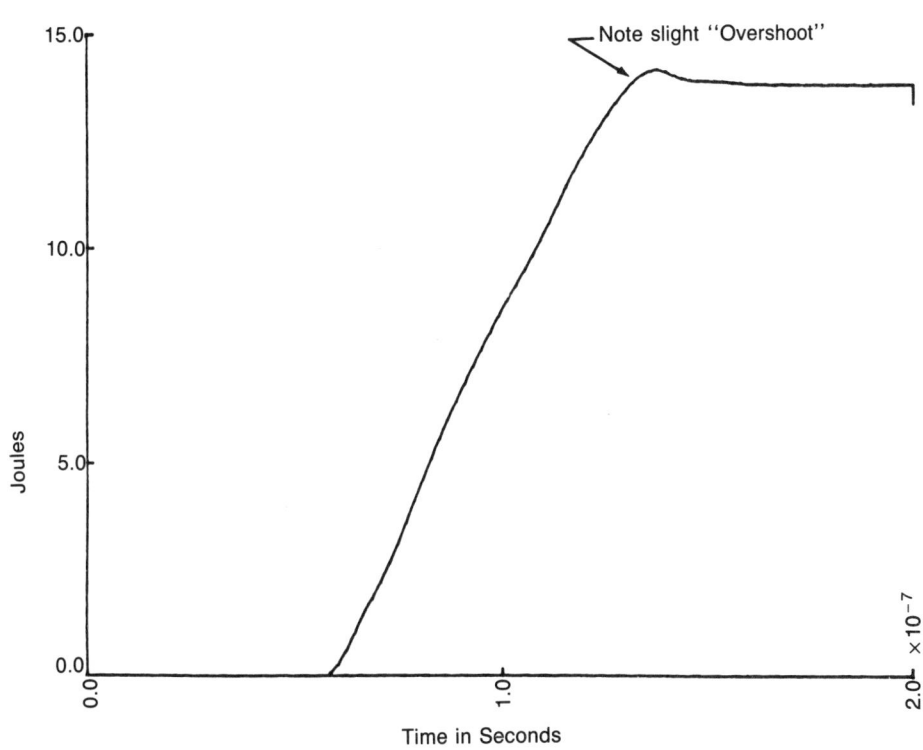

Fig. 11-8. Apparent energy deposited in laser.

Figure 11-9 is the time derivative of power, which is not constant at the time of peak current, but is reasonably well behaved in that region, having a value of about -5.7×10^{15}. The second derivative of the current with respect to time (Fig. 11-10) is not a constant at that time either, but again it is reasonably well behaved, and it turns out to be -4.7×10^{19}. The peak current is approximately 30 kA, which results in an inductance of 4 nH.

Recall that the measured voltage has two components to it: there is the resistive voltage (IR drop), and there is an L dI/dt term, which is the reactive component. This relationship can be rearranged to show that the resistive voltage is:

$$V_R = V_{meas} - L \frac{dI}{dt} \qquad (11\text{-}18)$$

The inductive voltage drop is shown in Fig. 11-11. There is a fairly appreciable voltage drop across the inductance due to the rate of change of current. The calculated resistive voltage pulse is shown in Fig. 11-12. The resistive voltage pulse appears to break down at a somewhat lower voltage. The pulse has more of a flat glow region in it, and it tends to be stretched out in time by the $L \, dI/dt$ term. Note the absence of the arc seen in the measured voltage (Fig. 11-4). The resistive power pulse is shown in Fig.

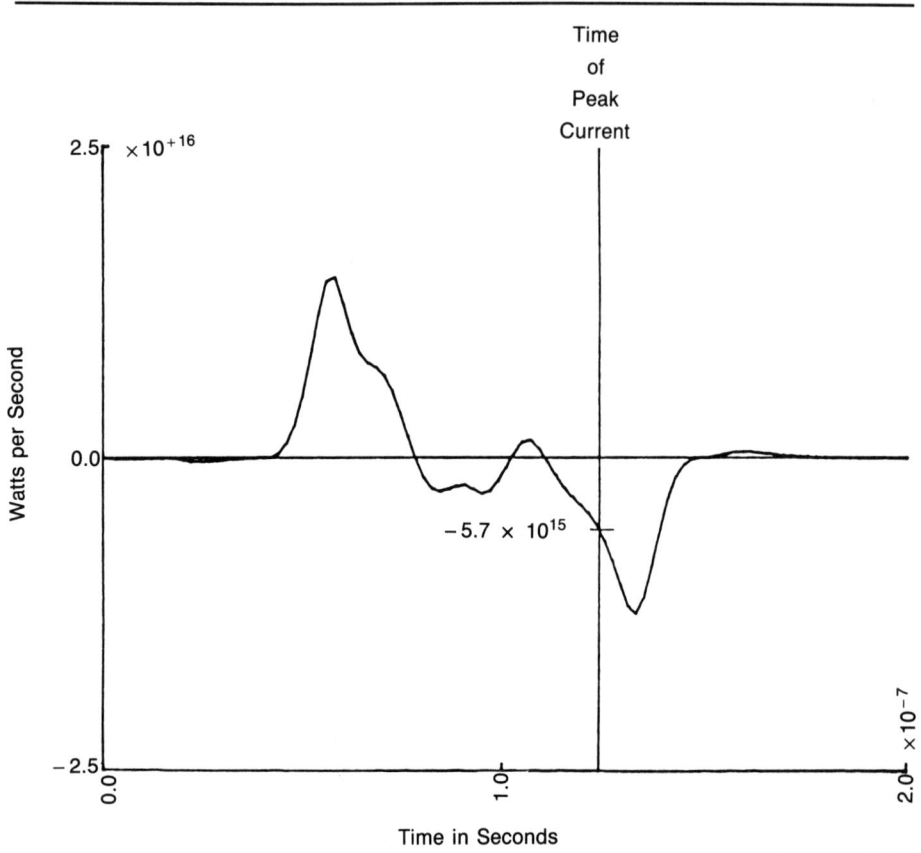

Fig. 11-9. Rate of change of power in laser.

11-13. The resistive power is the current times the resistive voltage. Comparing this to the apparent power pulse (Fig. 11-7), note that the resistive power pulse is slightly lower in magnitude and somewhat wider, and it is quite nicely missing the negative power. That is important because negative powers do not exist unless there is a reactive component. That is one way to tell if the inductance is a believable number. The minimum inductance required to eliminate the negative power is the lower limit. The resistive power pulse is also somewhat smoother than the apparent power pulse. The resistive energy pulse, that is, the energy that is really deposited in the resistive load, is the time integral of the resistive power (see Fig. 11-14).

Another way to calculate the inductance that sometimes works better than the rate of change of power technique is to use the energy that is returned to the circuit in the apparent energy plot, and use the relationship that inductance energy stored at the peak current is:

$$E = \frac{1}{2} LI^2 \qquad (11\text{-}19)$$

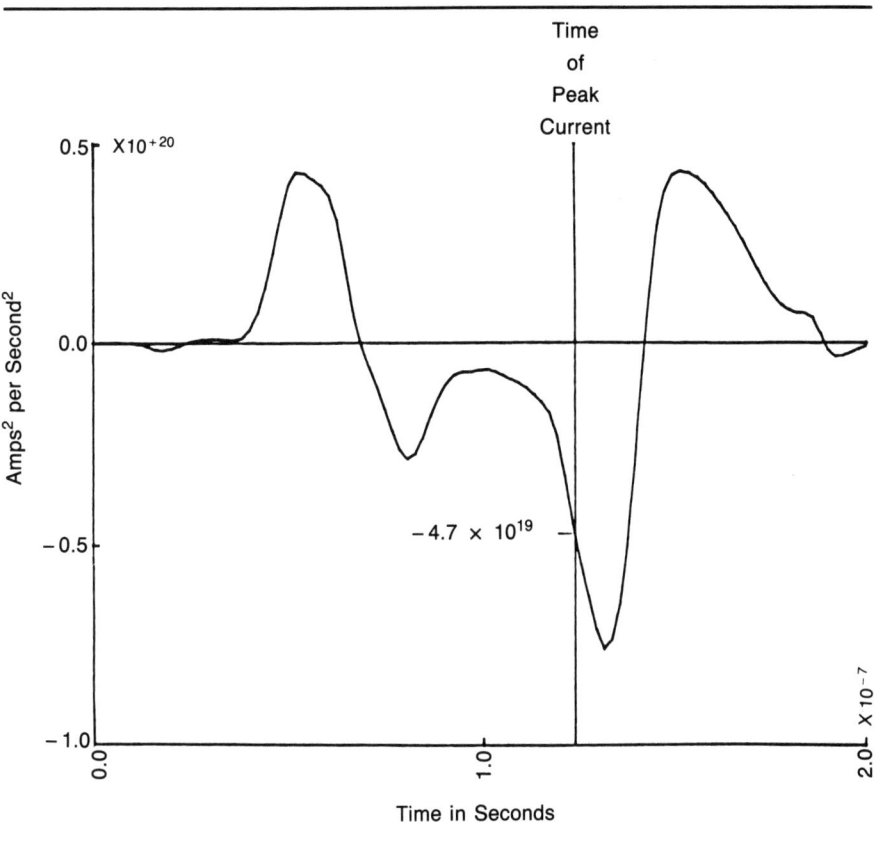

Fig. 11-10. Second derivative of current.

where E is the energy returned.

The inductance can be found from:

$$L = 2E/I^2 \qquad (11\text{-}20)$$

These are all somewhat smoother functions than those of Equation 11-17, but unfortunately the energy reversal is not always as large as might be expected, resulting in a low inductance value.

11.1.6 Laser Time-Varying Impedance

Consider the time-varying impedance of the laser. Figure 11-15 shows impedance (in ohms) as a function of time, (that is, the measured voltage divided by the measured current). It starts out at infinity up until the voltage breaks down; then it collapses below 1Ω quite rapidly. Notice that the impedance goes negative at about 130 nanoseconds, due to the current reversal in the inductor. In a similar plot (not shown) of resistance versus time (that is, the resistive voltage divided by the current), the resistance tends

328 Particular Applications

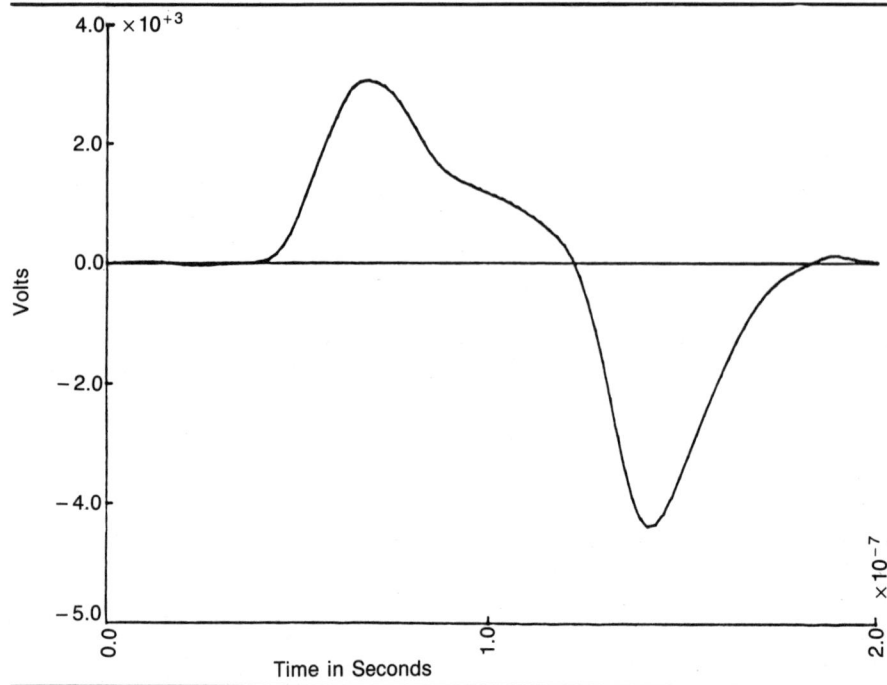

Fig. 11-11. *Inductive voltage drop in 4 nH inductance.*

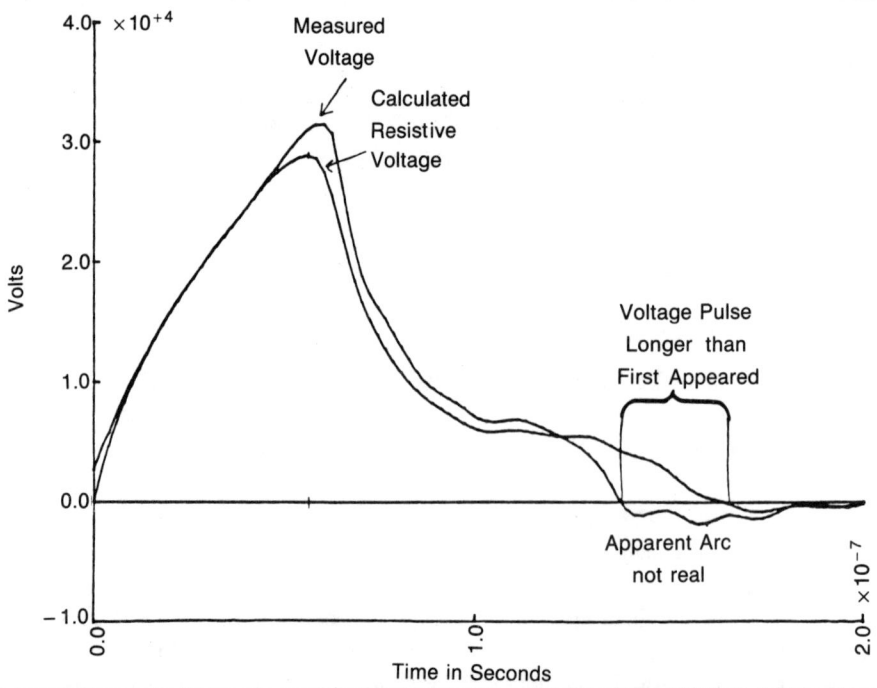

Fig. 11-12. *Comparison of measured and resistive voltages.*

11.1 Laser Loads

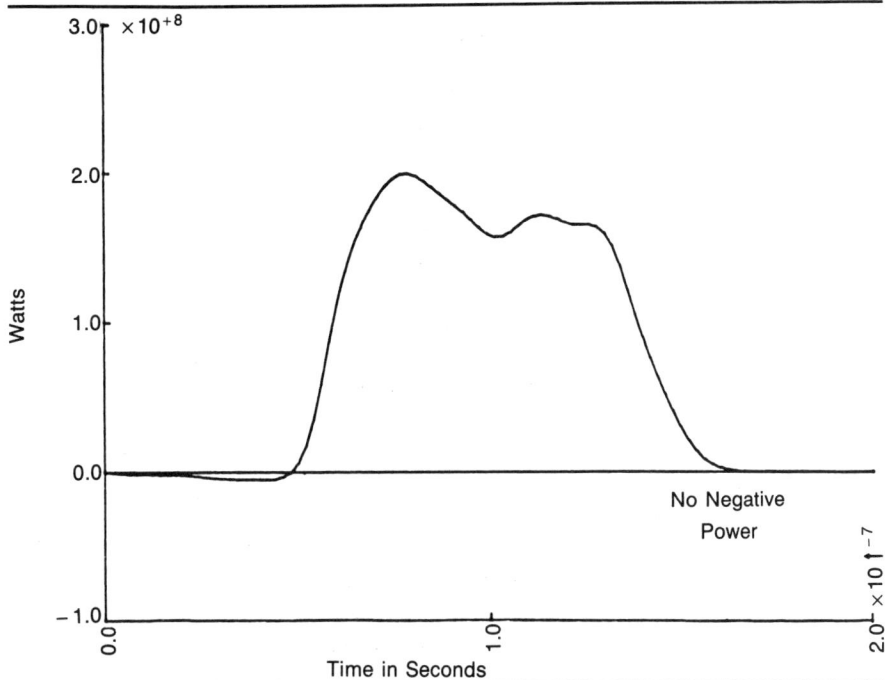

Fig. 11-13. Calculated resistive power.

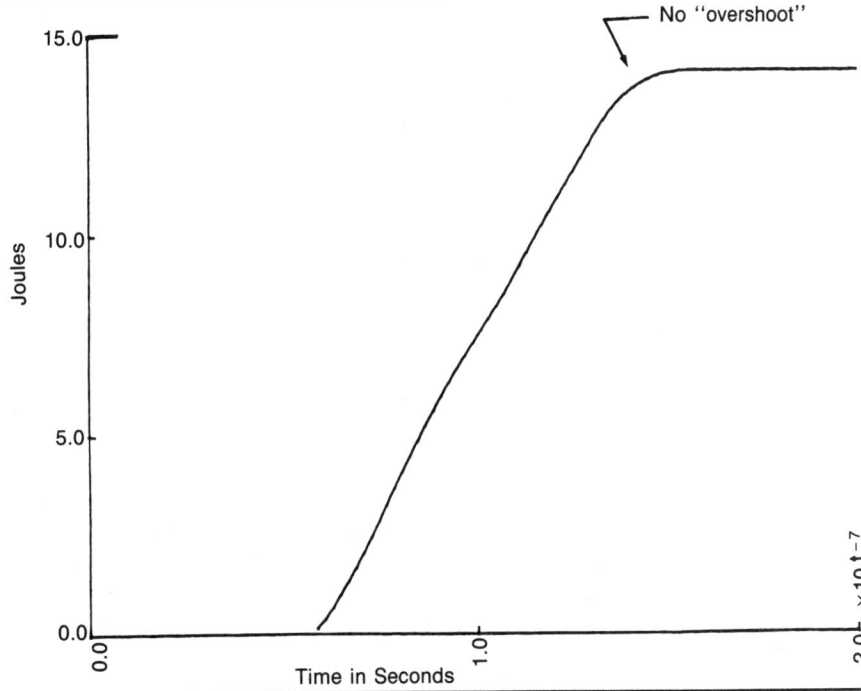

Fig. 11-14. Calculated resistive energy deposited in laser.

330   *Particular Applications*

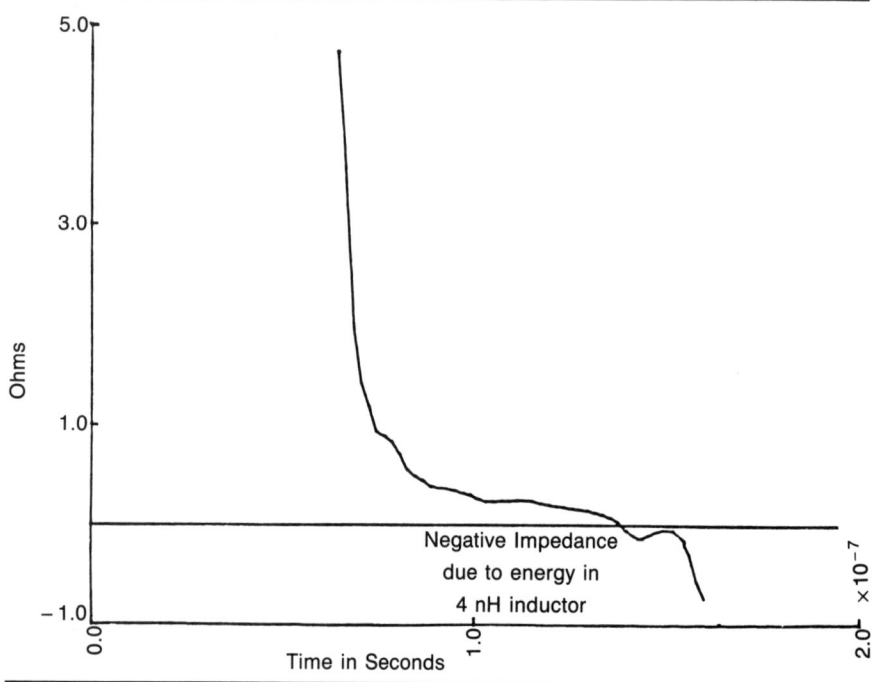

Fig. 11-15. Calculated impedance of laser gas.

to stay positive in the 130-nanosecond region instead of going negative. Referring to the voltage pulse and the current pulse (Fig. 11-4), the voltage at the time of breakdown looks something like an exponential decay:

$$V(t) = V_o \exp[-(t-t_o)/T_r] \qquad (11\text{-}21)$$

The current looks like an exponentially increasing current—the sort of current expected in an RL circuit:

$$I(t) = I_m[1 - \exp\{-(t-t_o)/T_r\}] \qquad (11\text{-}22)$$

The impedance can be found from the voltage divided by current:

$$Z(t) = \frac{V_o \exp[-(t-t_o)/T_r]}{I_m (1 - \exp[-(t-t_o)/T_r])} \qquad (11\text{-}23)$$

This function has an identical shape to the impedance shown in Fig. 11-15. A least-squares fit of measured points (the Xs) on the voltage curve compared to the least-squares fit of the voltage pulse is shown in Fig. 11-16:

$$V(t) = 32 \text{ kV} \left(\exp\left[-(t-57 \text{ nanoseconds})/23 \text{ nanoseconds}\right] \right) \quad (11\text{-}24)$$

Note that the 57 nanoseconds is the time of breakdown. The time constant, T_r, is the value of interest (23 nanoseconds). So it appears that the voltage across the gap is collapsing with a 23-nanosecond time constant. Some of this effect might be inductive, and some of it might be resistive. The resistive effect can be explained as a resistive phase of a spark gap. J.C. Martin [2] derived an empirical formula for the resistive phase time of a spark gap:

$$T_r = \frac{88(P/P_0)^{1/2}}{Z^{1/3} E^{4/3}} \quad (11\text{-}25)$$

This equation gives a resistive phase time of 38 nanoseconds using the density of the laser gas (mostly helium), the impedance of the cable PFN, and the electric field at the time of breakdown (16 kV/centimeter). This equation is for a single channel, however, and the laser is firing in many channels that tend to shorten T_r by increasing the effective Z.

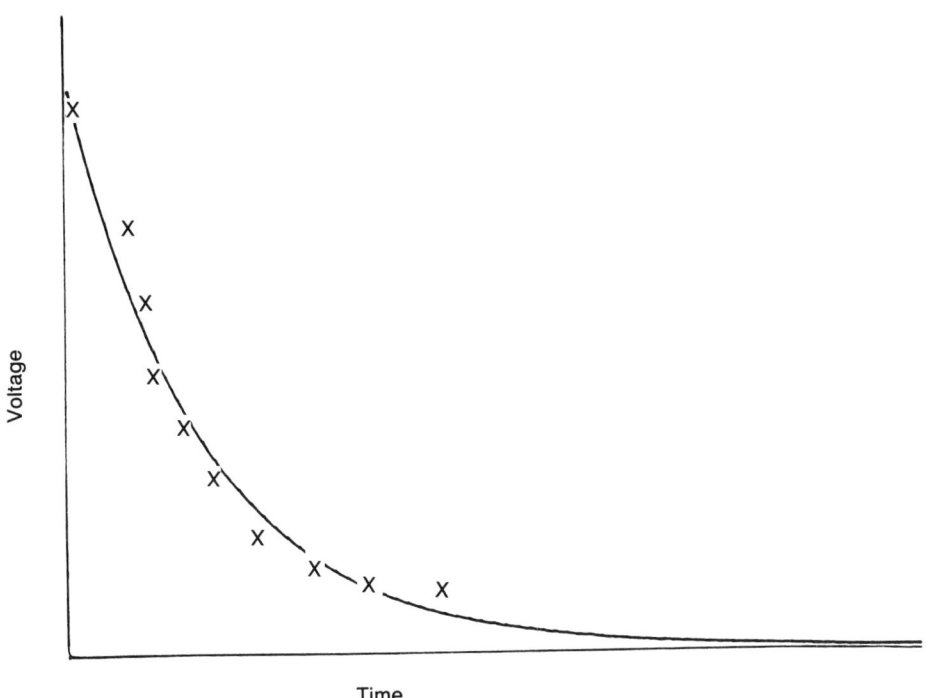

Fig. 11-16. *Least-squares fit to voltage during collapse.*

332 *Particular Applications*

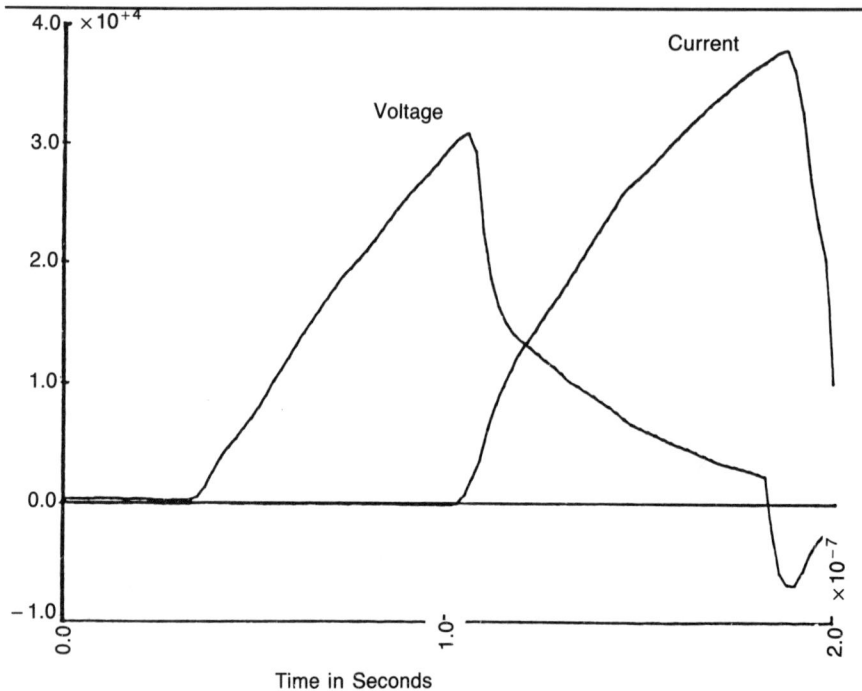

Fig. 11-17. Computer-predicted voltage and current.

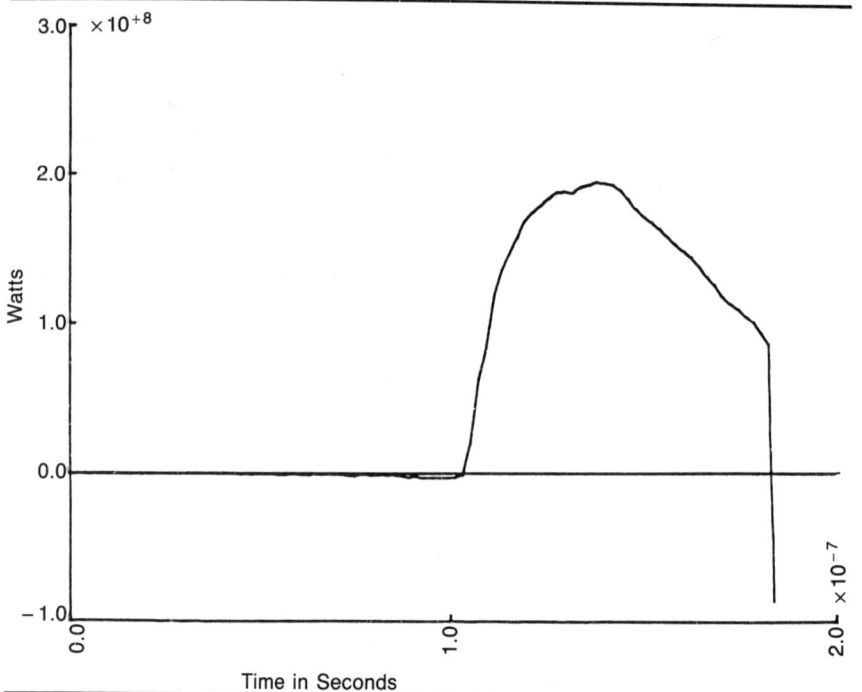

Fig. 11-18. Computer simulation apparent power.

11.1 Laser Loads

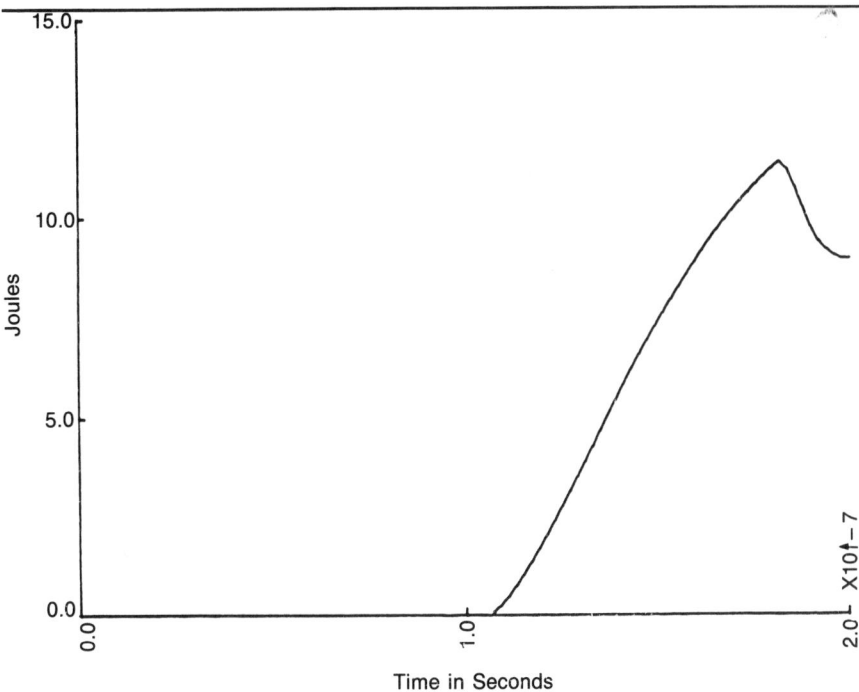

Fig. 11-19. Computer simulation apparent energy.

The resistance model Equation 11-23 can be checked with computer simulation using the 23-nanosecond time constant. Figure 11-17 shows the computer-generated voltage and current pulses. These pulses are very similar to those in Fig. 11-4, which indicates the model is reasonable. This simulation includes the total circuit: the Marx bank, the inductance in series of the Marx bank, the inductance in series with the laser (4 nH), the time-varying resistance of the laser gas. To simulate gas breakdown, the switch was closed when the voltage reached 32 kV. For a fixed-resistance model, the computer predicts drastically different current and voltage pulse shapes. So the time-varying resistance model for this particular circuit configuration appears to be valid. The computer-predicted power pulse is not too far from the real parameter (Fig. 11-7). The computer

Fig. 11-20. Program to Run Net-2.
```
RUNNER
RFL, 150000.
SETTL-400.
REDUCE, -.
ATTACH, METTWO/UM• 85213.
METTWO, DATA.
RETURN, METTWO.
REWIND, FILM.
REPLACE, FILM.
ATTACH, SCANgg/UM• 74136.
SCANgg.
REWIND, FILM.
END OF FILE
>?
```

```
Q
  1
0NET-a NETWORK ANALYSIS PROGRAM   RELEASE 9
  79/04/16  15.31.00
0        000000000111111111122222222223333333333444444444455555555556666666666
         1234567890123456789012345678901234567890123456789012345678901234567890
     1   *H-P67
     2   *UNITS-VOLTS, AMPS, OHMS, NANOFARADS, NANOHENRIES, NANOSECONDS
     3   *NAME IS CABLE
     4   V1 1 0 6E4 IE6
     5   S1 1 8 TIME-1
     6   Sa 5 6 F1(TIME)
     7   F1• N(4)-3a. 1E3 + 1E6*ABS(I(R1))
     8   F2• EXP(-(TIME-98)/23)
     9   F3• 1.9*F8(TIME)/(1.00001-Fa(TIME))
    10   L1 2 3 410 5E-4
    11   TLINE1 3 4 0 10 1.90 38 5E-4
    12   La 4 5 4 5E-4
    13   R1 6 0 F3(TIME)
    14   C1 10 75
    15   X?1-P(R1)
         PAGE 120 40
    16   MAXSTEP• .1
    17   STATE 1
    18     TIME 0 1 (100) 201
    19     PRINT N(4) I(R1) P(R1) X1 V(R1)
         FILM  1  0  0
    20   END
0NETWORK ELEMENT VALUE INVOLVED IN SIMULTANEOUS RELATION-  COMPUTATIONAL
 DELAY MAY OCCUR
```

Fig. 11-21. *Net-2 description of circuit in Fig. 11-2.*

actually predicted a much larger negative power pulse. Figure 11-19 is what the computer thinks the energy pulse should look like. This circuit is delivering a fair amount of energy back into the system. This is a case where the energy return to the system (proposed previously) could be used to calculate inductance. The result should agree quite well with Equation 11-20. The computer program used is NET-2, which is not a complicated program to use. Figure 11-20 is the program required to run NET-2. The NET-2 program to simulate this circuit, including the time-varying resistor and the switch that closes at 32 kV, etc., uses the 20 program lines in Fig. 11-21.

11.2 SUMMARY

Many interesting parameters can be calculated from voltage and current. Some circuit parameters can be derived from the data, and a digital computer can be used to check the results and perhaps to test circuit modifications for improved performance.

11.3 REFERENCES

1. R.R. Butcher, "A Comprehensive Study of Excimer Laser Systems," Los Alamos Scientific Laboratory, LA-7329-T, June 1978.
2. J.C. Martin, "Multichannel Gaps," Atomic Weapons Research Establishment, SSWA/JCM/703/27, March 1970, p. 8.

Chapter 12
Energy-Storage Capacitors

by
W.J. Sarjeant
(State University of New York at Buffalo)

ENERGY-STORAGE CAPACITORS (REFERRED TO AS *CAPACITORS* IN THIS CHAPTER) ALLOW electrical energy to be stored over a long charging time and then released as required over very short (nanoseconds to microseconds) periods under controlled conditions. In contrast to dc filter capacitors, pulse-discharge or energy-storage capacitors must provide large peak currents, often of an oscillatory nature, during this short release or discharge time. These large currents and their repetition rates can lead to significant internal power losses inside the capacitor, necessitating considerable care in design to achieve adequate heat-transfer rates for long capacitor lifetime. This chapter explores current capacitor technology and attempts to delineate some of the engineering considerations required to achieve user-specified lifetimes for the major classes of operation. A review of manufacturing techniques and their impact upon capacitor performance is presented and correlated with the known physical and chemical mechanisms responsible for observed capacitor electrical properties and lifetime.

12.1 INTRODUCTION

Webster's Dictionary defines a *capacitor* as: "A device giving capacitance, usually consisting of conducting plates or foils separated by thin layers of dielectric with the plates on opposite sides of the dielectric layers oppositely charged by a source of voltage and the electrical energy of the charged system stored in the polarized dielectric."

An illustration of the elements essential such a capacitor is shown in Fig. 12-1. Accept as an experimental fact that closing S1 with S2 open will place equal and opposite charges of magnitude q on the upper and lower plates. For the parallel-plate geometry shown, the constant of proportionality between q and V_{dc}, (namely C) is:[1]

$$q = C \cdot V_{dc} \qquad (12\text{-}1)$$

336 Energy-Storage Capacitors

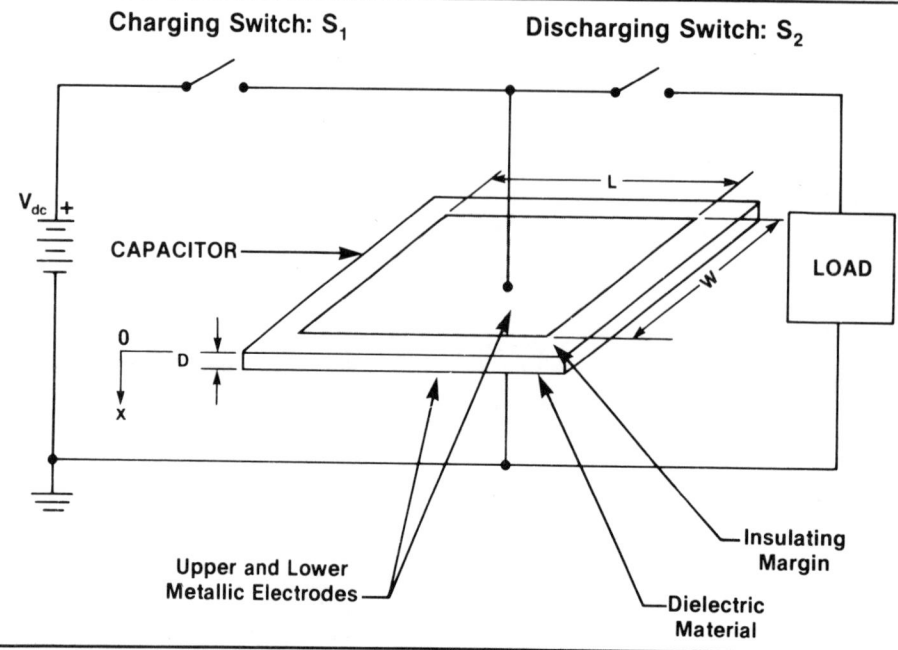

Fig. 12-1. Parallel-plate capacitor.

where:

$$C = \epsilon \cdot \ell \cdot \frac{W}{d} \quad \text{farads} \tag{12-2}$$

$$C = \epsilon_o \cdot K \cdot \frac{W}{d} \cdot \ell \tag{12-3}$$

where K is the relative dielectric constant or coefficient and ϵ_o is the dielectric constant of free space:

$$\epsilon_o = 8.854 \cdot 10^{-12} \text{ C}^2/\text{N}-\text{m}^2 \tag{12-4}$$

Table 12-1 lists some representative materials, giving the K and breakdown voltage or *Dielectric Strength* V_b, under an impressed dc voltage. The dielectric strength is expressed in volts per mil of dielectric thickness, at a voltage stress level where breakdown is virtually certain within 1 minute of application of voltage. Note that the dielectric strength for thicker pieces of material is considerably less as a result of bulk impurities, voids, and mechanical flaws.

As an example, consider a capacitor 1 × 1 meter with a plate spacing of 1 centimeter and a K of 1. Then:

$$C = 8.854 \cdot 10^{-12} \cdot \ell \cdot 1/10.01 = 8.854 \cdot 10^{-10} \text{ farads} \tag{12-4}$$

Table 12-1. Dielectric Constants, K, and
Breakdown Strengths, V_b, of Selected Insulators.

INSULATOR	K	V_b (V/mil)
Air	1.000585	75
Aluminum oxide	7.0	300
Bakelite (general purpose)	6.0	300
Castor oil	3.7	350
Ceramics	5.5–7.5	200–350
Ethylene glycol	39	500
High-voltage ceramic (barium titanate composite and filler)	500–6000	50
Kapton (polyamide)	3.6	7000
Kraft paper (impregnated)	6.0	2000
Lucite	3.3	500
Mylar	2.5	5000
Paraffin	2.25	250
Polycarbonate	2.7	7000
Polyethylene	2.2	4500
Polypropylene	2.5	9600
Polystyrene	2.5	500
Polysulfone	3.1	8000
Pyrex glass	4–6	500
Quartz, fused	3.85	500
Reconstituted mica	7–8	1600
Silicone oil	2.8	350
Sulfur hexafluoride	1.0	200 (per atmosphere)
Sulphur	4.0	--
Tantalum oxide	11.0	100
Teflon	2.0	1500
Titanium dioxide ceramics	15–500	--
Transformer oil	2.2	250–1
Water	80	500*

*Pulse charged in 7 to 10 microseconds.

Because a farad is a large unit rarely encountered in high-voltage capacitors, smaller subdivisions have been created for convenience in calculations:

$$1 \text{ microfarad} = 10^{-6} \text{ farad} = \mu F$$
$$1 \text{ nanofarad} = 10^{-9} \text{ farad} = nF$$
$$1 \text{ picofarad} = 10^{-12} \text{ farad} = pF$$

Note in Table 12-1 the relatively high K of reconstituted mica, which combined with its high dielectric strength of 1.6 kV/mil, makes it a rather interesting candidate for compact pulse capacitors.

Consider the work done in charging a capacitor of capacitance, C, to a potential difference of V_{dc} across the upper and lower metallic conductors. The incremental work done, dW, in adding an increment of charge, dq, to the capacitor when the plates are at a potential of V volts between them is:

$$dW = V \cdot dq \qquad (12\text{-}5)$$

338   Energy-Storage Capacitors

The total work done, in joules, in charging the capacitor to a potential difference of V_{dc} volts is the integral from o to q_{dc}, where q_{dc} is the charge on the capacitor plate (+q on the top and −q on the bottom plate) at V_{dc}. That is,

$$W = \int_0^{q_{dc}} V\, dq = \int_c^{q_{dc}} \frac{8}{c} \cdot dq = \frac{1}{2} \cdot C \cdot (V_{dc})^2 \tag{12-6}$$

where

$$q = C \cdot V \tag{12-7}$$

This energy is stored in the dielectric between the capacitor plates and is released into the load when the discharging switch in Fig. 12-1 is closed. It is important to note that the discharge of this stored energy gives rise to mechanical forces inside the capacitor, which can cause dielectric fracturing. In repetitive operation, the discharge can excite mechanical resonances in the capacitor structure, eventually resulting in destructive damage and electrical breakdown. For example, in the parallel-plate capacitor, the potential difference between the plates varies linearly with the distance from the bottom plate, and the electric field E is constant:

$$E = \frac{V_{dc}}{d} \cdot \wedge, \tag{12-8}$$

and the energy in the volume $\cdot w \cdot x$ is:

$$U = \frac{\epsilon}{2} \int E^2\, dv \tag{12-9}$$

$$= \frac{\epsilon}{2} \cdot \frac{V_{dc}^2}{d^2} \cdot \ell \cdot w \cdot x$$

This is the energy stored in a volume $\ell \cdot w \cdot x$ of the capacitor dielectric where the base of this volume is situated on the upper electrode of the capacitor. Similarly, the energy W' stored in a volume contained between x_0 and x_1 (any two distances between the plates) can be calculated and found to be:

$$W'\,(x = x_0 \text{ to } x = x_1) = \frac{C \cdot V_{dc}^2}{2 \cdot d} \cdot (x_1 - x_0) \tag{12-10}$$

In discharging this energy, there is a mechanical force that the dielectric exerts upon the mechanical conductors and container in which the capacitor is housed. For the nonconducting dielectric, that is, the very high dc resistance dielectrics, being considered here, this force, F_x, is equal to minus the rate at which the stored energy is changing in the x direction:[1]

$$F_x = -\nabla_x U \tag{12-11}$$

$$= -\frac{1}{2} \cdot C \cdot \frac{V_{dc}^2}{d} \cdot \hat{x} \tag{12-12}$$

Thus, discharging the capacitor causes the force in the dielectric to be opposite in sign to the direction in which the current is flowing through the dielectric. This action indicates that the dielectric is being compressed and pulled away from the metallic conductors. In fact, these disruptive forces can be large enough to cause rupture damage inside the dielectric and destroy the intimate bond required between the dielectric and rigid metallic conductors. Consider the effects the dielectric would experience with small impurities:

1. Not ionized
2. Not ionized but polarized
3. Ionized
4. Ionized and polarized

This chapter discusses impurity effects as well as the effects that isolated packets of charge at the edges of capacitors have on lifetime for both dc and very short pulse-charged methods of bringing the capacitor potential up to V_{dc}. This potential, V_{dc}, used throughout this note to designate either type of charging conditions. Any differences these conditions might make, either to capacitor lifetime or restrictions on modes of operation in the discharge loop, are explained in this chapter.

12.2 GENERAL PROPERTIES OF CAPACITORS

Whenever a capacitor is selected to perform in an electronic circuit, its characteristics have been optimized to provide the designer with a well-defined level of reliability of the component throughout the design lifetime of the circuit. Table 12-2 lists a number of the more important parameters that effect the capacitor designer's choice of geometry, impregnant, and materials.

The selection of a capacitor design requires the matching of available capacitor characteristics and parameters to the application needs. In addition to the basic capacitance value and voltage rating, specifying all the characteristics presented in Table 12-2 allows the supplier to provide the most cost-effective capacitor for the given application. The fundamental parameters open to control by the designer, listed on the left of Table 12-3, are controlled to a large degree by the environmental factors, such as temperature range, voltage, wave shape, and duty cycle. Essentially all these environmental factors affect the life expectancy of the capacitor. It is here that the user can effect considerable cost savings by providing the designer with all the data listed in Table 12-2 and an accurate assessment of the lifetime of the equipment into which the capacitor is to be placed.

Table 12-2. Capacitor Design Parameters.

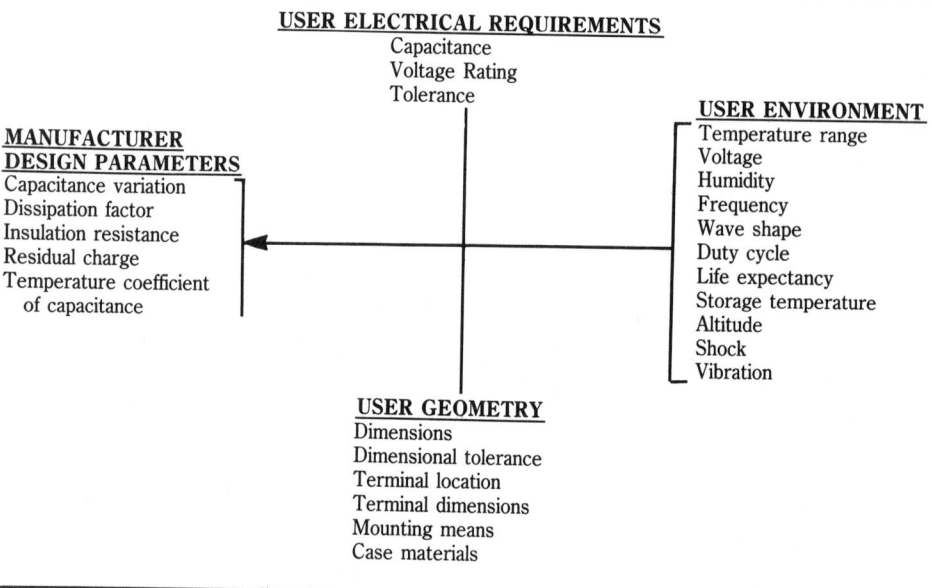

Table 12-3. Capacitor storage temperatures and capacitance decreases.

CAPACITOR TYPE	STORAGE TEMPERATURE RANGE—°C	CAPACITANCE DECREASE RATE WITH TIME IN PERCENT PER YEAR
1) Castor oil/Kraft paper	+10 to +40	nil
2) Plastic film/proprietary impregnant	−20 to +60	nil
3) Ceramic type DHS from Murata	−40 to +125	−3% and equilibrium at 2,000 hours
4) Ceramic type 715 from Cera-Mite	−55 to +125	−20% in first year, decreasing logarithmically. Can be pre-stabilized by mfr.
5) Ceramic type 720 from Cera-Mite	−55 to +125	As above
6) Reconstituted mica	−65 to +125 storage and operating—no derating needed.	nil

It should be noted that very long lifetimes, in excess of 100 million shots, can often take the capacitor designer into regions where lifetime data are not available. Thus, it might be advantageous to allow for capacitor replacement at fixed time intervals as a function of allowable equipment downtime, utilization factor, and the specific cost-benefit ratios of the capacitor designs for longer lifetimes. The user who is faced with very long lifetime requirements will almost always be restricted in utilizing capacitor designs that have been based upon the highly reliable insulation systems for power distribution systems. Low inductance and very fast discharge times at high repetition rates do not find an adequate data base in this wealth of dc and 60 Hz test information. With the advent of the pressing needs for high repetition rate capacitors, emphasis in this chapter is oriented in this direction. Those who require single-shot devices (that is, 1 pulse/minute or fewer) will find no difficulty in acquiring capacitors of high reliability for more than 100,000 shots at modest reversal (less than 30 percent), peak currents of 50,000 A and voltages up to 100,000 V. Suffice to say the enormous data base available for designs utilizing this performance profile surface is very well documented, especially in the very thorough studies by G. Boicourt[2], and it would be presumptuous to attempt to summarize them here. A word of caution: it is worthwhile asking any supplier who provides a design that is smaller, cheaper, and has a longer lifetime than all the competition, where similar devices have been made and used before. Then discuss with these users the performance they obtained in service.

The balance of this section addresses most of the individual headings in Table 12-2 in detail. At this point, assume that specific recommendations concerning capacitor compositions, impregnants, and lifetime will be incorporated under the umbrella of these headings. Thus, there might be some need to digress within each area while discussing the physical, geometrical and chemical processes responsible for affecting each of these parameters.

As a preamble to this discussion, the basic characteristics of capacitors will be summarized and then related to the unique requirements of pulse capacitors.

12.2.1 Dielectrics

The capacitance relation presented earlier states that the capacitance C varies directly with the dielectric constant, directly with the area of the plates, and inversely with the distance d between the plates. In general, the plate area is determined by the surface dimensions the user can tolerate, and the separation d is controlled almost totally by the lifetime the user specifies and the dielectric the manufacturer must use to provide the highest reliability for this lifetime. It is here that several iterations with the potential suppliers and careful specification of the desired mean time between failure (MTBF) at a high confidence level, approximately 90 percent, can often save the user money. As a general rule, the smaller the capacitor is specified to be for a given lifetime, independent of repetition rate, the more expensive it will be. For high reliability, the user is generally better off to place the minimum of size constraints upon the capacitor manufacturer and to avoid unique, sole-source designs at all costs unless there is no other alternative.

The dielectric constant of any material placed into an elementary parallel plate capacitor is a direct measure of the ability of the material to store electrical energy when charging of the plates to V_{dc} is completed. This energy is stored in the dielectric polarization, either permanent, as in water, or induced as in the case of aluminum oxide. Essentially, this polarization comes from one of two sources. The first source is the alignment of permanent dipoles in the dielectric parallel to each other and in line with the internal electric fields (the field applied during charging, in Fig. 12-1, is parallel with the x axis and its unit vector originates from the top [positive] plate of the capacitor[1]). The second source is the generation of an induced polarization by the presence of the applied electric field.

12.2.2 Induced Polarization

With induced polarization, the electric field is an external perturbation of a coulomb nature on the molecule or atom and tends to attract the electron cloud to the positive plate direction in the capacitor. On the other hand, the attractive forces between the electrons and the nucleus tend to resist this displacement of the electron cloud, giving rise to an equilibrium condition wherein the molecule has a finite dipole moment. This moment then is solely induced by the presence of an external (applied) electric field and is a result of an elastic displacement of the electronic charge distribution relative to the nucleus. To a first approximation, this polarization P_{ind} can be shown to be proportional to E, with the constant of proportionality being χ_{ind}, the susceptibility:

$$P_{ind} = \chi_{ind} \cdot E \qquad (12\text{-}13)$$

Note that both P_{ind} and E are vectors. For isotropic materials, they are parallel.

A discussion of the impact of anisotropic materials on any of the capacitor systems requires a tensor description for the susceptibility χ_{ind} and is beyond the scope of this chapter. Just for the sake of interest, such anisotropic materials tend to have rather large internal losses and low breakdown voltages. They might, however, be of considerable use in the field of electromagnetic shock lines. For induced polarizability caused by electronic displacements within atoms and molecules of dielectric under the influence of an applied external electric field, the polarization is a true elastic process so that no bound internal energy states exist. Thus, the induced dipole moment susceptibility, χ_{ind}, is constant up to frequencies corresponding to the ultraviolet portion of the spectrum.

In the case of ions, χ_{ind} can be much larger as a result of atomic or ionic displacements within the molecules (that is, changes in bond angles and interatomic distances caused by the application of E). It is constant for frequencies up into the infrared portion of the spectrum and is of concern primarily for ionic impurities in almost all capacitors. In this case, these ionic impurities experience a force proportional to the local electric field at their site and this force, constant during dc on-time and oscillatory during discharge, can be sufficient to cause mechanical rupture in the dielectric and electrical breakdown.[2,3] Note that the ratio of the forces upon the interstitial ionic impurity and dielectric molecules is in direct proportion to their relative dipole moments, in the presence of the applied electric field. Thus, if all impurities have interstitial or lattice site polarizations less than those of the host dielectric, little chance of mechanical damage will exist. The question of ionization and breakdown from voids in the dielectric is rarely dominated by this class of polarizations and is discussed later along with other corona processes.

12.2.3 Permanent Polarization

Dielectrics can also possess a permanent dipole moment μ, which is antiparallel to the applied electric field E for isotropic dielectrics. The external field will tend to exert a torque on each little dipole and try to orient it in the direction of the field.[3] This orienting and ordering influence of the field is counteracted by the thermal motion of the particles and their local interaction of one dipole moment with another. Analysis shows that, the application of the field E to such a dielectric results in a net orientation of some fraction of all these dipoles parallel to the field, thus:

$$P = \chi_e \cdot E \qquad (12\text{-}14)$$

where P is the polarization per unit volume of dielectric resulting from the dipole orientation caused by E; χ_e is the permanent dipole moment susceptibility generally referred to as K_p (relative dielectric constant) and $K_p = (1 + \chi_e)$, $\epsilon_p = K_p \epsilon_o$.

Because thermal equilibrium is the depolarizing perturbation on P, increasing the temperature, T, of the dielectric can be shown to result in a linear decrease in χ_e. This accounts for the dramatic decrease in capacitance observed in many polar dielectrics. Note that operation at voltages above the level where all the dipoles are oriented parallel with the applied field is of little value in increasing the energy stored because the dielectric constant then falls to ϵ_o. Recall that all the fields here are the external fields applied to the dielectric as a result of having a potential difference of V_{dc} volts across the capacitor electrodes. There exists no general qualitative theory of dipolar materials that can be applied to these problems in capacitors because of difficulties in evaluating the internal fields and because the dipole rotatability possesses complex thermal and fabrication-induced anisotropicities.[3] The discussions thus become qualitative.

In solids, the susceptibility decreases with increasing temperature, except for a large susceptibility increase at the melting point, where all the dipoles are free to rotate strongly in the field direction. If your capacitor is at this temperature, it has likely melted and shorted, so further comment is irrelevant. On the other hand, ceramic materials such as barium titanate possess large susceptibilities as they are spontaneously polarized, giving rise to a very large permanent dipole moment in the material. Hence, the dielectric constant is large and has a significant decrease with increasing temperature. Above a critical temperature known as the *Curie* temperature, this polarization all but disappears and the dielectric constant drops dramatically to near ϵ_o. The influence of E upon C of these types of capacitors is discussed later, and it is interesting to speculate how this property allows time-varying pulse-forming networks to be synthesized.[4] As a result of the binders employed in their manufacture, the power loss per pulse generally exceeds 5 percent at high frequencies (greater than 1 MHz) so that repetitive operation ought to be pursued with some care.

In summary, there are materials having large dielectric constants, K_p, which in general one desires to maintain as constant as possible as a function of E in order to store the maximum energy in the capacitor. Almost all the major contribution to the dielectric constant for capacitor dielectrics can be shown to be made by the permanent dipole moment in the dielectric. This moment has a temperature, field, and frequency of discharge and repetition rate class of dependencies that are not yet subject to accurate physical modeling from fundamental principles. Thus, it is necessary to adopt a parametric approach for each material of interest, which is a research problem that has been well addressed in the case of dc and single-shot dielectric tests and capacitor design. The field of high repetition rate capacitors does not have an extensive data base and is a very fertile field for further research.

12.3 CAPACITOR CHARACTERISTICS AND THEIR CHANGES

In circuit applications, the capacitor is often subjected to electrical, mechanical, thermal, and other environmental stresses, some 11 in all that are listed on the right-hand side in Table 12-2. The capacitance value of the capacitor is generally measured on a low-voltage bridge at a frequency of 1000 Hz. Assuming the capacitor to be mechanically well made, neither the area or plate separation will change in any significant degree during charge or discharge. Hence, all the variations in capacitance come about as a result of these environmental factors changing the dielectric constant of the material between the capacitor plates. The dielectric constant, K, is initially set by design in choosing dielectric materials and impregnants that are optimum for the class of capacitor being constructed. The major environmental factors causing a change in K are:

1. Temperature range
2. Voltage
3. Humidity
4. Frequency
5. Storage temperature and time

12.3.1 Temperature Range

Temperature variations in K, and thus the capacity value, are widely different for various dielectrics. The variations depend also upon the liquid impregnant utilized to fill voids between the capacitor plates and dielectric as well as upon any small voids in permeable materials such as paper. In the latter case, permeation of the impregnant into the paper results in a very significant improvement of the dielectric strength, primarily as a result of suppressing the corona that would have caused breakdown in voids in the paper, by filling the voids with an impregnant of nearly the same K and thus preventing corona formation until bulk breakdown occurs at a much higher voltage. Figures 12-2, 12-3, and 12-4 illustrate average curves of K against temperature and the corresponding percent change in capacitance from its value at 25 °C. Note that special processing of the dielectrics in manufacture of the films, as well as utilization of proprietary impregnants, can alter these curves very significantly and reduce changes in K at the higher temperatures. The slowly varying plastic films such as Teflon and polystyrene (Fig. 12-2) find their K is changed very little upon impregnation with several of the available silicone dielectric fluids.[5] In addition, these fluids possess very small power losses that are essentially frequency independent. To maintain these low losses, a paper wick cannot be employed in such units. There are also significant manufacturing process innovations to be made in assembly packaging and impregnation of nonwicked capacitors, which will require considerable further development for high-reliability applications. The conclusion is that temperature-induced variations in K for plastic materials can be made small enough to be insignificant for energy-discharge capacitors, by the selection of the film and impregnant that gives the required lifetime/frequency loss characteristics. These techniques are to be a large degree proprietary. The various selected combinations that manufacturers have chosen and tested are briefly discussed in this chapter.

Return to two particular classes of capacitors: paper-impregnated with castor oil and high-K ceramics. The paper capacitor is vacuum impregnated with, for example in

12.3 Capacitor Characteristics and Their Changes

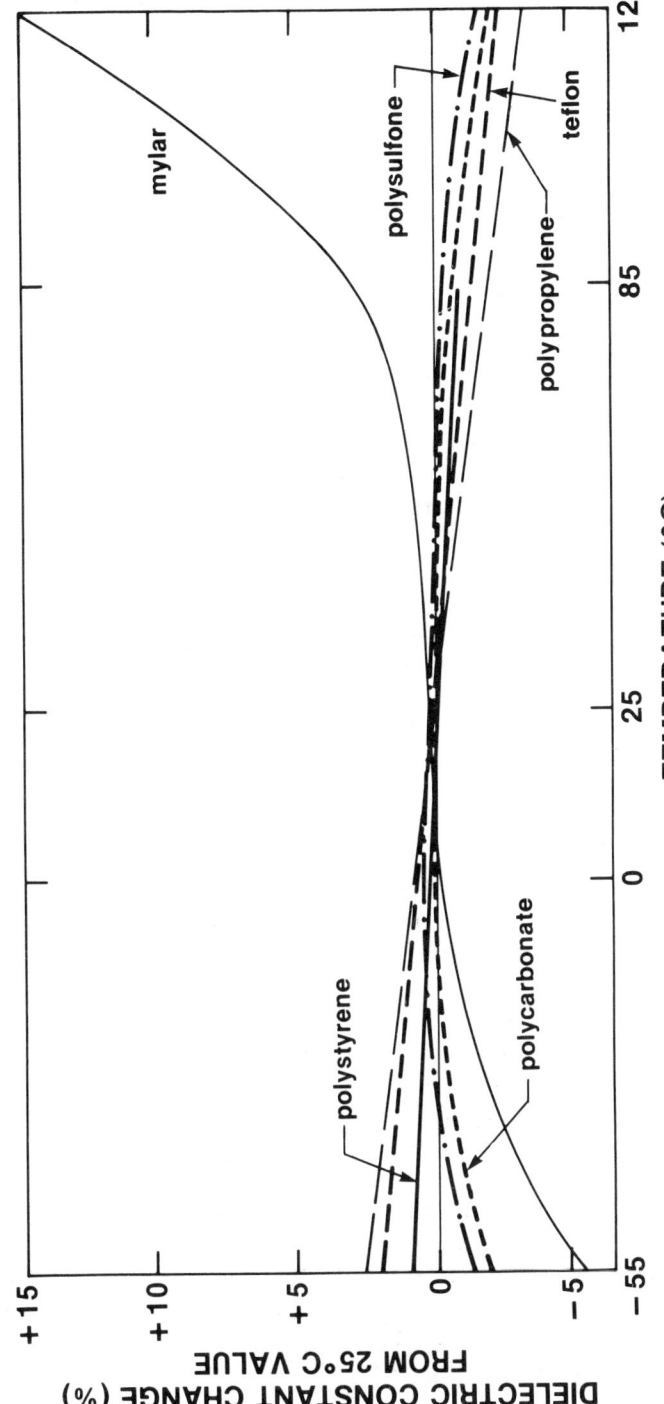

Fig. 12-2. K versus temperature—dry plastic film (after Electrocube Technical Capacitor Bulletin, with permission).

346 Energy-Storage Capacitors

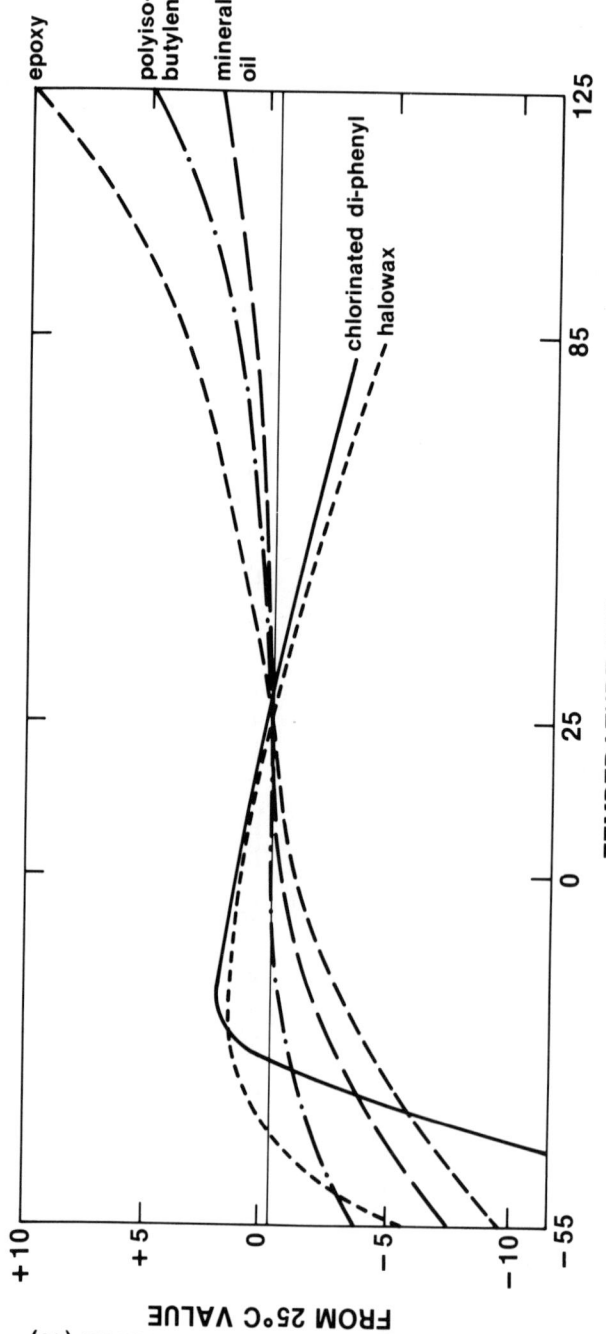

Fig. 12-3. K versus temperature—impregnated paper (after Electrocube Technical Capacitor Bulletin, with permission).

12.3 Capacitor Characteristics and Their Changes 347

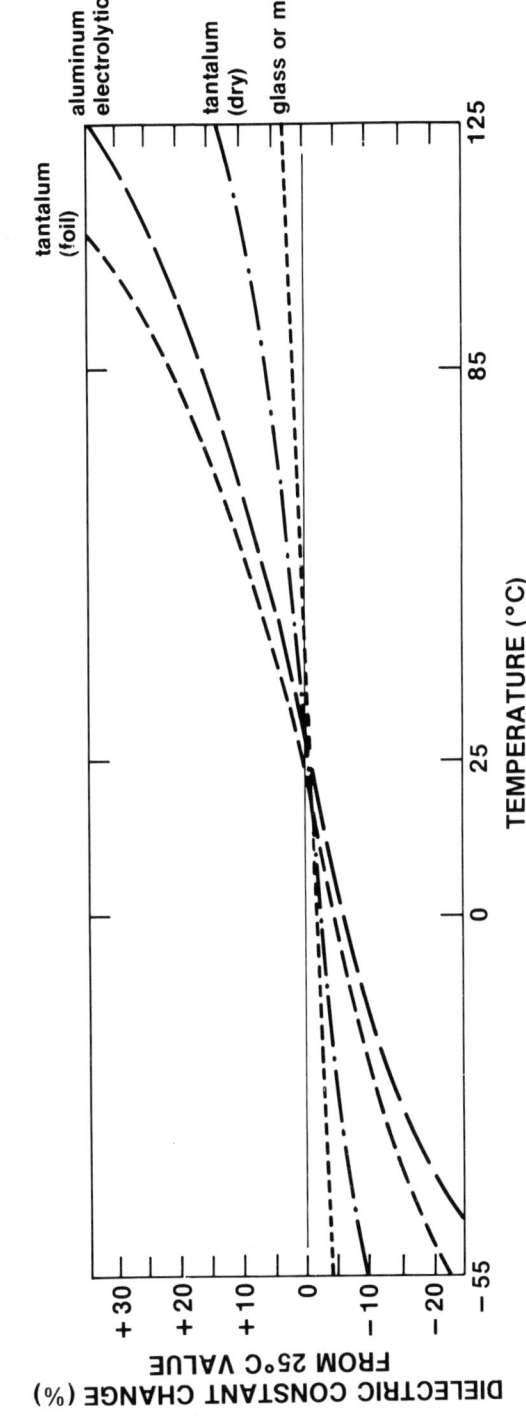

Fig. 12-4. K versus temperature—other dielectrics (after Electrocube Technical Capacitor Bulletin, with permission).

the case of Aerovox, a liquid named *Hyvol D* a specially processed and treated grade of castor oil specifically prepared for electrical applications. These capacitors exhibit temperature variations in K (Fig. 12-5) having a drastic decrease in the capacitance at low temperatures. This is also followed by very large increases in the internal losses, which are a result of increasing intermolecular losses as the temperature decreases. Castor oil capacitors used below 0 °C ought to be provided with appropriate heaters if constancy of capacitance is a requirement. If the capacitor is not heated, the increase in internal dissipation will, for high repetition rate operation, generally heat the dielectric sufficiently to increase the temperature, reducing the capacitance variation and as well the power loss.

Consider the classes of ceramic capacitors having published characteristics. The capacitance variation with temperature for the Cera-Mite 715 series illustrated in Fig. 12-6, is large indeed. Through gradual developments over the last few years, the manufacturer is now selling a very stable unit whose designation is the 720/722 Series (C-57 formulation). The improvement is clearly evident in Fig. 12-7 where the variation from 25 to 70 °C is less than 3 percent compared to 30 percent for the 715 Series as illustrated in Fig. 12-6.

Tests were run on a special version of the 720 Series having double silvering, half-inch studs soldered to the silvering after the silver has been hot-tin dipped. The edges are varnish coated, and there is a 1-millimeter nonsilvered margin for operation in Dow Corning 200 silicone dielectric fluid, 20-50 centistokes viscosity. Under these conditions, operation to 50 kV dc for single-shot applications is feasible provided fingers, smoke, solder flux, and especially silicone rubber are kept out of the oil. (The exception is the special silicone room-temperature vulcanizing (RTV) rubber made by Dow Corning for use on copper surfaces as no corrosive materials are given off during cure [Types 3140 and 3145]. This material requires the use of a rosin/ketone surface adhesion preparation available from Dow Corning known as Type 1201 primer. Note that cure times are 1 to 2 days for the RTV, which can be reduced somewhat through gentle heating.)

Pulse-charged (less than 100 nanoseconds) operation avoids corona limitations on lifetime. The tests showed that lifetime scales by about the 12th power of the ratio of the voltage stresses. At a datum point, 20 units, nominally 1800 pF, will have 10 failures at a pulse charge voltage of 190 kV and 30 percent reversal, with a discharge period of 50 nanoseconds full-width half-maximum(FWHM), after a total of 50 discharges. This correlated to a lifetime of two to three shots at 210 kV charging voltage at about a 50 percent reversal, obtained from a sample of three units that were not preconditioned at lower voltages. In contrast, the 715 Series was not usable above 130 kV and had a lifetime of about 100 shots at 110 kV and 20 percent reversal. There is virtually no application where the 715 cannot be replaced by the 720/722 Series for pulse duty.

An example of the temperature properties of a high-frequency, low-loss capacitor using plastic film impregnated with (as usual) the proprietary oil, is shown in Fig. 12-8. Polysulfone wins top marks, and these units from High Energy Corporation are rated to pass up to 47 kVar of reactive power at a RMS current of 25 A. From this figure it is also quite evident why Mylar capacitors are not suitable for high prf operation, from the standpoints of capacitance stability and increasing internal losses with temperature. These are impressive specifications for the polysulfone, but check with the users as to how they have performed in practice, particularly at the higher voltages, in the 25 kV

12.3 Capacitor Characteristics and Their Changes 349

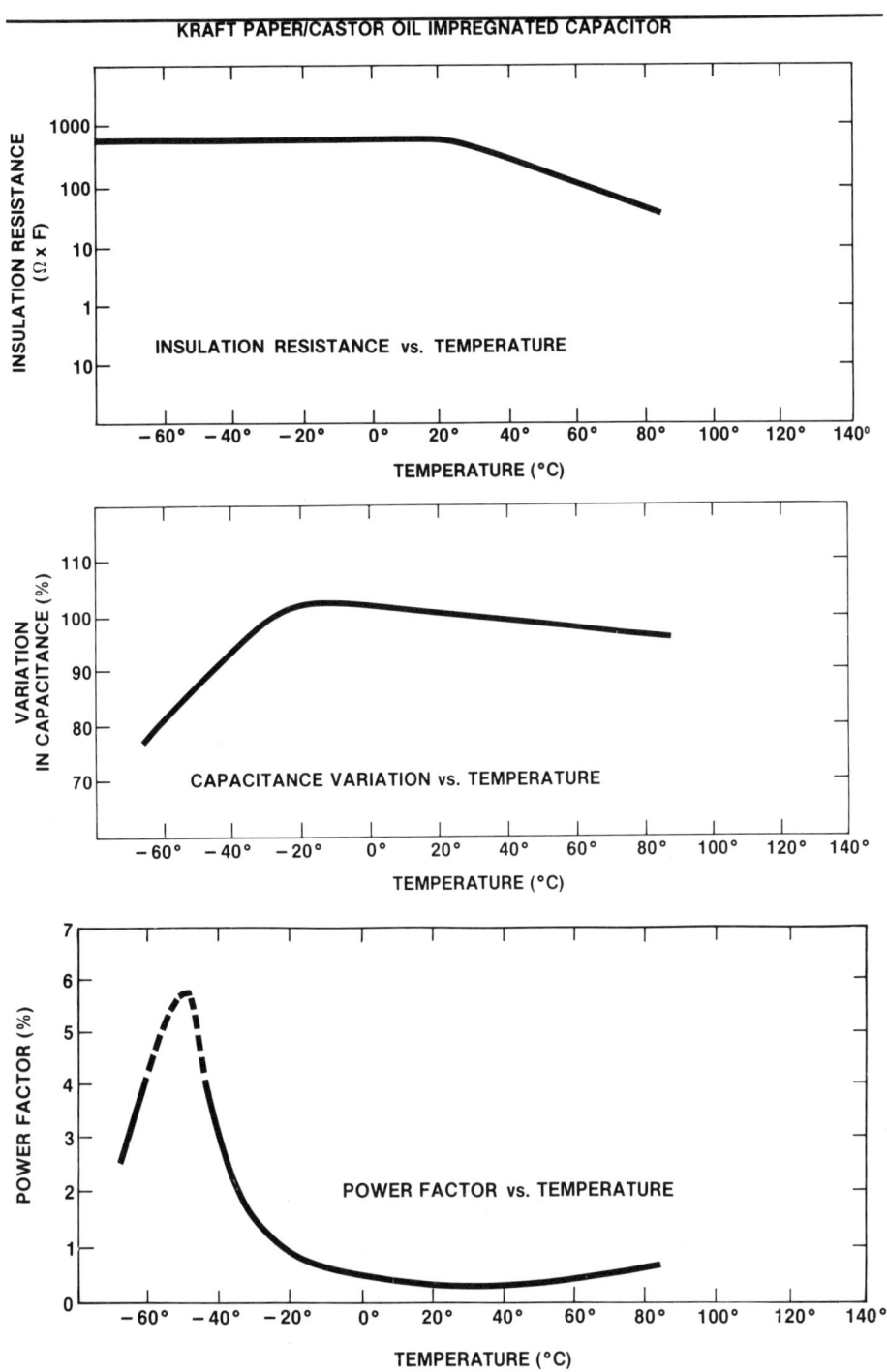

Fig. 12-5. *Typical characteristics of kraft paper/castor oil impregnated capacitor (after Aerovox, Technical Literature, with permission).*

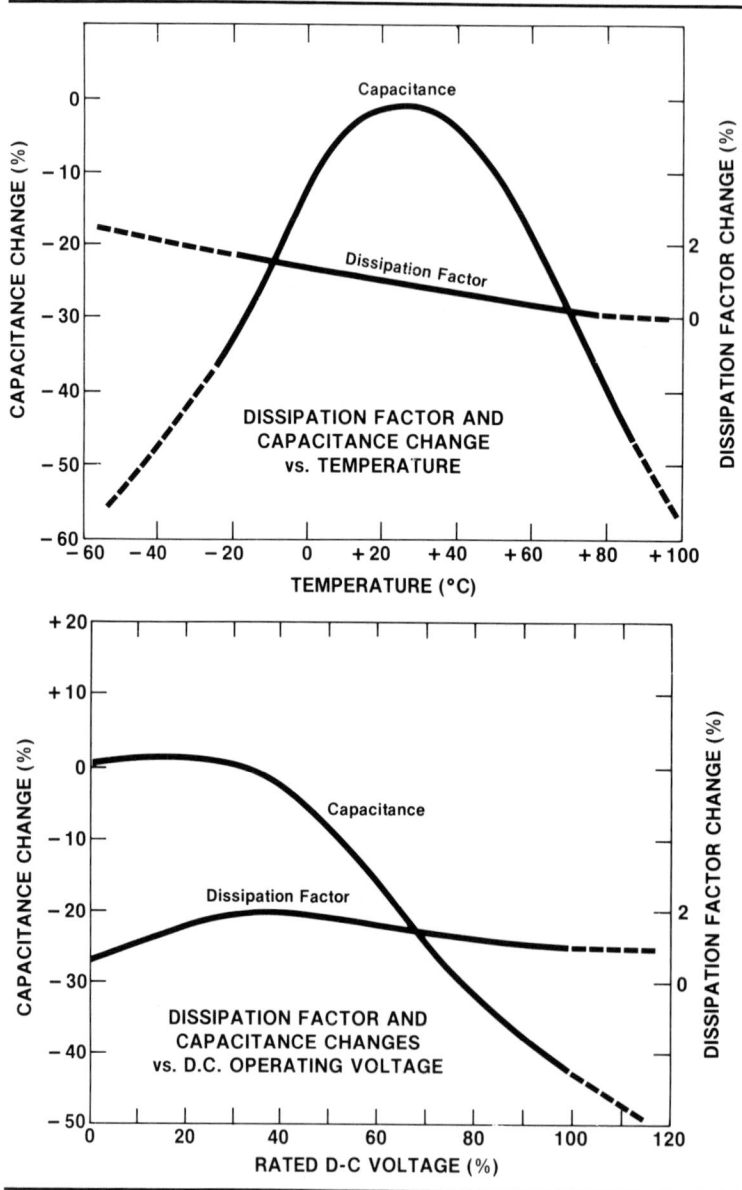

Fig. 12-6. Characteristics of Cera-Mite type 715 ceramic capacitors (after Technical Literature, Cera-Mite Corp., 1327 6th Ave. Grafton, WI 53024, with permission).

range. This should not be taken as a criticism of the company but rather as a suggestion based upon general past experience with all suppliers.

12.3.2 Voltage

The voltage rating of the capacitor, along with the other environmental factors will determine the lifetime the user can expect from the unit. For most dielectrics, except

12.3 Capacitor Characteristics and Their Changes 351

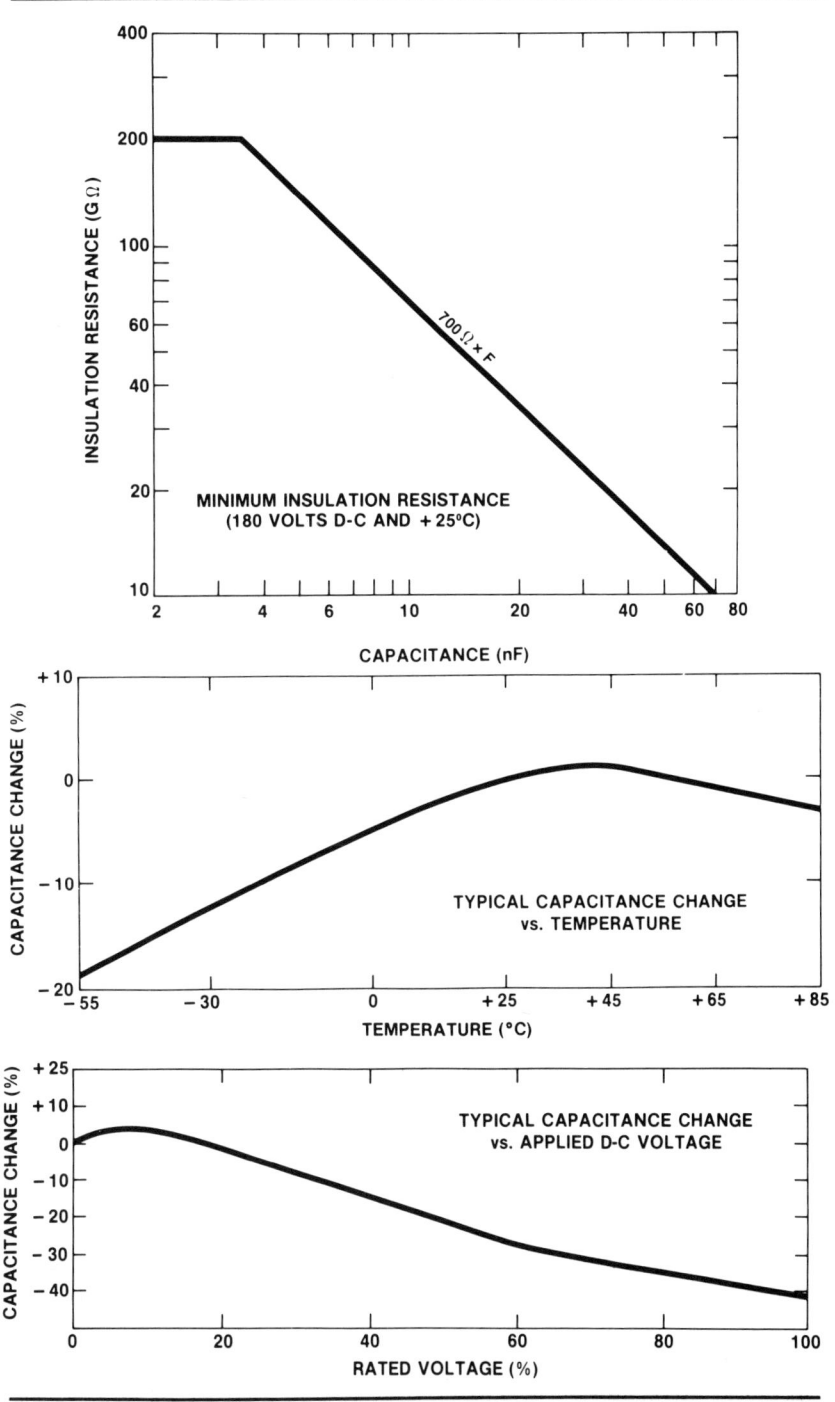

Fig. 12-7. Characteristics of Cera-Mite type 720/722 ceramic capacitors (after Technical Literature, Cera-Mite Corp., 1327 6th Ave. Grafton, WI 53024, with permission).

ceramics, there is an insignificant effect of the applied voltage upon the capacitance value of the capacitor. To illustrate the difficulties with ceramic capacitors, refer to Figs. 12-6, 12-7, and 12-9, wherein the percent change in capacitance as a percent of rated dc voltage is shown for Cera-Mite Type 715 and 720/722 capacitors, and Murata Type DHS capacitors. Suffice to say they are all about as poor so that the nameplate capacitance, measured at 1000 Hz and 1 Vac applied, must be reduced according to the curves as the dc voltage is increased. For those applications where pulse charging is used, Fig. 12-10 shows the decrease in capacitance for the 715 and 720/722 Series up to 120 kV

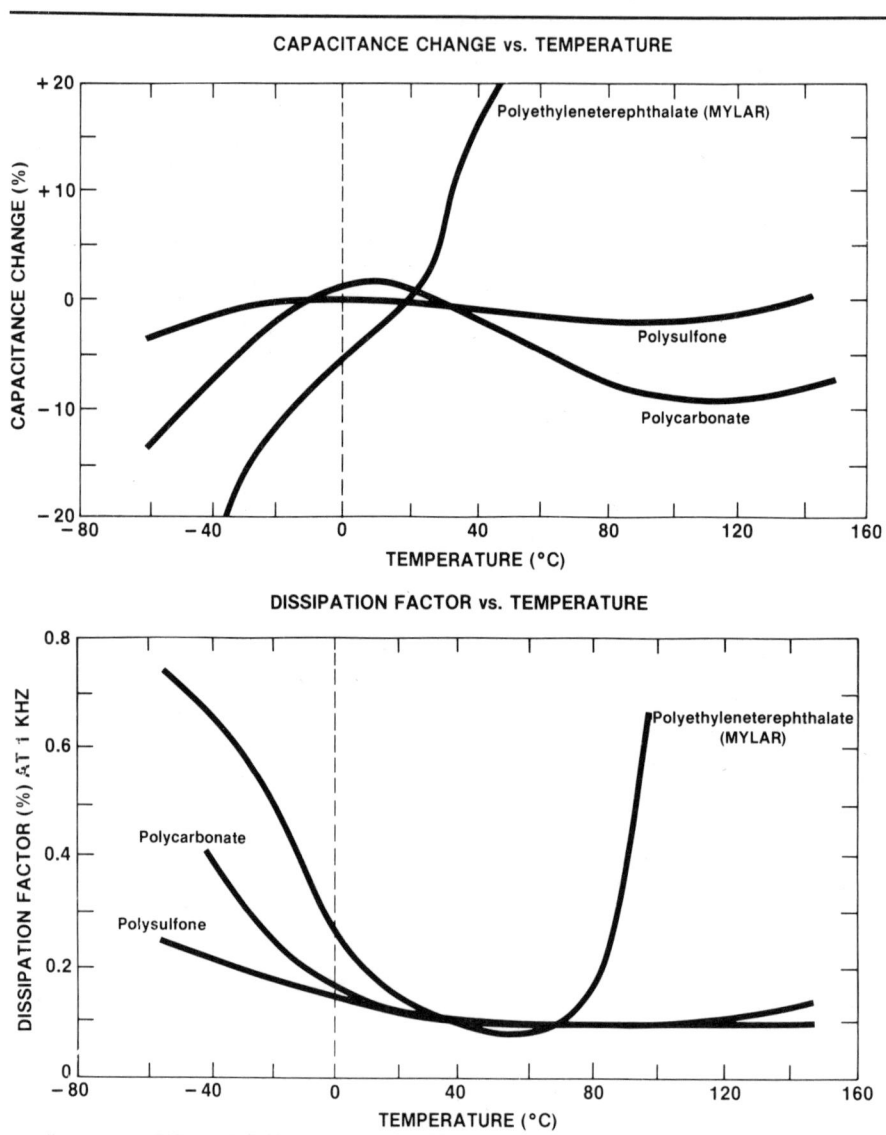

Fig. 12-8. Characteristics of type "CK" high-frequency, High-Energy Corporation capacitors (after Capacitor Technical Literature, High Energy Corporation, with permission).

12.3 Capacitor Characteristics and Their Changes

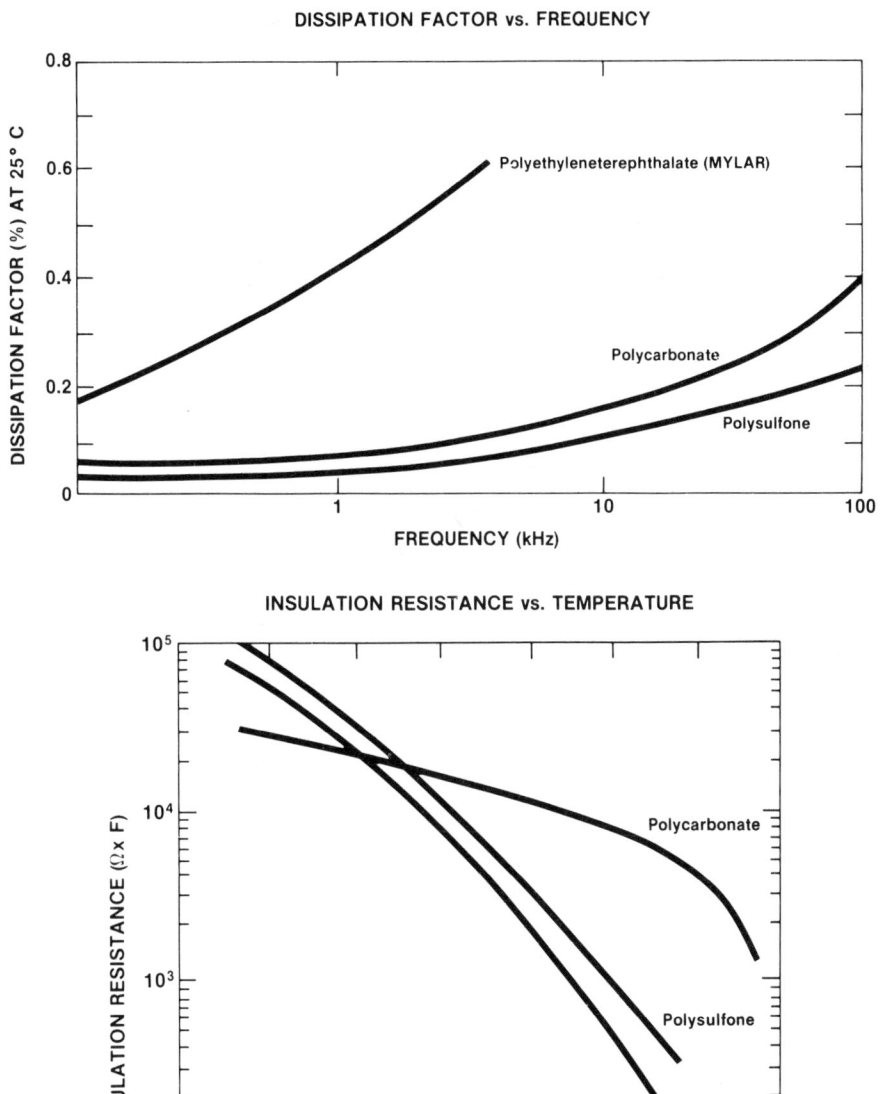

Fig. 12-8. Characteristics of type "CK" high-frequency, High-Energy Corporation capacitors (after Capacitor Technical Literature, High Energy Corporation, with permission) (continued).

354 Energy-Storage Capacitors

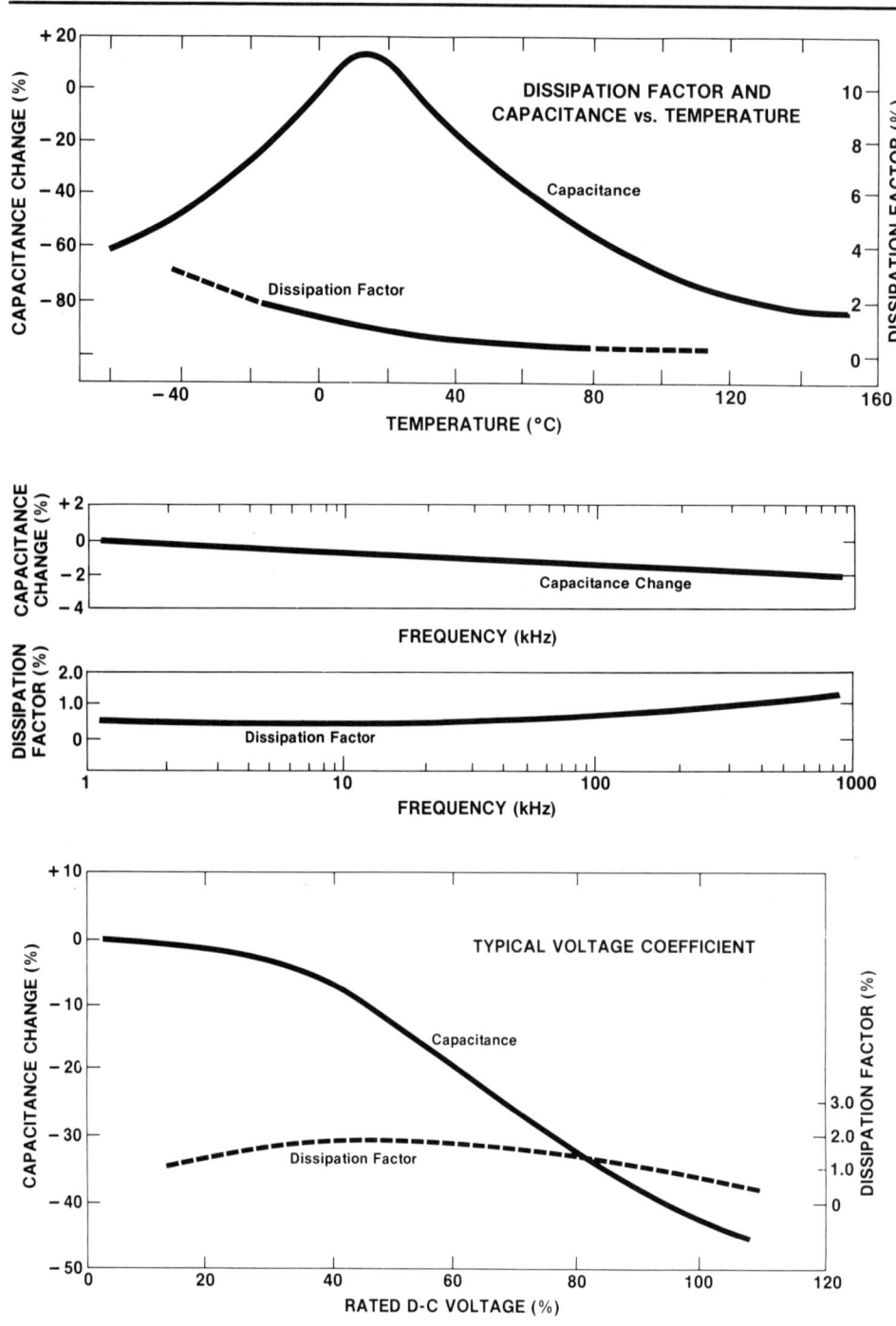

Fig. 12-9. Characteristics of Murata type "DHS" pulse discharge duty ceramic capacitors (Murata Technical Literature, with permission).

12.3 Capacitor Characteristics and Their Changes 355

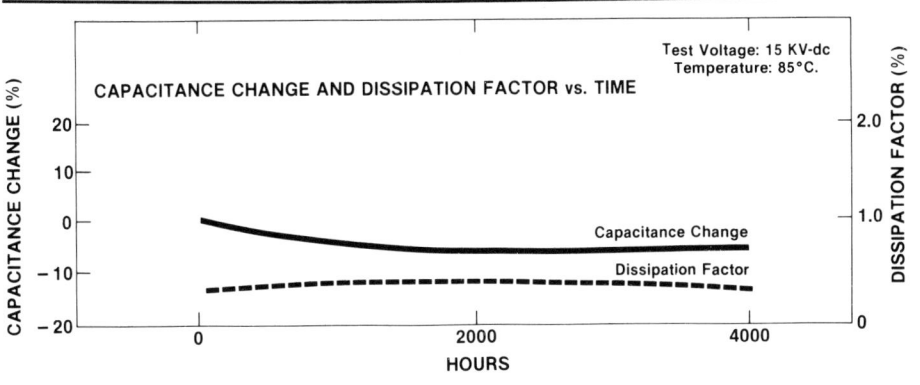

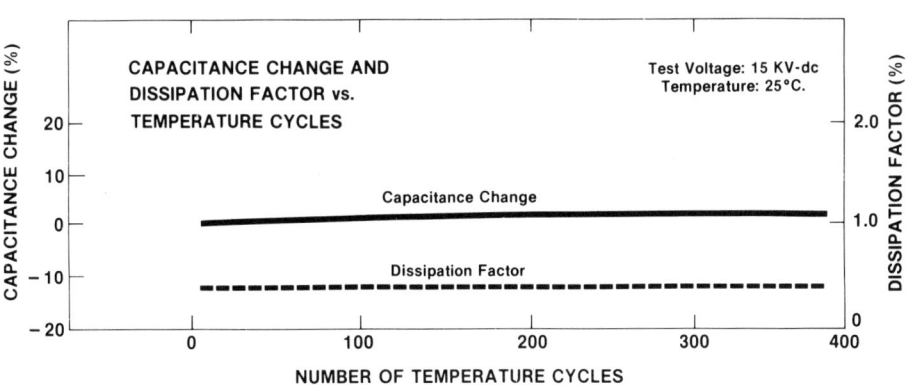

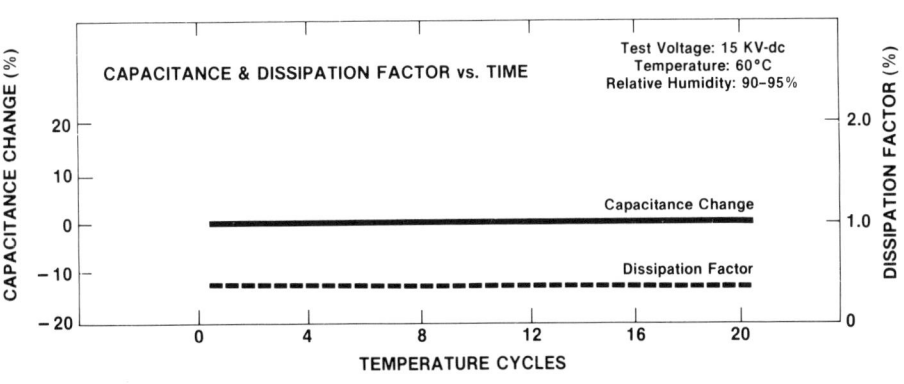

Fig. 12-9. Characteristics of Murata type "DHS" pulse discharge duty ceramic capacitors (Murata Technical Literature, with permission) (continued).

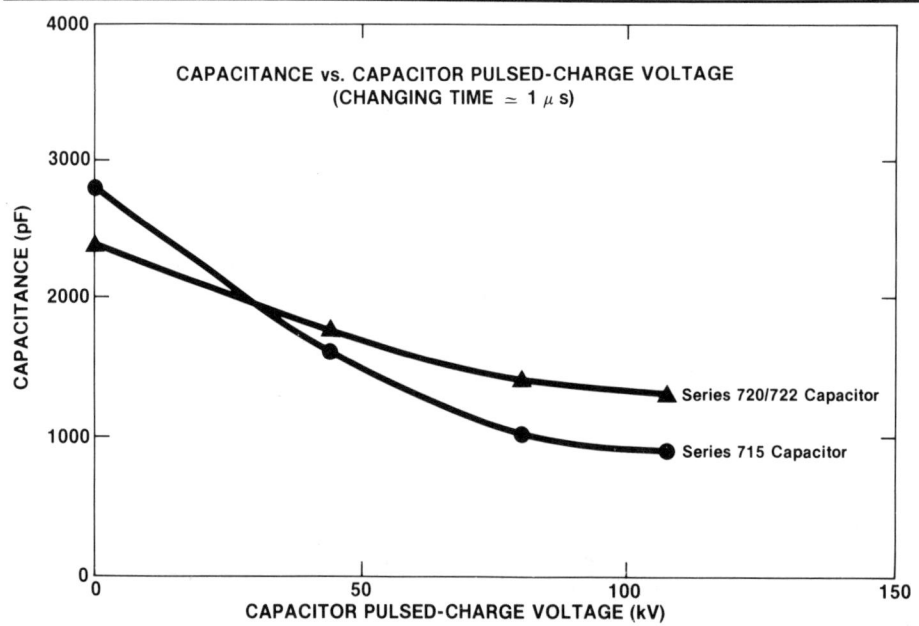

Fig. 12-10. Variation of capacitance with voltage of Cera-Mite series 715 and 720/722 ceramic capacitors.

charge voltage. Note that the capacitance decreased only by 30 percent at the nameplate voltage of 40 kV compared to the 45 percent reduction shown on the dc curves. There appears to be no straightforward explanation for the smaller decreases in capacitance under pulse charging. These measured decreases were obtained by calculating energy deposited into a low-inductance resistive load for each voltage, operating single shot, at a peak current of about 10,000 A at 120 kV. The 720 Series has a capacitance that drops to about 1200 pF, for the nameplate 1800 pF unit at about 150 kV. It was consistently found that the capacitance dropped to its nameplate value at 40 kV dc so the curves for these units are conservative. In contrast, the 715 Series was consistently somewhat worse in its capacitance drop at 40 kV dc, but this matters little because the advent of the Murata DHS and Cera-Mite 720/722 Series units allows the 715 units to be used where they ought to be: in dc filter capacitor service for which they were originally designed.

12.3.3 Capacitor Model

Only for ceramic capacitors is the amount of energy lost per pulse inside the capacitor a function of the applied voltage, as is shown in Figs. 12-6, 12-7, and 12-9 for both the Cera-Mite and Murata units. The maximum loss per pulse is 2 percent and varies little with frequency up to 1 MHz as shown in Fig. 12-9 for the DHS Murata capacitor. In order to discuss the dissipation factor (DF) in more detail, refer to the physical model

Fig. 12-11. Capacitor equivalent circuits.

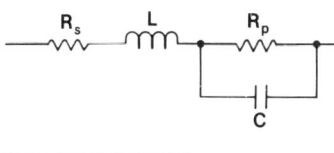

of the capacitor illustrated in Fig. 12-11. A practical capacitor possesses, in addition to capacitance, an inductance, and a resistance. In this equivalent circuit:

- R_s—The series resistance due to leads, contact terminations, and the electrodes on the capacitor dielectric. R_s is generally below 0.1 Ω and has a frequency and peak current dependence
- R_p—The leakage resistance of the capacitor due to the resistivity of the dielectric and case material as well as to dielectric losses. Generally the dc leakage resistance is several hundred megohms per microfarad of capacitance and is much larger than the frequency dependent dielectric losses.
- L —The measured inductance of the capacitor, generally having a negligible frequency or voltage dependence below the self-resonant frequency, as it is primarily a geometrically determined quantity.
- C —The capacitance of the capacitor, usually measured by the manufacturer on a bridge at a 1 Vac level and a frequency of 1000 Hz. Due note must be taken of the voltage coefficient of capacitance in ceramic units.

12.3.3.1 Equivalent Series Resistance (ESR). The ESR is the ac resistance of the capacitor, essentially R_s and R_p in parallel, at a given frequency so that the loss in the capacitor can be expressed as the loss in a single (generally frequency dependent) resistance, R, in the equivalent circuit, as shown in Fig. 12-11. Unless soldered or other eutectic bonds are made, this R will have a significant current dependence when peak currents rise into the multikiloamp region. In these cases, compression contacts have small oxide layers, which give rise to superheating where ohmic contact is made. Superheating increases the power lost per pulse over soldered connection configurations. Whether or not this is important depends upon the peak and RMS currents, as well as upon the repetition rate. The persistent usage of inserted tab designs at high repetition rates in liquid impregnated capacitors causes severe heating that can seriously lower the lifetime. In addition, the gases released in this process can cause dangerous pressures to build up inside the capacitor case, often causing explosions with the force of a small munitions shell. Flexible cases or solid cases with pressure relief vents are required. Experience to date with the latter has been inconclusive as to the protection level achieved. Only in the case of reconstituted mica capacitors (Fig. 12-12), where the tabs are under very high pressures as a result of the fabrication process, has the inserted tab design reliably sustained high RMS current loadings for far in excess of 10^{10} shots. The number of tabs per end connection determines the RMS and peak-current carrying capability of the capacitor connection, exact numbers depending upon the tab width and compression pressure.[6]

12.3.3.2 Capacitive Reactance. The reactance of the capacitor, X_c, at frequency f is given as:

$$X_c = -j/2\pi fC \text{ ohms} \tag{12-15}$$

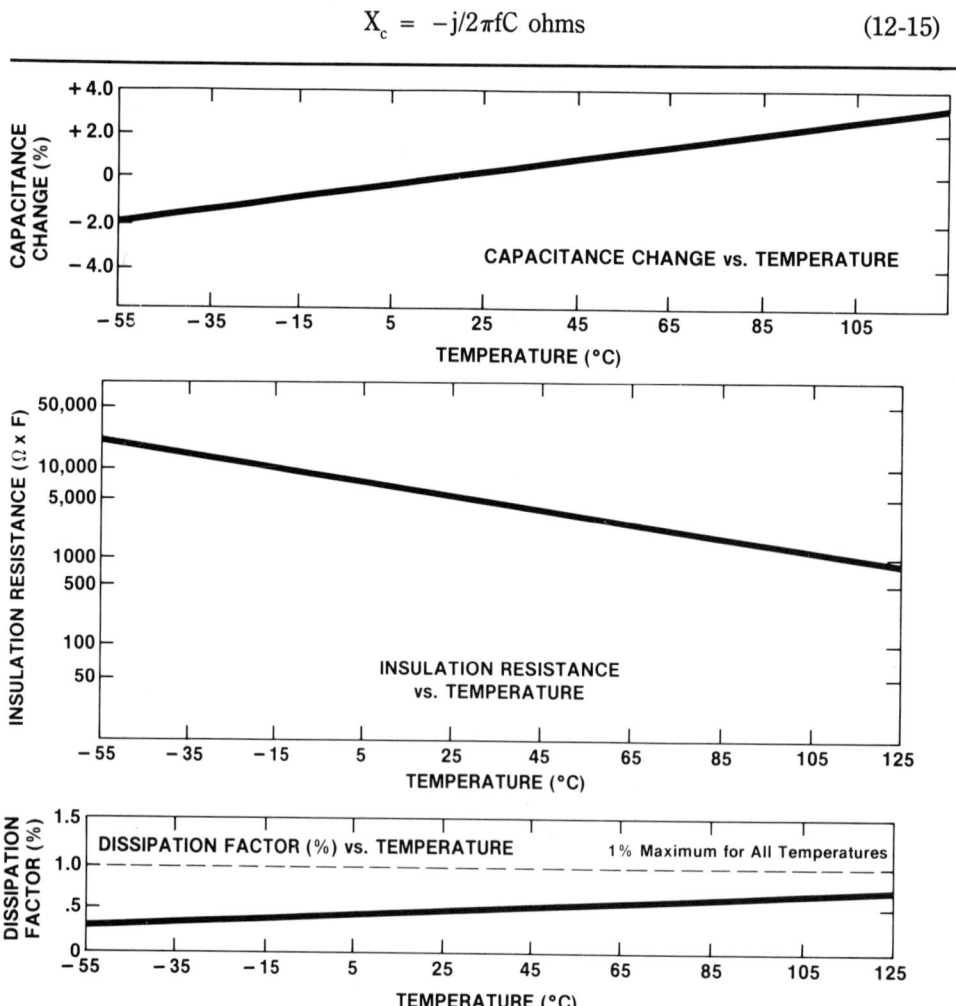

Fig. 12-12. *Characteristics of Axel mica pulse duty capacitors (courtesy of Axel Electronics, 134-20 Jamaica Ave. Jamaica, NY, 11491—A unit of General Signal Corporation).*

where $j = \sqrt{-1}$

$$X_c = -\frac{j}{\omega C} \qquad (12\text{-}16)$$

where $\omega = 2\pi f$

12.3.3.3 Impedance. At high frequencies, the inductive reactance component must be taken into account. From the equivalent circuit of this real capacitor (Fig. 12-11) the impedance, Z, at frequency f is:

$$Z = R + j(\omega L - 1/\omega C) \text{ ohms} \qquad (12\text{-}17)$$

and the magnitude of this impedance is:

$$= \left[R^2 + (\omega L - 1/\omega C)^2\right]^{1/2} \text{ ohms} \qquad (12\text{-}18)$$

At very high frequencies, the inductive term dominates. As the frequency decreases, a point is reached where the inductive and capacitive reactance are equal so that at this point $Z = R$, and the capacitor acts as a pure resistor at that frequency, generally referred to as the *self-resonant frequency* (SRF) of the capacitor.

It is best to stay far below the SRF because the very large RMS currents that can flow in a low-inductance discharge loop at the SRF, means large power losses inside the capacitor. Note that any capacitor can have more than one SRF, depending upon the construction geometry and the transmission line properties of each series section inside the capacitor. Detailed discussions on this matter are very complex and beyond the scope of this chapter. The only remark made is that a section inside a capacitor is generally a rolled-up, parallel-plate capacitor of some length, ℓ, which forms a transmission line and can only discharge in a time equal to the two-way transit time, t, along the line where:

$$t = \frac{2 \cdot \ell}{c} \cdot \sqrt{K} \qquad (12\text{-}19)$$

The term c is the velocity of light in free space, and K is the relative dielectric constant.

This time is independent of the lumped inductance of the capacitor and clearly shows why a supposedly 10 nH capacitor of this type with 0.1 μF capacitance can take, depending upon ℓ, much longer than the calculated 100 nanoseconds to totally ring up and down in current.[7] The only way to make a fast capacitor is to manufacture it with short-length sections whose discharge times are significantly shorter than the desired lumped element discharge time of the capacitor under construction.

Each capacitor manufacturer has its own bag of proprietary tricks, but just for fun, ask how the manufacturer would make the above capacitor for a 100,000 A peak current discharge application with the discharge time of 20 nanoseconds (zero to zero of current). Then ask how long each of the series packs are, and calculate t for yourself. There are ways of fabricating very low inductance capacitors utilizing many very fast-discharge, small capacitors, in parallel. However, a few moments of reflection on the required reliability of each of these capacitors in the assembly in order to achieve a specified system reliability, should give pause for thought.

In contrast to this requirement, there are a large number of very complex, single-shot systems in existence that work very well and have few reliability problems with their capacitors. It is only fair to point out that this is after significant amounts of money were spent to achieve high reliability capacitor designs and many research groups actively participated in and, in many cases, were instrumental in solving the major hurdles in these capacitor development programs.[2] This has not been the case for low-inductance, high repetition rate capacitor development.

The SRF of several ceramic capacitors has been measured in a recent experiment here.[8] For both the Murata Type DH and the Cera-Mite 720 of capacitance values 2890 and 2380 pf respectively, the SRF was 20 ± 0.5 MHz and the series inductance was 18 and 20 nH, respectively. Thus, many fast discharge circuits ($\cong$ 50 nanoseconds pulse width) employ these capacitors near their SRF, resulting in increased circuit losses and capacitor heating.

12.3.3.4 Dissipation Factor. The *dissipation factor*, DF, is defined as the ratio of the effective series resistance, R, to the capacitive reactance, X_c at a frequency f. It is normally expressed in percent:

$$DF = (R/X_c) \cdot 100\% \qquad (12\text{-}20)$$

$$= (R\omega C) \cdot 100\% \qquad (12\text{-}21)$$

The *DF* increases with frequency and is always a measure of the power loss inside the capacitor when multiplied by X_c times the RMS current[2] through the capacitor. Normally, this calculation of power loss is straightforward, since the discharge loop defines the discharge frequency for a half-sine pulse:

$$(I_{rms})^2 = (I_{peak})^2 \cdot \tau \cdot \frac{f^2}{2} \qquad (12\text{-}22)$$

where I_{peak} is the peak current and τ its duration, assuming no current reversal. The repetition rate is f hertz.

In complicated discharge systems where shaped pulses are required, a similar calculation can be effectively carried out by integrating the square of the discharge current over the discharge time and multiplying by the repetition rate. Typically, ceramic capacitors have dissipation factors of 2 to 3 percent in the 0.1 to 1 MHz frequency region, castor oil-paper around 5 percent, polysulfone-silicone and other proprietary impregnants about 0.5 percent, and reconstituted mica from 0.4 to 1.0 percent, depending also somewhat upon the geometry and compactness the user demands. The only capacitors to have compiled a long lifetime record at high repetition rates are the two low-loss type units described above. All in all, there are very sound arguments in favor of the reconstituted mica type up to 100 kV where high peak currents and nonflammable impregnants are demanded. High currents, up to 200 A RMS can be handled[6] by carefully fabricated mica units in low-inductance geometries, but they are somewhat more expensive than modules of liquid impregnated polysulfone capacitors; they also do not catastrophically fail, they just short and melt when overloaded.

Because the dissipation factor must be measured at the frequency of discharge of the capacitor, it is important that the supplier provide the user either with this number or the power loss at the specified repetition rate from which DF and hence R can be calculated. This R determines the ultimate Q of the circuit:

$$Q = \frac{1}{DF} = \frac{1}{\omega RC} \qquad (12\text{-}23)$$

for DF expressed as a pure number and not in percent.

Some manufacturers provide the power factor (PF) of the capacitor at the discharge frequency. The PF is defined as the ratio of R to Z at this frequency. When the PF is less than 10 percent, the DF and PF are essentially the same.

Recall that, except for reconstituted mica capacitors,[6] the internal resistance of a tabbed capacitor is current dependent. Thus, at short discharge times, high peak currents, and high repetition rates, Qs are found to be lower than shown by calculations based on low-level measurements. Once again, there is a strong argument for extended foil, soldered, and swaged geometries in all types of capacitors.

In addition to the frequency dependence of the dissipation factor, there is a temperature effect, which simply stated says: the higher the temperature, the higher the level of power lost inside the capacitor for the same stored energy and current waveform.[9] For ceramic capacitors, this is illustrated in Figs. 12-6, 12-7, and 12-9, for castor oil in Fig. 12-5, and for some low-loss plastic films in Fig. 12-8. For mica the curve is slowly varying (Fig. 12-12) from 0.5 percent at 25 °C up to 1.0 percent at 125 °C, at which temperature the capacitor is still happily operated without any derating. The variation in DF with temperature is illustrated in Fig. 12-13 for both dry dielectrics and liquid-impregnated kraft paper. The manufacturer can and does use the DF as a quality control tool. Variations in DF above normal values for a particular lot of the same capacitors generally indicates a possible loss of fabrication control or material control.

To sum up, the user operating a capacitor in ac or high repetition rate pulsed discharge applications is intimately concerned with the dissipation factor because the series resistance factor in DF is the heat producing element in these applications.

12.3.3.5 Humidity. For hermetically sealed capacitors, humidity is not a problem except that high humidity does cause some decrease in the surface breakdown voltage of the terminal insulators. Moisture in the capacitor dielectric will cause a large reduction in leakage resistance, and increases in the DF and internal gas pressures at high repetition rates. A long-term dc leakage test or corona-level test before and after operation will pick out such defective units.

12.3.3.6 Frequency. The variation of C and DF with frequency has been discussed in Sections 12.3.3.3 and 12.3.3.4. When utilizing capacitors in pulse-forming networks,

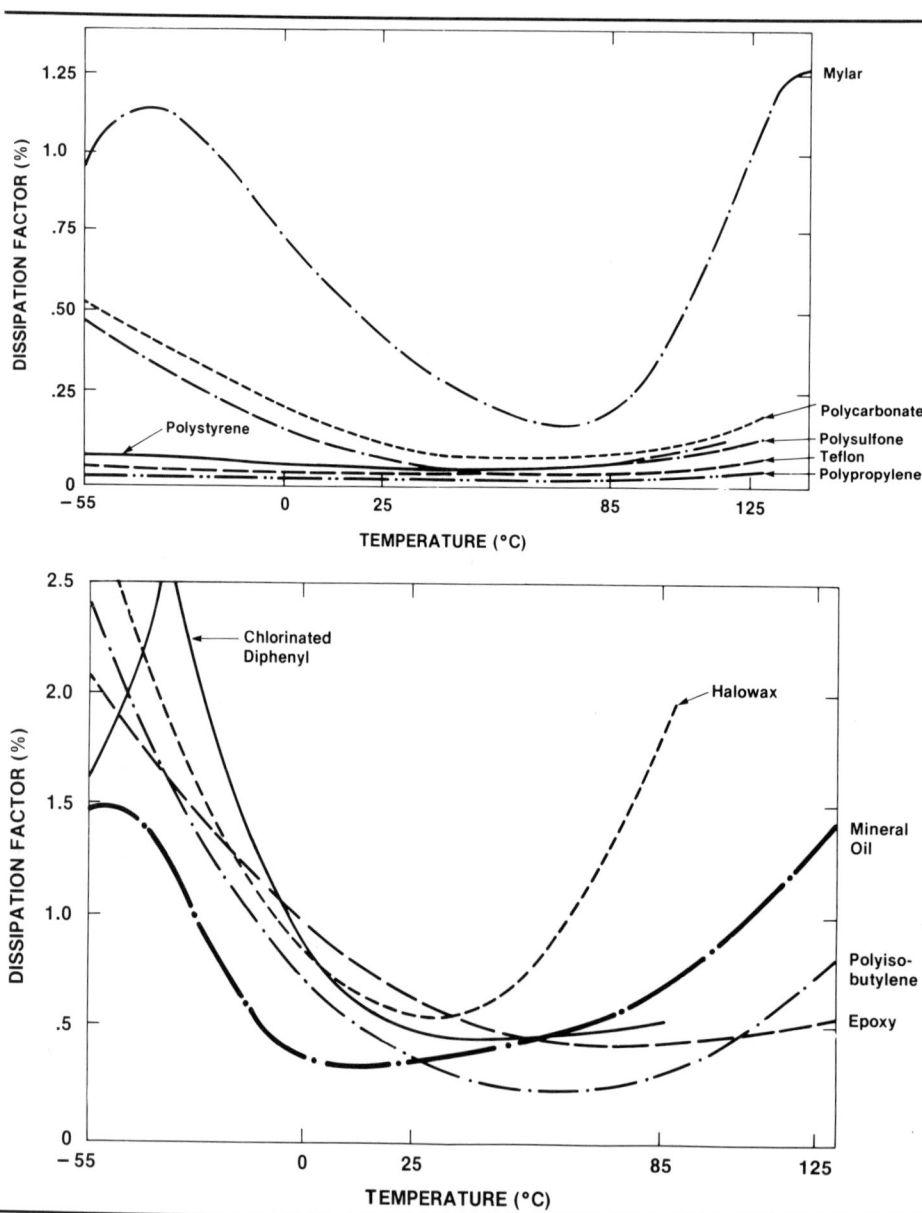

Fig. 12-13. Dissipation factors for film and impregnated papers as a function of temperature (after Electrocube Technical Capacitor Bulletin, with permission).

bear in mind that, particularly in tapered PFN designs wherein the impedance is constant but the capacitor values are staggered to achieve fast rise times, each capacitor sees a different Fourier frequency distribution in the currents that pass through. This must be taken into account in power-loss calculations and is best done by computer analysis.[4]

12.3 Capacitor Characteristics and Their Changes

12.3.3.7 Storage Temperature and Time. All capacitors have a range of storage temperatures beyond which either impurity migration, internal gasification from voids in the dielectric, or chemical changes in the impregnant will cause reductions in lifetime. Ceramic capacitors also suffer from depolarization with time, decreasing their capacitance. These variations are summarized in Table 12-3.

12.3.3.8 Lifetime. Lifetime is the main problem area in capacitor technology. If it is assumed that the manufacturer utilized proven devices from which specials are derived, then the lifetime for capacitors in general is limited by gasification from arcing at nonsoldered connections, resultant chemical reduction in dielectric strength, and corona-induced treeing in and on the dielectric, particularly at the edges of the capacitor sections. These points are illustrated schematically in Fig. 12-14. Arcing at tabs or between stacked, pressure-contacted sections can be eliminated with soldering, metalizing or cold-flow eutectic bonding. These are listed in order of increasing cost but will eliminate gas formation and arc-drop heating, particularly at high repetition rates. There is absolutely no excuse for the continued utilization of inserted tab constructions in very high peak current or/and high repetition rate capacitors where high reliability is desired.

Void ionization and impurity migration under the influence of the external electric field can be minimized through the use of high-quality dielectrics and the elimination of the unnecessary Kraft paper wick. This wick is retained in many film capacitor designs because of its very effective oil-wicking properties, which make impregnation of the capacitor faster and thus less expensive. Implementation of impregnation-while-winding procedures would eliminate this problem, but it requires substantial modification existing winding machinery. There are several new classes of silicone dielectric fluids of low loss that might be very good dielectric constant matches at higher frequencies. These, in conjunction with several of the new proprietary fluids,[10] could be combined with the above winding technique to eliminate almost completely the possibility of corona damage from impregnant voids between dielectric sheets or between dielectric edges and the electrode. Note that normally, several thin (<0.001 = inch) layers of dielectric are used rather than one thick sheet because the latter results in much higher breakdown vol-

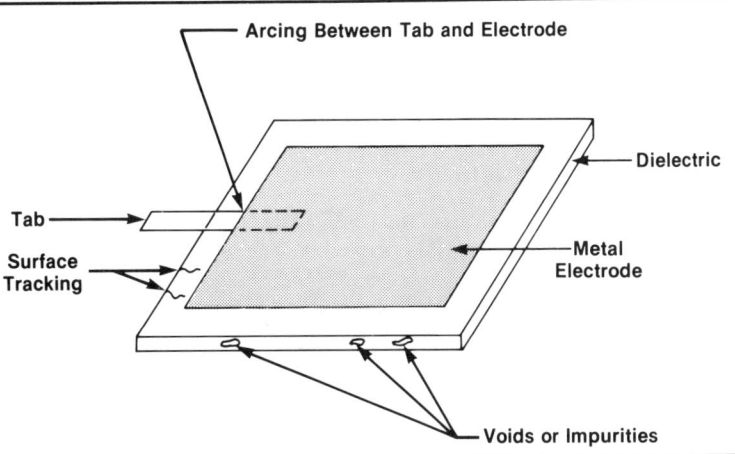

Fig. 12-14. Physical factors affecting capacitor lifetime.

tages because thicker sheets have a much higher percentage of imperfections, inclusions, and small voids. This is a technological limitation that could likely be overcome at some considerable cost.

The only problem remaining is that of surface tracking at the edge of the dielectric as sketched in Fig. 12-14. Charlie Martin at Aldermaston has a simple way of grading edges; namely, cover them with blotting paper impregnated with a dilute solution of copper sulphate in water. Having tried this technique on numerous occasions, it can be confirmed that it is very effective indeed. What the paper does is reduce the concentration of the electric field at the interface of the electrode edge and dielectric, essentially by providing a resistor-like layer from the electrode out to the edge of the dielectric. The results are dramatic, and stresses of much higher values than in commercial capacitors (which generally operate at less than 3,400 V/mil) are achievable. The difficulty with this technique is that such resistively graded systems must be pulse charged because of the dc resistance of the copper sulphate solution shunting the capacitor. Thus, some form of resistive grading will be necessary to minimize this effect, and this must be tempered by the increase in the loss factor that might follow, placing more power loss inside the capacitor.

When the user experiences failures in a pilot run of capacitors, the above notes might provide some help in identifying the problem areas and allow working with the supplier to rectify the problem. It is a general observation that larger size and very long lifetime are currently directly related.[9] The internal material cost is relatively a smaller portion of the price compared to the labor to make a unit very compact, often with special high-stress headers. Unless space constraints dominate, it is in the interests of reliability to allow the manufacturer to recommend a capacitor volume for that specific application. Often, the volume ratios of different supplier's capacitors provide a clue to the design margins each is using. If all suppliers utilize paper wicking, then the cost per unit capacitor element volume is a good selection factor to be included in the assessment of each potential supplier. Combine this with the knowledge of the stress level in each, the method of electrical connection, and the mechanical package pressure control upon the capacitor packs, and that provides most of the factors needed to make a cost-effective choice.

In dc and ac service, capacitor lifetimes of 30 to 40 design years have been achieved in the power industry and dc filter applications. These are obtained in most cases by using highly purified mineral oils with stabilizing additives; the capacitor is vacuum baked, dried, and then oil impregnated at an exceptionally low moisture level.[11] All these units originally used Kraft paper for historical reasons that are not of concern here. The oil-impregnated capacitor is hermetically sealed and might include a bellows or other thermal expansion device. In Fig. 12-15 is a curve of the internal impulse strength against years for this type of long-life design compared to a typical short-life design representative of what might be expected from a dc or ac filter or coupling capacitor for commercial service.[11] The differences are in the quality of manufacture and in the electrical stress levels in the dielectric. The long-life design might well operate below 300 V/mil (note that this makes for a large capacitor), and the short life might be in the neighborhood of 900 V/mil. The *base insulation level* (*BIL*) is the unity probability of failure impulse strength level of the capacitor, and all units must meet this standard after 30 to 40 years of service (at the end of life). In the case of film capacitors in the same duty, lifetimes of 10 to 15 years are likely, based upon accelerated life tests. A real advantage of film

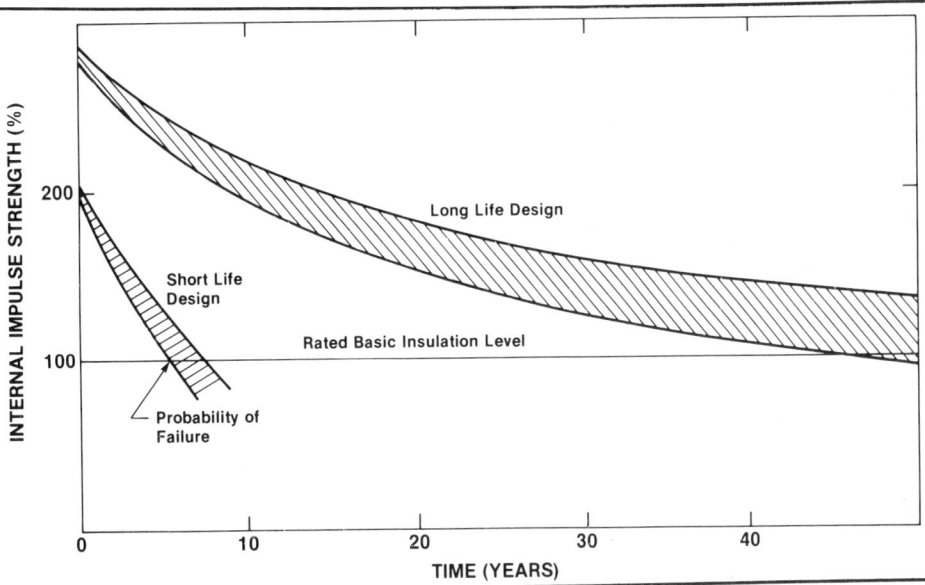

Fig. 12-15. Impulse strength as a function of time for very long lifetime capacitors for 60 Hz ac duty.

units is their very much lower power losses at higher frequencies and higher insulation resistances; both of these factors serve to reduce internal heating and corona currents. They are smaller, but considerably more expensive.

The question of lifetime of energy discharge capacitors must be addressed. Up to now, the discussion centers about capacitor parameters in general, with some notes on points of interest for pulse capacitors interjected where they related to the parameters under discussion. The rest of the chapter looks at the characteristics of discharge circuits having ideal switches and various types of loads. It is assumed that the discharge loop can be described by some lumped inductance and that the switch instantly closes. Although this is most clearly a gross oversimplification of the true state of affairs, it will serve to illustrate some of the salient features of the circuitry and of the currents and voltages the capacitor will likely see.

12.4 APPLICATIONS OF ENERGY-STORAGE CAPACITORS

The energy storage capacitor can be used singly, or with many units, in parallel, to provide short pulses of electrical energy up to many megajoules of energy. The electrical energy stored in a capacitor of capacitance C is $\frac{1}{2} CV^2$, where V is the final voltage across the capacitor at the end of the charging cycle. Applications for high peak current capability capacitors include particle accelerators, metal forming, laser drivers, and X-ray generators. In these situations, discharges are often highly oscillatory, and voltage reversals up to 90 percent are common. Flash tubes, some lasers, and numerous welding applications have capacitors discharging into resistors or other critically dampened loads, with relatively longer discharge times and small reversals compared to the previous case.

12.4.1 Discharge Circuit

A simplified discharge circuit is sketched in Fig. 12-16; this illustration shows a resistively charged capacitor being discharged into a load, which has a resistance R and an inductance L. The question of what happens when the load is a laser where a peak voltage must be reached before current flows is very complex.[12] In the circuit in Fig. 12-16, closing the switch discharges the capacitor into the load in a nonoscillatory, oscillatory, or overdampened manner depending upon the values of R, L, and C as illustrated in Fig. 12-17.[13] The terminology is defined just below the heading and differs slightly because E is substituted for V_c. Using these relationships, the ringing frequency, peak current, voltage reversal, and number of post-discharge rings in the oscillatory case can be calculated.

12.4.2 Ringing Frequency

For oscillatory discharges and small load resistances the ringing frequency is:

$$f = \frac{1}{2\pi \sqrt{LC}} \text{ or } \omega^2 LC = 1 \qquad (12\text{-}24)$$

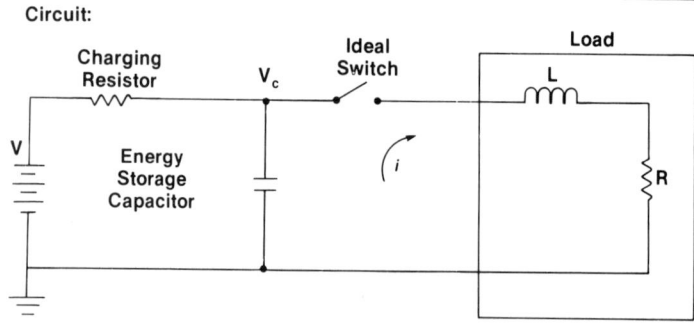

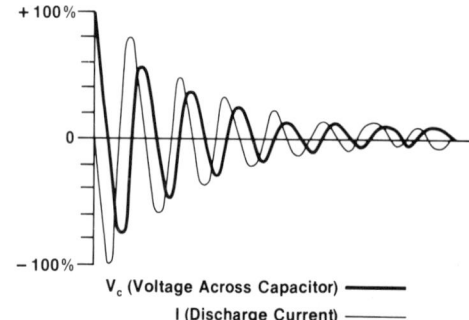

Fig. 12-16. *Simplified circuit diagram for charging and discharging an energy-storage capacitor.*

It is very difficult to make this larger than 100 MHz in lumped capacitor systems for high-power applications at the multikilovolt level. The reasons for this have been discussed before and are the lumped inductance limitations in high-voltage geometries inside the capacitor and the transmission line discharge time of each pack in the capacitor. There are some kilojoule or so, 100-kV systems capable of 25 to 50 nanosecond discharge times operating at low repetition rates (less than 10 Hz). These were developed a number of years ago for laser drivers, and they worked very well but are rather expensive.[14] High repetition rate versions would require pulse charging and grading development, particularly if the 85 percent reversal duty cycle was retained.

12.4.3 Peak Current

Resistance in the discharge loop represents a power loss term, which should be minimized for maximum energy transfer. The limit here is the resistance of the load, be it a laser, diode, or pure inductor. If the resistance is made small in order to increase the peak current and reduce the pulse width (Fig. 12-17) then the peak current I_p is approximately:

$$I_p = V \cdot \sqrt{\frac{C}{L}} \qquad (12\text{-}25)$$

Increasing the value of R, where:

$$Q \cdot R = \sqrt{\frac{L}{C}} \qquad (12\text{-}26)$$

gradually reduces the peak current, until, at critical damping, I_p is reduced by a factor of 0.4 from the oscillatory case. Note that the current waveform can be clamped when a unidirectional switch such as a thyratron is used. In this case the switch opens when the voltage across it turns negative and the device turns off. The RMS current is thus significantly reduced and the voltage reversal experienced by the capacitor is controlled primarily by the inverse voltage-limiting circuits placed across the switch. Each case here must be individually analyzed and this subject, for modulator and laser drivers, finds computer simulation a very useful tool.

12.4.4 Voltage Reversal

The voltage reversal has a first negative peak given by:

$$\frac{V_{reversal}}{V_{charge}} = \exp\left(-\frac{\pi \cdot R}{2} \cdot \sqrt{\frac{C}{L}}\right) \qquad (12\text{-}27)$$

Normally, this is expressed as a percentage of the charge voltage. For most high repetition rate capacitor geometries, 85 percent reversal is the cost-effective limit and should cost a small premium over the standard 20 percent reversal units. Even if these large

NONOSCILLATORY	OVER-DAMPED	OSCILLATORY
Critically damped, $R = R_c = 2\sqrt{\frac{L}{C}}$	$R > 2\sqrt{\frac{L}{C}}$	Under-damped, $R < R_c$
$i(t) = \frac{E}{L} t e^{-\frac{Rt}{2L}}$	$i(t) = \frac{E}{2j\omega L}\left[e^{-\frac{Rt}{2L} + j\omega t} - e^{-\frac{Rt}{2L} - j\omega t}\right]$	$i(t) = \frac{E}{\omega L} e^{-\frac{Rt}{2L}} \times \sin \omega t$
		Where $\omega = \sqrt{\frac{1}{LC} - \frac{R^2}{4L^2}}$
$I_p = 0.736 \frac{E}{R_c} = 0.368 E \sqrt{\frac{C}{L}}$	$I_p = 0.736 \frac{E}{R} = 0.368(E)\sqrt{\frac{C}{L}}$	$I_p \cong \frac{E}{\omega L} e^{-\frac{R}{4\omega L}}$ When $R \ll R_c$
Time to reach $I_p = \frac{2L}{R_c}$	Time to reach $I_p = \frac{2L}{R}$	$I_p = E\sqrt{\frac{C}{L}}$ Ideal case when $R = 0$
		$I_p = \left[1.00 + \frac{\%\text{REVERSAL}}{100}\right] \times \frac{\omega CE}{2}$
		For high values of voltage reversal.
		Time to reach $I_p \cong$
		$\frac{\pi}{2}\sqrt{LC}$ When $R \lll R_c$

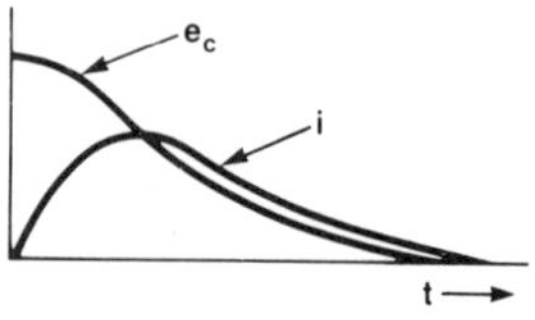

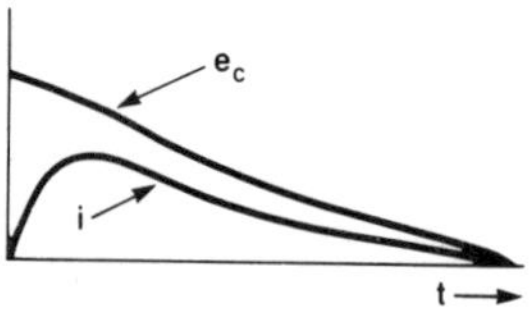

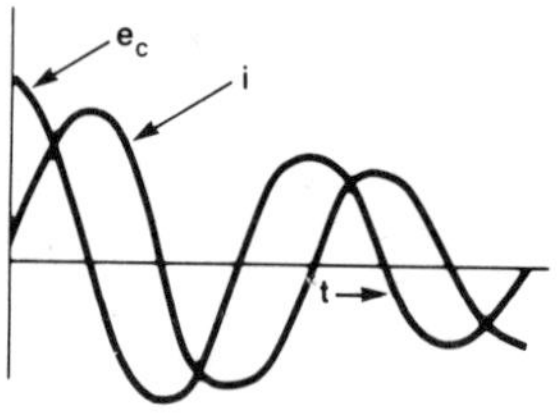

EQUATIONS Definitions of Symbols

- $i(t)$ — Current at any time, t
- I_p — Peak current (in amp)
- C — Capacitance (in farads)
- L — Total circuit inductance (in henries)
- E — Charge voltage (in volts)
- f — Ringing frequency of the circuit
- R — Total circuit resistance (in ohms)
- Q — Merit factor
- ESR — Equivalent series resistance of capacitor
- R_c — Capacitor ESR

1. Stored energy (in joules) = $1/2\, CE^2$

2. Voltage reversal (in percent) =

   $e^{-\frac{R}{4Lf}}$, where $f = \frac{1}{2\pi}\sqrt{\frac{1}{LC} - \frac{R^2}{4L^2}}$

3. Voltage reversal (in percent) = $(1 - \frac{\pi}{2Q}) \times 100$ (for high % reversals)

4. Q of the capacitor (at resonance) = $\frac{1}{ESR}\sqrt{\frac{L}{C}}$

5. $Q = \frac{X_c}{ESR}$

Fig. 12-17. Capacitor discharge circuit relationships for a load having a lumped resistance and inductance of fixed values.

reversals are experienced on a small percentage of the pulses, reliability can be significantly enhanced by requiring the unit to live with a 1-cycle reversal of 85 percent for all pulses. Reversal primarily causes corona damage at dielectric edges in a mechanically well-designed capacitor. This damage is currently minimized by designing such capacitors for ac duty stresses, making them rather large but at a slight increase in cost.

12.4.5 Lifetime

Lifetime for most liquid impregnated capacitors follows the relation:[2,5]

$$\text{Lifetime} = \left(\frac{V_{charge}}{V_{breakdown}}\right)^{-8} \cdot Q^{-2.2} \quad \text{in shots for DC charging} \quad (12\text{-}28)$$

For a given capacitor geometry, the manufacturer has accurate data on the unit probability breakdown voltage, generally after 1 minute at that dc charge voltage.[2] Note that this lifetime dependence upon stress results from corona and dielectric surface tracking, with occasional bulk breakdown. The exponent can be significantly increased through pulse charging. The same law holds true for all corona limited systems; this applies to almost all highly stressed discharge capacitors—paper, mica, ceramic and true liquid dielectrics. Ceramic capacitors follow the same law but with an exponent of 12, for pulse charging times less than a few hundred nanoseconds, which is shorter than the corona inception time:

$$\text{Lifetime} = \left(\frac{V_{charging}}{V_{breakdown}}\right)^{-12} \cdot Q^{-2.2} \quad \text{in shot for pulse charging} \quad (12\text{-}29)$$

It is tempting to hazard a guess that this relationship might hold for oil-impregnated and mica capacitors, but this remains to be shown. The factor Q is the circuit quality factor and is related to the percent reversal by:

$$\text{Percent Reversal} = \left(1 - \frac{\pi}{2Q}\right) \cdot 100 \quad (12\text{-}30)$$

12.5 CLOSING NOTE

Low-inductance kilohertz repetition rate mica capacitors are now being produced for operation at several hundred amps peak current, with measured lifetimes of 10^{10} shots at very low failure levels. Extension to multikilohertz operation appears feasible, but will require great care in keeping internal power losses sufficiently low to conserve this long life. Further developmental work is required to extend this class of capacitor into the tens of kiloamps of peak current at voltages in the region 100 to 150 kV.

12.6 REFERENCES

1. H.E. Duckworth, "Electricity and Magnetism," (Holt, Rinehardt, and Winston, New York, 1961).
2. G. Boicourt, "Problems in the Design and Manufacture of Energy Storage Capacitors," Los Alamos Scientific Laboratory report LA-4142-MS (January 1970).
3. A.J. Dekker, "Solid State Physics," (Prentice-Hall, Inc., 1965).
4. D.G. Ball and T.R. Burkes, "Pulse Generation for Time-Varying Loads," Texas Tech University Report.

5. W.G. Dunbar, "High Voltage Design Guide for Airborne Equipment," Boeing Aerospace Co. report AD-A029-268 (June 1976).
6. Axel Electronics Ltd., Sales literature and discussions with S. Zweig.
7. W.C. Nunnally, M. Kristiansen, and M.O. Hagler, "Differential Measurement of Fast Energy Discharge Capacitor, Inductance, and Resistance," IEEE Trans. on Inst. and Tech., Vol. 24, No. 2, pp. 112–114, June 1975.
8. W.C. Nunnally, unpublished.
9. "Technological Development of High Energy Density Capacitors," Hughes Aircraft report, 1976.
10. Maxwell Laboratories, for example, fabricates nonwicked, extended-foil, high repetition rate capacitors with long lifetimes utilizing a proprietary fabrication process. They are currently in use in a number of laboratories, for repetition rates up to 1 kHz. Inductances are in the range of 20 to 50 nH per unit.
11. Trench Electric technical memo, 1975.
12. Robert Ray Butcher, "A Comprehensive Study of Excimer Laser Systems," Los Alamos Scientific Laboratory thesis report LA-7329-T (June 1978).
13. DEL Engineering technical literature (1974).
14. Condenser Products module Type MXS-TA.

Chapter 13

Grounding and Shielding in High-Power Electronics

by
T.R. Burkes
(Texas Technological University)

GROUNDING AND SHIELDING OF ELECTRICAL COMPONENTS AND SUBSYSTEMS HAVE always been a source of nuisance problems. In systems where large amounts of energy are moved in short, high-power pulses, these problems might no longer be of nuisance value. They can become critical to the survival and well being of sensitive subsystems and adjacent electronics that might be functionally unrelated to the high power electronics system. As it happens, most of the problems associated with grounding and shielding turn out to be conceptually simple, but as in computer systems, the problems become lost in a maze of complexity. This chapter illuminates those practices that will lead to problems and removes some of the apparent mystery from the sources of electromagnetic interference (EMI). Also, a set of rules are presented that hopefully will help prevent or solve problems derived from EMI.

A word of caution is appropriate at the start of this chapter. It has been the experience of the many of those involved in pulse-power systems that the majority of the EMI problems are derived from preconceived notions, misconceptions, and oversights on the part of the system designer and those charged with system construction. The implication is that, although the problems are simple in concept, the solutions for their prevention must be employed from the outset of system conception and be continued through construction.

The usual approach to the solution and/or prevention of EMI problems is to use screen rooms, shielded cable and to employ a good, solid grounding scheme. Unfortunately, customary implementation of these practices might not result in the solution or prevention of EMI problems. At this point an element of mystery and confusion enters the picture, and the use of cut-and-dry practices usually results in an effort to remove the gremlins from the system.

13.1 ELECTRICAL GROUND: MEANING AND INTERPRETATION

A major source of confusion centers around the meaning of *electrical ground*. For instance, does a well-grounded system mean one in which a chassis (ground for the electronics) is connected to the relay rack, which in turn, is connected to the ac power ground, which is eventually connected to an earth ground? Clearly, the meaning of *ground* changes with the particular component. However, these grounds are not functionally independent and treating them as such can lead to the introduction of unwanted noise. In environments where high-power faults can occur, additional failure modes can be unconsciously introduced. Thus, it is important to understand the various ramifications of interconnecting grounds.

Connecting electrical systems to the earth has special meaning to systems exposed to natural phenomena. Ben Franklin started the practice of lightning protection with his lightning rods, which were solidly connected to the earth. Since most electrical systems are eventually connected to the earth grounds, the opportunity for unexpected system failure exists, especially if the system is earthed at more than one point.

Consider the simple model shown in Fig. 13-1.[1] The distributed capacitance between the cloud and earth is discharged by the lightning stroke. Because the earth is not a perfect conductor, the discharge results in a voltage distribution and results in a potential between adjacently connected points such as between the feet of a person in the vicinity of the stroke. Clearly, this can result in some discomfort to the individual and in failure for electronic systems using sensitive solid state components.

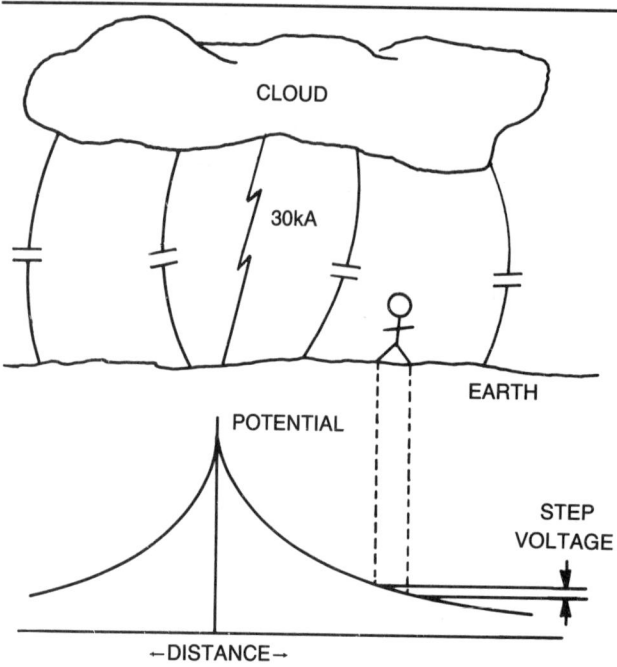

Fig. 13-1. A lightning stroke results in a voltage gradient along the Earth.[1] Systems connected to the Earth at more than one point may sustain very high voltages between Earth "ground."

13.2 LIGHTNING PROTECTION

One can seek protection by moving inside a building as shown in Fig. 13-2. However, the capacitance between the metal roof and earth, C_2, is in series with the capacitance between the roof and an approaching leader, C_1, forming a simple capacitance divider circuit. Thus, the roof can be raised to very high voltage, again endangering the individual seeking protection. If, however, the roof were connected to the earth forming a simple shield, the danger would be greatly reduced because, C_2 would be shorted by the earth connection.

In order to protect against lightning, power systems use extensive grounding schemes. A simple example is shown in Fig. 13-3. Because of the many earth grounds, a fault such as a live wire on the ground results in voltage distribution. Adjacently grounded systems, such as the telephone system, might be jolted.

Although many systems might not be exposed to nature in the above manner, they are interconnected such that crosstalk and interaction is likely to occur. Starting of large motors can, for instance, cause an error generation in a computer. Thus, grounding and shielding revolve around three primary objectives:

1. Safety of personnel
2. Protection of susceptible equipment
3. Measurement of the desired data

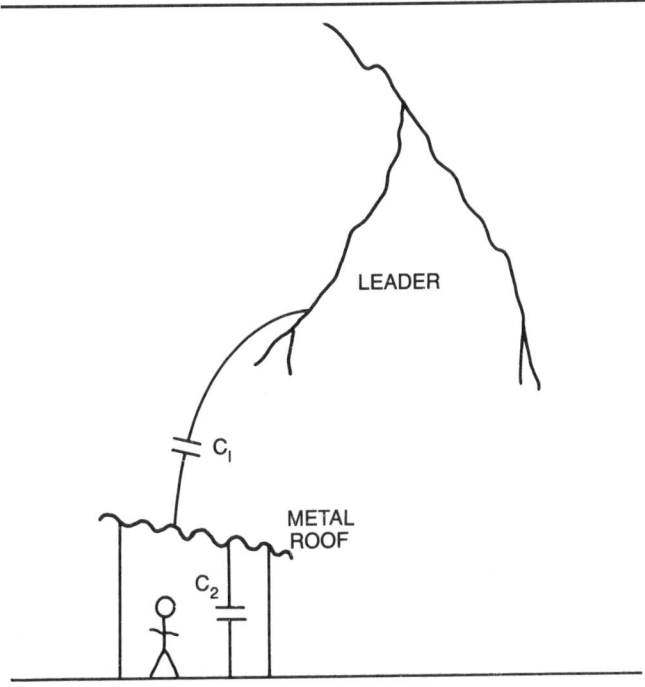

Fig. 13-2. The metal roof does not provide "shielding" from an approaching leader. The capacitive divider may raise the roof to a dangerous voltage unless the roof (shield) is connected to Earth, shorting out capacitance C_2.

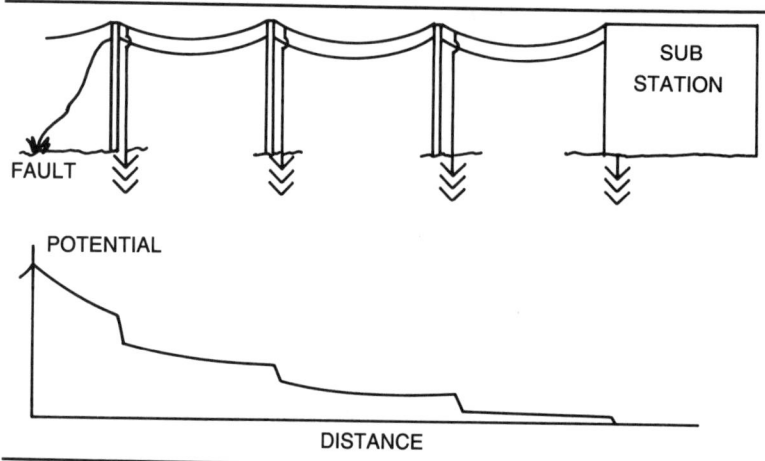

Fig. 13-3. Power system grounds to provide lightning protection cause voltage distributions in the earth whenever faults occur. The power grounds must be solid enough so that fault protection equipment is activated in the substation.

The methods available to achieve these objectives usually require the use of separate grounding schemes. These usually are the power, signal, and safety grounding systems. Keeping these systems separate results in reduced conducted noise and, when extended to shields, increased shielding effectiveness.

13.3 VOLTAGE DISTRIBUTION AND SKIN EFFECTS

Conducted noise is by far the most common of the evils. This conclusion follows from the assumption of a zero reference ground plane or bus wire to which all systems might be connected and not interact. Because no practical materials are available with infinite conductivity, no zero reference ground plane exists. Consider, for example, an infinite conducting plane with current injected and removed as shown in Fig. 13-4. Solving LaPlace's Equation for the situation indicated results in a voltage distribution similar to that shown in Fig. 13-5. For an infinite sheet of aluminum, approximately ¼ inch thick, voltages of 50 to 100 mV are easily obtained between equipotential lines for an injected current of 50 kA dc. Although the current in this example is large, the skin effect will greatly modify the value of current required to achieve the same values of voltage distribution. Recall that the depth of penetration is given by:

$$\delta = \frac{1}{\sqrt{\pi f \mu \sigma}} \tag{13-1}$$

which indicates that, at 1 MHz, two skin depths are only .006 inches for aluminum. This increases the magnitude of the voltage distribution by approximately 40 times for the same current.

Consider the example of the use of a ground plane as shown in Fig. 13-6. Clearly, the resulting voltage between the power amplifier (point C) and the system controller (point F) may be sufficient, especially under transient conditions, to cause malfunctions of overall system response. Also, note that a filter of the usual design in the signal con-

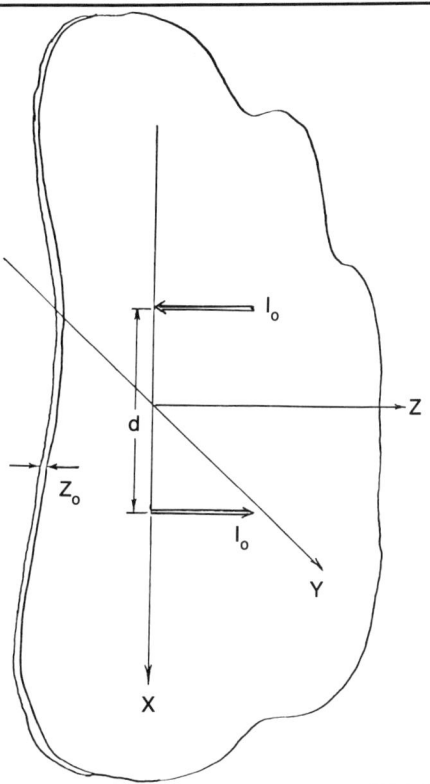

Fig. 13-4. An infinite conduction plane of finite conductivity $_o$ and thickness Z_o, with current I_o injected and extracted.

ductor lead between the amplifier and controller might not reduce the level of conducted noise introduced into the amplifier input. In fact, the noise level might be increased, depending on the type of filter and its location in the signal loop.

13.4 GROUNDING SCHEMES

The usual types of grounding schemes are shown in Fig. 13-7. The series ground is probably the most common but, all the systems can interact through the common impedance of the grounding strap or wire. This type of grounding scheme is best for safety or power grounding system. The parallel connection to a ground plane is also satisfactory for safety. It is likely to introduce unwanted noise if it is used for the power ground because of the likelihood of establishing ground loops. The single-point ground is usually the best of all, especially in those systems where circuit dimensions are a small fraction of a wavelength of the highest frequency. Using three separate grounding systems is the best strategy. It should be noted that if this strategy is to be successful, each of the grounding systems must be insulated from each other.

Examples of correct and incorrect implementation of the use of three separate grounding systems are shown in Fig. 13-8. The relay racks are strapped together and are strapped to the common ground point, forming the safety ground system. The signal

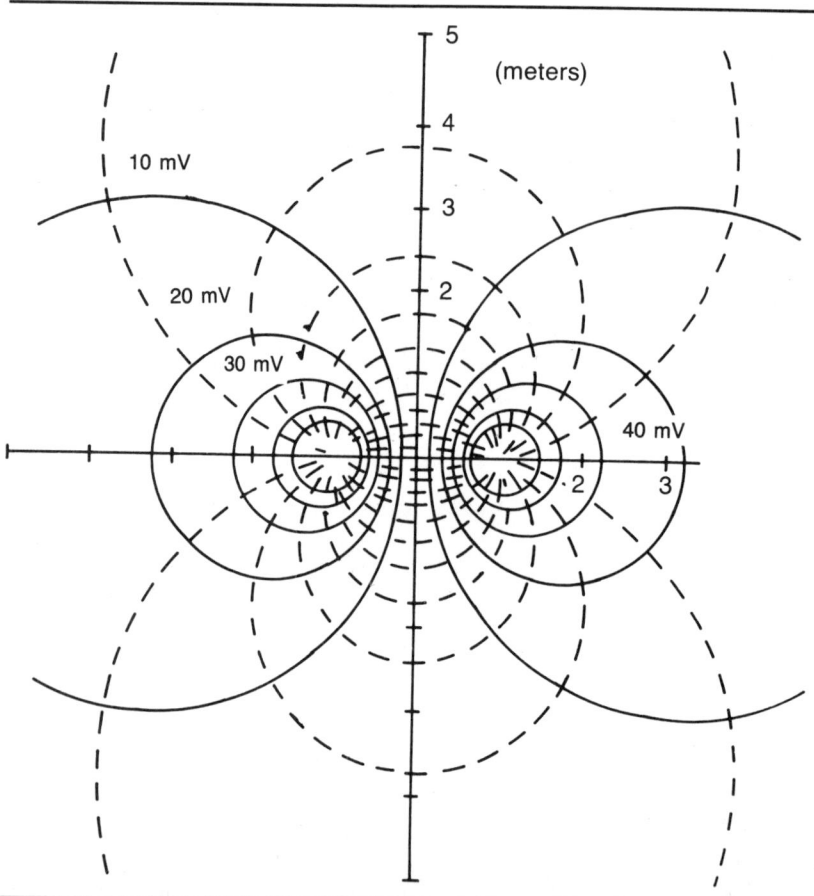

Fig. 13-5. Equipotential lines and current paths in an infinite aluminum ground plane of 5-millimeter thickness due to a 50 kA dc ground current.

ground for the amplifier in Rack 1 should not be connected to the chassis ground strap as shown. This connection can allow the introduction of unwanted noise to the amplifier through the signal input and its ground (not shown). If this type of connection is to be avoided, the signal ground must be insulated from both the safety ground and the power ground.

In some cases, it is impractical to avoid ground loops by using separate grounding systems. In these cases, various devices and schemes are available to break the direct ground connections (Fig. 13-9). Transformers have no direction internal connection so that they serve well to break any dc coupling that might be present. The primary and secondary of transformers are capacitively coupled, however, so that the ground loop is still completed through the primary to the secondary capacitance. Photocoupled devices are also useful to break ground loops and have much smaller feed-through capacitances than transformers. For very low frequency analog signals, the transformer and photocoupler might not be appropriate so that other means must be used to reduce the effect of the ground loop. The use of a one-to-one transformer inserted in the signal circuit

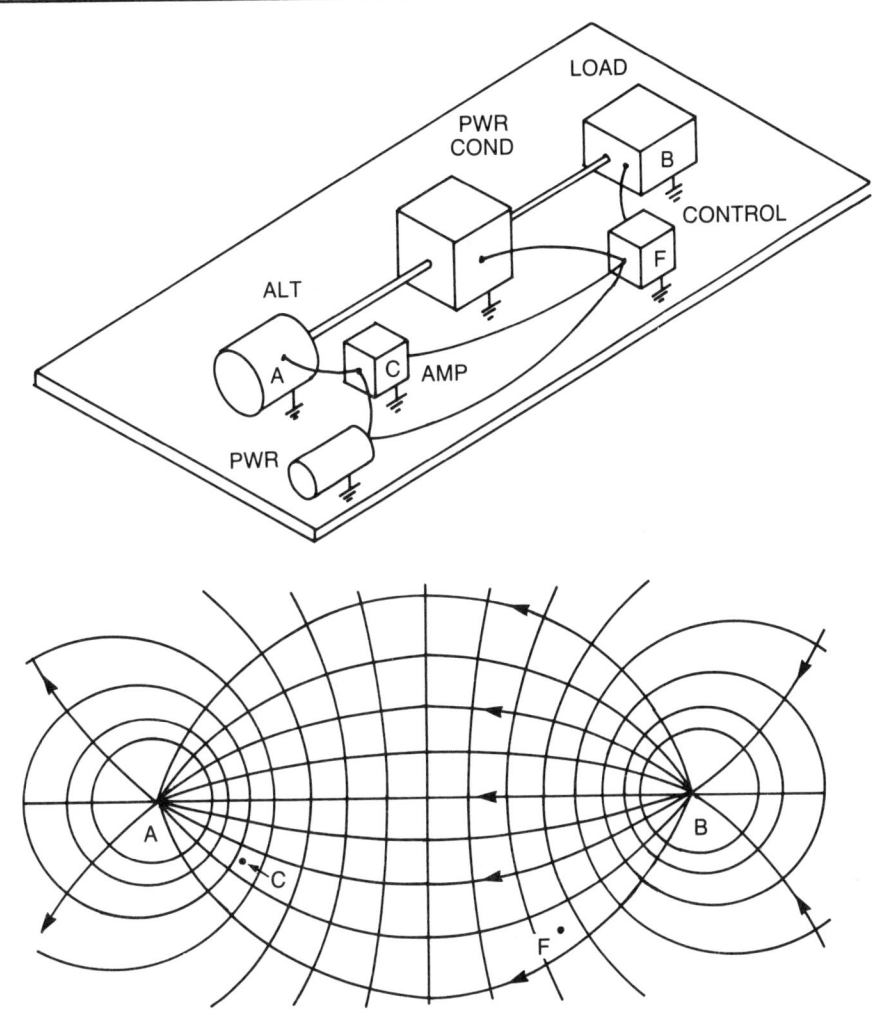

Fig. 13-6. High-power system mounted on a ground plane and approximate voltage distribution due to a ground current between points A and B.

as indicated in Fig. 9C will serve to reduce the effect of ground voltages by inducing an equal but opposite polarity signal in the signal leads. Only the desired signal, in the ideal case, would be unaffected. A differential amplifier is also useful to reduce the effect of ground voltages (often called common-mode noise), but in HPE systems, care must be taken that the ground voltages do not become so large that damage to the amplifier occurs.

13.5 RADIATED ELECTROMAGNETIC NOISE

Radiated noise is often blamed for problems that are more likely caused by conducted noise. However, radiation can be an important source of noise and system inter-

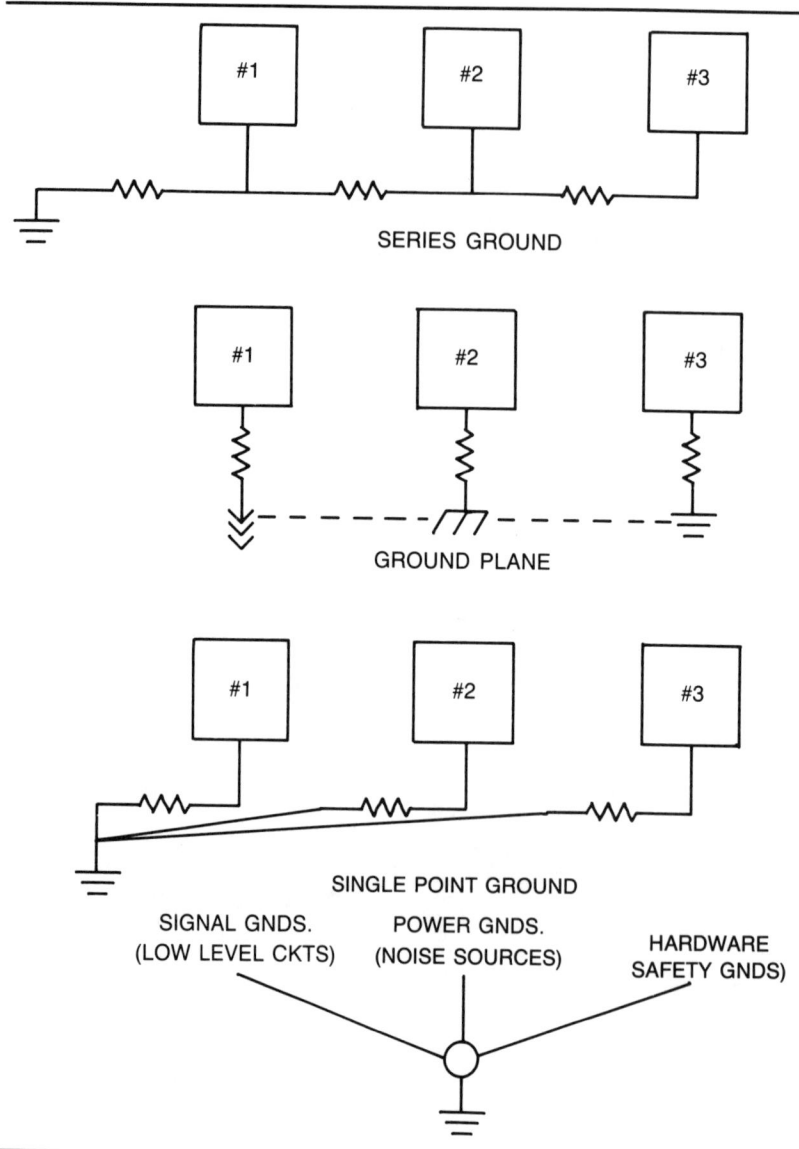

Fig. 13-7. Grounding schemes in common use.

action. Shielding of near fields is usually more appropriate than designing for far-field interference. Figure 13-10 gives the impedance character of near-field radiation that in the far field becomes the familiar 377Ω impedance of free space. Dipole-type radiation is characteristically high-impedance radiation or electric-field dominated and radiation from a loop antenna is low impedance or magnetic-field dominated. Very often the character of the noise source can be ascribed to one or the other; it is convenient to give separate treatment to the two cases.

13.5 Radiated Electromagnetic Noise

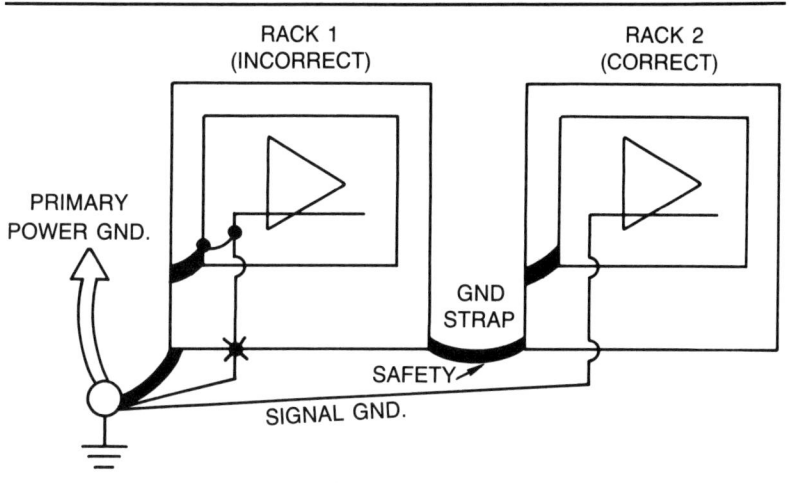

Fig. 13-8. Example of implementation of three separate grounding systems.

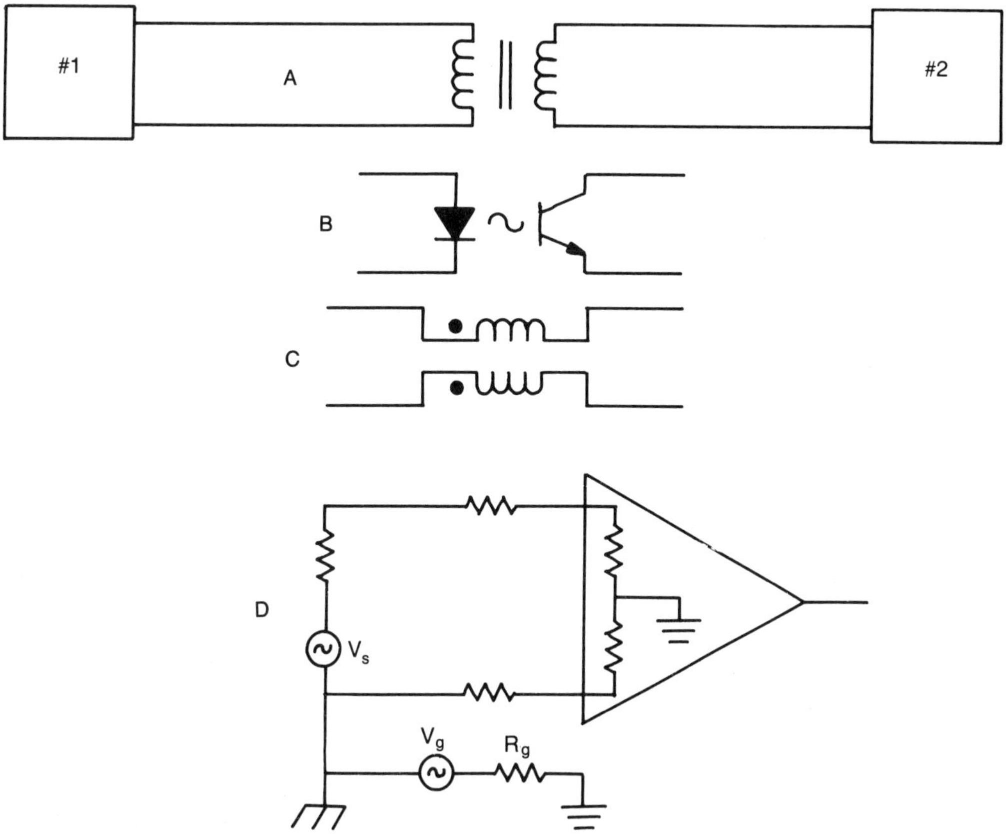

Fig. 13-9. Various devices for interrupting or reducing the effect of ground loops.

380 Grounding and Shielding in High-Power Electronics

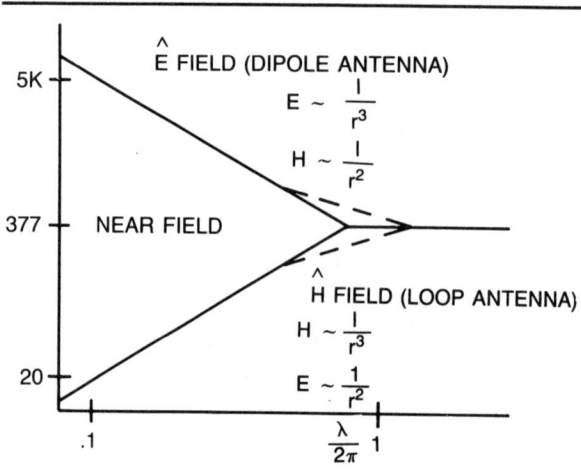

Fig. 13-10. *The character of near-field radiation as a function of distance from the source in wavelengths.*[2]

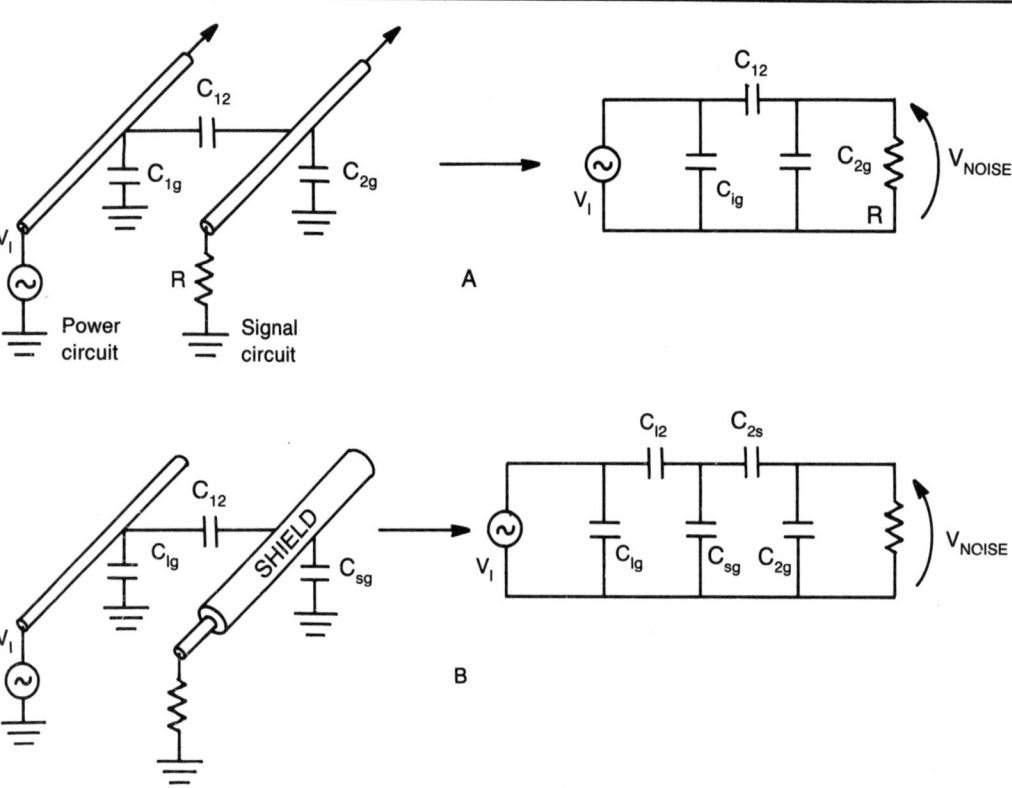

Fig. 13-11. *Making R small in circuit A will reduce the level of noise voltage. Using a shield will effectively eliminate the electronic coupling if the shield is properly grounded.*

Identification of noise sources may be more difficult than at first thought. For instance, a coaxial cable carrying large amounts of power might have the inner conductor off center so that a large magnetic field (low impedance) might exist outside the cable. An example of a high-impedance field can be generated by the rapid closure of a high-voltage switch (hydrogen thyratron). The rapid collapse of the associated electric field can result in unwanted currents flowing in signal leads exposed to the high field.

13.6 ELECTROSTATIC COUPLING

A simple example of electrostatic coupling between two circuits is shown in Fig. 13-11. If the signal circuit impedance is made small, then clearly the noise will be reduced, but it cannot be eliminated by this technique. Placing a conducting shield around the signal can greatly reduce the noise if the shield is properly grounded.

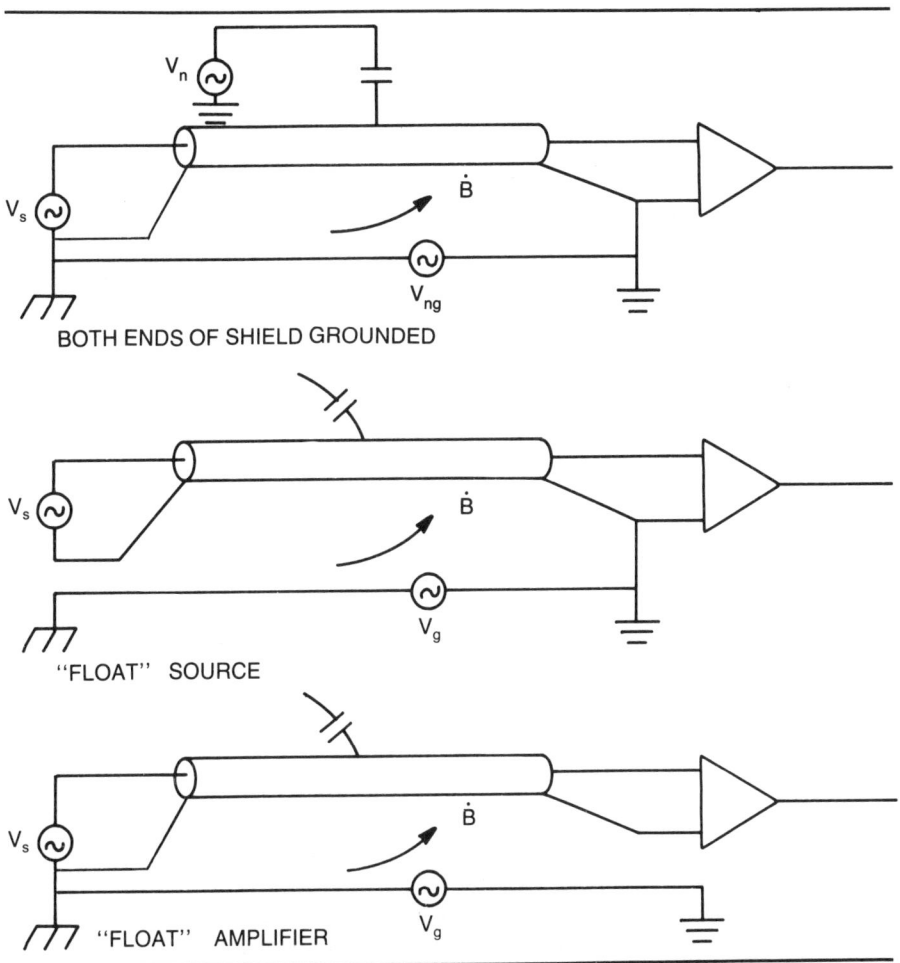

Fig. 13-12. Grounding both ends of shield results in ground loops and magnetic susceptibility. Floating either the signal source or the amplifier will eliminate the problem. However, some noise is still present because of shield impedance.

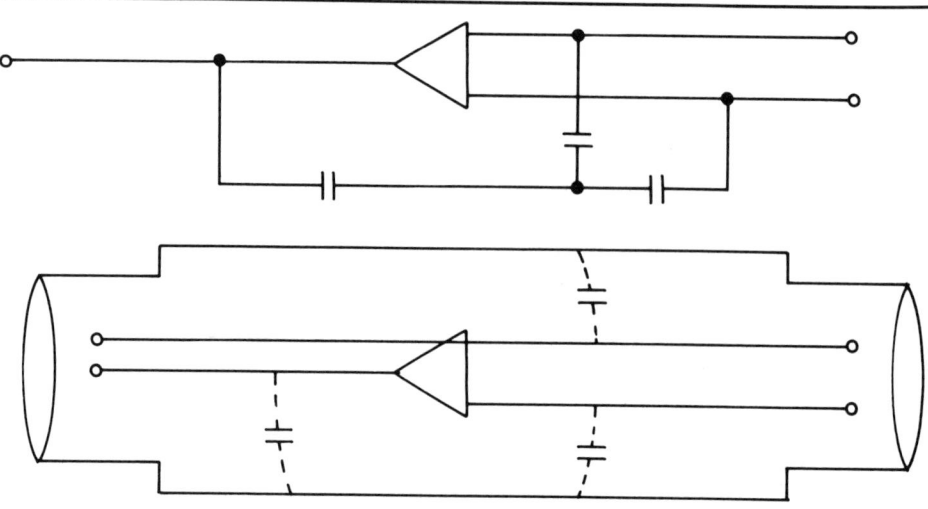

Fig. 13-13. *An electrostatic shield introduces feedback paths affecting frequency response and/or system stability.*

The usual procedure is to ground both ends of the shield. Clearly, this grounding procedure results in the introduction of a ground loop as well as a loop for the introduction of noise derived from magnetic fields. Both of these evils can be eliminated by electrically *floating* either the signal source or the receiver as shown in Fig. 13-12.[2,4] In either case, the shield is also a signal conductor, and electrostatically induced noise will still be present. Also, the implementation of this scheme is difficult because of long-established practices and availability of suitable connectors. It is evident that the standard coaxial cable is not, strictly speaking, a shielded cable.

To eliminate electrostatically induced noise effectively, a separate, electrically insulated shield should completely surround the signal processing system, similar to that shown in Fig. 13-13. However, the shield might introduce additional problems, affecting frequency response and system stability. Clearly, a tight shield will introduce feedback paths through the various capacitances and it is necessary to eliminate the feedback paths by connecting the shield electrically to the signal ground at some point. Multiple connections might result in reintroduction of ground loops or magnetic susceptibility.

13.7 SHIELD GROUNDING

Several methods of grounding the shield are summarized in Figs. 13-14 and 13-15. For the case of the amplifier input grounded (Fig. 13-14), connection A will result in shield currents being conducted to ground through a signal lead. Connection B will drain shield currents but will result in ground potentials being applied to the amplifier input through the shield to signal lead capacitances. Connection D reduces the ground noise problem, but Connection C will eliminate the ground potential problem and should reduce magnetic susceptibility that the other connections might introduce.

The case for the signal source grounded is summarized in Fig. 13-15. As before, the best connection prevents shield currents from flowing in the signal conductors, pre-

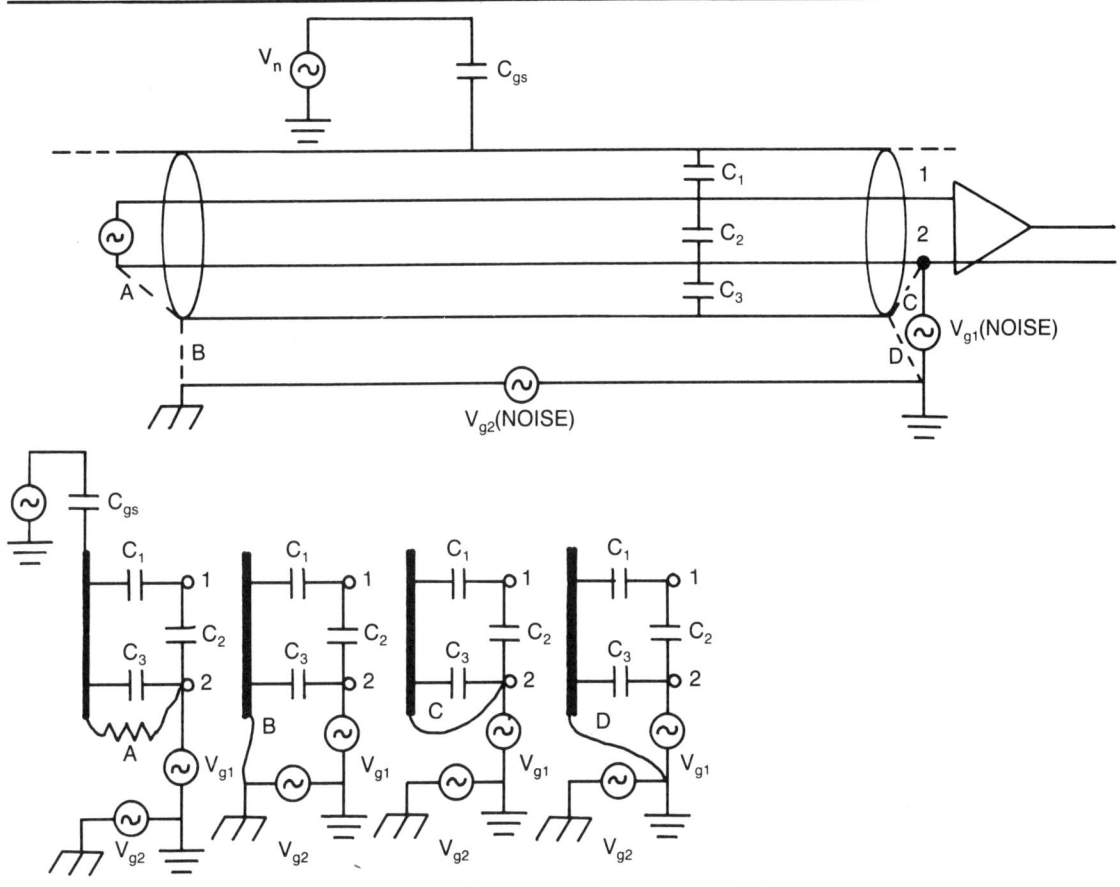

Fig. 13-14. *Various connections for connecting the shield to the signal ground for the amplifier grounded. Connection "C" results in minimum noise.*

vents introduction of ground potentials in the signal circuit, and prevents loops for the introduction of magnetically derived potentials. Connection A is the most likely means of preventing noise for the signal source grounded.

13.8 TRANSFORMER SHIELDING

To *float* a source, amplifier, etc; it is usually necessary to use isolated power supplies. Any carefully designed shielding scheme should consider the possibility of noise introduction through the power supply system. Most isolated power supplies utilize transformers, and it should be noted that the primary and secondary windings are capacitively coupled. Thus, noise signals can flow in the signal circuit by completing a path to ground for noise signals through the interwinding capacitances (Fig. 13-16). Electrostatic shields placed between the windings will reduce this possibility provided the shield is properly connected to the overall shielding system.[3] For the case of the singly shielded transformer, the interwinding shield should be connected to the overall electrostatic shield (Fig. 13-17). More complicated transformer shields are available and for

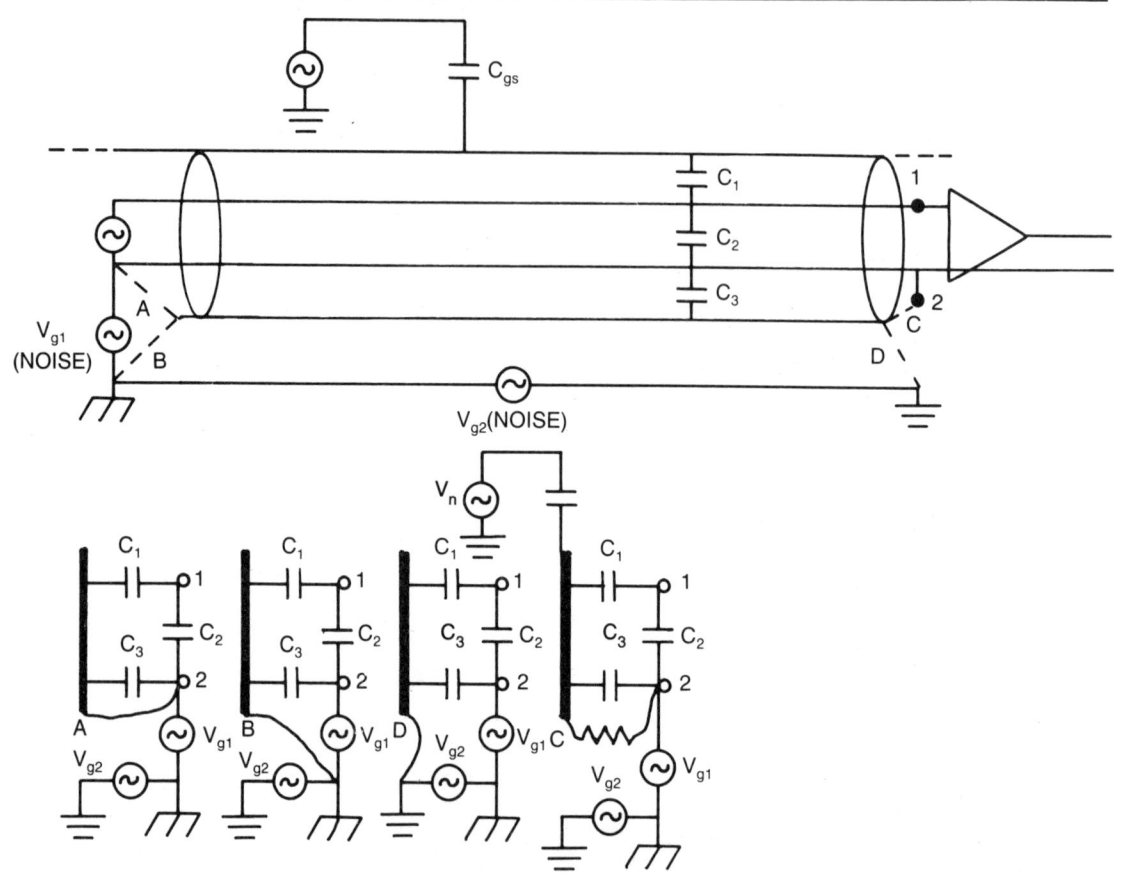

Fig. 13-15. Shield connections for the signal source grounded.

their proper connection; refer to Reference 3 at the end of this chapter. It should be noted that the usual power supply with any electronic regulator might not prevent the type of noise introduced through the primary to secondary capacitance of the transformer.

13.9 SHIELDING FROM MAGNETIC FIELD

Powerful magnetic fields are very difficult to shield, especially at low frequencies. Fortunately, most magnetically derived interference is sufficiently weak that the electronic shields and grounding practices previously discussed serve to eliminate magnetic effects. However, for those cases where magnetic fields are troublesome, either the source of the interference must be controlled, or the signal system must be shielded.

Where the source of magnetic interference is easily identified, such as exposed power leads with large currents, the use of large power coaxial cable might serve to reduce the amplitude of the magnetic field (Fig. 13-18). This strategy is effective only if the shield current is equal to or very nearly equal to the center conductor current. Quite often, the power source and the load for the source of magnetic fields are grounded through a common bus or equivalent ground plane. In this case, the shield of the power

13.9 Shielding from Magnetic Field

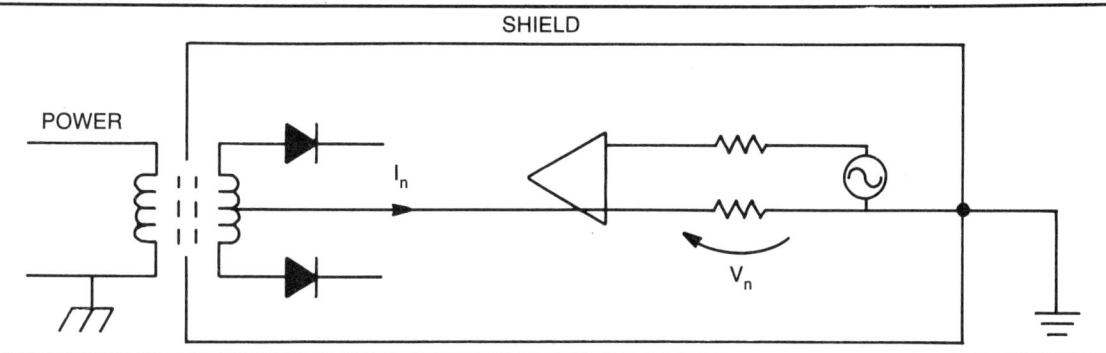

Fig. 13-16. Noise path through primary-secondary transformer capacitance.

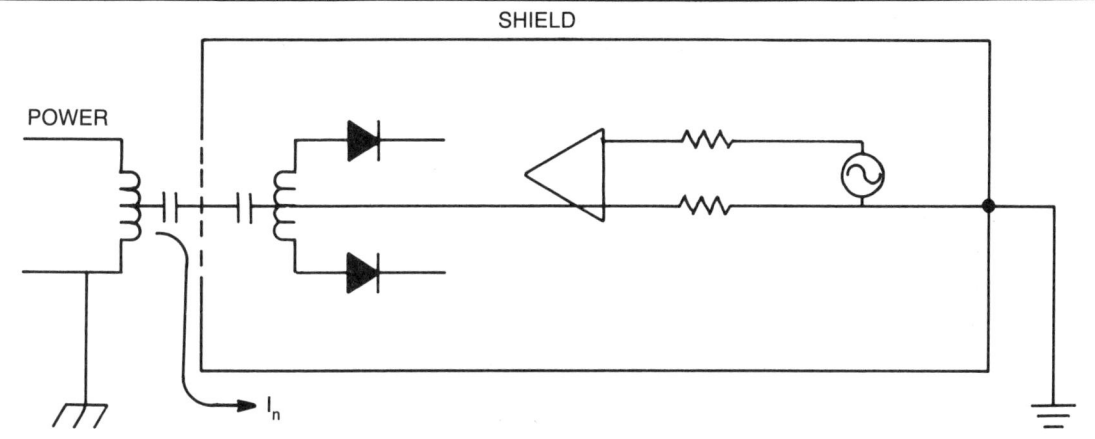

Fig. 13-17. Electrostatic shield placed between transformer windings and connected to overall shield breaks the noise path.

coaxial cable should also be grounded at both the sources and the load. Because of the mutual inductance between the shield and center conductor, this connection will result in a good portion of the ground return current flowing in the shield rather than in the ground plane. Thus, not only are ground potentials reduced, but a sizable portion of the magnetic fields are cancelled by the shield current.

For reduction of magnetic interference in signal circuits, the standard practices of reduction of loop size, elimination of ground loops that might couple magnetic fields, and the use of twisted-pair signal leads are useful and will not be discussed here. It should be noted that the foregoing practices concerning grounding and shielding should still be used to prevent reintroduction of noise derived from ground potentials and electrostatic sources.

Where the magnetic fields are sufficiently high, or of sufficiently low frequency to penetrate the electrostatic shield, additional protection can be achieved by using an additional layer of magnetic material in the shielding scheme. Ordinary steel is usually adequate. Care should be exercised that the magnetic shield does not become saturated or the frequency become so high that the shield has little or no permeability and thus

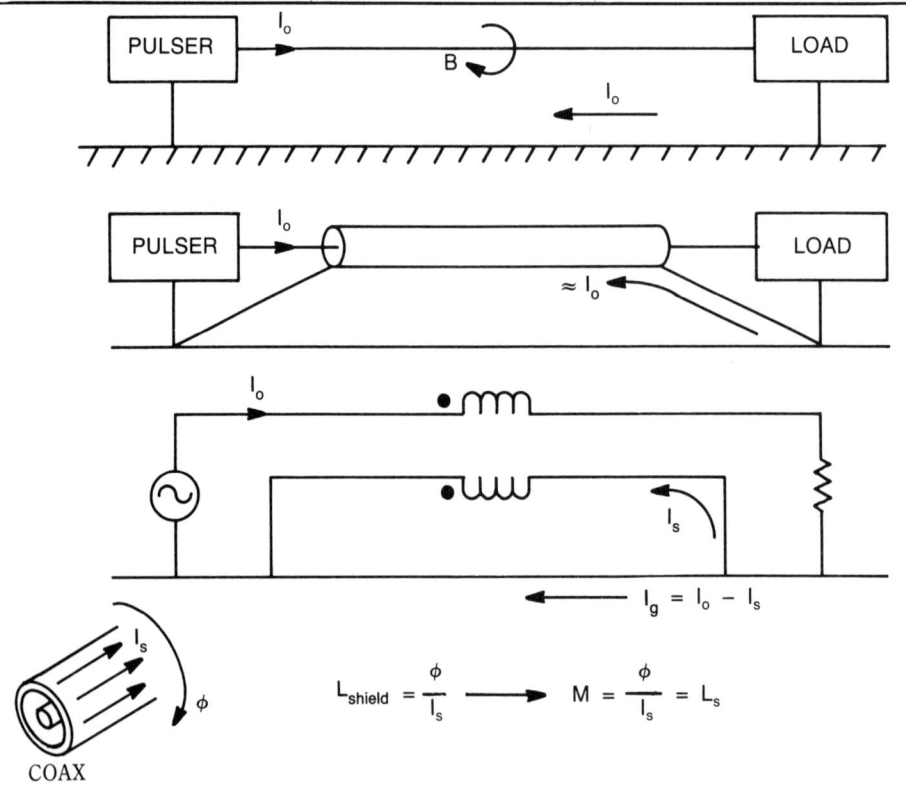

Fig. 13-18. Power coax can be used to reduce the magnitude of magnetic fields if the shield is used for the current return conductor. Note that assymetry in the coax will result in external fields.

no shielding effect. The electrical connection of the magnetic should follow the practices outlined for electrostatic shields. For extreme cases, magnetic shielding can become tedious and complex. For an understanding of the character of magnetic shielding required in pulse power, refer to Reference 5 at the end of this chapter. For implementation of multilayer magnetic shields refer to Reference 2.

13.10 REFERENCES

1. Golde, "Lightning Protection," (Chemical Publishing Co., Inc. New York, NY, 1973).
2. Ott, "Noise Reduction Techniques in Electronic Systems," (J. Wiley & Sons, New York, 1976).
3. Morrison, R., "Grounding and Shielding Techniques in Instrumentation," (J. Wiley & Sons, New York, 1977).
4. Fitch, R.A., "Salving Diagnostics in the Pulse-Power Environment," Proceedings 1st International Pulsed Power Conference, 1976.
5. Knoepfel, H., "Pulsed High Magnetic Fields," (North-Holland Publishing Company, Amsterdam, 1970).

Index

A

ac charging systems, 257-274
  average and RMS waveforms for, 262
  basic resonant charge circuit for, 258
  calculated resonant-charge efficiency and voltage transfer, 259
  charging diode for, 265-266
  charging inductor snubber, 266-268
  charging inductors in, 263-264
  charging switch snubber, 266
  command charge system, 260
  diode snubber for, 268-270
  diode-surge circuit and ratings, 272
  isolation network for, 270-271
  resonant circuit waveforms, 260
  transient suppression circuit design, 266
ac resonant charging systems, 24, 25
air core pulse transformers, 276-296
  breakdown from voids in, 290-291
  circuit tuning in, 285-287
  common problems in, 278
  corona-inception voltage, 291-294
  dual-resonance charging, 285
  eddy-current shorting in, 280
  electric stress in two-dielectric systems, 294-295
  erosion breakdown in, 290
  ideal relationships in, 283-285
  insulation in, 287-289
  liquid and solid insulation breakdown in, 287-289
  Martin's breakdown formula for, 291
  oil insulation breakdown for, 288
  shielding and, 281
  single-layer helical wound, 277
  spiral strip, 277
  streamer breakdown, 289
  thermal breakdown in, 290
  time-dependent breakdown model for, 289-290
  transformer circuit analysis for, 281-287
  types of, 277-278
  voltage-grading techniques in, 278-281
apparent power, laser loads, 322
applications, 317-334
asymmetrical fault current
  capacitor input dc power supplies, 55
  dc power supplies, 34

B

backbone circuit, Marx banks, 95
base insulation level, 364
bleeder resistors, dc power supplies, 43
blocking PRV, dc power supplies, 34
Blumlein device, 4, 105-107
  folded coaxial, 107
  thyratrons and, 231
  transient equivalent, 106
breakdown formula, Martin's, 291
Burkes, T.R., 79, 371
Butcher, R.R., 117

C

capacitive dividers, voltage measurements and, 297, 306-307
capacitive energy storage, hard tube pulsers and, 74
capacitive loads, 12
capacitive reactance, capacitors and, 358
capacitor input dc power supplies, 47-55
  actual and equivalent circuits of, 48
  asymmetric fault currents in, 55
  diode parameters, 50
  fault current in, 52, 53
  filter capacitor heating in, 53
  overload protection, 53
  ripple calculations for, 51
  transformer design, 50
  transformer leakage inductance, 53
capacitors, 3, 4, 6
  base insulation level, 364
  capacitive reactance in, 358
  characteristics and changes in, 344-365
  charging, 5

dc power supplies, 43
  definition and operation of, 335-339
  dielectric strength of, 336
  dielectrics in, 341
  dissipation factor, 360
  energy-storage, 335-370
  equivalent series resistance, 357
  frequency of, 361
  general properties of, 339-343
  high repetition rate, 4
  humidity effect, 361
  impedance in, 359
  induced polarization, 342
  lifetime of, 363
  model for, 356-365
  parallel-plate, 336
  permanent polarization in, 342
  storage temperature and time, 363
  temperature range for, 344-350
  voltage ratings for, 350-356
carbon dioxide laser loads, 166
Caristi, Bob, 227
cascade doubler, 30
cascade quadrupler, 30
charging diode, ac charging systems, 265-266
charging systems, 239-275
  ac resonant, 24, 25
  alternate current, 257-274
  alternative, 273-274
  command charging in, 24
  constant current efficiency, 248-249
  constant power, 249-251
  constant-current, 24
  direct current, 239-257
  flyback, 274
  linear, 23
  monocyclic constant, 246-248
  phase control, 254
  phase-control, 251-252
  pulse forming network, 21
  reactance, 245
  repetition-rate factors in, 24-25
  resistance, 244
  resistive and resonant, 23
  saturable reactor, 252-257
  three-phase monocyclic, 249
  voltage ramp, 246
chokes, 43, 53
  ac charging systems, 263
  dc power supplies, 39
clippers, 231
coaxial cables
  impedance matching and, 21
  transmission lines as, 118
coaxial generators, 114-115
Cockroft-Walton voltage measurements, 31
command charging systems, 24, 260
compensated pulse alternators, 2
compulsator, 5

constant power charging systems, 249-251
constant-current charging systems, 24
corona formation, 6
corona-inception voltage, 291-294
Coulomb's Law, 303
cross-field closing switch, 227
current measurement, 312-316
  Hall effect and, 315
  Rogowski pickup and, 314
  Townsend rotation and, 316
current viewing resistor, 312
current-fed pulser, 17
cycloinverters, 5

D

damping
  dc power supplies and, 33, 55-71
  equivalent circuits and, 59-63
  hard tube pulsers, 76, 78
  networks for, 56, 63-71
  primary, 66
  secondary, 65, 67, 69
  series-resonant circuit, 63-65
  snubbers in, 65-71
  thyratrons, 214
  transformers, 57-58
  two-frequency oscillatory discharge networks, 63
dc charging systems, 239-257
  ac line current surge in, 248
  charging circuits for, 241-243
  charging efficiency of, 245-246
  circuit time constants for, 240
  constant current charging efficiency, 248-249
  constant power, 249-251
  control methods for, 243-244
  impedance limited charge systems, 244-245
  monocyclic constant current, 246-248
  phase-control, 251-252
  resistance charge, 244
  saturable reactor, 252-257
  sizing for, 242
  three-phase transformer connections for, 241
  transformer primary-secondary connections, 242
  voltage ramp, 246
dc power supplies, 28-74
  ac voltage in, 34
  asymmetrical fault current, 34
  blocking PRV, 34
  capacitor design and voltage equalization, 43
  capacitor input (see capacitor input dc power supplies)
  choke input and ripple in, 39
  circuit operation in, 31-39
  component characteristics of, 33

current in, 34
current waveshape, effect of finite input inductance on, 41
damping networks, 56, 63-71
design data for resistive loads, 36-37
design PRV, 34
diode lifetime and reliability in, 45
diode parameters in, 34, 40, 49
diode stack, 43, 45
diode stack, transient voltage division in, 42
filtering in, 33
  cascade doubler, 30
  cascade quadrupler, 30
  Cockroft-Walton, 31
  elementary types of, 30
  full-wave, 29, 30
  full-wave bridge, 30-39
  half-wave, 29, 30
  major classes of, 29-31
  transients, 28
  voltage doubling circuit, 30, 31
KVA in, 34
oil insulation in, 70-74
output voltage in, 34
overload protection in, 33
peak output ripple in, 41
rated current in, 34
rated PRV, 34
rectifier circuits in, 35, 38, 46
regulation control with bleeder resistors, 43
ripple and filter calculations for, 39-40
schematic of, 32
short circuit current in, 34
short circuit KVA, 34
terminology of, 34
three-phase KVA in, 34
transformer parameters, 34, 40, 49, 61
transient damping in, 33, 55-71
transient PRV, 34
voltage in, 34
design PRV, dc power supplies, 34
dielectrics, capacitors, 341
diode stacks
  high-voltage RC compensated, 44, 45
  lifetime and reliability of, 45
  snubber, 270
  transient voltage division in, 42
  voltage grading in dc power supplies, 43
diode-surge circuit, ac charging systems and, 272
diodes
  capacitor input dc power supplies, 50
  charging, ac charging systems, 265-266
  dc power supplies, 34, 40, 49
  snubber for, 268-270
  surge ratings for, 51

transient voltages during commutation in, 58
discharge circuits and loads, 137-170
  general properties of, 137-139
  laser loads and, 160-166
  load impedance changes and, 150-152
  load-short circuits and, 155
  open-circuit protection in, 159-160
  parasitic capacitance and, 152-155
  parasitic capacitance effects and 140-141
  pulse cable interconnection to laser loads, 148-150
  pulse forming networks and, 139-140
  pulse transformers, 143-145
  switch recovery and resistive effects, 145
  switch rise time effects and, 145-148
  switches for, 141
  thyratron overvoltage protection, 155
  thyratrons, 141
  trigger generator for spark gaps and, 147
  voltage reversal after PFN discharge, 151
  voltage reversal during load faults, 152
dissipation factor, 360
Dollinger, R.E., 1, 28
dual-resonance charging, 285
duty factor, pulse shapes, 10

E

e-folding time, 8
eddy current, air core pulse transformers and, 280
electrical energy storage, 1, 3
electrical grounds, 372
electro-optic effect, voltage measurements and, 298, 311-312
electro-optical modulators, 9
electromagnetic interference (see grounding and shielding)
electromagnetic noise, radiated, 377-381
electron beam/light ion, 2
electrostatic coupling, 381-382
electrostatic fields, 1, 5
electrostatic voltmeters
  feedback system for, 305
  geometry of, 304
  linear design of, 305
  voltage measurements and, 297, 303-306
energy storage, 1-6
  capacitive, 74
  capacitors for, 335-370
  electrical, 1
  inductive, 74
  mechanical, 1
  problems and performance limitations in, 6-7

repetitive high-power electronics system technology, 3-7
  storage elements in, 5
energy-storage capacitors, 335-370
  applications for, 365-369
  base insulation level of, 364
  capacitive reactance in, 358
  characteristics and changes in, 344-365
  dielectric strength of, 336
  dielectrics in, 341
  discharge circuits using, 366
  dissipation factor, 360
  equivalent series resistance, 357
  frequency of, 361
  general properties of, 339-343
  humidity effect, 361
  impedance in, 359
  induced polarization, 342
  introduction to, 335-339
  lifetime of, 363, 369
  model for, 356-365
  peak current in, 367
  permanent polarization, 342
  ringing frequency of, 366
  storage temperature and time, 363
  temperature range for, 344-350
  voltage ratings for, 350-356
  voltage reversals in, 367
equivalent circuits
  ac charging systems, 263
  helical wire air core pulse transformers, 278
  isolation network, 271
  spiral strip air core pulse transformers, 279
  transformer, 60, 62
  transient damping and, 59-63
equivalent series resistance, capacitors and, 357
excimers, 2, 160
  laser driver circuitry, 161
  schematic for, 162

F

fall time, 8, 9
  hard tube pulsers, 75, 76, 78
fault current
  capacitor input dc power supplies, 52, 53, 55
  dc power supplies, 34
field-distortion gaps, 182
filters
  capacitor input dc power supplies, 53
  dc power supplies and, 33, 39
Fink, 62
Fitch, R.A., 95, 103
flashlamp loads, 163-166
flatness, pulse, 9
Fletcher, R.C., 88
flyback charging system, 274

flywheels, 1, 5
FRIZZ transformers, 278
Frungel, 312
full-wave bridge dc power supplies, 30-39
fusion technology applications, 2

G

Gap-Kap, 53
gas switches, 171-190
  arc recovery strength, 186
  electronegativity in, 172
  erosion in, 190
  gas breakdown in, 171-175
  inductive and resistive considerations, 186
  jitter in, 181
  maximum field strengths in, 176
  Maxwell 100kV gap geometry, 177
  Paschen curve for, 175
  Paschen curve, air, 178
  Paschen's Law, 172
  Penning effect and, 172, 174, 175
  radiation triggering in, 181
  recovery, 184-186
  relation between formative time and impulse ratio in, 174
  self break in, 175-180
  triggering, 180-184
  V-I characteristics for, 173
Gilmour, 5
Glasoe, 231
grounding and shielding, 371-386
  electrical grounds, 372
  electrostatic coupling, 381-382
  lightning protectors, 373-374
  radiated electromagnetic noise, 377
  schemes for, 375-377
  shield grounding, 382-383
  shielding from magnetic fields, 384-386
  transformer shielding, 383-384
  voltage distribution and skin effects in, 374-375
grounds, electrical, meaning and interpretation of, 372
Guillemin networks, 127-135

H

half-wave voltage measurements, 29, 30
Hall effect, 315
hard tube high-power electronics, 15-16, 25
  characteristics of, 17, 26
hard tube pulser, 72, 74-85
  capacitive storage system for, 74-76
  damping time in, 76, 78
  fall time in, 75, 76, 78
  high-power short pulse, 85
  inductive charging elements in, 76, 79

pulse shaping networks and, 76-80
resistive charging in, 76
rise time and, 75
shunt elements, inductive and diode, 77
switch tubes in, 79-84
systems using, 84-85
helical wire air core pulse transformers, 277
equivalent circuit for, 278
high-power electronics
categories and classes of, 14
energy storage in, 1-6
grounding and shielding for, 371-386
hard-tube vs. line-type, 15, 25
impedance matching and, 20-22
introduction to, 1-27
loads for, 12
pulse forming network charging systems in, 22-25
pulse shape characteristics in, 7-14
switch rise time effects in, 145-148

I

ignitrons, 6, 15, 231-237
commercial, specifications for, 234
cross sectional views of, 233
firing circuit, low-inductance housing for, 235
stacking, 236-237
standard characteristics of, 232-236
triggering for, 236
impedance, capacitors, 359
impedance matching, 20-22
coaxial cables and, 21
load mismatch and switch performance vs., 21
transformers for, 21-22
induced polarization, energy-storage capacitors, 342
inductive charging system, 3
hard tube pulsers, 76, 79
inductive energy storage, hard tube pulsers and, 74
inductive loads, 12
circuit information for, 161
inductors, charging, ac charging systems, 263-264
inversion generators, 103-105
circuit for, 104
inverters, 5, 6
charging systems and, 24
Ion Physics MiniMarx, 90
isolation network
ac charging systems, 270-271
equivalent circuit for, 271

J

jitter, gas switches, 181
Joule heating, 11

K

Keffel, E., 55
Kerr cells, 311, 316
klystrons, thyratrons and, 210-214
KVA, dc power supplies, 34

L

Lafferty, J.M., 192
laser loads, 160-166, 317-334
apparent power in, 322
carbon dioxide, 166, 167
circuit operation in, 317-319
direct-discharge pumped excimer, 160
discharge circuits and loads and, 160-166, 317
electrical characteristics of, 321-323
energy transfer in, 320
flashlamp, 163-166
hydrogen fluoride, 166, 168-170
inductance considerations in, 323-327
magnetron load characteristics for, 318
schematic and charging waveform of, 319
schematic for, 318
time-varying impedance in, 327-332
leakage inductance, capacitor input dc power supplies, 53, 54
Lebacqz, 231
line type high-power electronics, 15-20, 25
characteristics of, 26
current fed, 19
pulse forming networks and, 18
voltage- vs. current-fed, 17-18
linear charging, 23
liquid switches
spark gaps and, 190-191
triggering, 191
load faults
discharge circuits and loads and, 155-159
inverse voltage and control circuits for, 159
inverse voltage calculation during, 157
inverse voltage removal during, 156
short circuit, 158
load-short circuits, discharge circuits and loads and, 155-159
loads, 12-14
capacitive, 12
changing, discharge circuits and loads and, 150-152
discharge circuits and, 137-170
flashlamp, 163-166
general characteristics of, 137
high-power electronics, 12

impedance matching and, 20-22
inductive, 12
inductive, circuitry for, 16
klystron, thyratrons and, 210-214
laser (see laser loads)-
mismatched, switch performance and, 21
parasitic capacitance and, 138
PCS and, 12
positive mismatch in, 21
pulse cable interconnection to, 148-150
pulse shapes and, 11-12
resistive, 12
resistive, dc power supplies design data for, 36-37
time-varying, 13
losses, transmission lines, 124

M

magnetic fields, 1, 5
shielding from, 384-386
magnetrons, 10
Martin, J.C., 88, 99, 175, 186, 187, 191
Marx banks, 4, 87-103
backbone circuit for, 95
basic configurations for, 90-91
charging methods for, 91-93
circuit configurations for, 91
discharging of, 93-95
Erwin Marxes vs. Martin Marxes, 99
Ion Physics MiniMarx, 90
Kirchhoff's Law and, 95
layout and circuit relationships in, 96
Martin Marx, 100
Mesyats circuit in, 101
output considerations for, 100-103
Paschen's Law and, 101
reaction time in, 89
state of the art for, 88
switches, 90
transmission lines, 122
triggering and erection time in, 95-100, 95
triggering sequence in, 98
Marx, Erwin, 87, 99
measurement techniques, 297-316
current, 312-316
voltage, 297-312
mechanical energy storage, 1, 2
compulsator in, 5
Mesyats circuits, Marx banks, 101
Mesyats, G.A., 100
mixed dividers, 298
voltage measurements and, 308-311
mode hopping, 9
modulators, electro-optical, 9
monocyclic constant current charging system, 246-248
Morrison, R.W., 95

N

Nunnally, Bill, 146, 239

O

O'Rourke, Ray, 187
oil insulation
  air core pulse transformers and, 288
  antioxidants in, 71
  dc power supplies, 70-74
open-circuit protection, discharge circuits and loads, 159-160
overload protection
  capacitor input dc power supplies, 53
  dc power supplies, 33
  thyratrons, 155
overshoot, 8

P

parallel-plate capacitors, 336
parasitic capacitance
  discharge circuits and loads and, 140-141, 152-155
Paschen curve, air, 178
Paschen's Law
  gas switches and, 172
  Marx banks and, 101
peak pulse power, 11
peak-reading meter circuit, 306
Penning effect, gas switches and, 172, 174, 175
permanent polarization, energy-storage capacitors, 342
phase-control charging systems, 251-252, 254
Pockel's cells, 9
polarization, induced vs. permanent, capacitors, 342
positive mismatch, 21
power supplies, 5
  dc, 28-74
  ideal, 244
pulse forming networks, 16, 125-136
  charging systems, 22-25
  current fed line, 125
  discharge circuits and loads and, 139-140
  Fourier series approximation for, 126
  Guillemin networks as, 127-135
  line type high-power electronics and, 18
  thyratrons breakdown and recovery, 205
  thyratrons charging techniques for, 205
  thyratrons switching in, 202-214
  thyratrons triggering, 205-210
  voltage fed line, 126
pulse generators, thyratrons, 208
pulse modulator tubes, thyratrons, 215-217

pulse repetition frequency, 11
pulse shapes, 7-14
  derived parameters for, 10
  fall times, 9
  flatness in, 9
  Joule heating in, 11
  load characteristics and, 12
  load reproducibility effects, 11-12
  parameters for, 7, 8
  peak pulse power, 11
  pulse repetition frequency, 11
  rise times in, 8
pulse shaping networks
  constant-current, 81
  hard tube pulsers, 76-79, 80
pulse transformers, 68
  air core, 276-296
  discharge circuits and loads and, 143-145
  limitations of, 144-145,
pulse-charging transformers, 3
pulse-voltage circuits,g 87-116
  Blumlein devices, 105-107
  coaxial generators, 114-115
  inversion generators for, 103-105
  Marx bank, 87-103
  spiral generators, 107-114

R

radiated electromagnetic noise, 377-381
Radio Electronic Engineering, 62
rare-gas halogen lasers, 160
rated current, dc power supplies, 34
rated PRV, dc power supplies, 34
reactance charging system, 245
rectifier circuits
  dc power supplies, 35, 38
  resistive-load or choke input, 38
  solid-state, reliability of, 46
resistive charging systems, 23, 244
  hard tube pulsers, 76
resistive dividers, voltage measurements and, 298, 307-308
resistive loads, 12
  dc power supplies design data for, 36-37
resistors, bleeder, dc power supplies, 43
resonant charging systems, 23
  ac, 24, 25
ringing frequency, energy-storage capacitors, 366
ripple
  capacitor input dc power supplies and, 51
  dc power supplies, 39
rise times, 8
  hard tube pulsers, 75
Ristic, 186
Rohwein, Gerry, 21, 276
rope switches, 181

S

Sarjeant, W.J., 1, 28, 335
saturable reactor charging systems, 252-257
self-resonant frequency, capacitors, 359
series-resonant circuit, 63-65
shield grounding, 382-383
short circuit current, dc power supplies, 34
shunts, hard tube pulsers and, 77
simultaneous overvoltage, 182
single-shot systems, 3, 23
skin effects, grounding and shielding and, 374-375
Smith, A. M., 95
snubbers, 65-71
  ac charging systems, 266-268
  charging schematic for, 267
  diode, 268
  diode stacks, 270
  switch, 268
  thyratrons, 214
solid-dielectric switches, 192-196
  materials for, 196
  self-break curve for, 194
  stabbing, 193
  triggering for, 195
Sorensen, 186
spark discharge, 13
spark gaps, 6, 21, 53, 145, 171-197
  breakdown voltages, uniform field, 303
  calibration voltages for sphere, 299
  field-distortion, 182
  flashover voltages for, 300
  gas switches, 171-190
  hold-off strength and, 301
  liquid switches and, 190-191
  Maxwell 100kV gap geometry, 177
  model for, 188
  point angle vs. breakdown voltage in, 302
  solid-dielectric switches, 192-196
  spark distance vs. point-plane, 302
  trigatron, 182
  trigger generator for, 147-148
  vacuum gaps, 191-192
  voltage measurements and, 297, 298-303
spiking, 9
spiral generators, 107-114
  calculations for, 111
  inversion loss mechanism in, 110
  resistive loss, ideal switch, 114
  switch loss, 113
spiral strip air core pulse transformers, 277
  coaxial shield for, 280
  equipotential profile of, 280
  equivalent circuit of, 279
stabbing, solid-dielectric switches, 193

streamer breakdown, air core pulse transformers, 289
Strickland, 195
Swift, G.W., 92
switch tubes
  characteristics of, 83
  hard tube pulsers, 79-84
  high-vacuum, 82
  materials for, 82
switches, 4, 6, 14, 15, 21, 141
  corona formation in, 6
  cross-field closing, 227
  gas, 171-190
  high-power electronics, rise-time effects in, 145-148
  liquid, spark gaps and, 190-191
  Marx banks, 90
  mismatched loads and, 21
  modeling of, 146-147
  recovery, resistive effects and, 145
  rope, 181
  snubber for, 268
  solid-dielectric, 192-196
  thyratron, 4

T

tetrode tubes, 6
thyratrons, 4, 5, 6, 21, 141-142, 145, 198-201
  anode spacing in, 198
  applications for, 231, 232
  basic configuration and materials summary for, 201
  breakdown time history for, 206
  calculated resistive phase fall time against pressure, 228
  capability progress for, 225
  ceramic, 202, 203
  commercial and developmental overview, 215-225
  damping, 214
  developmental, 219-225, 230
  extended ratings, operating conditions for, 225
  geometry and pulse shapes for, 199
  glass, 201-202
  high dI/dt, 225-231
  hydrogen, 204, 217-218
  klystron load, 210-214
  long-pulse, 224

low-inductance, 226-227
MAPS-40 megawatt, 219-220, 222
overvoltage protection, 155
parameters for, 228-231
peak current-voltage profile for, 215
plane parallel, 203
pulse forming network switches using, 202-214
pulse forming network, breakdown and recovery, 205
pulse forming network, charging techniques, 205
pulse forming network, discharge circuit, 201-205
pulse forming network, triggering, 205-210
pulse-modulator tubes, 215-217
recovery for Rl greater than Zn, 213
recovery for Rl less than Zn, 212
snubbers, 214
switch resistive phase fall time relation, 227-228
tetrode, low-inductance housing for, 221
trigger-pulse shapes and pulse generators, 208
triode and tetrode, 198
tube-heating factors for, 217
wave shapes, charging voltage and current, 207
time-varying loads, 13
Townsend rotation, 316
Townsend, J.S., 171
transformers, 61
  air core pulse, 276-296
  capacitor input dc power supplies, 50, 53
  dc power supplies, 34, 40, 49
  equivalent circuit for, 60, 62
  FRIZZ, 278
  impedance matching and, 21-22
  primary damping for, 66
  pulse (see pulse transformers)
  pulse-charging, 3
  shielding of, 383-384
  transient damping and, 57-58
transient PRV, dc power supplies, 34
transients, 6
  ac charging systems and suppression of, 266

dc power supplies, damping of, 33, 55-71
diode stack and, 42
diodes, 58
voltage measurements, 28
transmission lines, 117-125
  capacitance and inductance calculations for, 117, 119
  charged, 123
  coaxial, 118
  current in, 125
  impedance calculation for, 119
  losses in, 124
  Marx banks, 122
  one-way transit time in, 118, 119
  open, 121
  reflection coefficient in, 121
  shorted, 122, 123
  strip-type, 119
  voltage and current waves in, 120
trigatron, 182, 192
trigger generators, 147
two-frequency oscillatory discharge networks, 63

V

vacuum gaps, 191-192
  plasma injection in, 192
  triggering in, 192
vector addition systems, 103
voltage doubling circuit, 30, 31
voltage measurement, 297-312
  capacitive dividers for, 306-307
  devices for, 297, 298
  electro-optic effect for, 311-312
  electrostatic meters, 303-306
  Kerr cells and, 311
  mixed dividers for, 308-311
  resistive dividers for, 307-308
  spark gaps for, 298-303
voltage ramp charging system, 246
voltage-fed pulser, 17-18

W

Willis, W.L., 87, 171, 297

Z

Zaengl, W., 55